FAST SOLAR SAILING

SPACE TECHNOLOGY LIBRARY
Published jointly by Microcosm Press and Springer

The Space Technology Library Editorial Board

Managing Editor: **James R. Wertz**, *Microcosm, Inc., El Segundo, CA, USA*;

Editorial Board: **Val A. Chobotov**, *Consultant on Space Hazards, Aerospace Corporation, Los Angeles, CA, USA*;
Michael L. DeLorenzo, *Permanent Professor and Head, Dept. of Astronautics, U.S. Air Force Academy, Colorado Spring, CO, USA*;
Roland Doré, *Professor and Director, International Space University, Strasbourg, France*;
Robert B. Giffen, *Professor Emeritus, U.S. Air Force Academy, Colorado Spring, CO, USA*;
Gwynne Gurevich, *Space Exploration Technologies, Hawthorne, CA, USA*;
Wiley J. Larson, *Professor, U.S. Air Force Academy, Colorado Spring, CO, USA*;
Tom Logsdon, *Senior Member of Technical Staff, Space Division, Rockwell International, Downey, CA, USA*;
F. Landis Markley, *Goddard Space Flight Center, NASA, Greenbelt, MD, USA*;
Robert G. Melton, *Associate Professor of Aerospace Engineering, Pennsylvania State University, University Park, PA, USA*;
Keiken Ninomiya, *Professor, Institute of Space & Astronautical Science, Sagamihara, Japan*;
Jehangir J. Pocha, *Letchworth, Herts, UK*;
Frank J. Redd, *Professor and Chair, Mechanical and Aerospace Engineering Dept., Utah State University, Logan, UT, USA*;
Rex W. Ridenoure, *Jet Microcosm, Inc., Torrance, CA, USA*;
Malcolm D. Shuster, *Professor of Aerospace Engineering, Mechanics and Engineering Science, University of Florida, Gainesville, FL, USA*;
Gael Squibb, *Jet Propulsion Laboratory, California Institute of Technology, Pasadena, CA, USA*;
Martin Sweeting, *Professor of Satellite Engineering, University of Surrey, Guildford, UK*

For further volumes:
www.springer.com/series/6575

Giovanni Vulpetti

Fast Solar Sailing

Astrodynamics of Special Sailcraft
Trajectories

Giovanni Vulpetti
International Academy of Astronautics—Paris, France
Rome, Italy

ISBN 978-94-007-4776-0   ISBN 978-94-007-4777-7 (eBook)
DOI 10.1007/978-94-007-4777-7
Springer Dordrecht Heidelberg New York London

Library of Congress Control Number: 2012943614

© Springer Science+Business Media Dordrecht 2013
This work is subject to copyright. All rights are reserved by the Publisher, whether the whole or part of the material is concerned, specifically the rights of translation, reprinting, reuse of illustrations, recitation, broadcasting, reproduction on microfilms or in any other physical way, and transmission or information storage and retrieval, electronic adaptation, computer software, or by similar or dissimilar methodology now known or hereafter developed. Exempted from this legal reservation are brief excerpts in connection with reviews or scholarly analysis or material supplied specifically for the purpose of being entered and executed on a computer system, for exclusive use by the purchaser of the work. Duplication of this publication or parts thereof is permitted only under the provisions of the Copyright Law of the Publisher's location, in its current version, and permission for use must always be obtained from Springer. Permissions for use may be obtained through RightsLink at the Copyright Clearance Center. Violations are liable to prosecution under the respective Copyright Law.
The use of general descriptive names, registered names, trademarks, service marks, etc. in this publication does not imply, even in the absence of a specific statement, that such names are exempt from the relevant protective laws and regulations and therefore free for general use.
While the advice and information in this book are believed to be true and accurate at the date of publication, neither the authors nor the editors nor the publisher can accept any legal responsibility for any errors or omissions that may be made. The publisher makes no warranty, express or implied, with respect to the material contained herein.

Printed on acid-free paper

Springer is part of Springer Science+Business Media (www.springer.com)

Dedicated to

Rosella

This book originated from my wife's constant inspiration to me, through her belief that solar photon sailing would eventually be realized in both its usefulness and beauty.

In memory of **Dr. Leslie R. Shepherd**

Leslie R. Shepherd was a great pioneer of the modern scientific studies on interstellar flight, and one of the promoters and founders of the International Academy of Astronautics.

Foreword

Once in a great while, a field of science or engineering will experience a revolution. I am not talking about the slow, methodical advancement of science—increasing our state of knowledge or engineering state-of-the-art by research, development, peer-review and publication. I am talking about the movement of a great idea from the realm of theory to practice and subsequently changing the way we do things, enabling new capabilities, and allowing future generations to look back and wonder how it could have ever been otherwise. I personally believe we are at that point in our exploration of space and the use of the fast solar sailing techniques discussed in this monograph may just enable us to break out of our snail's pace exploration of the inner solar system and enable us to take those first steps toward true interstellar exploration.

I think this monograph is important for many reasons. First, and foremost, and I say this because I work with traditionally trained aerospace engineers at one of the world's premier engineering organizations, fast solar sailing should become part of the standard education 'toolbox' of the next generation of aerospace engineers. As an advanced propulsion technology, Solar sails are now an engineering reality. Having a textbook/monograph available and in the modern engineering classroom will help educate this next generation as regards what is possible beyond traditional rockets. It is a step toward convincing them that solar sails are not just a scientific curiosity but a capability that they can use in tomorrow's deep space systems.

Secondly, fast solar sails are truly enabling. Educating those currently working in the field of space exploration as to their capabilities must begin now and not wait on the next generation. As NASA's propulsion lead for the early 2000s redefinition of the Interstellar Probe mission concept, it became apparent that there are very few viable propulsion technologies that will enable us to send a probe to the outermost portions of the solar system (200–250 AU) within the professional lifetime of the team that might develop such a system. If we are going to seriously study the outer solar system then we have to be able to get there and return data in reasonable periods of time. Granted, we are still getting some data from the venerable Voyager spacecraft some 35+ years after they were launched. I need to point out that we are now decades beyond their design life. Designing a spacecraft to provide operational data for 35–50 years is not an easy task and will be inherently risky, and expensive, for the foreseeable future. There are simply too many things that can go wrong over decades of operation in the space environment. Reducing trip times is essential and fast solar sailing is, in my opinion, the only technically viable and affordable approach open to us for truly deep space exploration.

Lastly, the monograph gathers into one place the fundamental astrodynamic underpinnings of fast solar sailing to serve as a starting point for those in or entering the field to make that next level of progress (revolution!) in their theoretical and practical application. Whether the interest is in solar photon sails or electric sails riding the solar wind, this monograph should become THE source for beginning serious study.

I had the privilege of working with Giovanni on the Interstellar Probe mission concept study, as a colleague on the scientific committee for a series of symposia dealing with deep space flight, and as a co-author of a popular science book about solar sailing. He has a passion for solar sailing and he has made many pioneering contributions to the field in his career. This, his latest, will help educate and inspire those who follow to build upon the foundation he has helped create and take us to the stars.

NASA Advanced Concepts Office   Les Johnson
George C. Marshall Space Flight Center

Preface

When the contract between Springer and the author was signed in 2007, the expected deadline for finishing the manuscript was March 31, 2010. In the Fall of 2009, when it appeared clearly to this author that a strong delay in writing the book was building up, he could not foresee that 2010 would have been a fundamental year for solar-photon sailing. As a point of fact, after many decades of waiting for the first experimental *full* missions with *proper* sailcraft, JAXA launched IKAROS (May 21, 2010) and NASA followed with NanoSail-D2 (December 19, 2010, with ejection on January 1, 2011). IKAROS' sail (about 200 square meters) consists of an aluminized polyimide, the commercially available APICAL-AH and the new material named ISAS-TPH, 7.5 μm in thickness. Although such a sailcraft is not a sail-ship capable to fly much faster than a chemical propulsion spacecraft, it however is of *basic importance* to solar sailing through all the technical items experimented with in space. In July 2010, JAXA announced two significant events: (1) using Doppler measurements, the orbit determination process indicated that the trajectory of IKAROS (after the second-stage sail deployment) was being perturbed by a solar-radiation thrust of about 1.12 mN, and (2) a slow attitude control maneuver was carried out on the spinning sail by using close-to-rim strips of voltage-driven liquid crystals, which can change their reflectance levels. Thus, thrust and sail attitude control torques were produced without any propellant consumption. Such things excited the international solar-sailing community, including this author. The historical IKAROS mission has affected this book in some notable points.

This book is a monograph on a branch of astrodynamics of solar sailing: *fast sailing*. This is the leading theme. This book has been arranged to be multidisciplinary to a certain extent. This was chosen for the following reasons: (1) *feasibility*: not always what in principle seems to exhibit no problem is realizable; (2) *challenge*: fast sailing is really a challenging endeavor, which could offer amazing potential to spaceflight; one should view sailcraft more generally than what may be got through the sole mathematical description of its trajectories; (3) *projection* to the future: building a future high-tech sail system could represent the realization key for systematically exploring the Kuiper belt, the "boundaries" of the solar system, the solar gravitational lens, and for going notably far beyond this. A fast sailcraft may be the core of an interstellar precursor mission.

Solar-Photon Sailing (SPS) is an old *in-space* propulsion concept that aims at getting thrust from the solar radiation pressure via a sufficiently large and light surface called *the solar sail*. This emphasizes the role of sunlight with respect to other propulsion concepts, which would utilize the momentum flux of the solar wind, namely the tenuous plasma consisting (mainly) of protons, electrons, and helium nuclei that the Sun ejects from its upper atmosphere. Both solar light and solar wind are omnidirectional and permanent, though variable, streams.

SPS, well beyond the denser zones of any planetary atmosphere, is one of the best candidates to fully become the 21st century in-space propulsion. One may wonder why future astronautics needs new propulsive systems, other than rockets and gravity assist via planetary flybys. After more than 50 years of space flight, they continue to do an excellent job indeed. Although very briefly, something has to be clarified in order to better understand this key-point. First of all, any gravity assist is not a real propulsion method, but actually a celestial-mechanics technique applied to space flight (very successfully). Many important historical missions would have been simply impossible without (multiple) planetary flybys. Second, rocket systems do not appear replaceable with regard to utmost-importance tasks such as launching, achieving some special orbits, and also soft landing on some massive celestial body.

Nevertheless, both rocket and gravity assist have their own bounds with respect to the future needs of robotic and/or human space exploration and expansion. Even though we shall expand this aspect in this book, we like to briefly mention some considerable drawbacks of them. By definition, any rocket system has to have a power source *onboard*; therefore, it is forced to eject a fraction of its mass-energy *directionally* for getting accelerated. There are quasi-rocket systems that could exhaust matter energized by some external power source via onboard photon collectors. Since the kinetic energy of such an ejection beam is normally very small compared with the beam's rest-mass, one can deal with the related energetics and dynamics as those ones of a rocket. A typical example is the solar-thermal engine where sunlight is used to heat some inert mass to be exhausted. Although rockets and quasi-rockets encompass as energy-limited as power-limited engines, they all obey the basic need to transport the propellant necessary to every trajectory and attitude maneuver. Many design features depend on this need; besides the complexity of a rocket vehicle, we highlight the fact that the vehicle's mass at departure may be really very high. This restricts the realizable space-mission classes very strongly.

Gravity-assist technique has aided spacecraft considerably, in general, and rockets, in particular. Without it, many significant space missions neither would have been accomplished in the past nor designed for the near term. Nevertheless, gravity assist relies on the relative positions of celestial bodies, because it is from their motions that the spacecraft will have to tap some (really tiny) amount of energy. This entails two important items: (a) missions have generally *narrow* launch windows, and the launch opportunities decrease with increasing the number of planetary flybys; (b) the transfer times to the final destinations are significantly long simply because the spacecraft have to follow (perturbed) Keplerian orbits between two consecutive flybys; if many flybys are required, then the ensuing long transfer time implies to use high-reliability technology and high cost. It is plain that the current or near-term status of in-space propulsion is no longer bearable because future highly desirable scientific, utilitarian, and even (some) commercial mission concepts have been increasing in number and payload mass. In addition, user communities normally request payloads with very long operational lives.

Although it is true that space-borne rocket engines will continue to be precious in normal and special situations entailing maneuver simplicity and rapidity, nevertheless a real revolution in space propulsion is mandatory for exploring the solar system and going beyond, but avoiding tremendous costs and the high risks of losing or strongly degrading high-return missions. We need not wait for a new frontier of fundamental physics. The point that should be grasped now is that hopefully we are approaching a long era where robotic/human exploration/expansion in space could be not only possible, but also systematic, reliable and cost-affordable. We are talking about something similar to what happened on the Earth oceans and seas for millennia *via* sails, which ultimately utilize solar energy through atmospheric winds. SPS could allow us to explore and utilize, widely and systematically, the Earth–Moon space, the Solar System and move our probes beyond the heliosphere in times acceptable and compliant with the human lifetime. Particularly interesting distant targets, which should be explored, are the many objects of the outer solar system known as the Edgeworth–Kuiper belt, the near and far heliopause, the near interstellar medium, the solar gravitational lens regions, and so on, in order of increasing distance from the Sun, from some tens to many hundreds of Astronomical Units. One will be aware that the technology developed for fast sailing could aid in accomplishing many missions—in the solar system—extremely expensive or practically unmanageable even via advanced rockets.

We are dealing with a special class of sailcraft and their related trajectories that should be capable to accomplish the deep-space exploration just mentioned. We will define *fast* solar sailing and study peculiar properties. Astrodynamics of fast solar sailing is relatively recent. The first technical paper on fast sailing was published in 1992, whereas a very preliminary quantitative investigation on nanotechnology-based SPS appeared in the specialized literature in 2008. An important step towards fast sailing missions has been the concept of Interstellar Probe by NASA (1999–2001, essentially). As said above, the era of space sailing has been opened very recently by IKAROS and NanoSail-D2; however, we are still far from missions systematically using sails for space exploration/utilization. Many space-ware countries should participate in such enterprise.

This book is organized as follows:

- Part I contains an introductory chapter on rocket dynamics in the context of very deep space missions. This chapter has a twofold purpose. The first one aims at strictly proving some basic rocket's properties that are not easy to find in the literature, especially for university students. The second objective is to show the considerable limitations a rocket has with respect to the growing needs of future spaceflight. For such combined reasons, some sections of Chap. 1 resort to basic concepts of special relativity. Although not recommended, skipping this chapter on a first reading does not affect the comprehension of the other chapters.

- Part II contains three chapters. For going beyond rockets, for instance by means of solar sails, one has to analyze the power sources in space.

  Chapter 2 describes and emphasizes the Sun's features the solar sailing relies on.

  Chapter 3 focuses on the concepts of sail-based spacecraft, or *sailcraft*, and which aspects of the complicated space environment may ultimately affect sailcraft design and trajectories.

  Chapter 4, in particular, deals with the problems related to an object such as a metallic sail moving inside the solar wind and the solar ultraviolet flow.

- Part III is the theoretical core of this monograph and consists of three chapters.

  Chapter 5 introduces the reader to the fundamentals of sailcraft trajectories, and a jerk-based analysis for finding all admissible types of thrust maneuvering, a set that includes sail axis changes.

  Chapter 6 describes in detail the very complicated problem of how radiation-pressure thrust can be modeled. One can see that, in addition to the first necessary condition from physics (i.e. the radiation pressure) for space sailing, there is a second necessary condition: optical diffraction, without which the control of the sail thrust would be quite limited.

  In Chap. 7, the theory of *fast* space sailing is explained in the framework of both two-dimensional and three-dimensional trajectories. The reader can also find how the 3D fast sailing is not a mere extension of the properties of 2D fast trajectories.

- Part IV will finally treat examples of fast missions through two chapters.

  Chapter 8 is devoted to the optimization and numerical calculation of sailcraft's fast trajectories. Two methods are discussed: (i) the variational approach, and (ii) the non-linear programming approach.

  Chapter 9 regards effects that should be considered in real designs of sailcraft and missions. It deals with (1) the influence of the variable solar irradiance on sailcraft trajectories, and (2) the problem of how sailcraft motion equations might be generalized *formally* for taking into account modifications of the sail's optical quantities caused by solar-wind ions and ultraviolet light.

References are included at the end of each chapter. The whole book has been designed by the author in LaTeX 2_ε with the document class provided by Springer, and other packages common to this language version. Most of the figures and tables, and almost all of the numerical examples have been prepared by the author expressly for this monograph by using a set of sophisticated (large) computer codes for astrodynamics and space mission flight design.

International System of Units (SI) is used throughout this monograph in general, a due choice not only for the sake of standardization, but also in order to make the different scientific areas related to solar sailing easily understandable in their own scales. A few exceptions are some well-known expressive units from Particle Physics and Astronomy, and the "solar units" for expressing distance, speed, energy, and angular momentum for heliocentric trajectories.

Preface xiii

Book prerequisites. The chapters of this book have been tailored for graduate students in Physics, Mathematics, or Engineering, especially the ones aimed at getting a Ph.D. in Astrodynamics and/or Space Propulsion. The book content is self-explanatory, in general. However, a limited understanding of the basic principles of Special/General Relativity is advisable for reading Chap. 1, and a few sections of Chaps. 5 and 6. The other sections use classical dynamics.

A few final words about the cut given to this book. The author strove to write a monograph containing *also* new aspects of solar-photon sailing not dealt with before in the specialized literature; also, he revised/enlarged heavily what he published in the past years, including computation codes. Of course, where applicable, results from very recent specialistic international journals and symposia are discussed and inserted in a wider context. Although the bibliography is always non-exhaustive in general, and the topic of this monograph is relatively recent in particular, many efforts have been made for indicating over four hundred references from various disciplines of Mathematics, Physics, and Engineering (until May 2012).

Rome, Italy Giovanni Vulpetti

Acknowledgements

First, I want to thank Mrs. Sonja Japenga, formerly at Springer Netherlands. During the 57th International Astronautical Congress, held in Valencia (Spain), October 2–6, 2006, I came in the Springer's Stand for buying some scientific books. There, I met Sonja, at that time Publishing Editor for Astronomy, Space Science and Space Technology at Springer. During the meeting, on Sonja's request, I gave some information about my work activities and membership rank at the International Academy of Astronautics (Paris). I also added that I was writing a co-authored popular book (published in August 2008) on solar sailing for Copernicus Book (an imprint of Springer Science + Business Media) and Praxis Publishing Ltd. She was very interested. Subsequently, she asked me to write a scientific book, possibly on solar sailing. Springer and I signed a publishing agreement a few weeks before Sonja left Springer.

Dr. Ramon Khanna took over Springer's previous projects in 2008, including this book. I have found immediate support from him. Then, he periodically checked my progressive drafts and suggested me a number of notable items for improving the book. In particular, I appreciated greatly when he encouraged me to continue to write this book in spite of my unexpected and severe health problems, and he was enthusiast with me about the IKAROS launch, namely, the beginning of a new astronautical era. I am significantly thankful to him for what he did as regards my book.

I do like to express special thanks for Dr. Charles L. Johnson, Deputy Manager, Advanced Concepts Office, NASA George C. Marshall Space Flight Center. He very kindly accepted my invitation to read the book and write the Foreword.

During the writing of this book, I had to face a number of serious troubles (not related to the book), which risked this project. I have received continuous help from my family in order to get over the various obstacles. Without such backing, I would not have been able to realize this book.

Contents

Part I   A Review of Rocket Spacecraft Trajectories

1   Some General Rocket Features 3
    1.1   Introduction 3
    1.2   Frames of Reference 4
    1.3   Modeling a Rocket 4
    1.4   Rocketship Dynamics in Field-Free Space 7
        1.4.1   Model Features 12
        1.4.2   Some Examples 13
    1.5   The Effect of Gravity on Rocket Dynamics 14
    1.6   Escaping from the Solar System by Photon Rocket 19
    1.7   Escaping from the Solar System by NEP Rocket 23
    References 31

Part II   Sailing in Space Environment

2   The Sun as Power Source for Spaceflight 35
    2.1   A Summary of Radiometric Quantities 35
    2.2   Monitoring Sunlight from Satellite 43
        2.2.1   The Total Solar Irradiance 47
        2.2.2   The Spectral Solar Irradiance 55
    2.3   Principles of Utilization of Sunlight for in-Space Propulsion 59
    References 62

3   Sailcraft Concepts 65
    3.1   Solar-Wind Spacecraft 66
        3.1.1   The Solar Wind: an Overview 66
        3.1.2   Magnetic and Plasma Sails 77
        3.1.3   Electric Sail 78
        3.1.4   Conceptual and Practical Issues 79

xvii

3.2	Main Elements of Solar-Photon Sailcraft	81
	3.2.1 Sail Shape	82
	3.2.2 Attitude Control	84
	3.2.3 Thermal Control	85
	3.2.4 Communication	86
	3.2.5 Payload	87
3.3	Micro and Nano Sailcraft Concepts	87
3.4	Additional Remarks	89
	References	90

4 Solar Sails in Interplanetary Environment ... 93

4.1	A Sail in the Solar Wind	93
	4.1.1 Mean Inter-Particle Distance	94
	4.1.2 Debye Length	95
	4.1.3 Electron's Gyro Radius	95
	4.1.4 Proton's Inertial Length	97
	4.1.5 Proton's Gyro Radius	98
	4.1.6 Collision Mean Free Path	98
	4.1.7 Comparing Solar-Wind Lengths to Sail Size	100
4.2	Ultraviolet Light onto Sail	101
	4.2.1 Concepts of Electric Spacecraft Charging	103
	4.2.2 Photoelectrons from a Solar Sail	108
	4.2.3 Evaluation of the Sail's Floating Potential	113
4.3	Solar-Wind Ions Inside the Sail	117
	References	120

Part III Sailcraft Trajectories

5 Fundamentals of Sailcraft Trajectory ... 125

5.1	Scales of Time and Frames of Reference	125
5.2	The Sun's Orbit in the Solar System	132
5.3	Equations of Heliocentric Motion	135
	5.3.1 Sailcraft Motion in Radius-Longitude-Latitude Chart	142
	5.3.2 Sailcraft Energy and Angular Momentum	143
	5.3.3 Sailcraft Trajectory Curvature and Torsion	148
5.4	Equations of Planetocentric Motion	153
5.5	Thrust Maneuvering: a Few Basic Items	156
	References	162

6 Modeling Light-Induced Thrust ... 165

6.1	Irradiance in the Sailcraft Frame	166
6.2	Photon Momentum onto Sail	173
	6.2.1 Approximating Irradiance and Input Momentum	175
6.3	What Does Surface Mean?	176

Contents   xix

6.4   Topology of a Solar-Sail Surface 180
    6.4.1   Concepts from Modern Optics 182
    6.4.2   BRDF and BTDF 184
    6.4.3   Describing Sail's Local Roughness 188
6.5   Calculation of Thrust 198
    6.5.1   Simplified Scheme of Sail-Photon Momentum Balance . . 198
    6.5.2   A Reminder of Optical Diffraction 202
    6.5.3   The Importance of Light's Polarization 207
    6.5.4   Diffraction-Based Reflectance Momentum 211
    6.5.5   The Twofold Effect of Absorption 219
    6.5.6   Transmittance Momentum 224
    6.5.7   Considering Large-Scale Curvature and Wrinkling 226
6.6   Total Momentum Balance and Thrust Acceleration 234
6.7   The Surface-Isotropic Flat-Sail Model 236
    6.7.1   Lightness Vector of Flat Sails 237
    6.7.2   Isotropic-Surface Scattering 240
6.8   Conclusions 247
References 249

7   The Theory of Fast Solar Sailing 255
7.1   Extending the Heliocentric Orbital Frame of Sailcraft 256
7.2   Two-Dimensional Motion Reversal 263
    7.2.1   Numeric Evidence 263
    7.2.2   *H*-Reversal Theory 274
7.3   Two-Dimensional Direct Motion 287
7.4   Three-Dimensional Motion Reversal 292
    7.4.1   Meaning of 3D Motion Reversal 293
    7.4.2   Numerical Issues 293
    7.4.3   Motion-Reversal Three-Dimensional Trajectories 295
    7.4.4   About the Definition of 3D Fast Sailing 316
7.5   General Comments 317
References 318

Part IV   Advanced Aspects

8   Approach to SPS Trajectory Optimization 323
8.1   Current Optimization Problem: Variational Approach 323
8.2   Current Optimization Problem: Non-Linear Programming 329
8.3   The Photon-Acceleration Technological Equivalent 336
8.4   Numerical Examples via Non-Linear Programming 338
    8.4.1   Using Results from Scalar Scattering Theory 342
    8.4.2   Lightness Number Smaller than Unity 345
    8.4.3   Lightness Number Greater than Unity 356
8.5   Other Studies on 2D and 3D *H*-Reversal Motion 369
8.6   Objections and Concluding Remarks 371
References 373

9 Advanced Features in Solar-Photon Sailing ... 379

9.1 TSI-Variable Fast Trajectories ... 379
9.2 Modeling Modifications of Optical Parameters ... 385
9.3 Conclusion ... 396
References ... 397

Index ... 399

List of Main Acronyms

ACE	Advanced Composition Explorer	72
ACRIM	Active Cavity Radiometer Irradiance Monitor	46
AFM	Atomic Force Microscope	190
AOCS	Attitude and Orbit Control System	235
ASTM	American Society for Testing and Materials	58
AU	Astronomical Unit	13
TCB	Barycentric Coordinate Time	4
BCRS	Barycentric Celestial Reference System	128
BE	Bose-Einstein distribution	36
BIPS	Barycentric Invariable-Plane System	133
BRDF	Bidirectional Reflectance Distribution Function	183
BSDF	Bidirectional Scattering Distribution Function	183
BSSRDF	Bidirectional surface-scattering Reflectance Distribution Function	185
BTDF	Bidirectional Transmittance Distribution Function	183
CAS	Computer Algebra System	216
CCBRDF	cosine-corrected Bidirectional Reflectance Distribution Function	208
CCBSDF	cosine-corrected Bidirectional Scattering Distribution Function	183
CH	coronal hole	66
CNT	Carbon Nanotube	395
CIE	Commission Internationale de L'Eclairage	35
CIR	Corotating Interaction Region	72
CVD	Chemical Vapor Deposition	179
D2FS	2D Fast Sailing	277
D3FS	3D Fast Sailing	316
DE405	JPL Ephemeris File DE405/LE405	30
DSFG	dual-stage four-grid	27
EHOF	Extended Heliocentric Orbital Frame	259
EMB	Earth-Moon barycenter	332
EMB/OF	heliocentric orbital frame of the EMB	332
EOS	Earth Orbital Speed	265

ERA	Earth Rotation Angle	126
ESA	European Space Agency	23
ESEM	Extra-Solar Exploration Mission	356
EUV	Extreme Ultraviolet	46
gHS	generalized Harvey-Shack	218
GOF	Geocentric Orbital Frame	153
GTC	Global Theory of Curves	278
GTT	Glass Transition Temperature	346
GR	General Relativity	127
HCS	Heliospheric Current Sheet	71
HIF	Heliocentric Inertial Frame	20
HOF	Heliocentric Orbital Frame	139
HMF	Heliospheric Magnetic Field	66
HPS	Heliospheric Plasma Sheet	71
HR	Hertzsprung-Russell diagram	44
IAU	International Astronomical Union	127
ICs	initial conditions	266
ICRF	International Celestial Reference Frame	128
ICRS	International Celestial Reference System	128
IEPS	integral equation of potential scattering	204
IERS	International Earth Rotation and Reference Systems Service	126
IF	Inertial Reference Frame	4
IKAROS	Interplanetary Kite-craft Accelerated by Radiation Of the Sun	33
ISO	International Standards Organization	35
IUPAP	International Union of Pure and Applied Physics	35
JAXA	Japanese Aerospace Exploration Agency	23
JPL	Jet Propulsion Laboratory	129
l.h.s.	left-hand side	26
LM	Levenberg-Marquardt	339
LME	Levenberg-Marquardt-Moré	340
LMM	Levenberg-Marquardt-Morrison	339
LVLHF	Local-Vertical Local-Horizontal Frame	155
M2P2	Mini Magnetospheric Plasma Propulsion	78
mBK	modified Beckmann-Kirchhoff	217
MHD	Magnetohydrodynamics	51
MHJT	Mean Human Job Time	21
MHOF	Modified Heliocentric Orbital Frame	257
MNHB	Mission to Near-Heliopause and Beyond	347
MSFC	George C. Marshall Space Flight Center	263
MWCNT	Multi-Walled Carbon Nanotubes	86
NASA	National Aeronautics and Space Administration	23
NASA-ISP	NASA InterStellar Probe	263
NEP	Nuclear Electric Propulsion	18
NIP	Nuclear Ion Propulsion	23
NLLS	Non-Linear Least-Squares	338

NLP	Non-Linear Programming	21
NOAA	National Oceanic and Atmospheric Administration	54
ODE	Ordinary Differential Equation	262
OF	Orbital Frame	17
ODP	Orbit Determination Process	385
PCHs	polar coronal holes	67
PPU	power processing unit	25
PSD	Power Spectral Density	195
PTE	Photon Transport Equation	225
PVD	Physical Vapor Deposition	178
RCD	Reflectance Control Device	157
r.h.s.	right-hand side	153
RLL	Radius-Longitude-Latitude	142
rtn	heliographic reference frame	96
RR	Rayleigh-Rice	214
RTE	Radiative Transfer Equation	224
S/C	Spacecraft	4
SDO	Solar Dynamics Observatory	66
SF	Ship Frame	4
SGL	Solar Gravitational Lens	347
SI	International System of Units	xii
SOF	Spacecraft Orbital Frame	139
SOFA	Standards of Fundamental Astronomy	127
SOHO	Solar and Heliospheric Observatory	45
SPS	Solar-Photon Sailing	x
SR	Special Relativity	3
SRIM	The Stopping and Range of Ions in Matter	117
SSI	Spectral Solar Irradiance	55
SST	Scalar Scattering Theory	204
STEREO	Solar Terrestrial Relations Observatory	72
TAI	International Atomic Time	126
TDB	Barycentric Dynamical Time	130
TFD	Thin-Film Deposition	177
TCB	Barycentric Coordinate Time	4
TCG	Geocentric Coordinate Time	127
TS	thruster system	25
TSD	Thermal Spray Deposition	179
TIS	Total Integrated Scatter	194
TSI	Total Solar Irradiance	44
TT	Terrestrial Time	126
UTC	Coordinated Universal Time	127
UV	Ultraviolet	46
VST	Vector Scattering Theory	204
WSB	weak stability boundaries	332
XUV	soft X-ray wavelengths	58
ZTM	zero total momentum	5

List of Figures

Fig. 1.1	Mass-energy flow in a rocketship	6
Fig. 1.2	Example of finite-burn losses	17
Fig. 1.3	Photon rocket escaping from the solar system	22
Fig. 1.4	Nuclear ion drive to heliopause: the heliocentric branch	30
Fig. 2.1	Geometry of an infinitesimal surface receiving radiation from an infinitesimal source	41
Fig. 2.2	Light's diffusion momenta	42
Fig. 2.3	Total solar irradiance	48
Fig. 2.4	PMOD composite of total solar irradiance	49
Fig. 2.5	Sun's main layers	49
Fig. 2.6	Temperatures of solar magnetic regions	52
Fig. 2.7	Babcock-Leighton solar dynamo mechanism	54
Fig. 2.8	Cycle-24 sunspot number	55
Fig. 2.9	Spectral solar irradiance in solar cycle-23	57
Fig. 2.10	Relative changes in XUV/EUV irradiance in cycle-23	57
Fig. 2.11	SSI variations in the 100–100	58
Fig. 2.12	Sun's brightness temperature profile	58
Fig. 2.13	Integrated irradiance in cycle-23	59
Fig. 2.14	TSI in frequency domain	62
Fig. 3.1	Some images of solar coronal holes	67
Fig. 3.2	SDO images	68
Fig. 3.3	Scheme of fast-slow wind near the Sun	68
Fig. 3.4	Scheme of heliosphere with large shocks	70
Fig. 3.5	Heliospheric plasma sheet	71
Fig. 3.6	Schematic of a CIR	72
Fig. 3.7	Variations across a CIR	73
Fig. 3.8	Speed/pressure of solar wind—1	75
Fig. 3.9	Speed/pressure of solar wind—2	76
Fig. 3.10	Solar-wind dynamical pressure and TSI distributions	89
Fig. 4.1	Solar-wind characteristic lengths	100
Fig. 4.2	Distribution of electron gas in an electric field	102

Fig. 4.3	Thick-sheath floating potential	107
Fig. 4.4	Sail's potential contours	116
Fig. 4.5	Sail damage by solar wind	119
Fig. 5.1	Sun's barycenter trajectory in the invariable plane	133
Fig. 5.2	Sun's barycenter motion along the angular-momentum direction of the solar system	134
Fig. 5.3	Sailcraft deceleration	148
Fig. 6.1	Geometry of sunlight on sailcraft	167
Fig. 6.2	Examples of sail wrinkling	181
Fig. 6.3	Reflection from and transmission through non-ideal surface	184
Fig. 6.4	Thin-film surfaces via microfacets	185
Fig. 6.5	Surface images by AFM	191
Fig. 6.6	Macrosurface's joint probability density	193
Fig. 6.7	Simple model of sail/photon momentum balance	198
Fig. 6.8	Aluminum's dielectric function	210
Fig. 6.9	Approximating the sail swelling	230
Fig. 6.10	Approximating sail wrinkles	232
Fig. 6.11	Net diffuse momentum from non-ideal sail: example-1	243
Fig. 6.12	Net diffuse momentum from non-ideal sail: example-2	244
Fig. 6.13	Net diffuse momentum from non-ideal sail: example-3	244
Fig. 6.14	Net diffuse momentum from non-ideal sail: example-4	245
Fig. 7.1	2D motion reversal: trajectories	265
Fig. 7.2	2D motion reversal: hodographs	266
Fig. 7.3	2D motion reversal: energy, H-invariant	267
Fig. 7.4	Plain acceleration vs H-reversal mode	269
Fig. 7.5	Control regions for H-reversal mode	273
Fig. 7.6	Rocket/sail reference flights	287
Fig. 7.7	Fast sailing: the option of full direct motion	290
Fig. 7.8	Three-L control for 3D motion: case-1	302
Fig. 7.9	Three-L control for 3D motion: case-2	304
Fig. 7.10	Three-L control for 3D motion: case-3	307
Fig. 7.11	Three-L control for 3D motion: case-4	309
Fig. 7.12	Three-L control for 3D motion: case-5	311
Fig. 7.13	Three-L control for 3D motion: case-6	313
Fig. 7.14	Three-L control for 3D motion: case-7	315
Fig. 8.1	Trajectory optimization schematic	339
Fig. 8.2	Aluminium's specular reflectance	343
Fig. 8.3	Aluminium's diffuse reflectance	344
Fig. 8.4	Aluminium's absorption	344
Fig. 8.5	Optimal trajectory of MNHB via $\mathcal{L} < 1$	350
Fig. 8.6	Optimal L for MNHB via $\mathcal{L} < 1$	350
Fig. 8.7	MNHB via $\mathcal{L} < 1$: evolution of L	351
Fig. 8.8	MNHB via $\mathcal{L} < 1$: evolution of the control angles and sunlight incidence	351
Fig. 8.9	MNHB via $\mathcal{L} < 1$: evolution of $\mathcal{L}$ and $\eta_{(thrust)}$	352

Fig. 8.10	MNHB via $\mathcal{L} < 1$: behavior of distance and temperature	352
Fig. 8.11	MNHB via $\mathcal{L} < 1$: evolution of H and H_z	353
Fig. 8.12	MNHB via $\mathcal{L} < 1$: orbit of $\mathbf{H}$	353
Fig. 8.13	Hodograph of the MNHB sailcraft	354
Fig. 8.14	Sailcraft's E, H, V exhibited by MNHB sailcraft via $\mathcal{L} < 1$	354
Fig. 8.15	Sun-Earth-sailcraft aspect angle in MNHB	355
Fig. 8.16	Boom specific mass vs sail side	358
Fig. 8.17	Example of ESEM: sequence of events	359
Fig. 8.18	Optimal trajectory of ESEM via $\mathcal{L} > 1$	363
Fig. 8.19	ESEM via $\mathcal{L} > 1$: $\mathbf{L}$ orbit	363
Fig. 8.20	ESEM via $\mathcal{L} > 1$: evolution of $\mathbf{L}$	364
Fig. 8.21	ESEM via $\mathcal{L} > 1$: evolution of mass and lightness number	364
Fig. 8.22	ESEM via $\mathcal{L} > 1$: evolution of sail axis angles	365
Fig. 8.23	ESEM via $\mathcal{L} > 1$: rotation of the sail axis	365
Fig. 8.24	ESEM via $\mathcal{L} > 1$: evolution of m, R, and T_s	366
Fig. 8.25	ESEM via $\mathcal{L} > 1$: evolution of H and H_z	367
Fig. 8.26	ESEM via $\mathcal{L} > 1$: orbit of $\mathbf{H}$	367
Fig. 8.27	ESEM via $\mathcal{L} > 1$: the hodograph	368
Fig. 8.28	ESEM via $\mathcal{L} > 1$: E, V, H	368
Fig. 8.29	ESEM via $\mathcal{L} > 1$: Sun-Earth-SC aspect angle	369
Fig. 9.1	TSI time series in 2009–2011	381
Fig. 9.2	MNHB profiles comparison: TSI-constant	382
Fig. 9.3	MNHB profiles comparison: TSI-variable	384
Fig. 9.4	MNHB: optimal trajectory from optical degradation	392
Fig. 9.5	MNHB: hodograph from optical degradation	392
Fig. 9.6	MNHB: control with optical degradation	393
Fig. 9.7	MNHB: changes of optical quantities	393
Fig. 9.8	MNHB: temperature profile from optical degradation	394
Fig. 9.9	MNHB: new $E-H-V$ profiles	394

List of Tables

Table 1.1	Nuclear ion drive to heliopause: main propulsion parameters and spacecraft mass breakdown	28
Table 1.2	Nuclear ion drive: heliocentric thrust control for maximizing the terminal speed	29
Table 4.1	Some solar-wind mean quantities at 1 AU over cycle-23	94
Table 7.1	Speed increase intervals	284
Table 7.2	Examples of direct fast-sailing trajectories	291
Table 7.3	3D trajectory inconsistency	294
Table 7.4	3D motion induced by the three-$\mathbf{L}$ control: case-1	301
Table 7.5	3D motion induced by the three-$\mathbf{L}$ control: case-2	303
Table 7.6	3D motion induced by the three-$\mathbf{L}$ control: case-3	306
Table 7.7	3D motion induced by the three-$\mathbf{L}$ control: case-4	308
Table 7.8	3D motion induced by the three-$\mathbf{L}$ control: case-5	310
Table 7.9	3D motion induced by the three-$\mathbf{L}$ control: case-6	312
Table 7.10	3D motion induced by the three-$\mathbf{L}$ control: case-7	314
Table 8.1	Trajectory optimization with $\mathcal{L} < 1$: sailcraft data	348
Table 8.2	Trajectory optimization with $\mathcal{L} < 1$: optimal control	349
Table 8.3	Trajectory optimization with $\mathcal{L} > 1$: sailcraft data	360
Table 8.4	Trajectory optimization with $\mathcal{L} > 1$: optimal control	362
Table 9.1	MNHB with optical degradation: optimal control	391

Part I
A Review of Rocket Spacecraft Trajectories

Chapter 1
Some General Rocket Features

Rocket's Dynamics in Field-Free Space and Gravity Field   Before dealing with any model for computing the trajectories of solar-sail vehicles, we review a topic of basic importance for spaceflight: *the rocket propulsion*. This chapter aims at highlighting features peculiar to in-space rocket vehicles and, consequently, to the related mission designs. In particular, one can see that the exhaust speed *not always* is the main reference quantity in rocket dynamics, though it is important, of course. In addition, the presence of a gravity field may strongly reduce the performance expected from the simple field-free space equation. The properties of photon rockets and ion propulsion rockets are compared through mission cases.

1.1 Introduction

During four decades, from 1950's to 1980's, numerous space scientists and engineers investigated the properties of very different concepts of rocket space vehicle. Plenty of engine design concepts were proposed and carefully analyzed. Much of the known physics was applied to try to design rocket ships compliant with ambitious mission concepts. Special Relativity (SR) was fully used in many cases for very deep space mission concepts. (In the Nineties and in the first years of 2000's, new space propulsion concepts, using frontier physics, were introduced; some of them are still in progress.) One of the problems was to compare different propulsions, including rocket and non-rocket systems. Even restricting the comparison between rocket-based engines, a number of interesting general features could be carried out. At the end of the Seventies and in the Eighties, general mathematical models of rocket vehicle were published in the specialized literature. Some basic features of rocket-vehicle can be obtained from such models (which use few parameters) independently of both the particular engine and power source onboard. Here, we shall use a rather general model for describing different rockets and determining a number of properties oriented to photon-rocket and nuclear-electric engine rocket. All this should help the reader to better understand the difference in performance between rocket and solar-sail propulsions.

G. Vulpetti, *Fast Solar Sailing*, Space Technology Library 30,
DOI 10.1007/978-94-007-4777-7_1, © Springer Science+Business Media Dordrecht 2013

1.2 Frames of Reference

The title of this chapter contains the adjective "General". This means that, in order to prove rocket's basic properties in their generality, we shall use some concepts and equations from SR.

In order to write the mass-energy conservation laws, we shall refer to an Inertial Reference Frame (IF) and the set of instantaneously-at-rest Ship Frames (SFs) as usually defined in SR. Each frame in such set can be considered as inertial in the time interval $[t, t + dt]$, where all velocity-dependent equations hold at $\mathbf{V}(t)$, namely, the ship velocity with respect to IF. This is the conceptual tool by which SR can deal with accelerating objects.

Implicitly above, we have denoted the coordinate time by t. The vehicle's proper time will be denoted by τ.[1] For example, one may think of t as the Barycentric Coordinate Time (TCB).[2] We shall return on this points in Chap. 5.

It is important to realize that the basic properties of vehicle-related physical quantities should *first* be written with respect to SF, and *then* transformed to IF. This holds for any symmetry is being applied in the transformation (rotation, translation, boost, and so on).

1.3 Modeling a Rocket

A rocket system is one using only mass-energy stored or produced *onboard* for getting translational motion. It is almost a standard way to name *the spacecraft* (S/C) the assembly containing mainly payload, attitude control, small translation engine(s), thermal control and telecommunication, *but* the main propulsion system. Normally, this is an engineering unit completely distinct from the spacecraft itself. Here, we adopt the convention that a spaceship consists of the complete propulsion unit *and* the spacecraft. This will apply also to sail-propelled vehicles, although we shall use other more specific names for them.

In this chapter, we are concerned with the motion of the center of mass, or the barycenter,[3] of the rocket spaceship (which ejects mass-energy) with respect to IF.

Any spaceship endowed with a pure-rocket propulsion system (rocketship) can be mass broken down into three main systems: the active mass (M_a), i.e. the fuel[4]

---

[1]Let us remind the reader that the coordinate time is integrable, but not invariant. The converse is true for the proper time.

[2]As for many international terms in Astronomy, this acronym is derived from the French language.

[3]In the dynamics of multi-body systems, where the concept of *augmented* bodies is introduced, some authors define the center-of-mass of any constituting body in the usual way, whereas call the (instantaneous) center-of-mass of the augmented body as its barycenter. The interested reader could deepen such formidable concepts in excellent textbooks, e.g. [50] and [12] of Chap. 5. In any case, center of mass and barycenter should not be confused with the center of gravity, which depends on the gravitational field where the considered body is immersed.

[4]For chemical/electric, nuclear, or antimatter propulsion.

1.3 Modeling a Rocket

from which the propulsive energy has to be extracted, the inert mass (M_i) to which part of this energy is transferred and then exhausted, and the gross payload (M_{GL}). Depending on the specific engine, the exhaust or ejection beam (jet) may contain some fraction of the active mass. In addition, not all of the inert mass may be exhausted. Therefore, the overall ejection beam may consist of sub-beams originated from active and inert masses. Many parameters could be used for describing real systems in any spacecraft, of course. In the present context, we shall use the minimum number of parameters necessary to describe the essential propulsion characteristics of a rocketship. One should note that dM_i represents the amount of inert mass going from the tank subsystem to the engine subsystem in the proper time $d\tau$. In the same time interval, the amount dM_a of fuel is burnt. This means that (atomic, molecular, nuclear or particle) reactions run (in some volume also permeated by an external field) by yielding end products, generally endowed with mass, but not necessarily.

Before showing the mass-energy flow in a general rocket, we define the parametric entries to the model as follows:

- ε_a denotes the fraction of dM_a transformed into energy potentially utilizable for propulsion and other spaceship needs; therefore, $1 - \varepsilon_a$ represents the rest-mass of the final reaction products.
- ε_k is the fraction of $dM_a c^2$ that is actually utilizable by the propulsion system as a whole.[5]
- $\varepsilon_l = \varepsilon_a - \varepsilon_k$ is the energy fraction of the reaction products, which have negligible interaction or are lost anyway from the power generator; $\varepsilon_l \, dM_a$ is assumed to be lost at zero total momentum (ZTM) in SF.
- η_b is the efficiency in transforming $\varepsilon_k \, dM_a \, c^2$ into the kinetic energy of the ejection beam.
- ζ denotes the fraction of the residual fuel $(1 - \varepsilon_a) \, dM_a$ contributing effectively to the jet.
- ξ is the dM_i fraction exhausted away; in contrast, the amount $(1 - \xi) \, dM_i$ (that is not ejected) is assumed lost at ZTM in SF.
- χ is defined as the ratio dM_i / dM_a.
- η_s is the beam spread factor, which essentially is related to the way the jet is exhausted, the (generalized) nozzle, and the particle properties.
- $\mathbf{U}$; strictly speaking, it is the velocity—in SF—of the exhausting particles when the particle-nozzle interaction ceases. It is a derived quantity, as shown below. Its magnitude $U \equiv |\mathbf{U}|$ is the ejection or exhaust speed.

The reader should note that the model parameters have been defined with respect to elemental amounts of mass in the interval $[\tau, \tau + d\tau]$; each of them may in principle be proper-time dependent. We make the simplifying assumption that $-\mathbf{U} \parallel \mathbf{V}$ at any τ, $\mathbf{V}$ being the spaceship's velocity in IF. Let V the length of $\mathbf{V}$. As a result,

---

[5]Energy is required for power conditioning, propellant rate control, pre-heating, ion generation, confinement fields, etc.

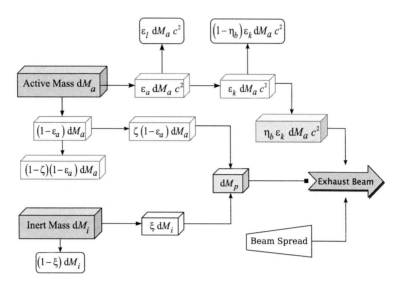

Fig. 1.1 Mass-energy flow in a rocketship. *Rounded rectangles* represent mass-energy wasted into space at zero total momentum in the spaceship frame

the Lorentz factor of the jet speed, as observed in IF with the value W, is equal to

$$\gamma_W = (1 - UV/c^2)\gamma_U \gamma_V$$
$$\gamma_x \equiv 1/\sqrt{1 - x^2/c^2} \tag{1.1}$$

Equation (1.1) will be useful in writing down the energy change due to the jet. In next section, we will show how U can be expressed as function of the efficiency parameters defined above. Thus, the current model contains seven independent parameters, which are sufficient for characterizing the main properties of the known rocket propulsion systems, including the photon rocket, as we shall see in the next sections.

Figure 1.1 shows the logical flow of the mass-energy available onboard in a rocketship. The active mass releases a certain amount of utilizable energy, a fraction of which is lost in the process necessary to transfer it to the ejection beam (regardless of the type of final particles the beam consists of). A part of the final reaction products or some inert mass (from tanks), or both, arrive to the beam generation system; there, such mass receives the above energy.

Note that there may be no rest mass in the beam, namely, the overall engine may consist of both an emitter of photons and a collimator. In the energy flow model, the values of two parameters $\varepsilon_{a,k}$ distinguish various power sources; in contrast, the different efficiencies typify the engine subsystems. In some cases, power and engine systems coincide (e.g. chemical or some envisaged fusion propulsion). It is important to realize that (1) in general, the exhaust beam can be characterized by a certain ratio χ, independently of the possibility to eject a fraction of the active mass; (2) a certain amount of mass-energy is unavoidably lost: the current assumption is

1.4 Rocketship Dynamics in Field-Free Space

that any losses occur at zero total momentum in SF. Apart from the total energy lost from a real rocketship, there is always a certain amount that does not contribute to the thrust. The related four-momentum in IF via Lorentz transformations (boost) result particularly simple and intuitive.

In the next sections, we will first analyze the dynamics of a rocketship in field-free space[6] and, then, in a gravity field. Among many things, this will produce the rocketship's basic equation in differential form.

1.4 Rocketship Dynamics in Field-Free Space

The four-vector formalism applied to the ship's energy-momentum, or four-momentum, is quite appropriate for carrying out the equation of motion of rockets. As explained above, we are interested in the basic rocket properties; therefore, we will focus on the two-dimensional (1-space + 1-time) motion equation.

> When the general four-dimensional space is used, the space-time metric is assigned a Lorentzian *signature*: either $\{+++-\}$ *or* $\{---+\}$. The first one is the preferred choice by relativity researchers, whereas particle physicists like the second one. Though the algorithm in this chapter appears independent of such a choice here, the underlying signature is $\{---+\}$, which produces the *invariant* $ds^2 = dt^2 - dr^2 = d\tau^2$, where $\mathbf{dr}$ is the vector position change (of a point-like body in IF) after dt, and $c = 1$.

In this case, the same dynamical information can be extracted from either the space-like part or the time-like part of the two-dimensional energy-momentum conservation law. We shall deal with *energy* and its flow, as shown in Fig. 1.1.

Unless otherwise specified, in all equations below, we shall use normalized units ($c = 1$, *year* $= 1$), where the year value may come from one of the usual definitions of year (we will use the standard year: 365.25 days). An appropriate unit mass is the initial mass of the spacecraft, or M_0. Note that, the unit acceleration, say, g^* takes on 9.5 m/s^2, approximately. The exact value depends on the chosen year value; in our choice, $g^* = 9.49985$ m/s^2.

Let us begin with expressing the energy conservation in the ship-frame. Reminding that, in the normalized units, the quantity $\gamma_U - 1$ represents kinetic energy per unit mass, we can write:

$$(\gamma_U - 1)\, dM_p = \eta_b \varepsilon_k\, dM_a \equiv K\, d\tau \tag{1.2}$$

$$dM_p = [\zeta(1 - \varepsilon_a) + \xi\chi]\, dM_a \tag{1.3}$$

---

[6]Field-free space is a strong idealization, considering the modern field theory implications. In our context, this means *no fields are considered but the local propulsive thrust*.

where we recall

$$\chi \equiv dM_i / dM_a$$

In Eq. (1.2), K represents the beam's kinetic power. Combining (1.2) and (1.3) results easily in

$$\gamma_U = 1 + \frac{\eta_b \varepsilon_k}{\zeta(1 - \varepsilon_a) + \xi \chi} \tag{1.4a}$$

$$U = \sqrt{1 - \gamma_U^{-2}} = \frac{\sqrt{\eta_b \varepsilon_k (\eta_b \varepsilon_k + 2(\xi \chi + \zeta(1 - \varepsilon_a)))}}{\eta_b \varepsilon_k + \xi \chi + \zeta(1 - \varepsilon_a)} \tag{1.4b}$$

Another basic relationship is the rocketship's rest-mass change, brought about by the processes schematized in Fig. 1.1, and expressed as function of the burnt active mass:

$$-dM = [\zeta(1 - \varepsilon_a) + \chi + \varepsilon_a] dM_a \tag{1.5}$$

As a point of fact, the fraction $(1 - \zeta)(1 - \varepsilon_a)$ of the active mass is retained onboard (e.g., a nuclear rocket where only part of the fuel reaction products is able to be exhausted away, and/or part of the released energy is re-absorbed). Obviously, only three of the previous five equations are *independent* relationships in the ship frame. Now, let us switch to the inertial frame.

In IF, during the coordinate time interval dt corresponding to the proper time interval $d\tau$ via $dt = \gamma_V \, d\tau$, the energy variations of the ship, the jet and the mass lost into space can be worked out by inserting the above ZTM assumption into the SR transformation between two frames of reference in relative motion.[7] The result can be cast into the following simple form:

$$dE_s = M V \gamma_V^3 \, dV + \gamma_V \, dM \tag{1.6a}$$

$$dE_p = \gamma_w \, dM_p = (1 - U V)\gamma_U \gamma_V \, dM_p \tag{1.6b}$$

$$dE_l = [(1 - \xi)\chi + \varepsilon_a - \eta_b \varepsilon_k]\gamma_V \, dM_a \tag{1.6c}$$

These energy changes are not independent, but they have to satisfy the conservation law:

$$dE_s + dE_p + dE_l = 0 \tag{1.7}$$

where the subscript l refers to as the mass losses (but the jet, of course). Inserting (1.2), (1.3) and (1.5) into (1.7) results in

$$\gamma_V^2 \, dV = -U_e \frac{dM}{M} \tag{1.8a}$$

---

[7] With ZTM in SF, the four-momentum of a point-like object observed in IF equals the object's energy in SF times the four-velocity of SF with respect to IF.

1.4 Rocketship Dynamics in Field-Free Space

$$U_e \equiv \eta_m U, \qquad \eta_m \equiv \frac{\eta_b \varepsilon_k + \xi \chi + \zeta(1 - \varepsilon_a)}{\chi + \varepsilon_a + \zeta(1 - \varepsilon_a)} \tag{1.8b}$$

In Eq. (1.8b), U is given by the rightmost equation of (1.4b), while η_m (or the mass utilization efficiency) is the ratio between the beam's total power and the overall power the vehicle gives out. Note that the parameter χ is the sole upper-unbounded quantity in η_m. For the conventional non-relativistic specific-impulse propulsion systems, the χ-terms are by far dominant so that $\eta_m = \xi$, namely, the fraction of the propellant accelerated up to the exhaust speed. On the other extreme, for *photon* rockets, both χ and ζ vanish, so the utilization efficiency results in $\eta_m^{(ph)} = \eta_b \varepsilon_k / \varepsilon_a \leq \eta_b$. Thus,

$$U^{(ph)} = 1, \qquad U_e^{(ph)} = \eta_b \varepsilon_k / \varepsilon_a \tag{1.9}$$

The *effective* jet speed may be rather lower than 1 in a rocket where the energy source releases some large amount of γ-rays and neutrinos.

Equation (1.8a) is the general (scalar) rocket equation in field-free space, in differential form, characterized by both the ship's total mass change and time dilation.[8]

> This equation, in various equivalent forms, has been re-discovered *independently* by many authors (including this author) in the 20th century and in the first years of the 21th century too. This may be explained in different ways. Until the first years of the Eighties, graduate students and young researchers in some countries found many difficulties in "discovering" old papers/books because the search in big libraries was time-consuming and expensive. Thus, some of them liked to carry out equations by believing to be the first in the history of spaceflight. In the last years, through the World Wide Web (WWW), search has become so a spread and common mentality that many students deem (much probably) that what is not in WWW simply does not exist. The result is the same: they ignore decades of scientific research. And, therefore, they re-discover many, many equations and related problems; but without the experience of elder investigators. Furthermore, many reviewers of current prestigious journals are used to not advise the young authors that a deeper bibliographic search—even of papers/books physically present in traditional libraries—should be done before proposing something as new.
>
> In addition to the fundamentals works by K.E. Tsiolkovsky, R.H. Goddard, F.A. Zander, and H. Oberth, the author likes to recommend some papers and books—in alphabetical order—which though do not exhaust all the pioneering works in those years. At least to the author knowledge, the following

---

[8]It was generalized to four dimensions in tensorial form [28], Eq. (31), involving any exhaust type. This is because energy conservation and momentum conservation give different pieces of propulsion information, in general. These ones coincide only in the one-space + one-time case.

ones (which contain many other references inside) are particularly worth citing, reading and remembering: [1, 4, 6, 7, 10, 13, 15–18, 21], and [14]. They truly belong to the history of Astronautics[9] with its profound meaning for the Humankind's future.

Were U_e constant, Eq. (1.8a) could be integrated easily to give:

$$\text{arctanh}(V(t)) - \text{arctanh}(V_0) = U_e \ln \frac{M_0}{M(\tau)} \equiv U_e \ln R \qquad (1.10)$$

where R denotes the propulsion mass ratio relative to the time interval $[\tau_0, \tau]$. If both $V_0^2 \ll 1$ and $V(t) - V_0 \ll 1$, then (1.10) reduces to the classical rocket equation with an effective jet speed, namely, a bit more general than Tsiolkovsky's equation. It is easy to see that time dilation, through the arctanh function, entails that more mass has to be ejected (to achieve a prefixed speed) with respect to the classical case with the same effective jet.

In the current $(1 + 1)$ case, Relativity tells us that a force, measured in IF, is equal to the corresponding force measured in SF. Therefore, the rocketship thrust, *as sensed onboard*, is equal to the thrust[10] measured in IF:

$$T^{(IF)}(t) = T^{(SF)}(\tau) \equiv T \qquad (1.11)$$

In contrast, the corresponding accelerations are related by

$$\left(\gamma v^3 a^{(IF)}\right)_t = a^{(SF)}(\tau) \qquad (1.12)$$

Equation (1.11) is very useful for calculating thrust straightforward. As a point of fact, the force stemming from the jet can be expressed as:

$$T = \gamma_U U \dot{M}_p = -U_e \dot{M}$$
$$= \sqrt{1 + \frac{2}{\eta_b \varepsilon_k}[\zeta(1 - \varepsilon_a) + \xi \chi]\eta_b \varepsilon_k \dot{M}_a} \equiv C_T K \qquad (1.13)$$

where the upper dot means differentiation with respect to τ. The first part of the second row of (1.13) has been carried out by means of Eqs. (1.5) and (1.8b). Note that, in the current natural units ($c = 1$), the second part contains the quantity K defined in Eq. (1.2): thrust equals the jet's kinetic power *amplified* by the presence of inert matter in the beam. Equation (1.13) is still partially ideal because, actually, jet's particles exhibit a non-negligible spread in speed and exhaust direction. Thus, in general, thrust should be written as

$$\tilde{U}_e = \eta_s U_e \quad \Rightarrow \quad \tilde{T} = \eta_s T \quad (0 \le \eta_s \le 1) \qquad (1.14)$$

---

[9]The new emerging science was named *Astronautics* by J.H. Rosny [7].

[10]In the general (3-space + 1-time) case, the relationship (1.11) holds provided that $\mathbf{T}^{(SF)} \parallel \mathbf{V}$.

1.4 Rocketship Dynamics in Field-Free Space

In Sect. 1.4.1, we will highlight some features of the current model of rocketship. For the moment, we will go on using the set $\{U, U_e, T\}$, which contains the main information relevant to the trajectory profile.

For any real propulsion system, the absolute minimum of thrust is achieved when $(\chi, \zeta) = (0, 0) \Rightarrow C_T = 1$, which characterizes a photon rocket:

$$T^{(ph)} = \eta_b \varepsilon_k \dot{M}_a = K \tag{1.15}$$

In other words, the thrust generated by a photon rocket engine is *the lowest one*—for the same amount of kinetic jet power—among all possible rocket systems. This is also confirmed by minimizing thrust from its general relativistic expression as function of the beam's kinetic power K and the (true) speed U:

$$K = (\gamma_U - 1)\dot{M}_p = \frac{\gamma_U - 1}{U\gamma_U} T = \frac{1 - \sqrt{1 - U^2}}{U} T$$

$$\Rightarrow \tag{1.16}$$

$$\min(T)\,|_{\text{given } K} = K \min\left(\frac{U}{1 - \sqrt{1 - U^2}}\right) = K$$

Perhaps, this might surprise a bit; however, it is the generalization of the elementary statement that, in comparing two bodies with the same kinetic energy, the lower momentum belongs to the lower-mass body.

Previous equations allow us to work out the thrusting proper time:

$$\tau_f - \tau_0 = \frac{M_0 - M_f}{-\dot{M}} = \frac{M_0 U_e}{T}(1 - R^{-1}) \tag{1.17}$$

R is given by Eq. (1.10). Equation (1.17) becomes particularly expressive for photon-rockets. To such aim, let us specialize the current ship mass breakdown model to photon-rocket:

$$M_0 = M_a + M_{GL} = M_a + M_W + M_L = M_a + \varepsilon_a \dot{M}_a / W_m + M_L \tag{1.18}$$

M_W denotes the whole photon propulsion system (i.e., the power and the collimator systems) and M_L is the ship's payload. Of course, there is no inert mass system in a photon rocket. In (1.18), W_m denotes the specific power of the power source. Thus we get

$$(\tau_f - \tau_0)_{(ph)} = \left(\frac{M_0 U_e}{T}\right)_{(ph)} (1 - R^{-1})$$

$$= \frac{1}{W_m} \frac{M_0}{M_W} \left[1 - \left(\frac{1 + V_0}{1 - V_0} \frac{1 - V_f}{1 + V_f}\right)^{\frac{1}{2U_e}}\right] \tag{1.19}$$

In Eq. (1.19), the identity $\text{arctanh}(x) = \frac{1}{2}\ln\frac{1+x}{1-x}$, Eqs. (1.5), (1.9), and (1.15) have been used. If one thinks of utilizing a photon rocket for getting $\Delta V \ll 1$ from

$V_0 \ll 1$, the acceleration time may be series expanded to give

$$(\tau_f - \tau_0)_{(ph)} \cong \frac{(V_f - V_0)M_0}{U_e W_m M_W}, \quad U_e \cong \eta_b$$

$$(t_f - t_0)_{(ph)} \cong (\tau_f - \tau_0)_{(ph)} = \frac{9.49985}{\eta_b} \frac{M_0}{M_W} \frac{V_f - V_0}{W_m} \quad \text{(years)} \tag{1.20}$$

after recognizing that $\varepsilon_k \cong \varepsilon_a$ in such context. In Eq. (1.20), $(V_f - V_0)$ is expressed in km/s, and W_m in kW/kg.

1.4.1 Model Features

In the previous section, the general one-dimensional (1-space + 1-time) rocket equation in field-free space has been carried out. The energetics behind such equation is general, namely, independent of the particular engine or class of engines one is being considered. It holds from chemical engine to photon rocket and for any type of power source. A seven-parameter model of energy flow virtually encompasses all known types of rocket (though, of course, the real specific design of propulsion systems entails models with many, many parameters). Thus, some general properties of the rocket propulsion have been pointed out; in particular, the photon rocket results to exhibit the absolute minimum in thrust for a given exhaust beam's power.

The above model, which is a generalization of [25], stresses the role of (1) molecular, atomic, nuclear, or particle exothermic reactions, (2) the macroscopic energy transformations, and (3) the loss of mass-energy that is unavoidable in real systems. As a point of fact, the first thing to do is to calculate ε_a and ε_k from the reactions inside the power system; it is not a simple task, but it is quite general, because is the amount of mass converted into exploitable energy that determines the primary output of the power source. Part of such output ($\varepsilon_l = \varepsilon_a - \varepsilon_k$) will be unavoidable lost, while the complement (ε_k) may be channeled to the thrusting system; but only a fraction η_b of such a complement will energize the exhaust beam. It is important to realize that the energy output from the source does not consist only of thermal or electric energy; depending on the leading reactions, high-energy particles (e.g. γ-rays) and/or weak-interaction particles (e.g. neutrinos) may spring out. In any case, the model parameters evaluation has to be accomplished in the sense of special relativity. The low-energy reactions (such as the chemical ones) fall within this context.

The physical origin of the beam spread factor is different from any other efficiency dealt with in the energy flow model. Beam spreading consists generally of two pieces: (1) beam's geometric divergence, and (2) non-uniform values of the jet's particle speed. In some real engines, either may prevail. In general, both factors

1.4 Rocketship Dynamics in Field-Free Space

come in computing thrust; but only factor-2 has to be taken into account in calculating the jet's kinetic power. Two different engines belonging to the same propulsion mode may exhibit significantly different spreads. Speed distribution function and/or geometric beam divergence depend not only on the engine type and design, but also on its degradation history, which, in turn, may be mission-dependent. In the special case of photon rocket, the spread is due to non-trivial configurations (in terms of energy flux and geometry) of extended sources in the collimation system, whereas the problem of the beam energy lessening can be included in the parameter η_b. Once the actual thrust is measured (directly or indirectly), the second expression in (1.14) will then give the actual spread factor, which can be used in any other equations. Although the detailed analysis of rocket's jet spread is not among the scopes of this chapter, nevertheless we will return to consider η_s explicitly in Sect. 1.7.

1.4.2 Some Examples

In principle, and in the ideal field-free environment, one may want to use a photon rocket since the allowed terminal speed would be the highest one (with the same R). However, that could have a significance for matter-antimatter annihilation systems, deeply analyzed in the Eighties, for which $\varepsilon_a = 1$, $\varepsilon_k \leq 0.63$. In those years, potential applications of the proton-antiproton annihilation to space propulsion were conceived; they followed (after many years) the historical papers by Shepherd [18] and Sänger [15], written with exceptional astronautical visions when antiproton was not yet discovered, and laser was far to be functioning. Recently, the basic energetics of annihilation rockets has been re-considered [30], and this one agrees with the full history of the antiproton-nucleon annihilation products as published in [24] and [26].

For instance, if one annihilated $0.1M_0$ per year (a huge value) in a rocketship with $\eta_b = 1$, the effective exhaust speed would be equal to $0.63c$, whereas the initial thrust acceleration (in IF) would amount to

$$T = -U_e \dot{M} = 0.63 \times 0.1 M_0$$
$$T/M_0 = 0.063 g^* = 0.6 \text{ m/s}^2 \tag{1.21}$$

namely, 101 times the solar gravitational acceleration[11] at 1 Astronomical Unit (AU).[12] If one wants to achieve $0.5c$ starting from rest, then equation (1.10) provides $R = 2.391$. The corresponding specific power is $W_m = 0.1 M_0 c^2 / M_W > 9 \cdot 10^{12}$ kW/kg! Using (1.19), one gets the acceleration *proper-time*, that is 5.818 years.

---

[11]This is the acceleration that allows a body test to orbit about the Sun in exactly 1 year; it is equal to 0.0059300835 m/s^2.

[12]This is one of the units, outside the SI, which are normally accepted for use with the SI. Other symbols for AU that one may find in scientific or popular books are ua, au or a.u.

14   1 Some General Rocket Features

In reality, the best annihilation-based rocketship would be one ejecting charged pions, not photons, with a true exhaust speed as high as $0.9c$.[13]

Any other rocket system would be quite inappropriate for achieving $\frac{1}{2}c$: simply put, the propulsion mass ratio would be hugely high.

On the other side, one may think about using a nuclear-based rocket for exhausting photons, starting from the Earth orbit ($V_0 = 30$ km/s), for getting, say, an increment of 70 km/s (i.e. a value decidedly beyond the current and near-term rocket standards). If we consider a specific power as high as 100 kW/kg (far from our current and near-term nuclear technology, indeed, and requiring both a highly-compact reactor and an efficient very-low-mass photon collimator), we may apply (1.20) with $M_0/M_W = 2$ and $U_e = \eta_b = 0.9$, namely, a well-designed rocket-probe. The acceleration time results in $(t_f - t_0)_{(ph)} = 14.778$ yr. Obviously, such value has to be compared to the lifetime of the nuclear power system onboard. Let us remind, once again, that such figures pertain to the *very favorable* field-free case. In the next section, we shall analyze how gravity affects the rocket performance.

1.5 The Effect of Gravity on Rocket Dynamics

In this section, we shall use classical dynamics for describing how the gravity affects a long maneuver by rocket engines. Many problems of astrodynamics regard large orbital transfers. Using rocket propulsion entails the minimization of the propellant or the total energy[14] necessary for achieving the target state. One might think that, since the gravitational field is conservative, one may ignore it if one keeps the product between thrust acceleration and burning time at a given value. However, apart from circular orbit or some particular points in a general trajectory, *both* thrust acceleration and gravity acceleration can have along-track components. The gravitational one could be negative for long time too, with its time integral affecting the final spacecraft state even when the engines are on.

We focus on the problem of *escaping* from a central body with gravitational constant μ. Although the strategy of directing the thrust along the track is not the optimal one for maximizing the rocket's terminal speed after a prefixed propulsion time, nevertheless it is close to the optimal thrust direction profile and simple to implement. That produces a two-dimensional trajectory. By the Euler-Lagrange equations with a non-conservative field such as the thrust, it is a very simple matter to carry out the equations of the in-plane motion in a central gravity. Their Cartesian form

---

[13] In the Eighties, many antiproton-nucleon annihilation relativistic mission concepts have been investigated; for some of them, pion-ejection systems were devised and analyzed in detail.

[14] In some envisaged relativistic missions, minimizing the propellant or the total energy spent by the power system does not give the same results.

1.5 The Effect of Gravity on Rocket Dynamics

can be used for high-precision numerical integration. For analytical purposes, we shall use their polar form as follows:

$$\ddot{r} = -\frac{\mu}{r^2} + r\dot{\psi}^2 + a\frac{\dot{r}}{\upsilon} \tag{1.22a}$$

$$\ddot{\psi} = -2\frac{\dot{r}\dot{\psi}}{r} + a\frac{\dot{\psi}}{\upsilon} \tag{1.22b}$$

where r and ψ denote distance and polar angle, respectively, whereas a denotes the thrust acceleration and υ is the vehicle speed. In general, a is time-dependent. A dot denotes differentiation with respect to the time scale as defined in IF. One can easily get the following meaningful relationship:

$$\mathbf{V} \cdot \dot{\mathbf{V}} = \upsilon\dot{\upsilon} = a\upsilon - \frac{\mu}{r^2}\dot{r} \quad \Rightarrow \tag{1.23a}$$

$$\frac{1}{2}\,d(\upsilon^2) - a\upsilon\,dt + \frac{\mu}{r^2}\,dr = 0 \tag{1.23b}$$

$$a \equiv a(t) = a_0 \Big/ \left(1 - \frac{a_0}{U_e}(t - t_0)\right) \tag{1.23c}$$

Equation (1.23b) represents the *scalar* equation, in differential form, of a rocket-vehicle. It generalizes the classical equation to any rocket working in a Keplerian field. One should note that (1.23a) follows formally from the Euler-Lagrange equations; in other words, it is not the consequence of some intuitive considerations, but a strict result. If $|\mathbf{g}| \equiv g > 0$ and $\upsilon > 0$, then (1.23a) can be cast into the more expressive form

$$\dot{\upsilon} = \left(\frac{a}{g} - \frac{\dot{r}}{\upsilon}\right)g \tag{1.24}$$

Equation (1.24) shows clearly that the evolution of the spacecraft speed is driven by the ratio a/g because $0 \leqslant |\dot{r}|/\upsilon \leqslant 1$, whereas a may differ significantly from g. This acceleration ratio, in turn, depends primarily on a_0 and, secondarily, on U_e. Given a_0, if one uses a specific impulse high, then the ratio changes more slowly than that stemming from a lower specific impulse.

For our aims, we need the integral form of (1.23a). It is easy to recognize that the rightmost term in the above equation is the scalar product of the gravity acceleration $\mathbf{g}$ and the velocity direction $\mathbf{V}/\upsilon$. Thus, integrating over the burning time interval t_b and using $a(t) = T/M = -U_e\,\dot{M}/M$ from Sect. 1.4 one gets

$$\Delta\upsilon(t_b) = \int_0^{t_b} a\,dt + \int_0^{t_b} \mathbf{g} \cdot (\mathbf{V}/\upsilon)\,dt$$

$$= -\int_{M_0}^{M_b} (U_e/M)\,dM - \mu\int_0^{t_b} (\cos\varphi/r^2)\,dt \tag{1.25}$$

where M_b is the vehicle's mass at burnout, and φ is the angle from the vector position $\mathbf{R}$ to the vector velocity $\mathbf{V}$. In the current context, $0 \leq \varphi \leq \pi$. In Eq. (1.25), the optimized effective jet speed may depend on M; here it is sufficient to assume it constant to get

$$\Delta v(t_b) = U_e \ln \frac{M_0}{M_b} + \int_0^{t_b} \cos(\pi + \varphi) \frac{\mu}{r^2} \, dt \tag{1.26}$$

Equation (1.26) describes the motion, in *integral* form, of a rocket craft moving in a Keplerian field. Some remarks are in order. *First*, let us note that the propulsion time appears explicitly in the rocket's speed change. *Second*, gravity losses, namely, the integral in Eq. (1.26) cannot be calculated a-priori because one should know the evolution of $[\mathbf{R}, \mathbf{V}]$, which is not known in general. *Third*, if one aims at a certain final rocket speed, though in a gravitational field, then the two terms in (1.26) cannot be computed independently of one another, since additional propellant has to be ejected resulting in a mass-ratio increase. *Fourth*, the previous equation is easily extended to the three-dimensional multi-body problem; as a point of fact, the key point to remember is the integration of the along-track component of the total vector gravity acceleration, as expressed in the first row of (1.25).

It appears from Eq. (1.26) that gravity losses vanish only if $\mu = 0$, i.e., no gravity is present (ideally). In the current framework, it is so. However, in a more general context, gravity losses are a particular case of *finite-burn* losses. For instance, let us suppose to fly—in the one-dimensional field-free case— with the following prefixed endpoints: $\{m_0 = 1, x_0 = 0, v_0 = 0\}$ to $\{0 < m_f < 1, x_f = l, v_f = 0\}$. Engines exhibit the set $\{U, a_0\}$, where the jet speed U is prefixed, whereas the initial acceleration a_0 may be varied. Flight entails three rectilinear pieces: acceleration arc, coasting arc, and deceleration arc. Curves of the flight time to target can be parameterized with $w \equiv a_0 t_b^{(a)}$, the product of the initial acceleration times the burning time of the acceleration arc. It is easy to prove that the overall propulsion mass ratio depends on U and w. Since U is prefixed, assigning w and varying a_0 means that we are comparing flights with different initial accelerations, but the same propellant consumption. The related transfer times are shown in Fig. 1.2 for a very high performance rocket with $U = 1000$ km/s. Note that a_0 spans three orders of magnitude. It is evident how the flight time increases by decreasing the acceleration, namely, by increasing the total burning time, even though (for each curve) the propellant amount is the same. Minimum time is achieved ideally at the impulsive conditions, which in this case are well approximated by $a_0 > 0.05g^* = 0.475$ m/s^2. The degradation of dynamical quantities is due to the rocket energetics: part of the energy released in any time interval goes to the jet, and the remaining part is "absorbed" by the propellant that will be ejected at a later time. The worst situation happens for a photon rocket.

1.5 The Effect of Gravity on Rocket Dynamics

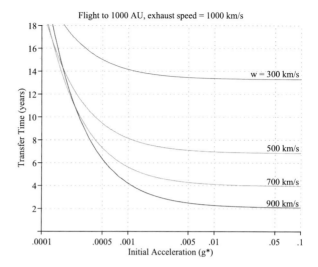

Fig. 1.2 Example of Finite-Burn Losses: a rocket traveling in field-free space with an ideal $U = 1000$ km/s. Quantity w is the initial acceleration times the acceleration-arc burning time. The vehicle leaves from and stops at rest

If we limit ourselves to a *spiral-like* trajectory about a central body, then we could use well-known approximate solutions in order to quantitatively show how the actual Δv depends on the gravity losses. If the rocketship is sufficiently far from the escape point (if any), its dynamical state is pretty described by

$$r(t) = \mu \bigg/ \left[v_0 + U_e \ln\left(1 - \frac{a_0 t}{U_e}\right) \right]^2, \qquad \psi(t) = \psi_0 + \int_0^t \left(\sqrt{\mu}/r(x)^{3/2}\right) dx \tag{1.27a}$$

$$\mathbf{R}^{(OF)} = \begin{bmatrix} r \\ 0 \end{bmatrix}, \qquad \mathbf{V}^{(OF)} = \begin{bmatrix} \dot{r} \\ r\dot{\psi} \end{bmatrix}, \qquad \cos\varphi = \dot{r}/v \tag{1.27b}$$

$$0 \le t \le t_b, \qquad 0 < 1 - a_0 t_b / U_e \le 1, \qquad t_b < t_{esc} \tag{1.27c}$$

In Eq. (1.27b), the superscript (OF) refers to as the spacecraft's orbital frame, here coincident with the polar-coordinates basis. Again, v_0 and a_0 denote the vehicle's initial speed and acceleration, respectively. In addition, t_{esc} denotes the escape time, namely when the spacecraft's orbital energy vanishes (provided that the product $a_0 t_b$ is sufficiently high). Since $a_0 t_b < U_e$ for any physical t_b, the integral Eq. (1.26) can be evaluated for the above spiral to give

$$\Delta v^{(g)}(t_b) = -a_0 t_b \left(2 + \frac{a_0 t_b}{U_e}\right) + O\big((a_0 t_b)^3\big), \qquad t_b < t_{esc} \tag{1.28}$$

from which one can see that an outward spiral trajectory undergoes gravity-losses that vary *parabolically* with the burning time! To complete the analysis, we note that the final arc of tangential-thrust escape trajectory is highly non-linear and there is no closed-form solution. Nevertheless, if the specific power of the rocket propulsion system is kept fixed, then an extensive numerical analysis for nuclear-electric and photon rockets can provide the following results:

$$t_{esc} \cong \varsigma a_0{}^{-(1+b)}, \quad 0 < b < \frac{1}{4}, \; \varsigma > 0 \qquad (1.29a)$$

$$\Delta v^{(g)}{}_{esc} \cong -\frac{\varsigma}{a_0^b}\left(2 + \frac{\varsigma}{a_0^b U_e}\right) \qquad (1.29b)$$

Parameters (ς, b) depend on the range of the specific power for a given type of engine. Equations (1.29a), (1.29b) do not hold past the escape point. Its immediate suggestion is that spiraling out with very low thrust acceleration is a bad strategy. In practice, either the propellant amount becomes prohibitive or the propulsion system working time exceeds some upper bound related to the system. Instead, it is wise (hopefully) to select propulsive devices for reducing the jet speed and increase thrust acceleration. Thus, the escape point can be achieved in significantly shorter time. Past this point, speed increases non-linearly. This is a good compromise between the two terms of the rocket equation (1.26). Obviously, a real transfer flight stems from a complicated optimization, but where such general considerations hold; resorting to a compromise like that just mentioned is tied to the rocket propulsion physics whereupon there are limitations in specific power, high-energy propulsion system lifetime, and the unavoidable onboard propellant.

Two sections have been arranged below. These will be particularly useful in other chapters, where we compare high-energy solar sailing to rocket trajectories. One is devoted to an example of computing the heliocentric escape trajectory of a nuclear-reactor-based photon rocket; the other deals with a similar trajectory by Nuclear Electric Propulsion (NEP) rocket. A model of vehicle's mass breakdown encompassing both cases is the following:

$$M_0 = M_a + \boxed{M_G}^{P/W_m} + M_T + M_i + \boxed{M_K}^{kM_i} + (M_S + M_{PL}) \qquad (1.30)$$

where M_G denotes the dry mass of the power generator, M_T is the mass of the thrusting system (or engine),[15] M_K represents the mass of the tanks system (if applicable), M_{PL} denotes the mass of the mission payload, and M_S is the total mass of the other structures. In the above expression, the boxed terms represent simple subsystem models, where k is the tankage factor, P is the released power (i.e. $\varepsilon_a \dot{M}_a$), and $W_m(P)$ denotes the specific power that is function of P, in general. (For many conventional power systems, M_a may be "embedded" in the overall power system mass.) The parameter W_m is quite important in designing a power system for space use; nevertheless, what matters in real rocket trajectory profiles is the ratio of the thrust acceleration on the local gravitational acceleration, as pointed out at (1.24).

---

[15]*Thruster* can be viewed as a device that channels energy into a beam of particles that are exhausted away from the rocket. Engine coincides with the collimator and its control system in the case of photon rocket.

1.6 Escaping from the Solar System by Photon Rocket

Use of (1.13) results in

$$\frac{a_0}{g_0} = \frac{C_T}{g_0}\frac{K}{M_0} = \frac{C_T}{g_0}\eta_b\frac{P}{M_0} \tag{1.31}$$

Thus, the primary quantity of interest is represented by the rightmost fraction in (1.31). Note that $\frac{P}{M_0} < W_m$ for any real space rocket. If W_m is bounded, then the related trajectory is power-limited, or acceleration limited.

1.6 Escaping from the Solar System by Photon Rocket

We saw that any photon rocket has $M_i = 0$ and $C_T = 1$ identically. Let us analyze what one may realistically do by such a system for escaping from the solar system, and starting from a circular orbit at 1 AU. To our aims, a two-dimensional analysis is sufficient. Re-formulating the equations of motion of a photon rocket, we will deal with a long burning *plus* a subsequent coasting in order to compute the overall effect for this escape mission class, which is very typical out of all fast missions. The system of equations can then be cast into the following form

$$\ddot{x} + \mu_\odot\frac{x}{(x^2+y^2)^{3/2}} = \frac{T_x}{M}s_b = \frac{T}{M}\frac{\dot{x}\cos\beta - \dot{y}\sin\beta}{\sqrt{\dot{x}^2+\dot{y}^2}}s_b$$

$$\ddot{y} + \mu_\odot\frac{y}{(x^2+y^2)^{3/2}} = \frac{T_y}{M}s_b = \frac{T}{M}\frac{\dot{y}\cos\beta + \dot{x}\sin\beta}{\sqrt{\dot{x}^2+\dot{y}^2}}s_b \tag{1.32}$$

In Eq. (1.32), we set

$$s_b \equiv s(t_b - t) = \begin{cases} 1, & t \leqslant t_b \\ 0, & t > t_b \end{cases} \tag{1.33}$$

and $\mu_\odot$ is the solar gravitational mass, the dot denotes time derivative, x and y are the components of the vehicle's vector position in the heliocentric inertial frame, and β is the angle measured (with the usual sign convention) from the velocity to the thrust direction.

Mass and thrust entering Eq. (1.32) are derived from the previous photon-rocket relationships:

$$M = M_0\left[1 - \frac{P}{M_0c^2}(s_bt + (1-s_b)t_b)\right]$$

$$T = (\eta_s\eta_b\varepsilon_k/\varepsilon_a)\frac{P}{c} \equiv \eta_{tot}\frac{P}{c} \tag{1.34}$$

$$P_M \equiv P/M_0 = W_m(M_G/M_0) \equiv W_m m_G$$

$T > 0$ for $t \leqslant t_b$, after which $T = 0$ and $M = M_b = $ constant. η_{tot} denotes the overall thrust efficiency of the whole propulsion system.

Some remarks about the above motions equations are in order:

1. Even in escape trajectories, the optimal control to maximize the terminal speed at some given distance from the Sun is different from a pure tangential thrust; for this reason, we indicated the thrust control angle explicitly. A good sub-optimal control, close to the optimal one, is $\beta(t) = \text{constant} \neq 0$.

2. From a dynamical viewpoint, a detailed mass breakdown, such as that given by (1.30), of a photon-rocket allows evaluating the two factors W_m and m_G with sufficient accuracy; as a point of fact, the trajectory profile depends on their product as well.

3. Maximizing η_{tot} is a complicated problem of system design, of course. The initial thrust acceleration equals $\eta_{tot} P/(M_0 c)$, whereas mass decreases by $P/(M_0 c^2)$. Because we will compare photon rocket with advanced solar sailing, for simplicity we may assume that the design of the whole propulsion system exhibits η_{tot} very close to unity.

4. The assumed Heliocentric Inertial Frame (HIF) for (1.32) may be built with starting from a standard inertial frame, rotating it in order to the reference plane coincides with the ecliptic plane at some standard epoch, and translating this intermediate frame to the Sun's center of mass; a precise definition will be given on p. 137. As shown in Sect. 5.1, the barycentric dynamical time can be adopted as the time parameter in the above equations.

5. We have restored the speed of light explicitly in (1.34). This is useful for emphasizing that the mass of a photon rocket is practically constant, unless a huge level of power is utilized.

6. There is no closed-form solution to the system of Eq. (1.32). Trajectory may be approximated with a spiral as long as the rocket is far from the escape point and thrust is tangential. However, we want to achieve and go across such point, whereupon the system is highly non-linear. Therefore, we need numerical integration. In addition, the problem class we are dealing with is actually an optimization problem.

In order to get realistically results, we have to exclude both matter-antimatter annihilation and nuclear-fusion power generators, although extensively analyzed in literature, just for their near/medium-term infeasibility. Thus, the only feasible low-mass photon-rocket conceivable hitherto consists of a *bare* compact nuclear-fission generator put in the focal zone of a paraboloid-like mirror. This would collimate the directly emitted electromagnetic radiation (namely, with no intermediate energy conversion) from the high-temperature reactor. Serious issues may come from the fission reactor lifetime. Without entering the many aspects of the nuclear-reactor technology for Space, we consider a life-time (before some unrecoverable failure) of 10 years at full power, although this does not exclude a factor 2 or higher in the medium-term advancement. In addition, only one thrusting arc is allowed: once the reactor is shutdown, the whole propulsion unit is jettisoned.

We have to do some other considerations for stating a realistic problem. In the present view of general money investment, it is highly improbable that some governments(s) approve space mission(s) requiring an excessive time to get appreciable

1.6 Escaping from the Solar System by Photon Rocket

returns. On the other side, when a scientific program is proposed, people from the proponent scientific communities will want to get scientific data before many of them retire. Such real-life aspects may be included in the concept of Mean Human Job Time (MHJT), which we may give a mean value of 36 years. Thus, in the current case for photon rocket, we assume that the *overall* time for (1) mission request and approval, (2) spacecraft and photon-rocket design, and (3) flight time to the minimum target distance should not exceed t_{MHJT}. The flight time may be assigned a reasonable value of $t_f = \frac{2}{3}t_{MHJT} = 24$ years.

Let us now state the optimization problem to be solved numerically:

Problem 1.1 Prefixing $r_0 = 1$ AU and $t_f = 24$ years, and given the effective specific power $\eta_{tot}P/M_0 \equiv w_0$, find the piecewise-constant control $\beta(t)$ that maximizes the terminal distance $R(t_f)$ with the constraint $t_b \leq 10$ years.

This statement reflects clearly that the dynamical driver of a photon rocket is the quantity w_0. It would be highly desirable that, after the nominal transfer time, the spacecraft will achieve ≈ 100 AU, that is of the order of distance of the near heliopause: an efficient escape mission should be capable to cross this "boundary" of the solar system in the prefixed flight time. That is also suitable for the mission prolongation,[16] if any. Considering the explicit values of this problem statement, one expects that

$$v(t_f) \cong v_\infty = \sqrt{v_b(w_0)^2 - 2\mu_\odot/r_b(w_0)} \qquad (1.35)$$

where we have emphasized that both distance and speed at burnout are determined by the specific power w_0. Of course, $r_b \geq r_{esc}$.

The above optimization problem has been stated as Non-Linear Programming (NLP) problem. Its formulation entails that the long thrusting arc of trajectory (under solar gravity) is segmented into a number of pieces where the angle β is constant; its values are among the optimization parameters. In practice, it is found that three segments are sufficient for the solution is close to that via the variational approach. Great advantages are a much simpler attitude control (to be implemented) and a trajectory design with general constraints, even with classical NLP methods. Figure 1.3 summarizes the results[17] relevant to the topic of this section. The first striking result is the lower bound of w_0, that is 20.67 kW/kg, below which 10 years of continuous thrusting are not sufficient to achieve an osculating parabolic orbit; this is the 10-year threshold for the effective specific power. Such a value is enormous inasmuch as even a photon rocket-vehicle of 1000 kg of mass would need a nuclear-fission reactor of over $(20/\eta_{tot})$ MW. The second aspect regards the obtainable terminal speeds; if one wants to recover the speed at 1 AU (about 29.785 km/s, or 2π AU/yr) after 24 years, one would need 37.75 kW/kg specific power. The

---

[16]Of course, some scientific instruments can send data before reaching such distance.

[17]These numerical results have been carried out by the author specifically for this book.

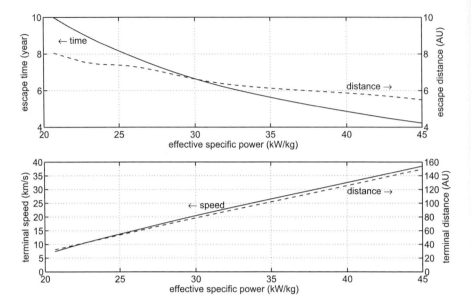

Fig. 1.3 Photon rocket escaping from the solar system starting from the Earth's orbit: [*top*] distance and time at the escape point ($E = 0$), [*bottom*] terminal distance and speed after 24 years as function of $\eta_{tot} P/M_0$, the effective specific power for trajectory shaping

specific power of the reactor would take on $W_m = 37.75/(\eta_{tot} m_G)$ kW/kg, according to Eq. (1.34), a huge value even for an advanced nuclear reactor and spacecraft technology. The terminal distance is 114.7 AU.

The third feature obtained from the current results is that, terminal speed and terminal distance can be analytically related to the effective specific power w_0 via simple functions. For instance, the following function proves a very good regression to the numerical curve of the terminal speed shown in the lower part of Fig. 1.3:

$$V_f \cong 54.2934 - 231.019/\sqrt{w_0} + 0.0092063 w_0^2 \qquad (1.36)$$

As above, w_0 is expressed in kW/kg whereas V_f is in km/s. The maximum error in using (1.36) is $\cong$77 m/s.

At first glance, one might expect that a photon rocket, which owes the highest exhaust speed, should deliver a very high terminal speed. However, as explained in Sect. 1.5, a sufficiently low thrust acceleration in a gravity field induces small gain of speed because of the long transfer time. In the current case of photon rocket escaping from the solar field, the ratio between the thrust acceleration and the local gravity acceleration is given by

$$a/g = w_0 \frac{M_0}{M} \frac{1}{cg} \qquad (1.37)$$

At 1 AU, the above threshold of 20.67 kW/kg entails $a/g = 1.16 \times 10^{-2}$. During the thrusting, the first fraction in (1.37) is almost constant and the second one scales as R^2. However, even during the last year of propulsion, the ratio a/g keeps less than unity. In a photon rocket, one has only one way to increase the thrust acceleration: designing the nuclear reactor with higher specific power. However, even doubling/tripling such power does not induce a big gain in the context of extra-solar missions. Such an unfavorable situation worsens significantly in the case of a photon rocket spiraling outward from a low-altitude orbit about a planet.

In contrast, other *rocket* systems are much more efficacious than the photon rocket for utilizing a high specific power reactor, as we shall see in the next section.

1.7 Escaping from the Solar System by NEP Rocket

Contrarily to the photon-rocket concept and feasibility, ion propulsion has been the object of intense theoretical, numerical and experimental work since 1960's, even though the idea of propelling a space-vehicle by ejecting charged particles can be traced back to Robert Goddard in the early 1900'. Only in the last 15 years, space missions such as Deep Space-1 (NASA, 1998, main target = Comet Borrelly), SMART-1 (ESA, 2003, target = the Moon), Hayabusa (JAXA, 2003, target = asteroid Hayabusa), and Dawn (NASA, 2007, target = asteroids Ceres and Vesta) have been accomplished, thus showing that ion-propulsion is a viable *in-space* technology. In these missions, solar arrays are the power generator.

In this section, we discuss only one meaningful example of ion propulsion for escaping from the solar system, and start just from the key-point for thrusting efficiently in a gravity field, namely, a sufficiently high a/g ratio. This reduces considerably the propellant loss due to the long finite burning inside the gravity of a massive celestial body.

In the current context, the vehicle's propellant and mass time rates are given by

$$\dot{M}_p = \eta_m \chi \dot{M}_a = \eta_m \dot{M}_i = \dot{M}_i^{(ion)} \tag{1.38a}$$

$$-\dot{M} = \chi \dot{M}_a = \dot{M}_i = \dot{M}_i^{(ion)} + (1 - \eta_m)\dot{M}_i \tag{1.38b}$$

where $\dot{M}_i^{(ion)}$ denotes the flow rate of ionized mass (ions and electrons) accelerated and exhausted, η_m denotes the propellant utilization efficiency, and $\dot{M}_i$ is the flow rate of (neutral) propellant from tank(s) to thruster(s). The previous relationships can be carried out from Eqs. (1.3) and (1.5) as soon as one realizes that a (fission) Nuclear Ion Propulsion (NIP) system is characterized by

$$\varepsilon_a \cong 0.0009, \qquad \chi \gg 1, \qquad \zeta = 0, \qquad \xi = \eta_m \tag{1.39}$$

The quantity $(1 - \eta_m)\dot{M}_i$ of neutral atoms emerges from the thruster at ZTM; it does not contribute to thrust, but contributes to the vehicle's mass decreasing.

Now, let us generalize a bit for getting more realistic expressions of thrust and beam's kinetic power. Since the jet speed does not consist of photons, the ejected particles will have a spread in speed. Therefore, we can write

$$d\dot{M}_i^{(ion)}(u) = \dot{M}_i^{(ion)} f(u)\,du \tag{1.40}$$

where $f(u)$ is a normalized probability density function: the r.h.s. of (1.40) gives the mass flow of the charged particle with speed in the range $[u, u + du]$ independently of the direction of $\mathbf{u}$. For simplicity, we assume that this probability density does not change with time. Therefore,

$$\dot{M}_i^{(ion)} = \int_{u_{min}}^{u_{max}} \dot{M}_i^{(ion)} f(u)\,du, \quad 0 \le u_{min} < u_{max} \tag{1.41a}$$

$$\langle q(u) \rangle \equiv \int_{u_{min}}^{u_{max}} q(u) f(u)\,du \tag{1.41b}$$

In Eq. (1.41b), $q(u)$ denotes a generic function of the particle ejection speed.

At the start from the heliocentric branch of flight, the magnitude of the thrust can be expressed as follows:

$$a_0 M_0 = \kappa g_{1\,AU} M_0 = \tilde{T}$$
$$= \dot{M}_i^{(ion)} \langle u_{\parallel} \rangle = (\eta_m \dot{M}_i)(\eta_d \langle u \rangle) \equiv \dot{M}_i \tilde{U}_e \tag{1.42}$$

where $\tilde{T}$ and $\tilde{U}_e$ have the meanings (apart from the relativistic aspect) explained in Sect. 1.4, $0 \le \eta_d \le 1$ denotes the geometric beam divergence factor, and $u_{\parallel}$ is the projection of $\mathbf{u}$ along the beam's symmetry axis.[18] In Eq. (1.42), κ is the *mission-dependent* ratio between the initial thrust acceleration (a_0) and the solar acceleration at 1 AU, or $g_{1\,AU}$.

The exhausting-beam power is given by

$$\tilde{K} = \frac{1}{2}\int_{u_{min}}^{u_{max}} u^2 \dot{M}_i^{(ion)} f(u)\,du = \frac{1}{2}\dot{M}_i^{(ion)} \langle u^2 \rangle$$
$$= \frac{1}{2}\dot{M}_i^{(ion)} (\langle u \rangle^2 + \sigma_u^2) \tag{1.43}$$

where σ_u^2 is the variance of the random variable u. Note that

$$U = \sqrt{\langle u \rangle^2 + \sigma_u^2} = \langle u \rangle \sqrt{1 + \frac{\sigma_u^2}{\langle u \rangle^2}} \tag{1.44}$$

---

[18]The ion propulsion community is used to call $\tilde{U}_e/g_0$ the *specific impulse*, g_0 being the standard acceleration gravity value ($9.80665\ m\,s^{-2}$). It is normally expressed in *seconds*. Of course, gravity has nothing to do with the specific impulse concept; g_0 is only a scaling factor. As a consequence, the specific impulse should instead be reported as a speed.

1.7 Escaping from the Solar System by NEP Rocket

is the exhaust speed that a jet with flow $\dot{M}_i^{(ion)}$ and kinetic power $\tilde{K}$ would have if all particles were endowed with the same speed. By inserting (1.42) into (1.43) and rearranging, one gets

$$\tilde{K} = \frac{1}{2} \frac{\tilde{T}\tilde{U}_e}{\eta_m \eta_d^2} \left(1 + \frac{\sigma_u^2}{\langle u \rangle^2} \right) \tag{1.45}$$

It is easy to recognize that both $\tilde{T}$ and $\tilde{K}$ return to their ideal expressions if and only if $f(u) = \delta(u - \bar{U})$, $\bar{U} > 0$, and there are both complete neutral-to-ion conversion and no geometric divergence. $\delta(\cdot)$ is the Dirac's distribution.

The beam spread factor η_s, formally used in (1.14), and discussed in Sect. 1.4.1, can be applied to the thrust expression via the exhaust speed spread:

$$\eta_s^{(jet)} = \eta_d \frac{\langle u \rangle}{u^{(ref)}} \tag{1.46}$$

where the denominator denotes a reference value of the exhaust speed. By choosing $u^{(ref)} = U$, we get

$$\eta_s^{(jet)} = \eta_d \Big/ \sqrt{1 + \frac{\sigma_u^2}{\langle u \rangle^2}}, \qquad \tilde{U}_e = \eta_m \eta_s^{(jet)} U \tag{1.47}$$

In contrast, the beam's actual kinetic power is affected by spread via

$$\eta_s^{(power)} = \eta_d^2 \Big/ \left(1 + \frac{\sigma_u^2}{\langle u \rangle^2} \right) = \left(\eta_s^{(jet)} \right)^2 \tag{1.48}$$

Thus, if a real mission require certain values of $\tilde{T}$ and $\tilde{U}_e$, which both affect trajectory dynamics, then the non-perfect ejection process comes in non-linearly the general engine inefficiency.

Note that, among many parameters, even a simple numerical code for trajectory computation requires the following quantities in input: the initial vehicle mass M_0, the actual thrust $\tilde{T}$, and the *effective* exhaust speed $\tilde{U}_e$, or an *equivalent* set, for computing both spacecraft mass and acceleration profiles.[19]

Now, what is the actual power of the nuclear reactor corresponding to the pair $\tilde{T}$ and $\tilde{U}_e$, namely, to the baseline kinetic power $K^* \equiv \frac{1}{2}\tilde{T}\tilde{U}_e$? We work it out by going back from the jet up to the reactor through a few power/propulsion subsystems[20] endowed with efficiencies denoted by $\eta_{(\cdot)}$. Thus, the electric power input to the thruster system (TS), including the power processing unit (PPU), amounts to

$$P_{(TS)} = \tilde{K}/\eta_{(TS)} = K^*/\eta_{(tot,\ TS)}, \qquad \eta_{(tot,\ TS)} \equiv \eta_{(TS)} \eta_m \eta_s^{(power)} \tag{1.49}$$

---

[19]This set of independent quantities lends itself to be easily generalized for including secondary effects on trajectory stemming from the attitude control system.

[20]In real space designs, the current systems could be dealt with as either systems or sub-systems or sub-subsystems, depending on the design context. Here, we are using a few simple terms.

PPU design depends on the type and class of the electric engine to be supplied; a PPU has its own efficiency that, in the current model, we embedded in $\eta_{(TS)}$. Thus, power $P_{(TS)}$ actually comes in the PPU-subsystem of the TS. Only a fraction $\eta_{(TS)}$ of it emerges as beam power (because of many power needs: atom ionization chamber, radio-frequency source, magnet, neutralizer, grids, etc.). Assuming that a fraction $\eta_{(NR \to TS)}^{(el)}$ of the electric power from the nuclear reactor goes to the TS, the reactor's total electric power amounts to

$$P_{(NR)}^{(el)} = P_{(TS)}/\eta_{(NR \to TS)}^{(el)} \tag{1.50}$$

Finally, the thermal power[21] of the NR is calculated via

$$\varepsilon_a \dot{M}_a c^2 = P_{(NR)}^{(th)} = P_{(NR)}^{(el)}/\eta_{(NR)}^{(el)} \tag{1.51}$$

Using these chain-system efficiencies, the reactor's full power can be written as a function of thrust, effective exhaust speed and unambiguous efficiencies:

$$P_{(NR)}^{(th)} = \frac{1}{2}\tilde{T}\tilde{U}_e/\eta_{tot}, \quad \eta_{tot} \equiv \eta_{(NR)}^{(el)}\eta_{(NR \to TS)}^{(el)}\eta_{(TS)}\eta_m\eta_s^{(power)} \tag{1.52}$$

Therefore, using the l.h.s. of Eqs. (1.42) and (1.52) results in the following formulas of the thermal and electric specific power of the vehicle, and the specific power of the whole propulsion system

$$P_M^{(th)} = P_{(NR)}^{(th)}/M_0 = \frac{1}{2}\frac{\kappa g_1 \, _{AU}\tilde{U}_e}{\eta_{tot}} \tag{1.53a}$$

$$P_M^{(el)} = \eta_{(NR \to TS)}^{(el)}P_M^{(th)} \tag{1.53b}$$

$$\alpha^{-1} = \frac{P_{(TS)}}{M_G + M_T} = \frac{\tilde{K}}{\eta_{(TS)}(M_G + M_T)} \tag{1.53c}$$

$$W_m = \frac{P_{(TS)}}{M_G} = \frac{\tilde{K}}{\eta_{(TS)}}\frac{1}{M_G} \tag{1.53d}$$

One should note that $\eta_{tot}P_M^{(th)}$ is be compared to the w_0-values relevant to the photon rocket (Sect. 1.6), whereas α, named the *specific mass*[22] of the whole propulsion system, is often used in the ion-propulsion literature for characterizing the overall technology of the power-plant *and* the thruster subsystem for a given class of NIP systems and missions.

---

[21] Apart from a small amount of neutrinos, we may consider that the energy released by nuclear fission will eventually go all to heat.

[22] This value, comprehensive of the generator and thruster subsystems, is normally split into subsystem-peculiar values.

1.7 Escaping from the Solar System by NEP Rocket 27

The NIP's three-dimensional motion equations in the solar inverse-square gravity field can be cast into the form:

$$\ddot{\mathbf{R}} + \mu_\odot \frac{\mathbf{R}}{R^3} = \frac{\tilde{\mathbf{T}}}{M} s_b$$

$$-\dot{M} = \dot{M}_i s_b = \frac{\tilde{T}}{\tilde{U}_e} s_b \tag{1.54}$$

where we have used the switching function (1.33). The last equation of (1.54) should be compared to (1.34). Considered the aims of this section, we omitted perturbative accelerations from asymmetric exhaust beam, and non-ideal attitude control torques. The extension to three dimensions is necessary here for showing that, much more easily than a photon rocket, a long thrusting by NIP could deliver the scientific payload to distant targets out-of-ecliptic.

Let us apply the previous equations to a modern concept of ion engine, such as that proposed in [8] and [9], with the mass breakdown model given by (1.30). This ion propulsion system consists mainly of a number of dual-stage four-grid (DSFG) thrusters, which should exhibit some advantages with respect to conventional ion engines. For instance, a DSFG system should allow *separating* the four basic processes necessary to get a ion jet of high power: (a) in-chamber plasma production, (b) ion extraction, (c) ion acceleration, and (d) space-charge neutralization. As a result, there are favorable features from design and operational viewpoints: (i) utilizing the same plasma chamber and neutralizer, thrust *and* effective exhaust speed may be varied within somewhat broad ranges, (ii) geometric beam divergence may be as low as a few degrees, so avoiding ions impacting on other spacecraft structures, and (iii) thruster lifetime is longer. DSFG may be designed with a high perveance utilization factor.[23]

Although the vast subject of optimizing an ion drive is not within the aims of this chapter, however—in our next example—we could *not* use some significant classical results [3, 21, 23] for getting an escape mission from the solar system with enhanced terminal distance and speed. Substantially, if the jet's kinetic power were kept constant, the best exhaust speed program in free-space would be $U \propto M^{-1}$. In addition, because the current mission concept uses engines of the same type and characteristics, we cannot use the concept of variable specific impulse via two different types of electric engine [23]. Due to the presence of the solar gravitational field, instead of using the thrust program with constant jet's power, the transfer should be performed at constant thrust to reduce the gravity losses strongly.

Table 1.1 contains the main parameters regarding the assumed propulsion system, and the mass breakdown of a (reasonable) spacecraft for the near heliopause.

---

[23]This *dimensionless* parameter is a measure of the non-uniformity of the ion current density at the screen grid. If the current density distribution, as function of the radius, were rectangular, then it would be equal to 1, its maximum (and ideal) value. There is a *dimensional* parameter, called the *perveance*, which is characteristic of the ion accelerator subsystem. Two ion engines performing a given specific impulse with both the same propellant and accel-to-decel grid voltage ratio exhibit different thrusts if the product of these two perveance factors is different.

28   1 Some General Rocket Features

Table 1.1 Nuclear ion drive to heliopause: main Propulsion Parameters and Spacecraft Mass Breakdown. (The above symbol 'u', in SI, denotes the atomic mass unit, formerly shortened in literature by amu or a.m.u.)

Description	Current-symbol	Value	Unit
Dual-Stage Four-Grid Engine			
propellant: Xenon (natural mix)	$_{54}\mathrm{Xe}$	131.293	u
tank-to-engine mass flow	$\dot{M}_i$	9.5	mg/s
thrust	$\tilde{T}$	1.274	N
actual jet speed	U	149.9	km/s
specific impulse	$\tilde{U}_e/g_0$	13682	s
ion beam voltage		15.3	kV
max grid voltage		33.6	kV
ion current density		6.0	$\mathrm{mA/cm^2}$
total ion current		12.5	A
perveance utilization factor		0.5	
mass utilization efficiency	η_m	0.9	
jet divergence half-angle		2	degree
jet speed spread	$\sigma_u/\langle u \rangle$	0.1	
thruster's total efficiency	$\eta_{(tot,TS)}$	0.712	
lifetime	τ_{engine}	30000	hour
Nuclear Reactor			
lifetime	$\tau_{reactor}$	>1	τ_{engine}
reactor/radiator + thruster(s) specific mass	α	25	kg/kW
thermal power	$P_{(NR)}^{(th)}$	650	kW
electric power	$P_{(NR)}^{(el)}$	130	kW
power to thruster(s)	$P_{(TS)}$	120	kW
total efficiency	η_{tot}	0.1315	
Spacecraft Mass Breakdown			
scientific-payload mass	M_{PL}	200	kg
total propellant	M_i	1026	kg
reactor/radiator + thruster(s)	$M_G + M_T$	3000	kg
tanks + telecom + other structures	$kM_i + M_S$	240	kg
mass injected at 1.0260 AU	M_0	4465	kg
mass at burnout	M_f	3439	kg

Although the table is self-explanatory, a number of remarks are in order. The main purposes of the mission can be summarized as follows: (a) to deliver a certain set of scientific instruments to the near heliopause, ~ 100 AU; (b) to hopefully extend the flight up to approximately 200–250 AU for measuring the basic properties of the near interstellar medium.

1.7 Escaping from the Solar System by NEP Rocket 29

Table 1.2 Nuclear Ion Drive: heliocentric thrust control for maximizing the terminal speed at 140 AU along the near heliopause direction. The spacecraft-fixed reference frame for getting the optimal control is the RVH-system given by the directions of the vehicle's $\mathbf{R}$, $\mathbf{V}$, and $\mathbf{H} \equiv \mathbf{R} \times \mathbf{V}$

thrusting time	arc No.	1	650	day
		2	450	
		3	150	
azimuth from $\mathbf{V}$	arc No.	1	$-0.17°$	
		2	$0.26°$	
		3	$0.10°$	
elevation from $\mathbf{RV}$-plane	arc No.	1	$0.70°$	
		2	$9.38°$	
		3	$12.43°$	
Mission Performance				
injection date: 2021 06 24 00:00:00				
injection speed *wrt* Earth	v_{in}		1.073	km/s
injection distance *wrt* Earth	r_{in}		2.803	10^6 km
heliocentric initial speed	V_0		30.337	km/s
heliocentric initial distance	R_0		1.026	AU
accelerations ratio	$(a/g)_0$		0.0481	
heliocentric-branch thrusting time	τ_h		1250	day
speed after τ_h	V_h		32.713	km/s
heliocentric gravity losses			32.649	km/s
total flight time	τ_f		8516	day
terminal speed	V_f		30.445	km/s
terminal distance	R_f		140.00	AU

The whole propulsion system (power generator + thrusters) considered here features both advanced and more conservative values, although even the latter ones are beyond the current technology. The main advanced value is the specific mass of the reactor-thruster complex, which includes radiator, radiation shield, and power processing unit. The assumed value of 25 kg/kW, though not very low in literature, is considerably better than the SP100[24] technology [5]. Another key factor is the ion thruster lifetime, in general. Although set of engines may work serially, we assumed an overall value of thrusting time equal to 30,000 hours (or 1250 days), also considering aspects of engine redundancy. The propellant choice is of fundamental importance, of course. Using Xenon has advantages, though its current cost is high; however, an increasing demand may lessen cost significantly. The overall efficiency of DSFG thrusters, computed from subsystem efficiency values, appears pretty good. Other design parameters, which satisfy the basic equations of ion engines and are listed in Table 1.1, are a bit conservative, though comparable or su-

---

[24]US Space Nuclear Reactor Design.

Fig. 1.4 Nuclear Ion Drive to the Near Heliopause: projection of the 3D trajectory onto the ecliptic plane. Thrusting arc lasts 1250 days, whereas only a piece of the coasting arc (7266 days) is shown

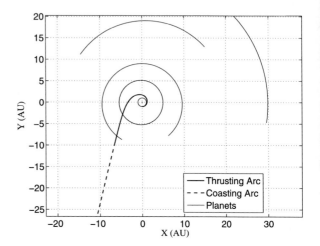

perior to the engine technology at the time of this writing. Finally, the value of 200 kg for the scientific payload should look very good for a mission beyond the heliopause, also considering the capability of the emerging technologies. As a result of the discussed items, the mass of the full spacecraft, injected into the heliocentric field, is less than 4.5 metric tons, well within the performance of some current powerful launchers delivering heavy deep-space spacecraft.

Table 1.2 reports the best attitude history[25]—in the sense of Non-Linear Programming—for the considered mission via a three-arc piecewise-continuous thrusting phase. This control enables the values of mission performance in the bottom part of the same table. Figure 1.4 shows the projection of the three-dimensional trajectory onto the reference plane of the HIF. The orbits, from JPL Ephemeris File DE405/LE405 (DE405), of the Earth-Moon barycenter, Jupiter, Saturn, Uranus, and Neptune are projected as well. Perturbations from such planets have been included just for showing that the heliocentric transfer in the chosen periods of time is practically unperturbed. Spacecraft points to the nose of the heliopause: ecliptic longitude 254.5°, ecliptic latitude 7.5°. One can note the high value of gravity losses, i.e. over 32.5 km/s. This means that, in field-free space, the terminal speed would amount to $V_0 + \tilde{U}_e \ln(M_0/M_f) = 65.37$ km/s. This is the direct effect of a really long thrusting with an initial $a/g = 0.048$.

One could relax (i.e. increase) either the payload mass or the power/propulsion system's specific mass, or both, significantly in order to meet high requirements; however, currently, a much more advanced spacecraft would be im-

---

[25]Once fixed the overall thrusting time, the dependence of the terminal speed on the single arc times is fairly weaker than the attitude angles, which here have been specified in the RVH frame of reference.

plied in terms of thrust and electric power, otherwise M_0 would increase so much that no launcher could inject it directly into the full solar field. Many-tonne NEP/NIP missions would entail long spiraling about the Earth with additional problems. *Both* planetary and solar gravity-losses would arise. Nevertheless, beyond doubt, NEP/NIP is a great advance with respect to chemical rocket solutions for deep space. Of course, one may think of a hybrid propulsion (e.g. nuclear-thermal + NIP) for dealing with the problem of very massive payload transportation. In literature, past and current analyses are not only very numerous, but also broad in proposal type and technological implications/cost, including the need of multiple launches via a heavy-lift vehicle from ground to low Earth orbits.

Probably, one of the greatest advantages from NIP-based very deep space vehicles is to be independent of any planet other than the Earth. For instance, our envisaged flight to the heliopause has a launch window *every* year. Of course, many advanced missions have been proposed and analyzed, by combining NIP with gravity assist for increasing the total velocity change; however, the launch window rarity is not an advantage. Among the things above remarked, the mission discussed in this section highlights some very important items, and—in this sense—may be considered representative for some class of fast missions beyond the heliosphere:

- In order to get a good escaping from the solar system, NIP systems needs a specific power three order of magnitude *lower* than that relevant to photon rockets. This is roughly the ratio of the NIP's exhaust speeds to the speed of light. In NIP systems, the specific impulse may act as a control parameter.
- Non-chemical systems are power-limited. This has two consequences: (1) if the specific mass is too high for fast missions, then the mass of a deep-space rocket-vehicle may result unacceptable, and (2) if the power-plant lifetime is too short, then the terminal speed may be low though the propulsive power is high.
- In escaping from the solar system, a high ratio a/g is somewhat more important than high specific impulse; a_0 and $\tilde{U}_e$ should be optimized simultaneously.

Such properties will be recalled when, in the subsequent parts of the book, we highlight the features of sailcraft missions.

References

1. Ackeret, J. (1946), Helvetica Physica Acta (in German), and Journal of the British Interplanetary Society, 6, 116–123 (1947).
2. Angelo, J. A., Buden, D. (1985), Space Nuclear Power. ISBN 0894640003.
3. Brewer, G. R. (1970). Ion Propulsion: Technology and Applications. New York: Gordon and Breach.
4. Bussard, R. W. (1958), Concepts for future nuclear rocket propulsion. Jet Propulsion, April issue.

32   1 Some General Rocket Features

5. Demuth, S. F. (2003), SP100 space reactor design. Progress in Nuclear Energy, 42(3), 323–359. doi:10.1016/S0149-1970(03)90003-5.
6. Esnault-Pelterie, R. (1928), L'Exploration par fusée de la très haute atmosphère et la possibilité des voyages interplanétaires. L'Astronomie, Souplément (March).
7. Esnault-Pelterie, R. (Ed.) (1930), L'Astronautique.
8. Fearn, D. G. (2006), Ion propulsion: an enabling technology for interstellar precursor missions. Journal of the British Interplanetary Society, 59(3–4), 88–93.
9. Fearn, D. G. (2008), Technologies to enable near-term interstellar precursor missions: is 400 AU accessible? Journal of the British Interplanetary Society, 61, 279–283.
10. Koelle, H. H. (Ed.) (1961), Handbook of Astronautical Engineering. New York: McGraw-Hill. ASIN: B0000EFZCO, 1867 pages.
11. Loeb, H. W., Bassner, H. (1987), Solar and nuclear electric propulsion for high-energy orbits. In IAF Congress, Brighton, UK.
12. Martin, R., Bond, A., Bond, R. A. (1988), Ultimate performance limits and mission capabilities of advanced ion thrusters. In 20th International Electric Propulsion Conference, Garmisch, Germany. IEPC-88-084.
13. Ruppe, H. O. (1966). Introduction to Astronautics, Vol. 1. New York: Academic Press.
14. Ruppe, H. O. (1967). Introduction to Astronautics, Vol. 2. New York: Academic Press.
15. Sänger, E. (1953). Zur Theorie der Photonraketen. Ingenieur-Archiv, 21, 213–226 (in German).
16. Shepherd, L. R., Cleaver, A. V. (1948), The atomic rocket 1–2. Journal of the British Interplanetary Society, 7(5–6).
17. Shepherd, L. R., Cleaver, A. V. (1949), The atomic rocket 3. Journal of the British Interplanetary Society, 8.
18. Shepherd, L. R. (1952), Interstellar flight. Journal of the British Interplanetary Society 11, 149–167.
19. Shepherd, L. R., Vulpetti, G. (1994), Operation of low-thrust nuclear-powered propulsion systems from deep gravitational energy wells. In IAF Congress, Jerusalem, October 1994. IAA-94-A.4.1.654.
20. Shepherd, L. R. (1999), Performance criteria of nuclear space propulsion systems. Journal of the British Interplanetary Society, 52(9–10), 328–335.
21. Stuhlinger, E. (1964), Ion Propulsion for Space Flight. New York: McGraw-Hill. ISBN B0000CM8P7.
22. Vulpetti, G. (1973), Considerazioni su consumo di propellente nelle traiettorie di fuga di veicoli spaziali con propulsione elettrica e chimica. In XXI International Meeting of Communications and Transports, Genova, Italy, October 8–13 (in Italian).
23. Vulpetti, G. (1982), More about the pulse-on-bias propulsion performance, academy transactions note. Acta Astronautica, 9(11), 687–688.
24. Vulpetti, G. (1984). A propulsion-oriented synthesis of the antiproton-nucleon annihilation experimental results. Journal of the British Interplanetary Society, 37, 403–409.
25. Vulpetti, G. (1985), Maximum terminal velocity of relativistic rocket. Acta Astronautica, 12(2), 81–90.
26. Vulpetti, G. (1986), Antimatter propulsion for space exploration. Journal of the British Interplanetary Society, 39(9), 391–409.
27. Vulpetti, G., Pecchioli, M. (1989), Considerations about the specific impulse of an antimatter-based thermal engine. Journal of Propulsion and Power, 5(5), 591–595.
28. Vulpetti, G. (1990). Multiple propulsion concept for interstellar flight: general theory and basic results. Journal of the British Interplanetary Society, 43, 537–550.
29. Vulpetti, G. (1999), Problems and perspectives in interstellar exploration. Journal of the British Interplanetary Society, 52(9–10), 307–323.
30. Westmoreland, S. (2010), A note on relativistic rocketry. Acta Astronautica, 67(9–10), 1248–1251.
31. Wilbur, P. J., Beattie, J. R., Hyman, J. (1990), Approach to the parametric design of ion thrusters. Journal of Propulsion and Power, 8(5).

Part II
Sailing in Space Environment

– The introductory chapter has highlighted how rocket propulsion could be inappropriate for accomplishing a large number of the increasing-in-performance future space missions. Although desirable, high-energy *and* high-power rocket vehicles are very difficult to design, sometimes even in principle, because of the intrinsic properties of the fundamental interactions from which the propulsive power should be extracted. Even in the positive case, by definition a rocket must carry propellant, and this provokes a number of problems every time the objectives of the rocketship mission are pushed forward significantly. Thus, one has to search for propulsion methods with power sources *external* to the vehicle.
– Whatever the power source may be located, either in space or on ground, it should exhibit certain properties in order to be utilized as the primary source for translational propulsion. In this second part of the book, many considerations are made with regard to the two major candidates for space solar sailing: the solar irradiance, and the solar wind. All known concepts of sailcraft, but the laser-pushed one, are based on such sources of free momentum. The irradiance-based sailing has been proved by Interplanetary Kite-craft Accelerated by Radiation Of the Sun (IKAROS), and NanoSail-D2 (that ended its successful mission on September 17, 2011).

Chapter 2
The Sun as Power Source for Spaceflight

In-Space Use of the Continuous Sunlight Pressure   If we were in the pre-IKAROS scenarios of space sailing, the main aim of this chapter would be to show that one *may* utilize the radiant energy from the Sun as external-to-spacecraft source of thrust. Now, as the Japanese sail-based spacecraft proved that solar sailing is a reality (even though at the level of orbit perturbation, for the moment), this chapter aims at discussing more in detail the principles of sunlight utilization for efficacious in-space propulsion.

We start with a summary of radiometric quantities, and then we point out the immense contribution of space-era to the solar physics, in particular through the high-precision records of the total and spectral solar irradiances. The time series of such quantities will be important in designing fast solar-sail trajectories. In particular, we highlight the principles of utilization of the solar irradiance for in-space propulsion.

2.1 A Summary of Radiometric Quantities

In this section, we expound on radiometric quantities involved in space trajectories, when the radiation pressure produces either the main thrust acceleration or a perturbing acceleration, as in many real situations. Radiometry is standardized by the Commission Internationale de L'Eclairage (CIE), or International Commission on Illumination, International Standards Organization (ISO), and International Union of Pure and Applied Physics (IUPAP).

Radiometry is the science aiming at measuring physical quantities relevant to electro-magnetic (e.m.) radiation, in principle from radio waves to γ-rays. However, in practice, one often refers to as radiometry in the context of infrared, visible and ultraviolet bands of the e.m. spectrum. By saying that, one implicitly assumes that light is viewed as classical waves, not as quantum objects—the photons—though a net separation is neither possible nor required for propulsion purposes. The twofold amazing nature of light is well known; however, some further clarification is necessary in the framework of this book. Going straight to the problem's core, for a

G. Vulpetti, *Fast Solar Sailing*, Space Technology Library 30,
DOI 10.1007/978-94-007-4777-7_2, © Springer Science+Business Media Dordrecht 2013

general photon-induced space sailing (i.e. not only solar-photon sailing), we need light incident on some (sufficiently large) spacecraft's surface (the *sail*), and emerging from it after having interacted with the sail's material. Well, we can picture both incident and reflected light *classically*, namely, as the full wave theory states. However, the photon-surface interaction should be modeled by treating the material's atoms and molecules as quantum objects. Even in this semi-classical approach, things look somewhat complicated, and so they are indeed. In-space propulsion needs high flows or streams of photons because radiation pressure has to be transformed into sufficiently high thrust.

If we utilized a single-mode-operated laser, with a given frequency and well above the threshold, for pushing a sail, then the photon statistics of this light would be close to a Poissonian inasmuch as this operational type of lasing is a good approximation to perfectly coherent light, which obeys the Poisson distribution.

In general, the statistics describing the fluctuations of the number of photons from a *single-mode* of the radiation field is known as the Bose-Einstein distribution (BE). Let us consider *thermal light*, namely, the light emitted from a blackbody. A single mode of thermal light is super-Poissonian (i.e. the variance is higher than the mean). However, blackbody light has a continuum of modes. As a result, the statistics of it exhibits a Poissonian character, as measured in many experiments. As a point of fact, the super-Poissonian variance tends to become Poissonian if the number of modes is high.

Quantum optics re-interprets the usual energy expression of the quantum harmonic oscillator in the n-state, namely, $E_n = (\frac{1}{2} + n)\hbar\omega$ as n photons *excited* at angular frequency ω. The very quantum nature of light can be detected in special experiments where photon statistics is sub-Poissonian [13] (i.e. the variance is lower than the mean). Beams of light can be classified not only with respect to the photon statistics, but also via the second-order correlation function, which quantifies the time fluctuations of intensity. In other words, there may be streams of light (the so-called *antibunched* light), where the photons are "separated" by regular gaps.

Sunlight originates from a stellar atmosphere with different brightness temperatures (as we will see later in this chapter); it is, as a whole, a classical electromagnetic radiation delivering pressure. (Where appropriate, we will view sunlight as photons with energy and momentum). The practical consequence of these facts is that macroscopic **radiometric** quantities represent an appropriate tool for dealing with the generation of radiation-pressure thrust, a topic that shall be developed in many details in Chap. 6.

Like radiometry, *photometry* regards measurements of electromagnetic radiation, but that one detectable by the human eye. Quantitatively, a photometric quantity is strongly affected by the eye's spectral sensitivity, which is restricted from 380 nm to

2.1 A Summary of Radiometric Quantities 37

800 nm, approximately, in wavelength. Although human eye exhibits photopic vision, mesopic vision, and scotopic vision, CIE's spectral luminous efficiency function regards the eye's average response under daylight conditions. When the behavior of the human eye under dark-adapted conditions (or also in intermediate lighting conditions) becomes important for some measurable quantity, one will replace the photopic function by the scotopic (mesopic) sensitivity function. We are not concerned with photometry here since a solar, laser, or microwave sail can be viewed as a special detector of e.m. radiation as a device utilizing the pressure of light, and in a range well wider than the human eye's.

Now, we can list and briefly comment on some radiometric definitions. The list is incomplete; fundamental radiometric functions are to be defined in Chap. 6 where we shall model the solar-sail thrust. Here, for each concept, *spectral*-quantity is stated first, then the *integral* one will follow. We will make only a few changes in symbols, but **not** in concepts, with respect to the international standard. Units follow strictly the International System of Units (SI). Each item is arranged as follows: quantity name, symbol, units, definition, and comments (if any).

- **Source of Light**: a source of electromagnetic radiation can be the surface of a proper photon emitter (e.g., stars, lamps, gases, plasmas, living bodies, etc.) or any surface reflecting/scattering some fraction of the received light. When the source does not appear as point-like to some observer, it can be partitioned in infinitesimal, or differential, surfaces, each of which is endowed generally with its own radiation characteristics.

- Let dA be a differential *directed* surface,[1] which emits or receives electromagnetic energy. Let also $\mathbf{n}$ be the positive normal to dA. A direction $\mathbf{d}$, along which radiation can be emitted or received, forms an angle θ with $\mathbf{n}$; $d\omega$ denotes the solid angle about $\mathbf{d}$. Then, the orthogonal-to-$\mathbf{d}$, or *projected*, area is equal to $dA_\perp = \cos\theta\, dA$, where dA denotes the magnitude of $d\mathbf{A}$. Note that an infinitesimal surface may be a pure geometric object *across* which light can go. In any case, light is assumed to impinge on or leave a surface element in the semi-space containing *its* $\mathbf{n}$.

  The concepts we are discussing have to be eventually applied to real-life surface elements. Each element will be sufficiently small to be dealt mathematically with as infinitesimal area, but also sufficiently large to contain a high number of atoms/molecules, and to exhibit two sides that may have some different optical properties.

- **Spectral Radiant Power** or spectral radiant **Flux** (Φ_λ, $\mathrm{W\,nm^{-1}}$). Let $d^2 Q(t, \lambda)$ denote the amount of electromagnetic energy (as established via Maxwell equations) that stems from a source of light in the wavelength band $[\lambda, \lambda + d\lambda]$ and during the interval $[t, t + dt]$. According to the two previous definitions, such elemental energy may be emitted by the surface of any real *active* body, reflected/diffused by, or transmitted across any real surface, or may "impinge" upon

---

[1] In the language of geometric algebra, $d\mathbf{A}$ is a bivector.

38 2 The Sun as Power Source for Spaceflight

a purely geometrical surface. The following expression defines the spectral radiant power

$$\Phi_\lambda = \frac{d^2 Q(t, \lambda)}{d\lambda\, dt} \qquad (2.1)$$

where function $Q(t, \lambda)$ is supposed differentiable with respect to t and λ.[2]

- **Radiant Power** or radiant **Flux** (Φ, W) is the spectral radiant power integrated over the full wavelength spectrum

$$\Phi(t) = \int_0^\infty \Phi_\lambda\, d\lambda \qquad (2.2)$$

In practice, the integral (2.2) is restricted from some $\lambda_1 > 0$ to some finite $\lambda_2 > \lambda_1$. Though it may be a time-valued function peculiar to the emitting body, normally Φ contains no other source-related information.

However, both radiant-power concepts can be extended to emitting/receiving surface and solid angle about a propagation direction based on $d\mathbf{A}$; for instance, one could mean

$$\Phi = \Phi(t, \lambda, A, \omega) \qquad (2.3)$$

for a more general function of *independent* variables to be used in additional radiometric definitions, as shown below.

- **Spectral Radiant Exitance** (M_λ, $W\,nm^{-1}\,m^{-2}$) is the spectral radiant power emitted per source's unit area

$$M_\lambda = d\Phi_\lambda/dA = d^2\Phi/d\lambda\, dA \qquad (2.4)$$

- **Radiant Exitance** (M, $W\,m^{-2}$) is the spectral radiant exitance integrated over the full wavelength spectrum

$$M = \int_0^\infty M_\lambda\, d\lambda \qquad (2.5)$$

In practice, the integral (2.5) extends from some $\lambda_1 > 0$ to some finite $\lambda_2 > \lambda_1$.

- **Spectral Radiant Intensity** (F_λ, $W\,nm^{-1}\,sr^{-1}$) is the spectral power, per unit solid angle, emitted by a radiation source. By definition one writes

$$F_\lambda = d\Phi_\lambda/d\omega = d^2\Phi/d\lambda\, d\omega \qquad (2.6)$$

where $d\omega$ is an infinitesimal solid angle about the direction $\mathbf{d}$ from $d\mathbf{A}$. Therefore, additional information regards the angular distribution of the source's spectral

---

[2] In the framework of this book, such assumption is always satisfied.

2.1 A Summary of Radiometric Quantities

power. Spectral intensity and spectral exitance are interrelated by the following equation:

$$F_\lambda \, d\omega = M_\lambda \, dA \qquad (2.7)$$

which follows from (2.6) and (2.4).

- **Radiant Intensity** (F, $W \, sr^{-1}$) is the source's spectral radiant intensity integrated over the full wavelength spectrum

$$F = \int_0^\infty F_\lambda \, d\lambda \qquad (2.8)$$

In practice, the integral (2.8) extends from some $\lambda_1 > 0$ to some finite $\lambda_2 > \lambda_1$.

- **Spectral Radiance** (L_λ, $W \, nm^{-1} \, sr^{-1} \, m^{-2}$) is the spectral power, per unit solid angle and per unit *projected* area, emitted by a radiation source. In formulas,

$$L_\lambda = d^3 \Phi / d\lambda \, d\omega \, dA_\perp = dF_\lambda / dA_\perp \qquad (2.9)$$

This radiometric quantity contains information about power distribution as function of the emitting area element, emission direction, and wavelength at some time t.

- **Radiance** (L, $W \, sr^{-1} \, m^{-2}$) is the spectral radiance integrated over the full wavelength spectrum

$$L = \int_0^\infty L_\lambda \, d\lambda \qquad (2.10)$$

In practice, the integral (2.10) extends from some $\lambda_1 > 0$ to some finite $\lambda_2 > \lambda_1$. We then have a *broadband* radiance.

- **Spectral Irradiance** (I_λ, $W \, nm^{-1} \, m^{-2}$) is the spectral radiant power incident on or crossing a surface of unit area; this power can come from any direction in the semispace containing $\mathbf{n}$ according to the above convention, in particular from a finite number of single directions simultaneously:

$$I_\lambda = d\Psi_\lambda / dA \qquad (2.11)$$

where Ψ_λ denotes the spectral power arriving at dA.

- **Irradiance** (I, $W \, m^{-2}$) is the spectral irradiance integrated over the full wavelength spectrum

$$I = \int_0^\infty I_\lambda \, d\lambda \qquad (2.12)$$

In practice, the integral (2.12) is restricted from some $\lambda_1 > 0$ to some finite $\lambda_2 > \lambda_1$. We then have a *broadband* irradiance. One should note that the cosine law is automatically taken into account in computing the total rate of energy deposited on dA from different directions. Therefore, if one or more light's beams arrive orthogonally to the considered surface, total irradiance is at its maximum. Spectral irradiance should not be confused with spectral intensity.

We have discussed radiometric functions starting from the concept of infinitesimal radiant energy, and radiant power has been defined for the first. As a point of fact, normally is power (not energy) involved in the characterization of a source of light (in the general sense explained above).

Definitions of the radiometric quantities do not mention any relative motion between surfaces which emit/receive/transmit electromagnetic radiation. They implicitly refer to some frame attached to the considered surface. However, when one deals with at least one energy source and at least one receiving surface, aberration of light can, for example, no longer be neglected if their relative velocity is sufficiently high.

Some further notes about the above definitions are in order.

1. Normally, the adjective *radiant* may be suppressed during statements, sentences etc, when no ambiguity is generated since the sole energy form in the context is the radiant one.
2. Both (spectral) exitance and (spectral) irradiance can be grouped conceptually into (spectral) *flux density*, but with the distinction between the 'leaving from' and 'the arriving at' possibilities, respectively.
3. Depending on the problem at hand, either wave frequency or energy may be used instead of wavelength.
4. Although many international efforts to achieve a world-wide standard in defining radiometric quantities, however some scientific communities prefer to have their own definitions. The problem is that some words refer to different concepts. One of the terms causing confusion is *intensity*, often referred to as power per unit solid angle or power per unit area. Another term of misunderstanding is *flux* sometimes used as shorthand for flux density. The two terms have different meanings and should *not* be interchanged. Thus, one should be particularly careful when comparing results coming from documents issued by different scientific communities.

The concept of completely isotropic source of light implies a spherical source whose *intensity* is direction independent. A distant star can be considered an isotropic and point-like source from our vantage points in the solar system. Not so the Sun, as we shall see in the next section.

A flat radiating surface, either active or passive, may in principle exhibit a *radiance* independent of the emission direction over its whole hemisphere. Such an ideal surface is called Lambertian. If the surface is passive, uniform diffusion of light takes place. The concept is applicable to each of the differential areas in which a finite curved surface can be partitioned. It's easy to show that the exitance of a Lambertian surface, either infinitesimal or finite, is related to its radiance via a simple but meaningful relationship:

$$M = \pi L \tag{2.13}$$

2.1 A Summary of Radiometric Quantities

Fig. 2.1 Geometry of an infinitesimal surface receiving radiation from an infinitesimal source

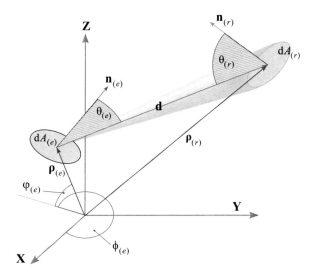

The factor π steradians comes from the definition of radiance that relies on the area projected along a given direction. A Lambertian surface is also known as the ideal diffuser. A blackbody radiator is Lambertian as well. In Chap. 6, we will analyze the blackbody sail.

Now let us work out the irradiance of an elemental surface brought about by the radiance of an infinitesimal emitting surface. This will be useful in other chapters for setting the algorithm giving the solar irradiance onto an arbitrary oriented sail at an arbitrary distance from the Sun. Figure 2.1 shows the geometry we are using; subscripts (e) and (r) refer to as the emitting and receiving areas, respectively. One has

$$\mathbf{d} = \boldsymbol{\rho}_{(r)} - \boldsymbol{\rho}_{(e)}, \qquad \|\mathbf{d}\| \equiv d \tag{2.14a}$$

$$\cos\theta_{(e)} = \mathbf{n}_{(e)} \cdot \mathbf{d}/d, \qquad \cos\theta_{(r)} = -\mathbf{n}_{(r)} \cdot \mathbf{d}/d \tag{2.14b}$$

$$\cos\theta_{(e)} \geq 0, \qquad \cos\theta_{(r)} \geq 0 \tag{2.14c}$$

The differential power received from the surface of area $dA_{(r)}$ can be expressed as

$$d^2\Phi_{(r)} = L_{(e)}\, dA_{(e)} \cos\theta_{(e)}\, d\omega_{(r)}$$

$$= L_{(e)}\, dA_{(e)} \cos\theta_{(e)} \frac{dA_{(r)} \cos\theta_{(r)}}{d^2} \tag{2.15}$$

Therefore, from the definition of irradiance and (2.15), the irradiance onto the passive area element results in

$$dI_{(r)} = \frac{d^2\Phi_{(r)}}{dA_{(r)}} = L_{(e)} \cos\theta_{(e)} \cos\theta_{(r)} d^{-2}\, dA_{(e)} \tag{2.16}$$

Fig. 2.2 Light's diffusion momenta: the z-axis of the reference frame is along the normal to the surface element. In general, each differential solid angle hosts a different 2nd-order momentum rate $d^2\Gamma$ of leaving light, as indicated by different-length arrows

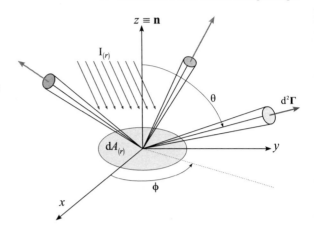

Inequalities in (2.14a), (2.14b), normally taken for granted, mean that the receiving and the emitting surfaces can see one another *directly*. The second (or back) *side* of a passive real surface can see the radiating surface indirectly, namely, via some reflection(s) (of the emitted light) from other surfaces, or (partial or total) transmission of light through the first (or front) side: for example, an arbitrary (concave or convex) finite surface radiating energy onto another arbitrary (concave or convex) finite surface. Things are still more complicated when either the emitting or the receiving surface, or both, change their orientation in space. Furthermore, the source of e.m. radiation may be a hot plasma that the receiving surface sees at different depths. Whereas (2.16) holds in general, its integral in various real configurations can be complicated to evaluate, even numerically.

Now, let us introduce the concept of coefficient of diffuse momentum[3] related to a surface that scatters light. We will use the same area element labeled by (r) in Fig. 2.1. Let us suppose to know the irradiance onto the area $dA_{(r)}$ coming from one (arbitrary) direction and *scattering* in the same hemisphere (as shown in Fig. 2.2); we are referring to the bidirectional-reflection function, detailed in Sect. 6.4.2.

The 2nd-order momentum rate $d^2\Gamma$, associated to the reflected radiance in the direction (ϕ, θ), can be expressed by

$$d^2\Gamma_{(re)} = \frac{L_{(re)}(\phi, \theta)}{c} dA_{(r)} \cos\theta \begin{pmatrix} \sin\theta\cos\phi \\ \sin\theta\sin\phi \\ \cos\theta \end{pmatrix} d\omega \qquad (2.17)$$

where the subscript *(re)* refers to reflection, and we have emphasized the fact that the radiance from $dA_{(r)}$ is not isotropic, in general; (this concept will be made more precise in Chap. 6). As a result, the direction of the integrated momentum rate can be different from either $\mathbf{n} = (0\ 0\ 1)^\top$ or the specular-reflection direction. If we focus

---

[3]This one should not be confused with the coefficients related to particle diffusion in gas or plasma.

2.2 Monitoring Sunlight from Satellite

on the total momentum rate due to *diffuse* reflection, then we have formally

$$d\Gamma_{(diff)} = \int_{2\pi\,\text{sr}} d^2\Gamma_{(diff)} \tag{2.18}$$

Let us note that the scattered exitance from the considered surface can be written

$$M_{(re)}\,dA_{(r)} = d\Phi_{(re)} = (\mathcal{R}_{(diff)} + \mathcal{R}_{(spec)})I_{(r)}\,dA_{(r)} \equiv \mathcal{R}I_{(r)}\,dA_{(r)} \tag{2.19}$$

$\mathcal{R}_{(diff)}$ denoting the diffuse reflectance of the surface, namely, the fraction of the received energy that is reflected diffusely, whereas $\mathcal{R}_{(spec)}$ is the specular fraction. The coefficient of diffuse momentum is defined as follows

$$\chi = \frac{\|d\Gamma_{(diff)}\|/dA_{(r)}}{M_{(diff)}/c} = \frac{\|d\Gamma_{(diff)}\|/dA_{(r)}}{(\mathcal{R}_{(diff)}/\mathcal{R})M_{(re)}/c} \tag{2.20}$$

In this ratio, the numerator represents the radiation pressure associated with the vector sum of the diffuse momenta, whereas the denominator is the pressure that one would get if the diffuse exitance were turned into a directional beam. If one deals with a Lambertian surface, i.e. $L_{(diff)} = L_{(re)} \equiv L^{(\mathfrak{L})} = M^{(\mathfrak{L})}/\pi = M_{(re)}/\pi = M_{(diff)}/\pi$, then the integration of (2.17) with respect to ϕ and θ becomes easy and produces

$$d\Gamma^{(\mathfrak{L})} = \frac{L^{(\mathfrak{L})}}{c}\,dA_{(r)}\begin{pmatrix} 0 \\ 0 \\ 2\pi/3 \end{pmatrix} \tag{2.21}$$

where the superscript $(\mathfrak{L})$ denotes a Lambertian-surface quantity. As a result, $\chi^{(\mathfrak{L})} = 2/3$. Although not strictly, χ-deviations from such values may be interpreted as non-Lambertian behaviors. The usefulness of the coefficient of diffuse momentum comes from modeling the recoil thrust due to the scattering of light from a surface (Chap. 6), in particular a solar sail. In simple models with paraxial solar incidence, the diffuse-reflection thrust component is assumed directed along the surface normal and with a magnitude proportional to $\chi\mathcal{R}_{diff}$. In a quite similar way, one can define coefficients of diffuse-transmittance momentum. In Chap. 6, we will be concerned with models based on complicated diffraction theories and mathematical descriptions of real-surfaces.

2.2 Monitoring Sunlight from Satellite

In this section, we will describe features of the solar irradiance that are of particular relevance to the sailcraft dynamics and design. From chosen references of the specialized literature on solar irradiance and models of the solar photosphere, and the close atmospheric layers, we will recall a few items about the current knowledge of the Sun, but only for a more fluent description of what is of matter for solar sailing, in general, and fast sailing, in particular. We begin with mentioning the revolution in the measurements of solar irradiance started in 1978. Then, we will discuss about total and spectral solar irradiances in two separate subsections. Finally, in Sect. 2.3, we will draw some indications relevant to sailcraft trajectory and solar-sail design.

In building a standard in solar spectral irradiance, e.g. the tables in [3] from 0.1195 to 1,000 μm, not only measurements from spacecraft have been extensively employed, but also many, many observational data from high-altitude aircraft and ground devices; also, solar-irradiance models have been taken into account.

Before the beginning of the astronautical era, there was no universally-accepted evidence that the so-called *solar constant* might vary from a strict constant value. Ground measurements *sufficiently* accurate/precise to reveal any real deviation from a constant value were not possible, in practice, before space satellites (e.g. [6, 18]). Many historical ideas, concepts, and facts (from China, Europe, and United States) about the Sun and analyses of the solar objects affecting the Sun's electro-magnetic emission can be found in [32]. After the first direct measurements of sunlight carried out by padre A. Secchi (Vatican State) and the studies of S. P. Langley (USA) in the 19th century, C. G. Abbot (USA) was a pioneer in measuring the solar constant and its claimed variability for decades [1]. However, such measurements (though carried out from observation stations on mountains of United States, Chile, and Arabia, and by means of new instruments) indicated a too much high (some percent) solar change. A remarkable synopsis of his and his team's measurement campaigns are described in [2].

On November 16, 1978, the first cavity radiometer, named HF after Hickey-Frieden, was switched on aboard the U.S. satellite NIMBUS-7; this was followed by other satellites endowed with modern instruments for observing the Sun. One of the most striking results of the space era has been the discovery that the Total Solar Irradiance (TSI) (standardized at 1 AU) is variable indeed. Let us be a bit more precise: in the Hertzsprung-Russell diagram (HR), the plot of star's luminosity[4] *vs* star's surface blackbody temperature, each star occupies a dot. A main-sequence star like the Sun stays fixed in the diagram for a time ranging from hundred millions to billions of years, depending non-linearly and inversely on the star mass.

However, the extended measurement campaigns, started on November 1978, have been confirming that the Sun is a *variable* star, and new knowledge—also through advanced solar dynamo models—have been accumulated in the last three solar cycles (21–23), and is continuing in the cycle-24. (Sunspot cycles are counted conventionally from the cycle in the period 1755.03-1766.06, or the solar cycle 1).

Before going on, we have to clarify that, from an astrophysical viewpoint, the Sun is still a fixed-luminosity star because its variations with time are

---

[4]This is the radiant power emitted by a star.

2.2 Monitoring Sunlight from Satellite

> rather small. A variable star, in the astrophysical sense, is something of much different from and stronger than the Sun (e.g. Cepheid-type stars). Nevertheless, the actual small changes of the TSI are capable to affect the Earth's climate evolution in the long run, even though in the last decades the anthropogenic actions on climate have been becoming stronger and stronger, according to the Intergovernmental Panel on Climate Change (IPCC) of the United Nations.

In order to better grasp how and how much the solar-irradiance campaigns have so positively affected the current knowledge of the Sun, down here we will summarize the main instruments on some launched Sun-relevant satellites. We have arranged the main features by means of concise card-like information. Some pieces of information are related to the time of this writing.

Spacecraft: **NIMBUS-7** (launched in 1978)

- Experiment(s)/Instrument(s): Many experiments for Earth weather and total solar irradiance. Solar irradiance was measured by Hickey-Frieden (HF) radiometer.

NIMBUS-7 was the last of the NIMBUS-series of US satellites in Sun-synchronous orbits. The Hickey-Frieden cavity radiometer started measuring in November 1978. This type of radiometer replaced flat radiometers working on board former NIMBUS satellites.

Spacecraft: **Solar and Heliospheric Observatory (SOHO)** (launched in 1995)

- Experiment(s)/Instrument(s): Variability Irradiance Gravity Oscillation (VIRGO):
- measurements of total solar irradiance by two high-accuracy two-cavity differential absolute radiometers (DIARAD and PM06V)
- Monitoring of the solar spectral irradiance at 402, 500 and 862 nm with a 3-channel sun photometer
- Spectral radiance at 500 nm with a luminosity oscillation imager
- Continuous high-precision, high-stability and high-accuracy measurements of the solar total and spectral irradiance, and spectral radiance variation; long-term precision of the time series from VIRGO resulted in about 2 ppm/year ($1\,\sigma$)
- Experiment(s)/Instrument(s): Solar Ultraviolet Measurement of Emitted Radiation (SUMER):
- Spectrograph operating in either spectral scan (80–150 nm, 1024 pixels) or raster scan (360 spatial pixels), spatial resolution 1 arc-second (i.e. about 730 km on the Sun)

SOHO has been operating in a halo orbit about the (Lagrange) L1 point of the Sun-Earth gravitational system (about 1.5 million kilometers from Earth towards the Sun). Two absolute radiometers are onboard. This has been quite important in comparing and reconstructing data of total solar irradiance.

Spacecraft: **ACRIMSAT** (launched in 1999)

- Experiment(s)/Instrument(s): Measurements of total solar irradiance via ACRIM-III

46   2 The Sun as Power Source for Spaceflight

Instruments of the series Active Cavity Radiometer Irradiance Monitor (ACRIM) were put on various spacecraft designed for solar investigations. ACRIM-III, designed for NASA by the Jet Propulsion Laboratory, has continued to extend the database started with ACRIM I launched in 1980 on the Solar Maximum Mission (SMM) spacecraft. ACRIM II was on the Upper Atmosphere Research Satellite (UARS) launched in 1991. ACRIMSAT has operated in Sun-synchronous orbit.

Spacecraft: **ENVISAT** (launched in 2002)

- Experiment(s)/Instrument(s): Scanning Image Absorption Spectrometer for Atmospheric Cartography (SCIAMACHY)
- 240–2380 nm with resolution range from 0.2 to 1.5 nm (transmission, reflection, scattering of sunlight by either the Earth's atmosphere or surface)
- Experiment(s)/Instrument(s): Measurements of the Earth's reflectance (from surface and atmosphere) via Medium Resolution Imaging Spectrometer (MERIS):
- Earth's reflectance in the solar spectral range 390–1040 nm segmented in 15 spectral bands
- Experiment(s)/Instrument(s): Global Ozone Monitoring by Occultation of Stars (GOMOS)
- GOMOS looks to stars as they descend through the Earth's atmosphere and change color

ENVISAT, orbiting in a Sun-synchronous polar orbit, is completely oriented to radiation reflected by Earth, but data from it and other similar satellites are useful also to modeling radiation-pressure perturbations onto sailcraft spiraling about the Earth.

Spacecraft: **SORCE** (launched in 2003)

- Experiment(s)/Instrument(s): Total Irradiance Monitor (TIM), Spectral Irradiance Monitor (SIM), Solar Stellar Irradiance Comparison Experiment (SOLSTICE A, B)
- spectral range: 0.1–2400 nm (excluding the interval 40–115 nm)
- total-irradiance accuracy of 100 ppm with a long-term repeatability of 10 ppm/year
- daily measurements of the solar Ultraviolet (UV) irradiance in 120–300 nm with a spectral resolution of 1 nm; measurement accuracy less than 5 percent and long-term repeatability of 0.5 percent/year

SORCE (Solar Radiation and Climate Experiment) orbits about the Earth at an altitude of 640 km with 40° in inclination.

Spacecraft: **SDO** (launched February 11, 2010)

- Experiment(s)/Instrument(s): Atmospheric Imaging Assembly (AIA), Extreme ultraviolet Variability Experiment (EVE), Helioseismic and Magnetic Imager (HMI)
- AIA: Sun's images acquisition in 10 wavelengths every 10 seconds
- EVE/MEGS: multiple Extreme Ultraviolet (EUV) grating spectrograph (two Rowland-circle grating spectrographs) for spectral irradiance, operating in 0.1–7 nm (1-nm resolution), and 5–105 nm with (0.1-nm resolution)
- EVE/ESP: EUV Spectrometer (transmission grating spectrograph)
- HMI: continuous full-disk coverage at high spatial resolution

SDO (Solar Dynamics Observatory) is a 22-m^3 satellite with an initial mass of 3.2 tonne and has been designed for further understanding the Sun's influence on Earth and near-Earth space; it has been studying the solar atmosphere on small space and time scales, and in many wavelengths simultaneously. SDO is the first mission

2.2 Monitoring Sunlight from Satellite

for NASA's *Living With a Star* Program aimed at comprehending the causes of solar variability and its impacts on Earth. SDO circles Earth on an inclined geosynchronous orbit allowing for nearly-continuous and high data rate down to one dedicated ground station.

Spacecraft: **PICARD** (launched on June 15, 2010)

- Payload of PICARD consists of the following packages:
- SOVAP (SOlar VAriability PICARD): composed of a differential absolute radiometer and a bolometric sensor to measure TSI;
- PREMOS (PREcision MOnitor Sensor): a set of 3 photometers to study the ozone formation and destruction, helio-seismology, and 1 differential absolute radiometer for TSI;
- SODISM (SOlar Diameter Imager and Surface Mapper): an imaging telescope and a CCD aimed at measuring the solar diameter and shape with an accuracy of a few milli-arcsec; helio-seismological observations to probe the solar interior are also planned.

PICARD, from a French astronomer of 17th century, moves on a Sun-synchronous orbit at an altitude of 725 km.

2.2.1 The Total Solar Irradiance

This section is devoted to the discussion of some very notable results regarding the TSI measured in space. The panel-(a) of Fig. 2.3 shows the daily averages of the original TSI data coming from different radiometers, as described in Sect. 2.2, for solar observations from satellites: HF on spacecraft NIMBUS-7, ACRIM-I on SMM, ACRIM-II on UARS, either DIARAD or (DIARAD + PM06V = VIRGO) on SOHO, ACRIM-III on ACRIMSAT. (Data from TIM on SORCE have not be used in the construction of the following composites). Data in the same time intervals are dissimilar mainly for four reasons: different instrument characteristics, data acquisition interruption, different satellite attitudes, instrument degradation. Therefore, the (expected) problem arose from the need to compare the TSI time series coming from the various spacecraft in order to form a *composite* time series. Panels-(b,c,d) report the currently-available three composites named the PMOD, ACRIM, and IRMB composites, respectively, and covering three solar cycles, i.e. cycles 21–23. Such composites have been building in distinct ways: they use all the same original observational data-sets, but employ different procedures for removing sensitivity variations.

PMOD-composite is the only one which also corrects the early HF data for its degradation; in addition, it provides corrections to other satellite data (from HF and ERBS) available during the ACRIM's time gap (about 28 months) between the end of ACRIM-I operation and the begin of ACRIM-II campaign. The two other composites, similar to one another, do not include this correction procedure. The three composites are compared to the reconstruction of TSI from magnetograms and continuum images recorded at Kitt-Peak national observatory (USA/Arizona): the PMOD composite captures most of the variability variance, according to [11].

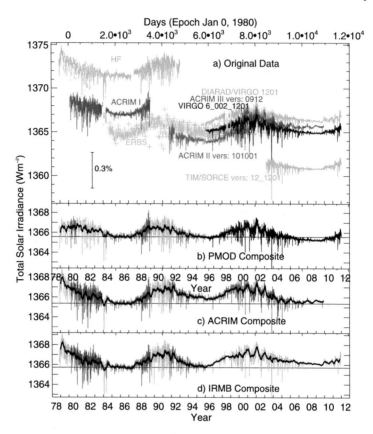

Fig. 2.3 Currently available composites of TSI since November 1978. They differ mainly on the considered satellite instruments and the way the raw data are corrected (Courtesy of C. Fröhlich, PMOD, World Radiation Center, Switzerland)

Figure 2.4 details the PMOD composite and reports additional information about the solar cycles 21–23; it marks the minima between two consecutive cycles. In particular, one can note that a proxy model [15], for explaining solar short and long-term behaviors during the last three cycles, has been successfully used for expanding the composite back to the cycle minimum occurred in 1976. Cycle-24 began in 2008.

Although highlighted at the beginning of Sect. 2.2 that—astrophysically speaking—the Sun has a constant radiant exitance, however it is notably variable for the solar system, in general, and for the Earth, in particular. Why is the Sun variable in TSI, though its variability is of the order of 0.1 percent (at least in the past decades)? Solar physicists have been advancing in explaining intrinsic phenomena in the upper layers of the solar atmosphere, in particular its "base" named the photosphere, a very useful picture of a spherical-like skin where astonishing phenomena take place.

2.2 Monitoring Sunlight from Satellite

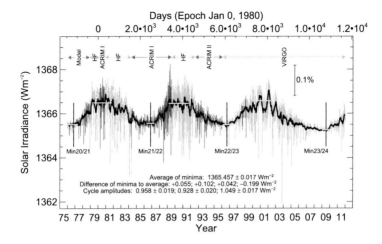

Fig. 2.4 Extended PMOD-Composite of Total Solar Irradiance. *Gray vertical segments* refer to the daily-means time series. Also reported are the differences between consecutive minima values, and the cycle amplitudes (Courtesy of C. Fröhlich, PMOD, World Radiation Center, Switzerland)

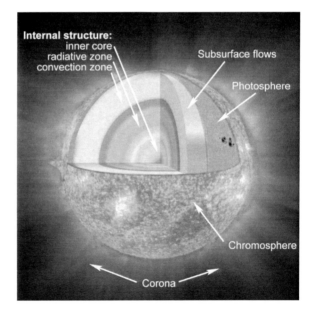

Fig. 2.5 Visualization of the solar structure via the shell concept (Courtesy of NASA)

Figure 2.5 shows a simplified shell model of the Sun. Substantially, the energy produced in the inner thermonuclear-reaction core (about $1/4$ of the solar radius $r_\odot$) diffuses outward by radiation (mainly γ and X-photons) through the radiative zone; the energy achieves the photosphere by fluid flows through the convection zone, extending in the last $0.3r_\odot$. Modern solar seismology has revealed a thin interface layer (not shown in the figure), called the *tachocline*, between the radiative zone

50   2   The Sun as Power Source for Spaceflight

(with constant spin rate) and the differentially-rotating convection zone; it is where the Sun's magnetic field[5] is thought to be generated. There are different estimates of the tachocline thickness, depending essentially on the specific ways it is defined by; for example, some authors calculate $\delta_{tc}/r_\odot \cong 0.039$ at the solar equator, and 0.042 at 60° of heliographic latitude; other authors estimate $\delta_{tc}/r_\odot$ ranging from 0.016 to 0.038, whereas other evaluations are notably more different.

To have even a rough idea of how much dense is the Sun in the above-mentioned shells, we can mention that, from the innermost core to the photosphere, density varies of about nine orders of magnitudes. A better impressive quantity is perhaps the time the energy produced in the core (in a given time interval) takes to reach the photosphere: at least some thousands of years (someone estimates this energy travel lasting for a considerable fraction of a million years), instead of the ideal $r_\odot/c \sim 2$ s.

This concept of energy transport time should not be confused with the time the Sun would take to get rid of all energy stored in its inner core if the Hydrogen fusion stopped now: the Sun would remain luminous for at least twenty million years, but without producing neutrinos.

Above the photosphere, there is a thin layer (about one Earth diameter thick) named the chromosphere, more transparent than the photosphere in the visible band. It is significantly hotter than the photosphere. Chromosphere can be observed during total solar eclipses or by means of spectrographs with filters that isolate particular emitted lines, e.g. the H_α (reddish chromosphere) or the K-line (violet chromosphere) of the ionized Calcium atom. Above the chromosphere, separated by a thin and irregular layer named the transition region, one can find the *solar corona*, a very low density extended region of temperature rising to over a million (Kelvin) degrees. The corona can be observed by special instruments (the coronagraphs) or during total solar eclipses; it shines in X-rays, and considerably too, because of its high temperature. Both photosphere and chromosphere host a number of important features revealing the interaction of surface plasma elements with the magnetic field originated in the tachocline. One can observe a very rich fine structure in the continuum, spectral bands and spectral lines.

Useful is the concept of *quiet* Sun, applied particularly to the photosphere, which encompasses all the elements (ionized-gas and magnetic flux) *weakly* affected by the 11-year solar cycle. These elements may be grouped according to their horizontal scales in magnetic-flux tubes (a few hundred kilometers), granules and mesogranules (some thousand kilometers), and supergranules (greater than 10 thousand kilometers). Supergranules may be described as a pattern of irregular bright regions or cells of horizontal plasma outflows, covering the photosphere, known as the *supergranulation* (with a lifetime of about 1 day). The boundaries of supergranules are outlined by local magnetic features, or the photospheric *network*. The network exhibits high magnetic fields; in contrast, the interior of the network pattern is not magnetized (or above some low threshold). The quiet Sun structure is very spread

---

[5]In this and subsequent chapters, we use the custom term "magnetic field" for actually indicating the *magnetic induction field* or magnetic flux density, which is denoted by **B** and measured in Tesla, or Wb/m^2, in the SI.

2.2 Monitoring Sunlight from Satellite

through the solar surface, so that its knowledge may be important for better understanding the photospheric magnetic evolution. Modern photosphere models, e.g. those ones considered in [21–25] and [30, 31], take a five-element typology into account: the Quiet Sun, sunspot umbrae, sunspot penumbrae, faculae, and the network. In addition to the network, sunspots and faculae are regions of high magnetic field, a typical sunspot extending in area much more than a facula. In the sunspot's darker core or umbra, the magnetic field is higher than in the peripheral regions or penumbra. Faculae appear notably brighter than sunspots, indicating higher temperatures of their magnetized plasma flows. Faculae are seen mainly at the limb of visible-band images of solar disk. Over them, images in the ionized Ca or H-α light show solar bright and large regions in the chromosphere, the *plages*.[6] During a typical sunspot cycle, faculae are usually close to sunspots; in period of intense magnetic activity, the area covered by both faculae and the (photospheric) network rises considerably. In addition, sunspot's mean-life is shorter than facula's, and sunspot "decays" into facula.

There is a more profound mechanism responsible for the fact that, when the number of sunspots increases, TSI increases (instead of decreasing as one would expect on sunspot basis only). This is explained in [11], and appears to be related to the local magnetic structure around sunspots and faculae, according to modern Magnetohydrodynamics (MHD) models. Very briefly, the sunspot darkness is ascribed to vertical and intense magnetic fields, which deflect the heat flow rising from deeper atmospheric zones. The blocked heat amount is not transferred or shunted nearby for re-emitting; instead, after horizontal diffusion, this energy flux goes back down rapidly into the Sun.

In contrast, faculae are bright because the magnetic field acts on smaller scales, and forms small depressions in the photosphere; thus, radiation from the hotter (because deeper) atmospheric layers is emitted with low attenuation.

Although not detectable in the periods of "classical" astronomy and astrophysics, namely, before the space era, however the fact that the net solar exitance can change should not surprise qualitatively too much once one considers the temperatures of the above-mentioned solar magnetic "objects". Figure 2.6 shows such temperatures versus the radial distance from the photosphere bottom. A big problem is to full understand quantitatively both short-time and secular trends in TSI variability (as also requested from many other areas of scientific investigation).

Observing the Sun radially from a sufficiently distant point in space (e.g. the Earth at 1 AU) at a chosen wavelength, the related optical depth $\tau(R)$ of such point can be set to zero. Then, one can define the *bottom* of the "solar surface" as the distance $r_\odot$ (from the Sun's barycenter)

---

[6]The French name standing for beach, because a plage resembles light-colored "sand" against the darker background.

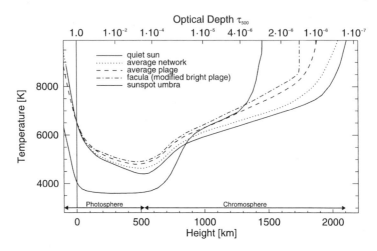

Fig. 2.6 Temperatures of solar magnetic regions *vs.* the height above the photosphere bottom. Height is defined in terms of the optical depth at 500 nm ($\tau_{500\,nm} = 1$): the optical thickness at 500 nm and the corresponding distance in kilometers are shown (From [15], Courtesy of C. Fröhlich, PMOD, World Radiation Center, Switzerland)

such that $\tau(r_\odot) = 1$. This depends on the wavelength, of course. One can choose the wavelength of 500 nm (one of the standard lengths in Optometry), very close to the 501.34 nm of the wavelength maximum of a blackbody at 5780 K (from Wien's displacement law, or $\lambda_{max}T = 2.897768 \times 10^6$ nm K) and sufficiently close to the Sodium doublet 589/589.6 nm (i.e. energy separation of 0.0021 eV). The top of the photosphere corresponds to the minimum temperature of the *quiet* Sun as above defined. In Fig. 2.6, photosphere is "extended" about 100 km below in order to show how temperature increases rapidly. A "negative thick" solar shell of −40 km is radiation-accessible around 1.6 μm, where solar opacity achieves a minimum.

In the last years, a significant work has been carrying out for reconstructing TSI over the past centuries in the contest of Earth's climatology studies, especially between the end of the Maunder minimum to the present time. Since measurements of solar magnetic flux were not available before the second half of the 20th century, proxy quantities have been using, e.g. the records of the Zurich or group sunspot number, the sunspot area, the facular area, and the production rate of cosmogenic isotopes. Current models of TSI are shown to explain 90–95 percent of irradiance variability, at least with regard to reconstruction. An extended and excellent review can be found in [11].

Prediction of TSI exhibits additional difficulties. We limit ourselves to few considerations, some of which are oriented to solar sailing. Some models for explaining the various aspects of the solar activity and its time variations resort to what happens (or is supposed to happen) in the convection zone and tachocline. There, an electrically conducting fluid evolves under the action of the solar magnetic field that, in turn, is affected by this plasma. The solar field acts as the privileged channel of transport of energy to be ultimately released from the star. In order to describe

2.2 Monitoring Sunlight from Satellite

the *coupled* evolution of such field and matter (both exhibiting a broad spectrum of spatial/temporal scales), one has to use the equations of MHD.

—The following gray-box may be skipped on a first reading.—Put very simply, the magnetic field (if any) of a body with finite electric resistance would eventually decay if there were no internal source of electric field. Referring to celestial bodies, decay time is usually very brief compared to the body lifetime. If the body is observed to show persistent and/or cyclic magnetic fields, then some process has to be responsible for maintaining these fields. Such a process is named a *dynamo*, which acts against Ohmic losses. With regard to the Sun, there are various kinds of dynamo models aimed at explaining the solar cycles. Because the full MHD equations are extremely complicated, researchers are used to resort to reduced MHD equations; for instance, a family of dynamos is represented by the mean-field hydromagnetic models. In this set, the so-called *flux-transport* dynamos have been receiving considerable attention.

In a nutshell, according to the flux-transport dynamo models, the Sun should exhibit a complicated circulation, in the convective zone, symmetric with respect to the solar equatorial plane. If one cuts the solar sphere by a plane passing through the rotation axis, one may think of observing one of the four sections of the convective zone, as sketched in Fig. 2.7, where the Babcock-Leighton solar dynamo concept is illustrated. This is based on a special form of the magnetic field, as inferred from solar macroscopic features such as Hale's polarity law (which puts into evidence that solar cycle lasts 22 years), the sunspot *butterfly* diagram, synoptic magnetograms, and the shape of the solar corona during the minima of solar activity. As a result, the typical large-scale magnetic field in the convection zone (including the photosphere) may be pictured with two components: one longitudinal component and one component parallel to the local meridians; they are usually termed *toroidal* and *poloidal*, respectively.

According to Babcock-Leighton mechanism, the poloidal magnetic field, accumulated on the polar surface region-A at cycle n, will be moved down to the tachocline neighborhood; then, the toroidal field for cycle $n + 1$ can be produced along the path B–C. The magnetized plasma rises to region-D exhibiting a poloidal field again. A slow polarward surface motion, related to the new cycle, will follow. The characteristic times of such plasma movements are of considerable importance inasmuch as they may induce a chaotic behavior in the magnetic-field evolution, and this may be of great importance for the Earth's climate.

The previous considerations may be summarized by asserting that the Sun "remembers" its magnetic-field history; such memory intervenes in the cycle. If one takes the polarity of sunspots into account, the full basic solar cycle lasts about 22 years. A flux-transport dynamo model, also called the calibrated

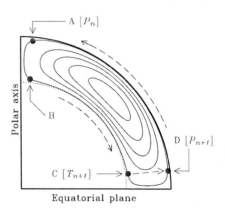

Fig. 2.7 Solar cycle model based on the Babcock-Leighton mechanism. The *closed lines* represent plasma streamlines on the solar meridian planes, known as the meridional circulation, inside the convective zone (From [8], courtesy of P. Charbonneau)

dynamo, has recently been setup including differential rotation (or the Ω-effect), Babcock-Leighton mechanism, meridional flow, and twisting/lifting of toroidal flux tubes (or the α-effect) on the surface and at the tachocline for regenerating poloidal fields. A model of such type (e.g. [10]) has been applied to the prediction of the cycle-24.

Very difficult is the problem of forecasting TSI. Figure 2.8 shows the evolution of the sunspot number from 2000, and the prediction for the current cycle-24. Thus, according to the Space Weather Prediction Center of National Oceanic and Atmospheric Administration (NOAA), maximum of sunspot number should occur in May 2013.

From the viewpoint of solar sailing, we have to stress some points:

1. The analysis of solar-cycle-related time series (in particular TSI) has not yet produced reliable predictions; some meaningful additional information is plainly required. However, for trajectory purposes, short-term proxy-based prediction might reveal to be sufficient for the guidance process. This point may represent an open research area for practical Solar-Photon Sailing (SPS).
2. Solar dynamo theories have been progressing considerably, but there is much to be done. Large-scale numerical simulations, supported by many additional data from solar-physics satellites, should eventually turn into results useful for SPS.
3. The time variations of the solar irradiance are described by means of the so-called *filling factors*, namely, the fraction of the solar surface covered by the elements (or components) of the model, e.g. sunspot umbrae, faculae, etc. Estimating such factors for each element represents a set of sub-problems to be solved. Quiet Sun's filling factor is the complement to unity of sunspots, faculae, or network.
4. TSI appears to exhibit several amplitude modulations. Some theoretical considerations suggest a superimposed chaotic behavior.

In Sect. 2.3, we will conceptually process the pieces of information of this section together with those of the following section devoted to spectral solar irradiance; the

2.2 Monitoring Sunlight from Satellite   55

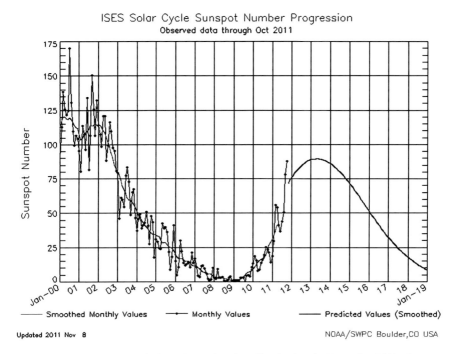

Fig. 2.8 Sunspot number evolution: observed and predicted values in November 2011. Courtesy of NOAA/SWPC

aim is to arrange a number of guidelines on the problem on how using the solar irradiance for a reliable in-space propellantless propulsion.

2.2.2 The Spectral Solar Irradiance

Total Solar Irradiance enters the thrust acceleration of any sailcraft directly (Chap. 6). Its relevance to solar sailing is therefore intuitive too. But also considerably important in designing a solar sail is the Spectral Solar Irradiance (SSI).[7] In order to show this in a simple way, let us recall the blackbody's main properties, which will be useful also to stress an important feature of the sailcraft thrust acceleration.

A blackbody is an ideal body absorbing electromagnetic radiation of every wavelength; it is in perfect thermodynamical equilibrium inside itself and with the ambient. Thus, it is also a perfect Lambertian emitter with spectral radiance $B(\lambda, T)$

---

[7]TSI represents the actual energy rate at the top of the Earth's atmosphere. Even SSI is notably important because, among various things, its large fluctuations in the UV band cause variations in the physical/chemical properties of the upper atmosphere.

depending only on wavelength λ and (absolute) temperature T:

$$B(\lambda, T) = \frac{2hc^2}{\lambda^5}\left[\exp\left(\frac{hc}{k}\frac{1}{\lambda T}\right) - 1\right]^{-1} n(\lambda)^2$$

$$= \left(\frac{1.03558434(4) \cdot 10^4}{\lambda}\right)^5\left[\exp\left(\frac{1.4387751(26) \cdot 10^7}{\lambda T}\right) - 1\right]^{-1}$$

$$\times\, n(\lambda)^2 \; W/(m^2\, nm\, sr) \tag{2.22}$$

h and k denote Planck and Boltzmann constants, respectively. The figures in parentheses after the values represent one standard-deviation in the last corresponding digits. The second row of Eq. (2.22) can be plotted directly by inserting wavelength and temperature in (nm, K), respectively. $n(\lambda)$ is the index of refraction, at λ, of the medium surrounding the blackbody. For our practical purposes, we will set such quantity to unity for any wavelength. The blackbody radiance $B(T)$ and exitance $\mathcal{M}(T)$ are therefore given by

$$B(T) = \int_0^\infty B(\lambda, T)\, d\lambda = \frac{2\pi^4 k^4}{15c^2 h^3}T^4 \tag{2.23a}$$

$$\mathcal{M}(T) = \pi\, B(T) = \mathfrak{s} T^4 \tag{2.23b}$$

where $\mathfrak{s} = 5.670400(40) \times 10^{-8}$ W m^{-2} K^{-4} is the Stefan–Boltzmann constant. Real bodies depart from the blackbody abstraction via different characteristics related to the complexity of the emitting surface; e.g. one can consider a celestial body endowed with variable atmosphere. In laboratory, there are very good realizations of the blackbody, or of the related concept known as the *hohlraum*, which is a cavity with the walls in thermal equilibrium with the inside radiant energy. A thermodynamical concept, which expresses any non-blackbody behavior in a simple way, is the *brightness* temperature. Suppose one has measured the spectral radiance, say, $\mathcal{L}(\lambda)$ of some body in a certain wavelength range $[\lambda_1, \lambda_2]$; at a given λ_0 in this interval, there is *one* blackbody spectrum passing through the point $(\lambda_0, \mathcal{L}(\lambda_0) \equiv \mathcal{L}_0)$ in the $(\lambda, \mathcal{L})$-plane. The corresponding temperature, the brightness one, can be easily obtained from the first row of Eq. (2.22). Denoting it by T_b, one gets

$$T_b(\lambda_0, \mathcal{L}_0) = \frac{hc}{k\lambda_0}\left[\ln\left(1 + 2\frac{hc^2}{\lambda_0^5 \mathcal{L}_0}\right)\right]^{-1} \tag{2.24}$$

We will apply Eq. (2.24) here down.

Nowadays, via some professional distribution of solar-physics data, it is possible to download solar spectral irradiance data adjusted at 1 AU. Such adjustment is necessary because of the variable distance from the Sun of the solar instruments onboard Earth satellites, including those ones around the Lagrange point L1 of the Sun-Earth system. When one computes the solar brightness temperature distribution, one has to transform irradiance into radiance data multiplying the measured

2.2 Monitoring Sunlight from Satellite

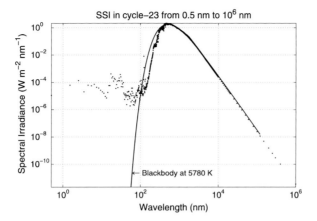

Fig. 2.9 The *dots* represent the solar spectral irradiance (daily data sets), averaged on all days of cycle-23, from 0.5 nm to 1 millimeter. The *solid line* is the irradiance at 1 AU caused by a blackbody with 5780 K

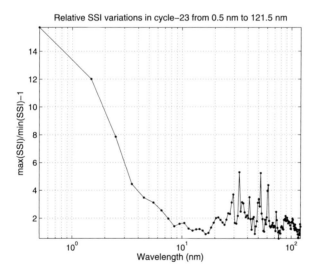

Fig. 2.10 Relative Changes in XUV/EUV irradiance during solar cycle-23: distribution is built by means of 1-nm bins. This 121-nm band is not covered by the ASTM-E490 tables

spectral irradiance by the factor $(AU/r_\odot)^2/\pi \cong 214.9^2/\pi$, which entails a quasi point-like Sun (an apt assumption in the context of this chapter). Of course, the reciprocal of this number has to be used in transforming radiance to irradiance.

Figures 2.9 and 2.10 regard SSI in cycle-23. We have processed these solar irradiance data provided by Space Environment Technologies, CA, [38, 39].

In Fig. 2.9, dots represent daily data sets, averaged on all days of cycle-23, from 0.5 nm to 10^6 nm. (The current solar cycle is 24). The solid line is the irradiance

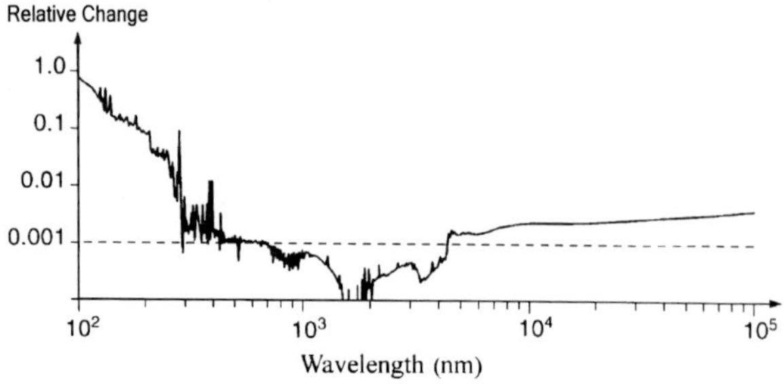

Fig. 2.11 Solar spectral variations over a solar cycle in the 100–100,000 nm broadband. A further finer structure, i.e. a higher resolution SSI in this band should not concern fast solar sailing. Adapted from [34], courtesy of Springer

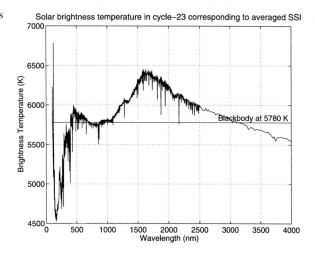

Fig. 2.12 Profile of the Sun's brightness temperature in cycle-23. Temperatures correspond to the means of SSI shown in Fig. 2.9; the reported band has been limited to 100–4,000 nm

at 1AU caused by a fictitious solar blackbody with 5780K. Roughly, the Sun looks like such a blackbody; however, SSI departs quite impressively from the ideal emitter in the EUV band down to the soft X-ray wavelengths (XUV). Not only, but these bands exhibit strong variations during the solar cycle, as shown in Fig. 2.10, where we plotted the relative change max(SSI)/min(SSI)-1 as function of the wavelength in 1-nm bins. This 121-nm band is not included in the E490 tables from American Society for Testing and Materials (ASTM), as documented in [3]. In particular, in the XUV region, changes are very strong. This situation has not been characteristic of cycle-23 only. As a point of fact, Fig. 2.11 shows SSI typical change over solar cycle in the 100–100,000 nm broadband. In the visible band and beyond up to the thermal and far infrared region (FIR), variations are negligible. In contrast, as wavelength goes below the visible region, fluctuations increase considerably. This is quite important, in particular, for the chemical-physical properties of the terrestrial thermosphere/ionosphere.

Figure 2.12 shows the brightness temperature profile for the spectral irradiance of cycle-23. The horizontal line refers to the 5780 K blackbody, which is crossed by the brightness temperature profile in a small number of points. Finally, in Fig. 2.13 we plotted the $\int_{0.5\,nm}^{\lambda} \mathcal{J}(\ell)\,d\ell$, namely, the mean SSI (shown in Fig. 2.9) integrated over wavelength. With regard to the visible region, different definitions are found in

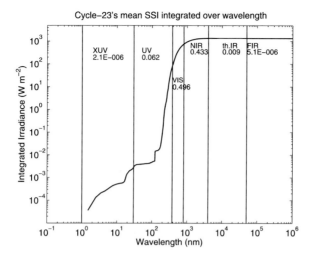

Fig. 2.13 SSI integrated over wavelength. Shown are the main electromagnetic bands, bounded by vertical lines, and their related fractions of TSI

literature (with different justifications). We have adopted the 380–800 nm band in Fig. 2.13, namely, the interval of wavelength of the spectral sensitivity of the human eye, which includes both the photopic and the scotopic visions. CIE has defined the spectral luminous efficiency functions relatively to this range.

Though we detailed cycle-23 only, the following considerations regard solar sails in general:

1. Approximately 99 percent of TSI falls in the 200–4000 nm range; this is important since the thickness of most of the all-metal solar sails already studied for fast sailing is close to 100 nm.
2. The UV-fraction of TSI might appear negligible; however, it amounts to about 48 times (at least) the full TSI relative fluctuations in a cycle, and we shall analyze the impact of the variable TSI on fast sailcraft trajectories.
3. UV photons may also have a cumulative effect on trajectory by altering the reflective layer of the sail (at least, with regard to sails conventionally conceived).
4. XUV and FIR bands do not seem to be a problem as long as the sailcraft does not spend much time very close to the Sun.

2.3 Principles of Utilization of Sunlight for in-Space Propulsion

Since the first half of the 20th century, sunlight has been taking seriously as the basic external source of momentum flux for non-rocket spacecraft. In particular, Tsiolkovsky and Zander wrote explicitly that very thin sheets, pushed by the sunlight pressure, should be able to achieve high speeds in Space. Garwin coined the term *solar sailing* [17]. From what discussed in Sects. 2.2.1 and 2.2.2, in particular at the end of those sections, it is possible arrange a number of items oriented to solar sailing. Sun releases plenty of energy per unit time. This is in form of photons, particles, and magnetic fields filling the heliosphere. Emission can be either continuous (e.g. the normal sunlight and solar wind) or time-localized (e.g. flares and coronal mass ejection); of course, *continuous* does not mean constant in any case. Also, ejection of matter and energy may involve a large fraction the whole solar surface, or be space-localized. When a space mission of any kind is designed, there is an implicit basic assumption that a mission planner/designer makes: the trajectory/orbit of the spacecraft has to be computable in essentially deterministic way. This does not mean that transfer trajectories and operational orbits—including any maneuvers—are free of uncertainties. Fatal errors may happen in delicate mission phases because something goes wrong, of course. In contrast, uncertainties come from sources of errors such as inadequate force field models, insufficient knowledge of relevant parameters, noised measurements of the observables, intrinsic fluctuations of engines, biases in numerical methods for propagating the spacecraft state, and so forth. The spacecraft state evolution with control and stochastic noise is the general problem to be dealt with in a real space mission. In all cases, however, one is confident that things (apart from accidents and/or quite unknown situations unforeseeable in advance) go sufficiently close to what calculated in the design phase and/or updated in the course of flight as soon as measurements (from ground and/or spacecraft instruments) become available.

The design and construction of space rocket engines have become highly accurate and reliable; so are the navigation, guidance and control processes. Now, if we want to use a non-rocket propulsion for many advanced missions, especially those ones impossible for any practical rockets, the first thing to do is to search for an external source of power *sufficiently* intense, stationary, and predictable. Let us discuss these three items, in the order, applied to the Sun. If one wants to utilize sunlight for space propulsion, there are two general options: (1) to absorb solar power and transfer it to some *onboard* reaction mass which shall be ejected, (2) to transform the radiation pressure into a force acting on the spacecraft.

The first option resembles a rocket, although conceptually it is not; such spacecraft does not need a primary source of power onboard, and this represents a strong advantage. Nevertheless, because some matter has to be ejected for getting thrust, we would go back to the problems of mass characterizing rocket spacecraft.

The second option entails that the electromagnetic radiation has to interact with some material surface, or the sail, in order to get a propulsive acceleration. This sail can be well approximated with a two-dimensional (more or less curved) surface with a supporting subsystem. Thus, the utilization of TSI appears natural, even though not so obvious as one might think because of the properties of SSI. We have to search for a low-density material, mechanically and thermally apt to be extended, and long-lasting thin space films operating in a wide range of distances from the Sun, and optically reflecting at least in the range 200–4,000 nm. This range is a good reference band; however, in the next chapters, we will considered different values in dealing with specific effects.

One needs an external source with a particular feature, namely, a sufficient *predictability*; it deserves special attention. In general, a propulsion concept based on

2.3 Principles of Utilization of Sunlight for in-Space Propulsion 61

a highly unpredictable, though existing, power source may turn into an unmanageable space system. A good policy should be to resort to a novel propulsion system that can solve the unsurmountable problems of the old one *without* introducing new big problems. Said differently, the novel system should to be able to deliver payloads at new targets, unrealistic for the old propulsion, with the desired accuracy and precision.

As it is known, mission success relies on high-quality spacecraft/mission design, including the processes of navigation, guidance and control. Nevertheless, a manageable flight depend primarily on the features of the momentum flux source. Let us clarify this point by means of an example. Suppose we want to design a sailcraft leaving some high Earth orbit for reaching Mars, putting the payload on the desired orbit about the planet, and going back to the departure orbit about the Earth. Suppose that the sailcraft will be launched in December 2019, and that the design of the transfer trajectory begins in January 2015. The envisaged flight should then start around the minimum between cycles 24/25 (if cycle-24 lasts 11 years). During the feasibility analysis of the flight design, assuming that the Earth-to-Mars arc and the Mars-to-Earth arc may last about one year each, it is reasonable to use the mean value of TSI throughout 2007 and 2008, respectively, in the trajectory calculations (see Figs. 2.4 and 2.8). For cycle-24, there is a non-negligible uncertainty about its maximum. For cycle-25, nobody knows today, even though a profile similar to that of cycle-24 appears a reasonable starting point. In any case, in the subsequent phases of the design, calculations should be refined by considering different levels of TSI; eventually, more appropriate TSI data—related to the time interval of the *real* flight—will have to be input to the mission analysis codes. Periodically updating the solar database via onboard measurements may be one of the key-points in designing sophisticated sailcraft missions.

One expects that high-frequency changes of TSI should not affect trajectory because of sailcraft inertia, whereas too smoothed TSI data should introduce some target missing higher than some prefixed mission-dependent tolerance(s). After the launch is performed, the actual solar irradiance the sailcraft will receive during its flight is not known deterministically. In Sect. 9.1, we will preliminarily analyze TSI-variable trajectories. This solar variability may be considered as consisting of three components: one is a secular trend, the second one regards cyclic changes, whereas the third one might be represented by stochastic fluctuations.

- What renders SPS attractive, even from this point of view, is the combination of the small amplitude (about 0.15 percent) of the TSI fluctuations about the mean (see also Fig. 3.10), and the long-period high spectral power, as Fig. 2.14 shows very clearly.
- Space sails will have to be controlled, of course. This should be possible with no insurmountable difficulty essentially because solar sail is just two dimensional, not a large volume that would act as a "reflecting sail". Nevertheless—since the TSI thrust acceleration can be continuous

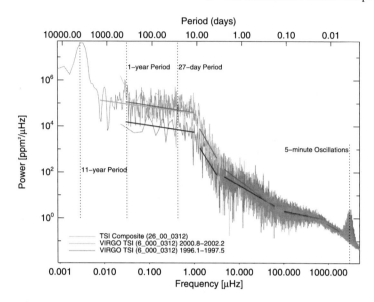

Fig. 2.14 Total Solar Irradiance in the frequency domain via Fourier analysis. Period is from 1978 to 2002. Temporal scales from few minutes to a solar cycle are evident. The 1996–1997 spectrum includes the minimum between cycles 22/23. The 2000–2002 spectrum pertains to high solar activity in cycle-23. Fourier-processed data come from SOHO/VIRGO radiometer system (From [15], courtesy of C. Fröhlich, PMOD, World Radiation Center, Switzerland)

> for many years—controlling, stabilizing and maneuvering a structure by far surpassing in size all other sailcraft systems represents one of the key points to manage in order to guide a sailcraft exactly as one wants.

Considering TSI fluctuations is one way of computing sailcraft trajectories; in Chap. 9, we will show that such time variations may cause deviations of sailcraft trajectories, if not dealt with in the flight design.

Either spin-stabilized or three-axis stabilized sail control may be appropriate, depending on the mission. The meaningful point is that—among the propellantless attitude control methods—TSI could be utilized even for attitude control especially when sailcraft draws close to the Sun in future missions. IKAROS inaugurated this method successfully.

References

1. Abbot, C. G. (1911), The Sun. New York: Appleton and Company. Read online at http://www.archive.org/stream/thesunab00abbouoft#page/n7/mode/2up.

References 63

2. Abbot, C. G. (1966), Solar variation, a weather element. Proceedings of the National Academy of Sciences of the United States of America, 56(6).
3. ASTM International (2006), Standard Solar Constant and Zero Air Mass Solar Spectral Irradiance Tables, E490, 16 pages, PA 19428-2959, USA.
4. Bennet, J. M., Mattsson, L. (1999), Introduction to Surface Roughness and Scattering (2nd edn.). New York: Optical Society of America.
5. Berrilli, F., Del Moro, D., Viticchiè (2008), Magnetic field distribution in the quiet Sun: a simplified model approach. Astronomy and Astrophysics manuscript No. 9683aph, August 4th.
6. Calisesi, Y., Bonnet, R.-M., Gray, L., Langen, J., Lockwood, M. (Eds.) (2007), Solar Variability and Planetary Climates. Berlin: Springer. ISBN 978-0-387-48339-9.
7. Charbonneau, P. (2007), Flux transport dynamos. Scholarpedia, 2(9), 3440.
8. Charbonneau, P. (2005), Dynamo models of the solar cycle. Living Reviews in Solar Physics.
9. Dikpati, M. (2005), Large scale organization in the solar dynamo and its observational signature. Astronomical Society of the Pacific, 346, 61–76.
10. Dikpati, M., De Toma, G., Gilman, P. A. (2006), Predicting the strength of solar cycle 24 using a flux-transport dynamo-based tool. Geophysical Research Letters, 33, L05102.
11. Domingo, V., Ermolli, I., Fox, P., Fröhlich, C., Haberreiter, M., Krivova, N., Kopp, G., Schmutz, W., Solanki, S. K., Spruit, H. C., Unruh, Y., Vögler, A. (2009), Solar surface magnetism and irradiance on time scales from days to the 11-year cycle. Space Science Reviews, 145, 337–380. doi:10.1007/s11214-009-9562-1.
12. Foster, S. S. (2004), Reconstruction of solar irradiance variations for use in studies of global climate change. Ph.D. dissertation at University of Southampton, School of Physics and Astronomy. ftp://ftp.pmodwrc.ch/pub/Claus/Publications/.
13. Fox, M. (2007), Oxford Master Series in Physics: Quantum Optics, an Introduction. Oxford: Oxford University Press. ISBN 978-0-19-856673-1.
14. Fröhlich, C. (2004), Solar energy flux variations. In Solar Irradiance Variability, Geophysical Monograph: Vol. 141. Solar Variability and Its Effect on Climate (pp. 97–110). Washington: American Geophysical Union.
15. Fröhlich, C., Lean, J. (2004), Solar radiative output and its variability: evidence and mechanisms. The Astronomy and Astrophysics Review, 12, 273–320. doi:10.1007/s00159-004-0024-1.
16. Fröhlich, C. (2006). Solar Irradiance Variability Since 1978, Space Science Review. Dordrecht: Kluwer Academic.
17. Garwin, R. L. (1958), Solar sailing: a practical method of propulsion within the solar system. Jet Propulsion, 28, 188–190.
18. Golub, L., Pasachoff, J. M. (2002), Nearest Star: The Surprising Science of Our Sun, Harvard: Harvard University Press. ISBN 0-674-01006-X.
19. Hathaway, D. H., Wilson, R. M. (2006), Geophysical Research Letters, 33, L18101.
20. Hoyt, D. V., Kenneth, H. Schatten (1997), The Role of the Sun in Climate Change. Oxford: Oxford University Press.
21. Krivova, N. A., Balmaceda, L., Solanki, S. K. (2007), Reconstruction of total solar irradiance since 1700 from the surface magnetic flux. Astronomy & Astrophysics, 467, 335–346.
22. Krivova, N. A., Solanki, S. K., Floyd, L. (2006), Reconstruction of solar UV irradiance in cycle 23. Astronomy & Astrophysics, 452, 631–639.
23. Krivova, N. A., Solanki, S. K. (2005), Modelling of irradiance variations through atmosphere models. Memorie Della Societa Astronomica Italiana, 76, 834–841.
24. Lockwood, M. (2002), An evaluation of the correlation between open solar flux and total solar irradiance. Astronomy & Astrophysics, 382, 678–687.
25. Mekaoui, S., Dewitte, S. (2008), Total solar irradiance and modelling during cycle 23, Solar Physics, 247, 203–216. doi:10.1007/s11207-007-9070-y.
26. NASA/MSFC Solar Group (2008), http://solarscience.msfc.nasa.gov/papers.shtml (a set of downloadable PDF-format papers), http://solarscience.msfc.nasa.gov/presentations.shtml (a set of downloadable .avi and .ppt-format presentations).
27. NASA, Solar Dynamics Observatory (SDO), http://sdo.gsfc.nasa.gov/.

64 2 The Sun as Power Source for Spaceflight

28. NASA, Solar Heliosperic Observatory (SOHO), http://sohowww.nascom.nasa.gov/.
29. Space Weather Prediction Center, http://www.swpc.noaa.gov/index.html.
30. Solanki, S. K., Krivova, N. A., Wenzler, T. (2004), Irradiance models. Advances in Space Research, 35, 376–383.
31. Solanki, S. K., Krivova, N. A. (2005), Solar variability of possible relevance for planetary climates. Space Science Reviews, 125, 25–37.
32. Wei-Hock, Soon W., Yaskell, S. H. (2003), The Maunder Minimum and the Variable Sun-Earth Connection. Singapore: World Scientific. ISBN 981-238-275-5.
33. Parr, A. (2005), Experimental Methods in the Physical Sciences: Vol. 41. Optical Radiometry. New York: Academic Press.
34. Rozelot, J. P. (Ed.) (2006), Lecture Notes in Physics: Vol. 699. Solar and Heliospheric Origins of Space Weather. Berlin: Springer. ISBN 3-540-33758-X, ISSN 0075-8450.
35. Rozelot, J. P., Neiner, C. (Ed.) (2008), Lecture Notes in Physics: Vol. 765. The Rotation of Sun and Stars. Berlin: Springer. ISBN 3-540-878300
36. Schatzman, E., Praderie, F. (1990), The Stars. Berlin: Springer. Translator King, A. R. ISBN 3-540-54196-9.
37. Wolfe, W. L. (1998), SPIE Tutorial Texts in Optical Engineering: Vol. 29. Introduction to Radiometry.
38. Tobiska, W. K., et al. (2000–2008), http://www.SpaceWx.com, many downloadable papers.
39. Solar Irradiance Platform (professional grade) (2009), Space Environment Technologies, version 2.35, http://www.SpaceWx.com.
40. Wenzler, T., Solanki, S. K., Krivova, N. A., Fröhlich, C. (2006), Reconstruction of solar irradiance variations in cycles 21–23 based on surface magnetic fields. Astronomy & Astrophysics, 460, 583–595.

Chapter 3
Sailcraft Concepts

Space Vehicles That May Utilize In-Space Natural Momentum Fluxes The concept of sea sailing can be extended to space traveling. The basic idea is to utilize some energy already present in space for avoiding the main drawback of a rocket vehicle: to be forced to carry all necessary reaction mass onboard. The more energetic is the transfer mission the more massive is the spaceship. In Chap. 1, we have seen the various equations that govern the motion of a rocket in field-free space, in a gravitational field, and even under relativistic laws. Space vehicles have augmented significantly in mass and complexity; since the 1950's, investigators have been searching for much more efficient ways to travel in space, especially for the future when space missions are expected to increase in number, purpose, and energy. It should be clear that any new good method for practically enlarging human exploration and expansion does not mean quitting rocket propulsion; it is sufficient to consider the problem to launch something from ground into space (at least to within our current knowledge of the physical laws). On the other hand, there are a high number of missions that are impossible to rockets, not strictly in mathematical terms, but because of the very large mass and complexity of the involved systems, including space infrastructures. Many things, allowed in principle, are completely impractical. Others, which seem feasible if an advanced technology were available, are in reality physics-forbidden; as a point of fact, one should not forget that complexity is not an engineering problem alone: it is first of all a conceptual problem of basic physics.

In this chapter, we deal with three sailing modes for traveling in space: actually, only the third one is a "strict" sail, namely, a *two-dimensional* object through which an *external*-to-vehicle momentum flux can be captured and translated into thrust. Because of what the term *external* implies, such a vehicle is **not** a rocket. After a summary of the solar-wind properties, the two subsequent sections describe concepts regarding generation of thrust, but involve large *volumes*. In principle, this is not a limitation: problems would come from other features, as we will see below. Finally, all other section/subsections of the chapter describe sailcraft and its main systems, at least as they are conceived today and how presumably may evolve.

G. Vulpetti, *Fast Solar Sailing*, Space Technology Library 30,
DOI 10.1007/978-94-007-4777-7_3, © Springer Science+Business Media Dordrecht 2013

3.1 Solar-Wind Spacecraft

In this section, we like to give the essential piece of information about solar-wind-based sail concepts, where the "sail" is actually a volume around the vehicle where some field is active. First of all, we describe some features of the solar wind, which may result in some potentials for in-space propulsion utilization. The second aim of such short description is to "alert" the mission analyst that the concept of trajectory of a wind-based sailcraft is not straightforward because of the very nature of the solar wind.

3.1.1 The Solar Wind: an Overview

The Sun has been progressively studied for centuries, as it is known. This investigation not only has been continuing, but also uses more and more sophisticated techniques and instruments on ground and onboard spacecraft. Solar spectral analysis via ultraviolet light and soft X-ray has shown special zones in the solar corona, regions where the plasma temperature and density are lower than the surroundings and the solar-corona's averages. These regions appear *darker* also for the lack of trapped electrons. Comparing solar spectral images to solar magnetic spectrograms has resulted in discovering that such regions exhibit magnetic lines open to the Heliospheric Magnetic Field (HMF)[1] (whereas other magnetic zones normally exhibit magnetic loops that close over short distances onto the solar surface). These so-featured regions have been named the coronal holes (CHs). Their shape, number, and extension depend on the solar cycle activity phase. Around the cycle maximum, there may be many such short-lived[2] regions spread in solar latitude and longitude; the associated magnetic fields resemble a sort of spikes. As the solar activity declines, both corona and magnetic field structures become "more organized", but still sufficiently asymmetric with respect to the Sun's rotation axis. Finally, around the cycle minimum, CHs are large Sun-corotating polar holes while the solar field looks like a magnetic dipole with its axis tilted with respect to the rotation axis. Figure 3.1 shows six images (selected by the author) at $\lambda = 195$ Å in chronological order from May 1996 to April 2011 in order to visualize what said above. Images come from the Extreme Ultraviolet Imaging Telescope onboard the spacecraft SOHO. CHs may also change significantly in a few days, as shown in Fig. 3.2, where two images of the Sun were taken at $\lambda = 193$Å by the Atmospheric Imaging Assembly on spacecraft Solar Dynamics Observatory (SDO). The tilt angle between the solar rotation axis and magnetic dipole changes slowly with time, and may also overcome $90°$. During a solar rotation, the dipole rotates as seen from an observation point far from the Sun.

---

[1] It was formerly named the interplanetary magnetic field, or IMF. Famous is the Parker's theory; a subset of Parker's papers is given in [23–29].

[2] I.e. lasting less than one solar rotation period, typically.

3.1 Solar-Wind Spacecraft 67

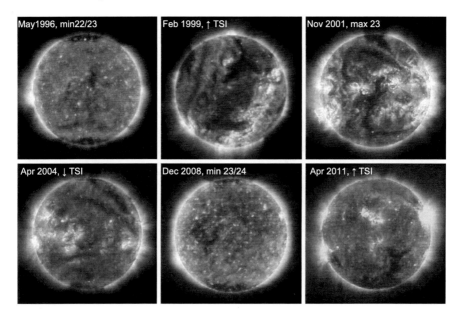

Fig. 3.1 Images of the Sun taken at λ = 195 Å by the Extreme Ultraviolet Imaging Telescope on satellite SOHO. From *top-left* to *bottom-right*, dates are in chronological order: May 1996 (minimum of cycles 22–23), February 1999 (increasing TSI), November 2001 (maximum of cycle 23), April 2004 (decreasing TSI), December 2008 (minimum of cycles 23–24), and April 2011 (increasing TSI). The sequence has been chosen by the author within the framework of Chap. 3. Courtesy of SOHO/EIT Consortium, NASA and ESA

The solar wind originates from CHs. Actually, it can be viewed as the hot *expanding* corona: as a point of fact, a hot corona cannot be confined like a hydrostatic-equilibrium (stationary) neutral atmosphere is. This wind, when viewed globally, is mainly a two-stream plasma flow: (1) the *slow* wind—with bulk speed typically in the 250–400 km/s range—originating from *low*-latitude CHs, and (2) the *fast* streams—with bulk speed ∼700–800 km/s (also sometimes beyond 1,000 km/s)—coming from (northern and southern) polar coronal holes (PCHs). Figure 3.3 shows a schematic of a model [30] well illustrating how an observer can see slow wind, and then—as the Sun rotates—encounter fast streams, and vice versa, depending on its heliographic latitude and angular speed with respect to the solar surface. Various configurations can arise; in particular, sufficiently close to the Sun, the observation of a wind change can be quick.

In a heliocentric inertial frame of reference, solar wind travels radially, with small transversal components (all fluctuating). Mostly, solar wind consists of electrons and protons, and ionized helium; however, there are about 30 heavier ion species[3] all endowed with the same bulk and thermal speeds, namely, independent of the ion's

---

[3]Many elements, one or several times or totally ionized, are present in the solar wind: for example, C, N, O, Na, Ar, Xe, Mg, Si, Ca, S, Kr, Fe. Energy/charge distribution tails up to about

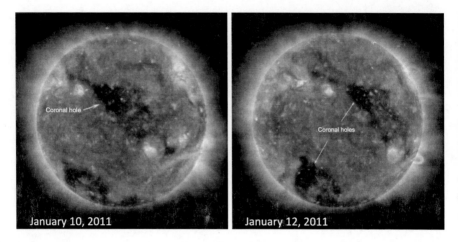

Fig. 3.2 Images of the Sun taken at $\lambda = 193$ Å by the Atmospheric Imaging Assembly on spacecraft SDO. As indicated, only two days temporally separate the images. Courtesy of NASA

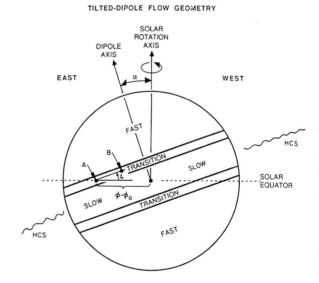

Fig. 3.3 Pizzo's model [30] of the solar dipole and the fast-slow solar wind transition near the Sun (see text). From [2], courtesy of Springer

ratio q^2/m, where q and m denote charge and mass, respectively. In the inner solar system, there may exist differential speeds due to interaction between particles and Alfvén waves; however, this speed non-uniformity dies away as the plasma recedes from the Sun.

Solar wind is a supersonic, superalfvénic, momentum-dominated, highly electrically conductive plasma with magnetic energy density small compared to the

---

60 keV/e have been measured by spacecraft Ulysses; ions exhibit energy/mass values up to about $60(Z^+_{ion}/A_{ion})$ keV/u, with a mean value of $\simeq 10$ keV/u.

3.1 Solar-Wind Spacecraft

bulk kinetic energy density. As a result, the solar wind transports the force lines of the HMF. Magnetohydrodynamics, which describes a fluid where electromagnetic forces affect its evolution, has been employing widely by solar physicists. Another approach is to consider the solar wind as a statistical ensemble of particles, which do not undergo collisions, or, more precisely, short-range encounters. This second view comes from the *kinetic* theory of gas and plasma. According to old exospheric models, electrons and free ions (mostly protons and α-particles) move in the solar gravitational field, the interplanetary electric field, and are guided by the HMF. Real things appear to be much more complicated. The profound difference between these two approaches lies on thermodynamics; the fluid method assumes a plasma in local thermodynamic equilibrium (therefore, the plasma is necessarily collisional), whereas the kinetic model entails a system of particles far from thermodynamical equilibrium as the collision rate decreases more and more. Kinetic theory also allows describing how collisions can cause "relaxation" towards the thermal equilibrium. In the solar wind there are rare collisions per characteristic time, and the particle speed distributions depart meaningfully from the Maxwellian. Not only the solar wind does not behave as one fluid, but also the single species (i.e. electrons, protons, and heavier nuclei) do not resemble a fluid. There are various approaches that try to explain a twofold puzzle: the heating of the solar corona and the acceleration of the solar wind up to super-alfvénic speeds, a half-century old problem. Perhaps, new data from sensors on solar probes traveling at few solar radii will be conclusive; however, things are complicated also on account of the solar-wind turbulence, which we will mention below.

Among various types of waves in a plasma, as described by magnetohydrodynamics, particularly important are the Alfvén waves, the speed of which represents the rapidity with which the whole *plasma system* is informed about a perturbation occurring somewhere in the system region (thanks mainly to the tension of the magnetic field lines). If the plasma has kinematic viscosity and magnetic diffusivity both negligible, then the wave dispersion relation simplifies considerably: $\omega = \pm \mathbf{k} \cdot \mathbf{B}_0 / \sqrt{\rho \mu} \equiv k_\| V_A$; the (scalar) Alfvèn speed is defined by $V_A = B_0 / \sqrt{\rho \mu}$, where B_0, ρ and μ denote the magnetic induction, the mass density and the magnetic permeability, respectively, of the unperturbed plasma. The existence of such waves can be carried out mathematically via *linearization* of the MHD equations. Alfvèn waves are anisotropic, in general, since their properties depend on the angle between the applied field and the local propagation direction. For the solar wind, the magnetic Mach number, or $M_A = V_{sw} / V_A$ (where V_{sw} is the wind's bulk speed) is notably greater than unity, and fluctuates considerably with time. Just for a reference example, as for the solar wind one can set $\mu = \mu_0 = 4\pi \times 10^{-7}$ H/m, using $B_0 = 10$ nT, $\rho = (1/4\pi) \times 10^{-19}$ kg/m^3, and $V_{sw} = 400$ km/s, one gets $V_A = 100$ km/s and $M_A = 4$. In the time frame 1995–2008 (which contained cycle-23), solar wind exhibited (daily averages of) M_A in the range $(1, 36)$, with a mean of $\lesssim 10$.

Fig. 3.4 Scheme of the Heliosphere with its three large shock waves. The termination shock has been crossed by two spacecraft: Voyager-1 (2004.12.15) and Voyager-2 (2007.08.30), thus giving confirmation of its existence. At the time of this writing, both Voyagers are crossing the heliosheath, where the solar wind is slowed down by the pressure of interstellar gas. Courtesy of NASA/JPL

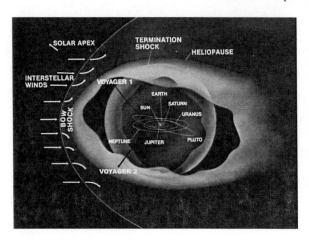

The solar wind fills a very large volume around the Sun, well beyond the Kuiper belt. This volume is called the *heliosphere*, but is neither a sphere nor of fixed size. The Sun (and the whole solar system) moves with respect to the interstellar medium, while the solar activity determines the "boundaries" of the heliosphere; thus, in a heliocentric frame, an interstellar wind is observed. The interaction between the interstellar wind and the supersonic solar wind should result in three shock waves: the termination shock, the heliopause, and the bow shock. The *heliopause* is expected and modeled as a "surface" where the pressure of the interstellar medium equals that of the solar wind. The nose of the heliopause is estimated of the order of 140–150 AU from the Sun (according to different authors). Finally, a *bow-shock* should take place if the solar system moves supersonically through the (local) interstellar medium; this may cause a shock wave in front of the heliopause, perhaps some tens of AU afar, across which the interstellar medium becomes subsonic with respect to the heliosphere. A bow shock (also called a detached shock) arises in general because a *supersonic* flow senses an obstacle in advance with respect to the usual sonic propagation; in the astronomical case, such an obstacle may be an object without magnetic field, or with a magnetic field around (e.g. the Earth with its magnetosphere). In the latter case, the obstacle consists of the whole (interaction-modified) field region. Of course, here the supersonic flow is a plasma and bow waves are more complicated than bow shocks in a neutral atmosphere. Figure 3.4 shows an artist drawing of the heliosphere and its shock waves. In any case, the important property is that the flow, before a bow shock, is unperturbed by the obstacle.

Downstream of *termination shock*, which received confirmation by Voyager-1 (94 AU) on December 15, 2004, and Voyager-2 (84 AU) on August 30, 2007, the solar wind becomes subsonic (similarly to a fluid across supersonic normal shocks in gasdynamics). At the time of this writing, Voyager-1 (about 119 AU from the Sun) and Voyager-2 (about 97 AU from the Sun) are crossing the so-named *heliosheath*, where the solar wind is slowed by the pressure of interstellar gas. There is more: measurements of low-energy charged particles show that the solar-wind's

3.1 Solar-Wind Spacecraft

Fig. 3.5 Shape of the Heliospheric Plasma Sheet: the ballerina skirt. The orbits of the first five planets are shown. Courtesy of NASA

radial speed has progressively slowed down to zero, a very important *unexpected* fact [5, 16]. To within a few years, Voyager-1 is expected to cross the heliopause, according to a JPL estimate of the heliosheath thickness.

Applied to the interstellar wind, a space mission more advanced than Voyager flights may be conceived for reaching the near interstellar medium, just where its original properties are not disturbed by the heliosphere. We will return to such an intriguing mission concept at the end of the book with a chapter devoted to possible fast missions to the bow shock and beyond via solar photon sailing.

Let us go back to the Sun and continue to mention other solar-wind properties that may result useful in feasibility analyses of spacecraft concepts using such particle flow as source of momentum. In the solar polar regions (as northern as southern), magnetic field lines are *open*; they first divert toward the equatorial regions, then go deeply into space to set up the HMF. Now, let us assume that the solar magnetic axis forms an acute angle (e.g. a few tens of degrees) with the solar rotation axis (or spin axis, for short, with no ambiguity in the current context), so the field polarity is *plus* in the northern hemisphere, and *minus* in the southern hemisphere. Therefore, there has to be a "surface", separating these two opposite-polarity fields, hosting an electric current that causes the magnetic discontinuity. This is the Heliospheric Current Sheet (HCS) that represents the magnetic "equator" of the Sun. Of course, it is not a plane or other geometric surface. Its median thickness, on statistical basis, changes with the radial distance from the Sun, of the order of 1,500 km, or 10^{-5} AU, in the range 1–5 AU. At large distances from the Sun, the HCS shape is notably wavy: it warps several times and looks like a ballerina skirt, as Fig. 3.5 shows. HCS is embedded in a considerable larger volume of plasma, called the Heliospheric Plasma Sheet (HPS). With respect to the ambient plasma, HPS exhibits significantly higher plasma-β, higher density, and higher total (kinetic + magnetic) pressure. HCS is not a stationary, but fluctuating, structure that is continuously affected by the solar rotation. If the wind sensor is onboard a spacecraft, or is the spacecraft itself, it passes through the folds of the HCS as the Sun rotates. In time scales much longer than a solar rotation, HCS changes orientation with respect to the spin axis [2].

Fig. 3.6 Schematic of a Corotating Interaction Region from [2]. Courtesy of Springer

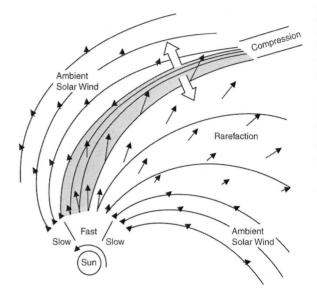

There is a further complication for the motion of a wind-based sailcraft. According to Lenz's law, fast and slow winds are prevented from penetrating into each other and a Corotating Interaction Region (CIR) originates; it deflects both flows. Again, solar rotation induces significant phenomena. As the Sun rotates, it can often occur that slow and fast flows are aligned radially inside the low-latitude zone of solar-wind variability. Faster flow runs into slower wind exhausted previously from another CH: a pile-up region with high density is brought about. Such parcels of plasma are passed through by different lines of HMF. Thus, a compression arises on the front side of the fast stream, and a rarefaction is induced on the trailing edge; of course, such regions co-rotate with the Sun. CIR is a transient phenomenon in the solar cycle evolution. Figure 3.6 shows a simplified diagram of a CIR. If a wind-pushed sailcraft crosses a CIR, then it undergoes an acceleration jump. This may be inferred from Fig. 3.7, which highlights proton density, bulk speed, and magnetic field as recorded by SOHO spacecraft (ESA-NASA), the two Solar Terrestrial Relations Observatory (STEREO) spacecraft (NASA), and Advanced Composition Explorer (ACE) spacecraft (NASA) in the time frame from 2008.01.29 to 2008.02.04 [46]. These spacecraft have been progressively crossed (counterclockwise) by a CIR; first STEREO-B (which trails behind the Earth in its orbit), then SOHO and ACE (which are at the Earth-Sun L1 point), and finally STEREO-A (which precedes the Earth in its orbit). Though the CIRs observed by STEREO are much different from the high-latitude CIRs usually considered in literature, however they should be considered in the feasibility analysis of large wind-based sailcraft. The case discussed in [46] exhibits a pulse density duration of 0.15 day, in which the number density starts with about 5 p/cm^3 and achieves a peak of about 25–30 p/cm^3. The increase of wind speed is significant, and delayed; in contrast, magnetic field decreases from STEREO-B to STEREO-A. All such profiles may have a non-negligible impact on wind-based sailcraft trajectory.

3.1 Solar-Wind Spacecraft 73

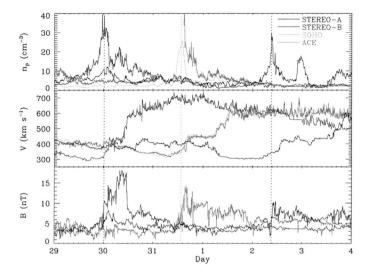

Fig. 3.7 Changes of three solar-wind quantities across the same CIR as measured by STEREO (A and B), SOHO and ACE spacecraft. *Vertical lines* represent the CIR density pulse arrival times at the four spacecraft. From [46], courtesy of The American Astronomical Society

Finally, a few words about the *turbulence* of the solar wind. Turbulence regard intriguing and very complicated aspects of the dynamical systems, from every-day fluid flows (e.g. regarding atmosphere in cities or over forests, rivers and river rapids, falls, air flow in lungs, blood in arteries and heart, etc.) to thermonuclear-fusion fields, and space plasmas like the solar wind. Turbulence is still a big unsolved problem in Physics, and although this topic is not among the aims of this books, nevertheless we like to mention just a few items to *emphasize* that solar-wind's irregularities should receive attention in investigating propulsion-oriented applications of this plasma.

Turbulence deals with a (wide) range of fluctuations that take place over various spatial and temporal scales. Strangely enough, they arise in the evolution of a dynamical system described deterministically. Statistical methods have to be used for analyzing such behavior. Essentially, energy is input at large spatial scales; it is transported progressively to smaller and smaller scales—an energy cascade—and eventually scales are achieved where energy dissipates, i.e. converts into heat, by viscosity. This cascade can be depicted conceptually by *large scales* ≫ *inertial scales* ≫ *dissipation scales*, where inertial scales are meant as the intermediate spatial range where large-scale forcing and viscous forces produce negligible effects. Fluid flow transport quantities are enhanced by turbulence. Such property may sometimes be favorable to the practical design of something (e.g. the elimination of gases injurious to the human health), or a strong obstacle to the achievement of some objective (e.g. the control of very complicated physical processes), or makes the human full comprehension of natural phenomena a very difficult problem (e.g. the solar corona). Osborne Reynolds [31] studied experimentally the different regimes

74   3 Sailcraft Concepts

of water flows in circular-section ducts; he discovered that the various regimes could be described via one parameter, which subsequently was named the Reynolds number, or *Re*, in Fluid Dynamics.

For a general fluid flow, there may be a huge number of parameters describing its time-variable global characteristics, and a statistical description is then preferred. Flows subjected to (fully developed) turbulence present a very high sensitivity to small perturbations, but exhibit statistical properties insensitive to perturbations. Another striking behavior of a turbulent flow is that its energy is spread over all scales admissible in the related dynamical system; the spectral power density analysis becomes quite important. A turbulent flow for which *Re* is very high (infinite, strictly speaking) is said to be in a *fully developed* state because the symmetries (which are broken as *Re* increases) are restored in statistical sense.

> The previous and the following paragraphs on solar-wind report concepts and results from, say, a classical interpretation of satellite data. As a point of fact, very recently
>
> "A LANL scientist examining the solar wind suggests that our understanding of its structure may need significant reassessment. The plasma particles flowing from the Sun and blasting past the Earth might be configured more as a network of tubes than a river-like stream, according to Joseph Borovsky of Los Alamos National Laboratory's Space Science and Applications group."
>
> LANL stands for Los Alamos National Laboratory (New Mexico, USA). At http://www.lanl.gov/science/NSS/issue1_2011/story2e.shtml, the reader can find the full article (2011), which followed the technical paper [4].

The solar wind expands radially and exhibits a feature of strong turbulence, which evolves towards a state similar to the hydrodynamic turbulence (e.g. see [7, 15, 19]). It appears to be in the fully developed state, though the solar-wind *Re* is not extremely large! In addition, the radial evolution of the fluctuations is different for slow solar wind and fast streams. The solar-wind's energy is distributed among ten orders of magnitude in frequency or, equivalently, on time scales from about one solar rotation period to less than a tenth of millisecond [34]. The power spectrum of the fluctuations depends also on the distance from the Sun, according to measurements from spacecraft. Power density decreases with increasing frequency, and increases as distance decreases (e.g. according to the spacecraft Helios-2 data that operated in 1976 in the range 0.3–1 AU). Different-scale structures interact with each other resulting in the generation of structures of other scales. At about 1 AU, typically the observed inertial range covers three orders of magnitude in frequency, i.e. from $\sim 2 \times 10^{-4}$ Hz to $\sim 10^{-1}$ Hz.

Solar wind exhibits fluctuation *intermittency*: put very simply, this is a remarkable turbulence feature, which means that the probability densities of the medium's various parameters change with time, in particular high-variability regions turn into "quiet" regions; the distributions depart from being Gaussian (at large scales) for becoming notably non-Gaussian (at small scales) with more and more prominent tails.

3.1 Solar-Wind Spacecraft 75

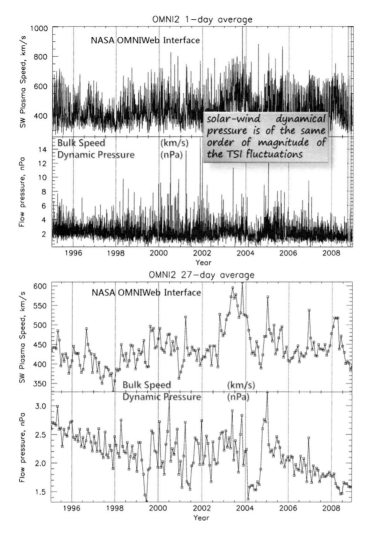

Fig. 3.8 Averages of speed and dynamic pressure of the solar wind at 1 AU in the 1995–2008 time-frame. (*Top*): One-*day* Means. (*Bottom*): One-solar-rotation Means. This plot is part of the HMF and plasma data obtained from NASA's OMNIWeb interface

The fluctuating quantities δn_p, δB^2, and $\delta(n_p V^2)$ regarding proton density, magnetic pressure, and dynamical pressure, respectively, show intermittent behavior as well.

On this though very short background, a question arises: what does a solar-wind-based sailcraft sense as it moves at different distances and through heliographic latitudes? What mainly matters to propulsion are dynamic pressure and speed (as seen from the spacecraft frame). Figure 3.8 shows the 1-day and 27-day averages, respectively, of the solar wind speed and dynamic pressure at 1 AU. In addition,

Fig. 3.9 Hourly averages of speed and dynamic pressure of the solar wind at 1 AU in October–November 2008. This plot is part of the HMF and plasma data obtained from NASA's OMNIWeb interface

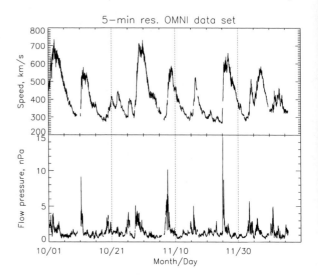

Fig. 3.9 reports the hourly averages in the October–November 2008; this period corresponds to the gray vertical rectangle on the rightmost of Fig. 3.8. Note that one-day averages smooth many plasma fluctuations, which nevertheless may affect sailcraft motion. An elementary (but sufficient here) statistical analysis on the time series of the dynamical pressure 1-day averages has produced the following values: *mean* = 2.15 nPa, *mode* = 1.93 nPa, *st.dev* = 1.22 nPa, *skewness* = 2.49, *kurtosis* = 16.05, *range* = 15.66 nPa. (Let us remind that for a Gaussian distribution, the skewness is 0, and the kurtosis is 3, exactly.) From a potential propulsion viewpoint, suppose for a moment that a sailcraft, utilizing the solar wind for getting thrust, is ready to fly in 2020 and its complete mission will last 14 years. What will the history of the dynamic pressure be in those years? It is possible that some statistical values of P_{dyn} in 2020–2033 be close to those ones related to 1995–2008 period, but the particular sample or realization could not repeat. One should realize that the instantaneous thrust acceleration of some types of wind-based sailcraft is proportional to the dynamic pressure of the solar wind; therefore, the vehicle trajectory, developing in the period $[t_0, t_1]$, depends on the realization of P_{dyn} just in the time interval $[t_0, t_1]$.

Analogous considerations can be done about the bulk speed of the wind impinging on the "sail". Contrarily to some misconception spread via World Wide Web, such quantity is important for "escape" propulsion as well; as a matter of fact, as the speed of a *very large* solar-wind-based radially-receding sailcraft is lower than the wind speed in any case, the solar plasma could be diverted by the obstacle represented by sail field, and by far gets past the space vehicle after the interaction. The subsonic receding sailcraft does not sense the plasma present in the space regions ahead it. In general, the instantaneous relative velocity between a fast sailcraft and the solar wind is important because it determines that plasma's dynamical pressure as observed in the sailcraft reference frame.

3.1.2 Magnetic and Plasma Sails

The previous section described the very general properties of the solar wind (and the heliospheric magnetic field), and emphasized the parameters that may be used for in-space propulsion, in particular its dynamic pressure and speed. At 1 AU, the solar radiation pressure, averaged over the 1995–2008 time frame, amounts to 4.56 µPa. In contrast, from the daily averages in the same time frame, the solar wind dynamic pressure amounts to 2.15 nPa, a reduced factor 2,100. Therefore, if one thinks of utilizing the solar wind pressure on some material sail (e.g. a particle absorber), there is no chance to get some useful acceleration. The first idea behind the "magnetic sail" (or *magsail*) concept was to use the volume field of the simplest magnetic lens: a current loop. In order to act as a lens, a high current should flow in a superconducting (multi-wire) loop such that the generated magnetic field "sees" every particle of the solar wind *individually*: each particle would undergo the well-known adiabatic invariants. Thus, most of them would be magnetically reflected. This particle model is particularly simple and, in principle, may work. However, one could not take this mode of operation for granted at any distance from the Sun. As a point of fact, the solar plasma compresses the magnetic field produced by the large-diameter ($\approx$100 km or more) coil until its kinetic energy density matches the (modified) magnetic pressure locally, giving rise to a vast "surface" bordering the field-particle interaction volume. In this situation, the interaction between the incoming plasma and the wire's extended magnetic field (which is not a dipole) would be complicated enough with a number of phenomena, which may dominate each other depending on the wind characteristics "at the impact". Considered the complexity of the calculations in the claimed situations of either particle or collective interaction, often in the past some authors resorted to the general laws of conservation after introducing macroscopic coefficients of particle reflection and transmission through the interaction volume. In so doing, the magnetic sail concept was generalized to that of *field sail*, dealt with relativistic dynamics for very deep-space flights. The related mathematical model included vehicle's acceleration by means of plasma beam (from a solar-system station) and deceleration via the interstellar medium. Although this approach bypassed the details of plasma and magnetic-field interaction, however it showed how much high the power of the plasma station should be, and how long the deceleration time is in various spacecraft/mission configurations.

A significant conceptual improvement, regarding the dynamical output of a magnetic-sail vehicle keeping the working temperature of the superconducting loop in the space environment, was investigated by analyzing in detail the *two-sail* concept. In the original layout, the first sail (facing the Sun) was a multi-layer solar photon sail annular in shape. Behind it, concentrically, a superconducting loop with radius equal to the mean radius of the annulus acted as a magsail. Since the loop should have a radius of some tens of kilometers, the annulus width may vary from a few meters to some tens of meters. The separation between the parallel planes of the annulus and the loop takes on the same values for optimization reasons. The role of the photon sail is twofold: (a) to protect the superconductor from the direct solar irradiance, and (b) to add acceleration for high-speed missions. The role of the field sail is likewise twofold: (1) to act as the main thruster, and (2) to protect the photon sail from the continuous bombardment of the solar-wind particles. The current wires are irradiated only by a very small fraction of the infrared energy coming from the photon-sail backside. In turn, the current loop wires are inserted into an asymmetric thermal-insulation sheath in order to allow the superconductors to be cooled by radiation. The detailed analysis of the systems involved in such sailcraft concept helped to better understand the level of complexity of realistic magnetic-field space vehicles, which was underestimated in the first papers about magsail. As an example of an extra solar mission, the mass of the whole vehicle at the heliocentric flight start resulted in about 8 tonnes with 100 kg of net scientific instruments, reaching 400 AU after 22 years from launch.

A variant of the original magsail is the so-called Mini Magnetospheric Plasma Propulsion (M2P2), according to which the solar-wind particle deflection surface may be notably increased (with respect to that envisaged for the original magsail) by using a strongly magnetized plasma—around the space vehicle—which is injected into the magnetic field produced by a suitable generator. The aim is to small-scale copy the Earth magnetospheric environment in a volume containing the vehicle and, in so doing, to cause the solar-wind to be diverted at the vehicle's magnetopause. This *should* produce thrust because the magnetic field of this artificial "magnetosphere" is attached to the spacecraft. Therefore, the plasma source is *onboard* as well. Noble gases are suitable in laboratory experiments, but carbon-hydrogen compounds are possible for space use. The considerable advantage of this concept should be the non-mechanical inflation of the spacecraft's magnetic field. Although early experimental tests at NASA/MSFC are inconclusive with regard to feasibility, nevertheless this concept—viewed beyond its intrinsic drawbacks (Sect. 3.1.4)—would deserve further attention.

3.1.3 Electric Sail

Instead of utilizing magnetic fields (no matter whether generated via circuit or plasma) for reflecting/deviating the solar wind (locally), one can think of using the electrostatic field produced by some generator *onboard*. This concept is new (2004) from the viewpoint of the field to be utilized for particle deflection propulsion, but it is not so with regard to the external source of momentum. The initial idea was to employ a large wire mesh, nominally planar and with a spacing of the order of Debye length of the solar wind, kept at a positive voltage by means of an electron gun onboard. The mesh would act as quasi-full reflector of protons, and as a quasi-full absorber of electrons. Of course, in the stationary case, this electron current has to be rejected into space through the electron gun. One possible show-stopper of such configuration is the extreme difficulty to deploy and keep the mesh planar and oriented. The geometric arrangement of the sail has been modified considerably.

To the author knowledge, a modified concept of electric sail, or *e-sail* for short, needs N multi-wire tethers of length l, all kept radial about the spacecraft body by rotation. N should be of the order of 100, whereas l should measure about 20 km, according to some analyzed configurations. Each tether might be a structure that should work in the case of breaking of one of the wires. Denoting the tether's linear density by ϱ, the moment of inertia about the (nominal) spin axis (i.e. the axis orthogonal to the plane of the tethers complex) is expressed by $N\varrho l^3/3$. The deployment of such system should not represent a show-stopper, even though it remains a very delicate matter for which approaches different from the conventional ones may be required. The e-sail systems analyzed to some extent hitherto are essentially: the multi-tether or particle deflection system (including tether materials and reels), deployment of the tethers, tether simulations, electron gun, and mission applications via parametric studies. It is claimed that an e-sail may reach the near heliopause in 15 years, and, more generally, may achieve cruise speeds approximately higher than 60 km/s (or about 13 AU/year).

3.1.4 Conceptual and Practical Issues

It is relatively easy to conceive of new rocket systems for space propulsion, even though their practical realization is a completely different matter, as it is well known. When one wants to face with quite new propulsion concepts and related systems, there is some risk to underestimate some necessary spacecraft systems, meet technological inconsistencies, and to take a conceptual problem for a technological one that may be eventually solved in the future. In a few words, one should be very careful not only with technological issues, but especially with *conceptual* pitfalls. The former ones are to be analyzed in the light of the aimed benefits. The latter ones are much more "dangerous" because they could invalidate the whole new concept; this situation may induce to ask funds for ground experiments and subsequent preliminary missions that will prove what could be found with some more accurate analysis of the new concept. Magnetic, electric and plasma sails—at least as they have been proposed so far—seem not to be an exception to this regard. Let us discuss briefly the main problems and vantages of solar-wind based sails.

Common Problems The first, and the most relevant issue, is to single out the type of interaction between the solar wind and the object—with its fields—to be put inside this solar plasma, which (as described in Sect. 3.1.1) is a supersonic super-alfvénic plasma flow. Normally, in a simple description, one pictures a charged particle entering a field region, computes its trajectory, its momentum transfer (and other quantities), and *extrapolates* the same features to a set of incoming particles differing only by the initial conditions, namely, such particles are independent of each other. This is the typical situation assumed in Monte Carlo computer code for particles coming in the system of interest. A magnetic/plasma/electric sail is a region of matter and field inside a collection of *interdependent* particles (the solar plasma). Information about disturbs propagates through the plasma via Alfvén waves. Is the single-particle interaction assumption sufficiently valid for such space

80 3 Sailcraft Concepts

propulsion? Where and when are models such as the kinetic theory or the fluid description apt? How to assess mission-related performances in the case of a trajectory that entails different interaction models, e.g. depending on the Sun-vehicle distance?

In Chap. 4, we shall discuss some criteria related to the interaction that an object, put in the solar-wind, may undergo. They will be applied to solar-photon sails, which are considerably smaller than the so far envisaged non-photon sails.

The second big problem is represented by the irregularities of the solar wind with respect to the (desired) aims of in-space propulsion. We briefly described the fluctuations of this plasma. Why such a worry in conceiving of a feasible and reliable propulsion system? Trajectory are generally endowed with unavoidable errors that can be determined and corrected, if necessary, in flight. The statistical treatment about nominal trajectories relevant to baseline and backup scenarios is an integral part of the mission analysis. However current astrodynamics is pervaded by the practical realization of trajectory and attitude maneuvers, and how to improve the vehicle's state knowledge via observables; guidance schemes and real controls are essential to the mission success. In turn, all such processes can be used to design and realize missions for greatly increasing the knowledge about special fields of scientific interest, e.g. the detailed structure of the gravitational field of a celestial body. In general how may one design space trajectories for which long continuous thrust accelerations are dominated by very large fluctuations, often notably higher than the mean values? In literature, it is customary to analyze trajectories driven by averaged values of the solar-wind's dynamical pressure and bulk speed, and with no fluctuations during the path. An analysis similar to, or even simpler than that done in [38] for relativistic rockets, might be appropriate in order to carry out a preliminary evaluation of the fluctuations impact on the actual dynamics.

Magsail This concept requires superconducting wires many hundreds or thousands of kilometers long. Once the superconducting circuit is deployed and the operational temperature is achieved, the loop has to be energized, usually some MJs, via a system of high-charge capacitors that add mass to the sail system. In the past, only low-temperature superconductors have been considered for magsails. In general, the major risk in designing a superconducting device is quenching, namely, any process that induces a region of the superconductor circuit to pass from the superconducting to the resistive state. Such transition is irreversible whereupon the magnetic energy will be dissipated as heat.

Two Sails Advantages and drawbacks of this concept have been described quantitatively and in detail in [40]. Although the cruise speed of extra-solar sailcraft was calculated in the 18–22 AU/year range for sailcraft of about 10 tonnes, however the solar wind was oversimplified (even though a rough calculation of plasma fluctuations compensation was issued). In any case, the deployment, the alignment and the achievement of the thermal regime of both sails is considerably complicated. The first two formidable problems appear to be technological in kind, but the last one is mainly a conceptual problem.

Plasma-Sail As said previously, this concept entails the remarkable advantage of no mechanical sail inflation. However, conceptually speaking, the plasma sail is not much different from a (high-temperature) rocket. As a point of fact, M2P2 requires a continuous flow of matter into space, ~ 0.5 kg/day, according to the analysis in [45]. Apart from the particular value of propellant consumption, this concept implies a *refueling* for each mission to be accomplished.

We pointed out in Sect. 3.1.2 that thrust on the spacecraft should be produced because the field—magnetizing the plasma around the spacecraft—originated in the vehicle itself. What remains to show is how much the thrust efficiency may be. As a matter of fact, any sail responds to the dynamical pressure on the interaction surface, on which—hence—a thrust is applied. One should calculate the *effective* momentum transfer from such input to the material vehicle, and the direction of the gained thrust with respect to the incident solar-wind's vector velocity. These two points would be important even if the solar wind were a monochromatic beam of particles.

E-Sail A strong vantage in the e-sail concept with respect to magsails is that the gained thrust per unit tether length depends on the applied positive voltage, which is a control parameter. In principle, might one reduce the effects of the large fluctuations of the solar-wind P_{dyn} through a smart control? Encouraging results come from the preliminary analysis in [37]; as the problem is of great importance, it would deserve a continuation of investigation. In general, large sails exhibit really very high moments of inertia about the sail's main axis of symmetry. This is particularly true for e-sails. The e-sail example given in [11] has $N\varrho l = 2.65$ kg with $l = 21$ km; therefore, the moment of inertia is $\cong 390 \times 10^6$ kg m^2. New materials from Nanotechnology, or new configurations, or both might hopefully relax the problem.

3.2 Main Elements of Solar-Photon Sailcraft

This section and its subsections are devoted to conceptual and technological considerations about the main systems an Solar-Photon Sailing (SPS) vehicle consists of. Differently from the plasma/field sail concepts of the two previous sections, the photon sail utilizes the pressure of electromagnetic waves, or the *radiation pressure*, to get thrust.

It is generally known that photon pressure and photon energy density are interchangeable terms; in addition, quantum physics confirms that any photon carries *both* energy and momentum. Perhaps, it is not known so much that the radiation pressure is one of the consequences of the second principle of thermodynamics. The proof of it was given by Italian physicist A. Bartoli in 1876, independently of J. C. Maxwell, via a conceptual experiment of thermodynamics.

The first immediate result is that one needs a physical surface interacting with the photons impinging on it. Because what matters is acceleration, the mass of this device—the sail—adds to the space-vehicle mass and has to therefore be as low as possible. This can be accomplished by decreasing the thickness of the sail and choosing low-density materials. Although such conditions are necessary, but not sufficient, to make an efficient sail, however (for the moment) one should realize that—almost similarly to the well familiar sails—a photon sail is virtually a two-dimensional material surface, i.e. a membrane, *large* with respect to the size of the other vehicle's systems.

One may think of the space sailcraft as merely composed of two structures: the proper spacecraft (i.e. the sail's payload) and the system consisting of sail with mast, spars, rigging, tendons, and a subsystem controlling the sail orientation in space. Such a picture, which may be common among the various types of envisaged sail, is simplified. Let us mention why: (1) the Earth orbits the Sun at about 1 AU, (2) the solar radiant exitance amounts to about 63.1 million watts per square meter, (3) the rate of linear momentum photons transport is scaled by the factor $1/c$ with respect to the rate of energy. As a result, a surface of one square meter, which is 1 AU distant from the Sun and perpendicular to the sunlight's direction, can receive about 1366 watts (on average during a solar cycle). What does it mean? Were this object a perfect mirror, it would experience a force equal to $2 \times 1366/c \cong 9.113 \times 10^{-6}$ N. If the mass of such a body were 9.113 grams, the ensuing acceleration would amount to 1 mm/s^2 at 1 AU; one should note that the solar gravitational acceleration at 1 AU is 5.93 mm/s^2. This simple example tells us two remarkable things. First, such an acceleration level would enable many, many space missions; second, if we aim at accomplishing very deep space missions, we have to significantly lessen the ratio between the *overall* vehicle mass and the (effective) sail area. This ratio is called the *sailcraft sail loading* and normally is expressed in g/m^2. This is the most important technological parameter out of the many ones affecting the performance of a sailcraft (for a given mission and payload mass).

In this chapter, we begin with adopting the following nomenclature: *Sailcraft = Sail System + Spacecraft*. In turn, we use the convention *sail system = bare-sail + sail deployment/support structures*. In the subsections below, we will highlight the interconnection between the sail system and the other main systems; this depends also on the specific mission at hand.

3.2.1 Sail Shape

After the whole sail system is manufactured on the ground (it should be so for a long time-span in the future), it has to be folded and placed in an apt box to be sent to orbit. Subsequently, sail will be unfolded in the sailcraft's early orbit and, then, some initial orientation has to be acquired. The sail architecture affects the thrust level.

There are a variety of responses of the sail material to mechanical and thermal stress, charge accumulation from space environment, etc. Such responses appear as wrinkles, creases, internal tears,[4] and so on. In particular, *wrinkles* represent

---

[4]During the flight, tearing may be caused by space debris and/or micro-meteorites.

3.2 Main Elements of Solar-Photon Sailcraft

elastic, or recoverable, response of the sail membrane to compressive stress; they appear as undulations spread over the surface. Wrinkles may induce multiple reflections of light and two effects: (1) locally, the sail can absorb more energy than in normal conditions (hot spot); (2) if wrinkles cover a large fraction of the sail, the solar-pressure thrust decreases with respect to what is expected for a flat smooth surface. Qualitatively, one might view wrinkles as macroscopic deviations from flat surface, which add to the sail's intrinsic roughness that is responsible for the surface scattering of light. Wrinkle distribution over a surface depends on geometry and disturbing/applied torques; for instance, a sail shaped as radially stretched annular disk with a central hub responds to an axial twisting moment with essentially two features: one interior *wrinkled* region, and one (significantly larger) exterior *taut* region.

In contrast, suppose that the sail is unfolded by means of telescopic booms, which slowly come out of the sail stowage box. The sail, either squared or polygonal in shape, should be normally divided into smaller (e.g., squared, triangular) sheets. These ones could be considered as membranes subjected to two-dimensional different tensions in their plane. Such membranes appear experimentally covered by wrinkles, well visible even with the naked eye. Some ground tests performed in the recent years have shown that thrust lessening is small.[5] Some other deployment method should apply to circular sails. For instance, the sail would be unfolded by a small-diameter inflatable tube attached around the sail circumference; once deployed, the tube has to be rigidized in the space environment for retaining its shape without the need of keeping the tube under pressure (a thing impractical for a long time). In general, reducing wrinkles entails some additional structural mass. This repeats for other problems such as tearing.

Nevertheless, a sail divided into a number (to be optimized) of pieces presents significant advantages from the viewpoint of construction and handling. This strongly affects the type and the structure of the sail system, which is tailored as function of the chosen deployment architecture and structural architecture.

By *deployment* architecture one means the technology of boom packaging and deploying. Inflatable, elastic, and rigidizable architectures are some example of deployment architecture. Also, booms have a *deployed structural* architecture, which regards a low-mass structure for keeping the sail in its operational configuration. In designing a boom subsystem, both architectures have to be specified.

Another problem of considerable importance is the scalability of the sail system, enhanced by the fact that the sail is a two-dimensional material device, not virtually a large empty region filled with some force field(s) (as in the case for magnetic sail). Thus, also sail distortions and vibrations under different loads have to be considered carefully in space mission designs dealing with larger and larger sails. Not only, scalability is also affected by multi-layer sails, which so far have been dominating the sail system concepts, at least with regard to the first sail generation.

---

[5]In Chap. 6, we will give this adjective a quantitative meaning together with equations that may allow the analyst to calculate the effect of wrinkling on thrust case by case.

3.2.2 Attitude Control

The problem of controlling radiation-pressure thrust will be introduced in Chap. 5, where we deal with which types of thrust maneuvers are admissible in SPS.

Qualitatively, here we highlight a few items that characterize a sailcraft. After the separation of the packed sailcraft from the launcher, the first maneuver and the related commands and procedures (or the first attitude acquisition) are performed in order to start the time sequence of the planned mission . The first critical event of the sequence is the sail deployment. After sail unfolding and checkout are completed,[6] the sail will be oriented stably toward the Sun. The sail's first orientation maneuver (which can be considered the second attitude acquisition) might be thought via some traditional equipment such as rotating wheels, extended booms, cold-gas thrusters, and so on, *if* the sail moment of inertia is sufficiently low; however, this will not be the case in general. Reaction wheels should be rather large for managing the high moments of inertia of a sail and, hence, considerably heavy. Booms may be long, but appear very promising though mass penalty is unavoidable; IKAROS employed a rotation-based deployment. It is important to realize that the architecture of sail-keeping can be used even as sail deployment.

Cold-gas engines, to be located at the tips of the sail spars, may result in an unacceptable propellant consumption, depending on the mission. For this reason, plasma-based high-specific-impulse micro thrusters have been considered in detail.

Another technique consists of using the rotation of two opposite panels of those ones the sail has been segmented into. In order to do so, the sail should have each (attitude) panel supported by two articulated booms gimballing at the sail mast. At the boom tips, these panels may be connected to small movable spars such a way panel edges could be independently lowered or raised with respect to the booms plane. As a result, full sail controllability could be achieved. However, the hardware enabling panel movements may be rather massive. For redundancy, a good design should include two panel pairs, equipped as described, but this increases the sailcraft mass to sail-area ratio. Consequently, the solar-pressure thrust decreases, and fast-sailcraft missions would not be allowed. For other mission types, though, this control technique exhibits two particular advantages: (1) attitude control should be still possible when, in some mission phases, the sunlight incidence angle is rather high, and (2) a priori, the spacecraft is not constrained to be located between the Sun and the sail, unless required for other purposes.

In general, at the very moment the sail is steered toward the Sun and the solar photons impinge on it (with some incidence angle), the center of pressure arises, similarly to the swelling of the sailboat sails. From that moment on, two objects, the spacecraft and the sail system, will move through the action of the sail on the spacecraft and the reaction of the spacecraft on the sail, because these bodies are interconnected. In general, the *sailcraft*'s barycenter (G) and the center of pressure (P) will not be aligned with the local sunlight direction. Since spacecraft and sail

---

[6]This monitoring may be effected via television cameras exploring the sail surface.

3.2 Main Elements of Solar-Photon Sailcraft

do not form a rigid body, one could carry into effect relative movements between G and P in order to change the sail orientation.

There is a significant literature on how to implement the mentioned attitude control methods (see Chap. 5). One could assert that control ways other than those ones usually employed for conventional spacecraft have to be considered, especially in sophisticated sailcraft missions entailing large sails.

3.2.3 Thermal Control

Spacecraft's thermal control might be expected to be of conventional type; space vehicles are designed to withstand the temperatures of space environments. However, in a sailcraft, spacecraft is necessarily close to the sail system; it should be strongly affected by the sail temperature. Controlling the temperature of the sail shall be of passive type, namely, via radiation balance. Sail temperature can be adjusted by changing its orientation with respect to the incident sunlight, but not too much, otherwise the sailcraft trajectory would change considerably. One has to design a trajectory satisfying the mission target(s) *and* the temperature constraints of the sail materials. This is not a new thing, of course. Nevertheless, one should note that the non-reflected photon energy is absorbed by the sail and then re-emitted over the whole solid angle; therefore, if the sailcraft is sufficiently close to the Sun, other sailcraft systems may be hit not only by part of the light diffused by the sail, but also by a significant amount of energy in the form of infrared radiation. The thermal control of such systems may require additional power in order to keep their operational temperature ranges.

The irradiance onto a solar sailcraft carrying out a mission in the planetary space comes from the Sun and the planet that sailcraft is being orbited. The planet irradiance may produce a negligible dynamical effect. In contrast, from a thermal viewpoint, a different situation occurs for such a sailcraft entering planetary shadows (penumbra and umbra). Since the sail is extended and very thin, the sail temperature immediately drops and adapts to the space environment. When the sailcraft returns to light, the sail temperature rises much quickly again. Although the space environment around a planet is very different from the interstellar medium, sail's temperature may jump up of hundreds of K-degrees (in some missions). It appears clear that sail materials have to be selected to withstand many high-low-high temperature cycles during their years of operational life.

There are a number of mission-dependent options in order to adopt a certain sail material for a given mission. In the near-term, sails should be a three-layer type, the supporting layer (or the substrate) consisting of a polyimide[7] thin film (e.g. Kapton, CP1, CP2[8]), or a polyester film (e.g. Mylar); advanced materials

---

[7] A polymer containing the so-called imide monomers, widely utilized in the electronics industry.

[8] A piece of 10 m × 10 m 2-micron sheet of CP1 was used by NASA for its experimental NanoSail-D [14], which did not achieve its orbit because of the failure of its launcher (August 2008). The

may come from space-born metallic monolayers (e.g. Beryllium membranes), bilayer sheets of Aluminium-Chromium (the so-called all-metal sail[9]), or the very advanced Multi-Walled Carbon Nanotubes (MWCNT) monolayer membranes. It is important to realize that, among different items, the sail materials affect the class of missions through the maximum allowed temperature before unrecoverable failures in the sail system. In particular, plastic-based sails will undergo unavoidable TSI-induced swelling as sailcraft moves toward the Sun. Each element of the sail may behave differently and two global effects could arise with respect to a flat sail: (1) an optical response (which depends also on the light's incidence angle) causing the total thrust to decrease, (2) temperature gradients bringing about thermal strains comparable or higher than the mechanical strains caused by the sail tension loading.

3.2.4 Communication

In general, communications between a space vehicle and the ground control center are fundamental for the mission success, though the control center is not the only base. As a point of fact, the spacecraft can be tracked periodically from other ground stations. Tracking stations and control center receive from and send electromagnetic waves to the spacecraft in different frequency bands. Plainly, the spacecraft has to be "electromagnetically visible" and the onboard antennas have to see the Earth.

Here is another implication of the relative size between sail system and spacecraft. The problem is where to put the onboard antennas in large deep-space sailcraft. This depends not only on the sail configuration, but also on the sail orientation along the sailcraft trajectory. On a spacecraft, there may be different types of antennas: high-gain antenna for scientific-data-return, telemetry/command antenna, emergency and low-gain antennas. IKAROS does not have a high-gain antenna; instead, it has two low-gain antennas oriented along the sail spin axis. IKAROS is not able to communicate with ground stations when the Earth angle is close to $\pi/2$.

Though very thin, the sail can cause obstruction of the antenna waves. To put antennas close to the sail rim would be wrong, as it could (1) cause mechanical and electrical problems, (2) induce sail instability, and (3) make the sail control much more difficult. A possible solution may be to use the structure that normally forms the "axis" of the sail; for each antenna type necessary for the mission, one may be placed on the front side of the sail and a twin on the back side. In future advanced missions beyond the solar system, a small part of a wide sail might be designed to function as a big antenna, so large amounts of scientific data may be downloaded to Earth-based or Moon-based receiving antennas from distances as large as hundreds of astronomical units. Nanotechnology may hopefully offer a solution.

---

flight spare—NanoSail-D2—was successfully launched in December 2010. It completed its mission on September 17, 2011.

[9]Only very preliminary studies have been performed in Italy in the Nineties.

3.2.5 Payload

In general, scientific mission payloads consist of a set of instruments for detecting particles and fields, analyzing e.m. radiation bands, receiving and sending signals, taking pictures of target objects, mapping surfaces, and so on. In a sailcraft, may the payload working be affected by the sail? Suppose we want to design a sailcraft circling a planet on a high (elliptic) orbit with the payload measuring the fine and evolving structure of the planet's magnetic field under the continuous action of the solar wind; this one, interacting with magnetic regions, continuously changes their shape and properties, even if the general configuration remains unchanged. One of the next sailcraft missions might be of a similar kind, about the Earth.[10] How is a sailcraft affected by a large region of magnetic and electric fields, and with flows of charged particles?[11] Since the sail size is normally larger than some characteristic plasma lengths, and smaller than others (see Chap. 4), one of the expected effects in the solar wind consists of space plasma surrounding the sail's front side by a positively charged sheath, whereas a wake of negatively charged flow extends beyond the sail's back side. Such a charge distribution can alter the local undisturbed properties of what the payload should measure. Therefore, in these cases, one shall locate the scientific instruments sufficiently ahead of the sail system, where the plasma is locally unmodified by the presence of the sail system. Of course, as each mission has its own features, the payload-sail configuration has to be analyzed on a case-by-case basis.

3.3 Micro and Nano Sailcraft Concepts

We limit our considerations to the following questions:

1. Were the ratio between the sailcraft mass and its effective area kept fixed (at the same sail orientation), would the motion of the vehicle remain unchanged regardless of the sail size?
2. How much realistically could we reduce the size of a sailcraft?
3. Is scaling a technological problem only or is there any physical limit that prevents having an (almost) arbitrarily small vehicle?

---

[10]Though space satellites such as the NASA IMAGE spacecraft (March 2000–December 2005) and the ESA four-satellite CLUSTER ($\sim$3–19 earth radii, still operational since August 2000) have discovered or confirmed fundamental phenomena in the Earth's magnetosphere, nevertheless the realization of a sailcraft-based mission like GeoSail would combine SPS technological/physical aspects with measurements in the Earth magnetosphere in the region $\sim$11–23 earth radii (http://sci.esa.int).

[11]Part of the solar wind penetrates into the magnetosphere and is channeled down to Earth, bringing about a number of phenomena, some of which may disturb human telecommunications and electric-energy transport (to cite the most common).

88   3 Sailcraft Concepts

Let us first recall that about 99 percent of the solar irradiance is due to photons with wavelengths from 0.2 to 4.0 µm, and the red-violet part of the spectrum carries about 50 percent of the total solar irradiance. Telecom systems operating in the microwave K-band have been widely employed. If one envisaged a future complete deep-space system transmitting information at significantly higher frequencies, the sole antenna could not be smaller than some millimeters. If one turns to telecommunication systems via laser for ultra-deep space missions ($\sim$200 AU, and more), minuscule-laser arrays would be nano-technologically possible, but there are other critical problems (e.g. pointing accuracy, ground receiving telescope, etc.) to be taken into account because of the enormous distances, in such missions, between spacecraft and Earth.

With regard to the scientific payload, interstellar spacecraft of about 1 kg have been proposed; however, if one thinks of accomplishing some high-performance deep-space science by tiny volume detectors, the probability of interaction between any space particle and the detector decreases dramatically. Even if we have one-event (large) detectors, however getting a sufficiently high number of events is fundamental for analyzing data compliant with the scientific aims. The minimum size of scientific instruments can vary significantly according to their types; it depends not only on technology, but mainly on the underlying physics. This should not surprise because measuring is a conceptual and basic problem involving interactions.

Although very briefly, such points may be considered an answer to the above third question. What about potential applications of nanoscience and nanotechnology to solar sailing, in general, and with regard to missions aimed at collecting precious data very far from the solar system, in particular? Here, we may note the suggestion of a *swarm* of many tiny spacecraft, or nanoprobes, which collectively behave like a large spacecraft. This is a very advanced concept indeed [17]; in principle, the spacecraft as a whole might look like many small antennas that act together as a very large non-constructible antenna, but much more intricate. However, at the moment, this fleet of interrelated tiny sailcraft is a pure conjecture.

In order to fix some order of magnitude useful to characterize sailcraft sizes, we may define micro-sail and nano-sail with respect to a fictitious sail of 1 km^2 in area. This is a factor 1.6 larger than the sail designed by JPL in the Seventies with the purpose to rendezvous the Halley comet on its 1986 passage close to the Sun.

In principle, a sailcraft with mass 1,537 kg with a *full* reflective sail 1 km^2 large may be oriented orthogonally to sunlight, thus *floating* in the solar field at any distance from the Sun greater than about 0.1 AU, and sufficiently far from the planets.

Well, a sailcraft with a sail of 1 m^2 may be called a *micro-sailcraft* if the whole vehicle had a mass 1.537 g. Such a sailcraft might be the typical element of a large swarm, because its payload should amount to 0.7 g at most. Strictly speaking, then a (perfectly-scaled) *nano-sailcraft* should have a sail of 0.001 m^2 and a mass of 1.537 mg.

3.4 Additional Remarks 89

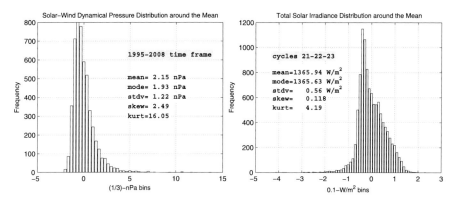

Fig. 3.10 Frequency distribution around the mean: (*left*) solar-wind dynamical pressure, (*right*) Total Solar Irradiance. Means and modes refer to as the unshifted distributions. Standard deviation, skewness and kurtosis are also reported. Note that 0.1 W/m²/c is very close to (1/3) nPa. These plots have been built by using HMF and plasma data obtained from NASA's OMNIWeb interface

3.4 Additional Remarks

In this chapter, we have reviewed the most important aspects of in-space sailing concepts. SPS has been detailed to a greater extent not only as the topic of this book, but also because the efforts regarding SPS system/subsystem analysis—made in the course of four decades—greatly surpass those ones relevant to the solar-wind-based concepts. We have to remark the following overall aspects that can be inferred from physics and geometry related to SPS concepts:

- Every mission envisaged by solar-wind sail could be got via photon solar sail.
- The two-sail and the e-sail concepts may, in principle, vary the "direction" of the interaction region with respect to the local velocity of the solar wind; instead, the interaction volumes of magsail and plasma sail (at least as conceived hitherto) appear to be hard to control in orientation. Controlling orientation with respect to the local momentum source is essential for enlarging the set of achievable goals dramatically. SPS has no practical limitation in this sense.
- Missions inside the planetary magnetospheres (including that of the Earth) cannot be performed by solar-wind sails; in particular such vehicles need another propulsion for escaping from the planetary gravitational fields. In contrast, a photon solar sail could perform any mission with the exclusion of maneuvers inside the dense atmospheric layers; for example, SPS cannot operate in Earth orbits lower than 700–800 km because of the stronger atmospheric drag. Nevertheless, one may utilize a sail for assisting the orbital decay of spacecraft into the low atmosphere, as experimented by NanoSail-D2.
- Above all, the problem of variability of the external energy/momentum sources. The TSI daily means fluctuate: $-0.0022 \lesssim \delta\widetilde{\text{TSI}/\text{TSI}} \lesssim 0.0015$ (we shall consider this problem of trajectory computation in Sect. 9.1). Solar wind P_{dyn} exhibits the changes $-0.93 \lesssim \delta P_{dyn}/\widetilde{P_{dyn}} \lesssim 5.1$. The related frequency distributions are shown in Fig. 3.10. This problem, perhaps the main one in solar-wind sailing,

should be *further* investigated because it may ultimately decide on the feasibility of the related sail concepts.

References

1. Andrews, D. G., Zubrin, R. M. (1988), Magnetic sails and interstellar travel. In 39th IAF Congress, Bangalore, India. IAF-88-553, also published on Journal of the British Interplanetary Society, 43, 265–272.
2. Balogh, A., Lanzerotti, L. J., Suess, S. T. (2008), The Heliosphere Through the Solar Activity Cycle. Berlin: Springer-Praxis. ISBN 978-3-540-74301-9.
3. Bavassano, B., Pietropaolo, E., Bruno, R. (2004), Compressive fluctuations in high-latitude solar wind. Annals of Geophysics, 22, 689–696.
4. Borovsky, J. (2010), Contribution of strong discontinuities to the power spectrum of the solar wind. Physical Review Letters, 105, 111102. doi:10.1103/PhysRevLett.105.111102.
5. Brumfiel, G. (2011), Voyager 1 reaches surprisingly calm boundary of interstellar space. Scientific American^{TM}, June 15, 2011.
6. Bruno, R., Carbone, V. (2005), The solar wind as a turbulence laboratory. Living Reviews in Solar Physics, 2.
7. Davidson, P. A. (2001), Cambridge Texts in Applied Mathematics. An Introduction to Magnetohydrodynamics. Cambridge: Cambridge University Press. ISBN 0-521-794870.
8. Ecke, R. (2005), The turbulent problem. Los Alamos Science, 29.
9. Goedbloed, J. P. H., Poedts, S. (2004), Principles of Magnetohydrodynamics: With Applications to Laboratory and Astrophysical Plasmas. Cambridge: Cambridge University Press. ISBN 0-521-62607-2.
10. Janhunen, P. (2004), Electric sail for spacecraft propulsion. Journal of Propulsion and Power, 20(4).
11. Janhunen, P., Sandroos, A. (2007), Simulation study of solar wind push on a charged wire: basis of solar wind electric sail propulsion. Annals of Geophysics, 25, 755–767.
12. Janhunen, P. (2008), The electric sail—a new propulsion method which may enable fast missions to the outer solar system. Journal of the British Interplanetary Society, 61(8), 322–325.
13. Jenkins, C. H. M. (Ed.) (2006), Progress in Aeronautics and Astronautics: Vol. 212. Recent Advances in Gossamer Spacecraft. Washington: AIAA.
14. Johnson, L., Whorton, M., Heaton, A., Pinson, R., Laue, G., Adams, C. (2011), NanoSail-D: a solar sail demonstration mission. Acta Astronautica, 68(5–6), 571–575. Special issue: Aosta 2009 Symposium.
15. Kolmogorov, A. N. (1941), The local structure turbulence in incompressible viscous fluids for very large Reynolds numbers. Doklady Akademii Nauk SSSR, 30, 301–305. Reprinted in Philosophical Transactions of the Royal Society of London, Series A, 434, 913 (1991).
16. Krimigis, S. M., Roelof, E. C., Decker, R. B., Hill, M. E. (2011), Zero outward flow velocity for plasma in a heliosheath transition layer. Nature, 474, 359–361. doi:10.1038/nature10115.
17. Matloff, G. L. (2005), Deep-Space Probes (2nd edn.). Chichester: Springer-Praxis. ISBN 3-540-24772-6.
18. Marsch, E. (2006), Kinetic physics of the solar corona and solar wind, Living Reviews in Solar Physics, 3(1).
19. Mathieu, J., Scott, J. (2000), An Introduction to Turbulent Flow. Cambridge: Cambridge University Press. ISBN 0-521-77538-8.
20. Mengali, G., Quarta, A. A., Janhunen, P. (2008), Considerations of electric sail trajectory design. Journal of the British Interplanetary Society, 61(8), 326–329.

References 91

21. Meyer-Vernet, N. (2007), Cambridge Atmospheric and Space Science Series. Basics of the Solar Wind. Cambridge: Cambridge University Press. ISBN 978-0-521-81420-1.
22. NASA/GSFC OMNIWeb Interface, http://omniweb.gsfc.nasa.gov/.
23. Parker, E. N. (1958), Dynamics of the interplanetary gas and magnetic fields. The Astrophysical Journal, 128, 664–676.
24. Parker, E. N. (1960), The hydrodynamic theory of solar Corpuscular radiation and stellar winds. The Astrophysical Journal, 132, 821–866.
25. Parker, E. N. (1991), Heating solar coronal holes. The Astrophysical Journal, 372, 719.
26. Parker, E. N. (1996), The alternative paradigm for magnetospheric physics. Journal of Geophysical Research, 101(A5), 10587–10626.
27. Parker, E. N. (1999), Space physics before the space age. The Astrophysical Journal, 525C, 792–793.
28. Parker, E. N. (2009), Solar magnetism: the state of our knowledge and ignorance. Space Science Reviews, 144, 15–24.
29. Parker, E. N. (2010), Kinetic and hydrodynamic representations of coronal expansion and the solar wind. AIP Conference Proceedings, 1216, 3–7.
30. Pizzo, V. J. (1991), The evolution of corotating stream fronts near the ecliptic plane in the inner solar system, 2: three-dimensional tilted dipole fronts. Journal of Geophysical Research, 96, 5405.
31. Reynolds, O. (1895), On the dynamical theory of incompressible viscous fluids and the determination of the criterion. Philosophical Transactions of the Royal Society of London, Series A, 186, 123–164. doi:10.1098/rsta.1895.0004.
32. Riazantseva, M. O., Zastenker, G. N. (2008), Intermittency of solar wind density fluctuations and its relation to sharp density changes. Cosmic Research, 46(1), 17. doi:10.1134/S0010952508010012.
33. Rozelot, J.-P. (2006), Solar and Heliospheric Origins of Space Weather Phenomena. Berlin: Springer. ISBN 3-540-33758-X, ISBN 978-3-540-33758-4.
34. Salem, C., Mangeney, A., Bale, S. D., Veltri, P. (2009), Solar wind magnetohydrodynamics turbulence: anomalous scaling and role of intermittency. The Astrophysical Journal, 702, 537–553. doi:10.1088/0004-637X/702/1/537.
35. Shepherd, L. R. (1990), Memorandum of the interstellar space exploration committee. In 41st Congress of the International Astronautical Federation, Dresden, Germany.
36. Smith, E. J., Zhou, X. (2007), Turbulence and Non-Linear Processes in Astrophysical Plasmas, Honolulu, Hawaii.
37. Toivanen, P. K., Janhunen, P. (2009), Electric sailing under observed solar wind conditions. Astrophysics and Space Sciences Transactions, 5, 61–69.
38. Vulpetti, G. (1980), Noise-effects in relativistic pure-rocket dynamics, Journal of the British Interplanetary Society, 33, 27–34.
39. Vulpetti, G. (1990), Dynamics of field sail spaceships. Acta Astronautica, 21, 679–687.
40. Vulpetti, G., Pecchioli, M. (1991), The two-sail propulsion concept. Paper IAA-91-721, 42nd Congress of the International Astronautical Federation, Montreal, Canada, October 5–11.
41. Vulpetti, G. (1994), A critical review on the viability of a space propulsion based on the solar wind momentum flux. Acta Astronautica, 32(32), 641–644.
42. Vulpetti, G. (2000), Sailcraft-based mission to the solar gravitational lens. In STAIF-2000, Albuquerque, NM, USA, January 30–February 3.
43. Vulpetti, G., Santoli, S., Mocci, G. (2008), Preliminary investigation on carbon nanotube membranes for photon solar sails. Journal of the British Interplanetary Society, 61(8), 284–289.
44. Wilson, M. N. (1986), Superconducting Magnets. Oxford: Oxford University Press.
45. Winglee, R. M., Slough, J., Ziemba, T., Goodson, A. (2000), Mini-magnetospheric plasma propulsion: tapping the energy of the solar wind for spacecraft propulsion. Journal of Geophysical Research, 105(A9), 21,067–21,077.

46. Wood, B. E., Howard, R. A., Thernisien, A., Socker, D. G. (2010), The three-dimensional morphology of a corotating interaction region in the inner heliosphere. The Astrophysical Journal Letters, 708, L89. doi:10.1088/2041-8205/708/2/L89.
47. Zubrin, R. M. (1989), Use of magnetic sails for Mars exploration mission. In AIAA-ASME-ASEE 25th Joint Propulsion Conference, July 1989. AIAA-89-2861.
48. Zubrin, R. M. (1993), The use of magnetic sails to escape from low Earth orbit. Journal of the British Interplanetary Society, 46.

Chapter 4
Solar Sails in Interplanetary Environment

An Introduction to Some Aspects of the Interaction of Ultraviolet Light and Solar Wind with Sail In this chapter, we deal with some aspects of the influence of the solar-wind particles and the ultraviolet/X-soft radiation to photon sail materials. In a certain sense, the following sections are preparatory to Chap. 9, where we will deal with a formal model to be inserted into the trajectory equations. This chapter consists of three main sections: the first one is devoted to the solar sail immersed in the solar wind, the second one regards the effect of the UV-XUV light onto the sail reflective film, and the third one considers the interaction of solar-wind protons and Helium ions on a sail moving with respect to the wind. As we shall see in such sections, there are some simultaneous processes that stem from the particle impact onto the solar sail; but ultimately one aims at regarding sail trajectory through the alteration of the optical properties of the sail surface. In addition to journal/book references and the author's personal investigation, plasma data information obtained from NASA's OMNIWeb interface will be utilized again.

4.1 A Sail in the Solar Wind

In Sect. 3.1 we emphasized the type of model (either particle, kinetic, or fluid) to be known for describing the interaction of the solar wind with a sail endowed with some *own* extended field. Conceptually speaking, the same thing holds for SPS. As a matter of facts, even though the solar-wind dynamical pressure is expected to produce a small perturbation on any reasonable SPS, nevertheless the sail is a physical obstacle inside the wind; therefore, we need a criterion to determine some appropriate model for calculating the fluxes of charged particles impacting the sail. To such an aim, we will follow the meaningful approach in Chap. 7 of [28], here adapted to the case for SPS. We are interested in a non-destructible object put inside the solar wind; in general, such an object interacts with the plasma according to:

(i) its own geometric sizes relatively to the plasma's characteristic scales;
(ii) its own magnetic/electric fields;
(iii) its own environment.

G. Vulpetti, *Fast Solar Sailing*, Space Technology Library 30,
DOI 10.1007/978-94-007-4777-7_4, © Springer Science+Business Media Dordrecht 2013

94 4 Solar Sails in Interplanetary Environment

Table 4.1 Values, averaged over solar cycle 23, of the proton number density (n_p), plasma temperature (T), and magnetic induction (B) of the solar wind at 1 AU. Time series of solar wind have been obtained from NASA's OMNIWeb interface

Solar Wind	symbol	value	unit
proton number density	$n_p(1)$	6.1	cm^{-3}
temperature	$T(1)$	0.11	MK
magnetic induction	$B(1)$	4.5	nT

Point (ii) can be excluded because SPS is not a solar-wind based object, which has to be endowed with electric or magnetic field (Chap. 3). A solar sail may have an electric field induced by energetic photons (Sect. 4.2), but the related potential is not so high to cause an effective size meaningfully larger than the geometrical size. SPS has no own environment, differently from many celestial bodies; therefore, point (iii) does not apply.

In the following subsections, we will calculate or report how the *mean* values of six characteristic lengths of the solar wind change with the distance from the Sun. Their numeric values will also depend on the mean values of the number density, plasma temperature, and magnetic induction, all at 1 AU. Using the data from [37] again, we will use the values averaged over cycle-23, and reported in Table 4.1.

In the following, we will emphasize the units of the quantities related to the solar wind. According to custom in plasma physics, many of the quantities called *frequencies* are actually angular speeds; thus, one has to divide by 2π in order to get strict frequencies.

4.1.1 Mean Inter-Particle Distance

After a parcel of solar wind leaves its coronal hole, it evolves along its streamline where, in particular, the plasma continuity equation holds (according to the one-fluid description). Since the parcel bulk speed may be considered insensitive to distances beyond ~ 0.1 AU and before the termination shock, then the parcel's number density n (i.e. n_e electrons $+n_p$ ions per cubic meter) scales as $1/R^2$, where R is expressed in AU. Every parcel has it own initial density, temperature, and so forth. If we average the observed fluctuations of the solar wind over cycle 23, then we may write the mean inter-particle distance as follows

$$\ell_{ip}(R) \sim \left(n(1)/R^2\right)^{-1/3} \cong 4.3 \times 10^{-3} R^{2/3} \text{ m} \qquad (4.1)$$

where $n = n_e + n_p$ with $n_p \sim n_e$, and we have used the value of Table 4.1 for $n_p(1)$. Therefore, even at large distance from the Sun, usual spacecraft linear sizes are much higher than ℓ_{ip}.

4.1.2 Debye Length

Put simply, Debye shielding length, say, ℓ_D is the minimum linear size of an ionized gas beyond which the plasma appears practically neutral. Said equivalently, a region of the order of $\ell_D{}^3$ is the maximum volume inside which local charge imbalance, ascribable to thermal fluctuations, may take place. In the current framework of solar wind, considering one flow of protons (i.e. neglecting heavier nuclei) and one flow of electrons, Debye length is given by

$$\ell_D = (\sqrt{\varepsilon_0 k}/e)\sqrt{\frac{T}{n}} \cong \frac{69}{\sqrt{n(1)}}\sqrt{T(R)}R \text{ m} \tag{4.2}$$

where $T \sim T_p \sim T_e$ is the solar-wind temperature assumed of the same order of the electron and proton temperatures. Using the two-fluid equations and neglecting viscous heating in the solar wind, the proton fluid would result to be adiabatic, namely, $T_p \propto n^{2/3} \propto R^{-4/3}$. In contrast, the electron fluid evolution appears to deviate from the adiabatic process. If one assumes that the electron heat flux follows a polytropic law of index γ_e (e.g. [28]), then the electron temperature scales as $T_e \propto R^{-2(\gamma_e-1)}$. However, in the specialized literature, one can read many values of γ_e ranging from about 1.2 to about 1.6, i.e. very close to $5/3$, and a variable polytropic index may not be excluded, e.g. [14] and [8]. Considering the current level of approximation here, we use a reference value of 1.4 that entails $T_e \propto R^{-0.8}$, which nevertheless is sufficiently far from the adiabatic value. We will use this function of the electron temperature for scaling the plasma temperature:

$$T(R) \cong T(1)R^{-2(\gamma_e-1)}, \quad \gamma_e \cong 1.4, \ T(1) \cong 1.1 \times 10^5 \text{ K}$$
$$\ell_D \sim 7R^{0.6} \text{ m} \tag{4.3}$$

Thus, Debye length is of the same order of many spacecraft sizes, but less than the typical lengths of sailcraft envisaged for studying the magnetosphere or the interplanetary space for several years.

Another important parameter in plasma physics is the number of particles in a Debye sphere, usually denoted by Λ:

$$\Lambda = \frac{4\pi}{3}n\ell_D{}^3 \tag{4.4}$$

In the plasma literature, various authors define Λ differently among them, but the corresponding values are not much dissimilar; these definitions do not change the order of magnitude of Λ. We will use this plasma parameter in Sect. 4.1.6.

4.1.3 Electron's Gyro Radius

The classical path of a particle of mass m and charge q in a uniform magnetic induction B is a circular helix, which has constant curvature and torsion (see Sect. 5.3.3).

96   4   Solar Sails in Interplanetary Environment

What matters in many problems is the *cyclotron* or gyro radius,[1] say, ℓ_q that is given by

$$\ell_q = \frac{mV_\perp}{qB} \rightarrow \boxed{electron} \rightarrow 5.686\frac{V_{e,\perp}\ (\text{km/s})}{B\ (\text{nT})}\ \text{m} \tag{4.5}$$

where $V_{e,\perp}$ is the electron's velocity component orthogonal to **B**. Now, this quantity and B have to be expressed as function of the distance R. With regard to the former one in the current framework, we may use the above temperature assumption and Eq. (4.3) to write

$$V_{e,\perp} \cong (2k/m_e)^{1/2}\sqrt{T_e} \cong 1790R^{-0.4}\ \text{km/s} \tag{4.6}$$

In order to quantify the magnitude B of the HMF, one can express **B** in heliographic coordinates, or the *heliographic reference frame* (*rtn*).[2] The first unit vector is the direction of the radial vector, namely, $\mathbf{r} \equiv \mathbf{R}/R$; if **s** denotes the Sun's rotation axis, the second unit vector is $\mathbf{t} = \mathbf{s} \times \mathbf{r}/\|\mathbf{s} \times \mathbf{r}\|$; the third one completes the positive triad as usually: $\mathbf{n} = \mathbf{r} \times \mathbf{t}$. HMF is detailed in technical books, e.g. [2] where theory and observations are widely discussed with a special emphasis on the excellent results from the instruments onboard spacecraft Ulysses. From what observed, we can draw the guidelines for a simple expression of B.

In the current astrodynamical context, we are concerned with the overall aspects of HMF in order to carry out an approximate function $B(R, \Box)$, where the slot stands for any additional parameter. Very shortly, this objective may be achieved on comparing HMF observations with Parker's model of the solar wind (e.g. [28, 39, 42]). Very close to the Sun, both the radial and transversal components of **B**, i.e. B_r and B_t, respectively, expand super-radially[3] and bend toward the solar equator; sufficiently distant from the lower corona, from several solar radii on, B_r becomes radial as predicted by Parker's model. The observed spiral angle, namely, $\tan\phi_{obs} = B_t/B_r$ shows fluctuations at all latitudes,[4] and about the theoretical spiral angle, i.e. $\tan\phi_{Parker} = -\omega_{Sun}R\cos\vartheta/V_r$, where ϑ denotes the heliographic latitude and ω_{Sun} is the angular speed of the Sun. With regard to B_n, it fluctuates as well; nevertheless, the observations (made in a wide interval of distances from

---

[1]This length is also reported as the Larmor radius in few textbooks and many websites. However, Larmor frequency does not regard gyration, but the precession of magnetic dipoles in an external magnetic induction field, and is a quantum phenomenon as well. The classical expression for Larmor frequency is very similar to the gyro-frequency but a factor $1/2$. The helix curvature is given by $\kappa = (\ell_q)^{-1}(V_\perp/V)^2$.

[2]Here, we adapted the customary RTN notation into *rtn* for avoiding confusion with the related capital letters that have different meanings in this book.

[3]According to Fisk's model of the solar wind and magnetic induction, the *super-radial* expansion of polar-cap field lines begins non-radially with large displacements in heliographic longitude and latitude with respect to the final radial direction.

[4]Histograms of $\phi_{obs} - \phi_{Parker}$, related to low latitudes, are double-peaked, one about $0°$ and the other about $180°$, revealing the existence of two magnetic sectors.

4.1 A Sail in the Solar Wind

the Sun) show that averages over many days and up exhibit vanishing values, according to Parker's theory.[5] Such global view of the HMF allows us to use Parker's expressions for $B(R, \Box)$ for our aims:

$$B(R, \vartheta) = B_{\mathrm{r}}(1) R^{-2} \sqrt{1 + (R \cos \vartheta)^2}, \quad B_{\mathrm{r}}(1) \cong \frac{4.5}{\sqrt{2}} \mathrm{~nT} \tag{4.7}$$

where, at 1 AU and $\vartheta = 0$, $\phi_{Parker} \cong \pi/4$ and, the reference value comes from cycle-23. Inserting Eqs. (4.6)–(4.7) into Eq. (4.5) results in the electron gyro radius as function of R:

$$\ell_e^{(th)} \sim 3,200 \frac{R^{1.6}}{\sqrt{1 + (R \cos \vartheta)^2}} \mathrm{~m} \tag{4.8}$$

4.1.4 Proton's Inertial Length

Proton's inertial length in a plasma is defined by:

$$\ell_p = \frac{c}{\omega_{p,\mathrm{p}}} = \frac{1}{e} \sqrt{\frac{m_p}{\mu_0}} \frac{1}{\sqrt{n_p}} \sim 0.92 \times 10^5 R \mathrm{~m} \tag{4.9}$$

$$\omega_{p,\mathrm{p}} = e \sqrt{n_p/\varepsilon_0 m_p}$$

where $\omega_{p,\mathrm{p}}$ is the *proton* plasma frequency. Note that, if we apply Eq. (4.5) to a proton with speed equal to the Alfvén wave propagation speed, we get ℓ_p exactly; ℓ_p is independent of the magnetic induction. However, in general, the proton's inertial length (or the proton skin depth) should not be confused with the proton's gyro radius that is the object of the next sub-section. Their physical meanings are different. For instance, the proton skin depth enters the interaction of the solar wind with a small magnetized celestial body such as Mars' satellite Phobos (e.g. [31, 32]). The solar-wind's response to this obstacle depends (also) on the ratio between ℓ_p and the body's effective size (which may be much larger than its geometrical size, as in this case).

Of far greater importance in plasma physics is the *electron* plasma frequency, which is defined as

$$\omega_{e,\mathrm{p}} = e \sqrt{n_e/\varepsilon_0 m_e} \tag{4.10}$$

normally denoted by ω_{p} (where the subscript p stands for plasma, not for proton). We will use such quantity in Sect. 4.1.6.

---

[5]In the theory, $B_{\mathbf{n}} = 0$ everywhere results from the assumption that, in an inertial frame, solar wind is perfectly radial.

4.1.5 Proton's Gyro Radius

When one substitutes m_p and e for m and q in Eq. (4.5), one gets the proton gyro radius provided that the speed is chosen appropriately. On the current assumption $T_i \sim T_e$, we do not get a really new information if we use the proton's thermal speed. Instead, let us insert the solar-wind bulk speed averaged over the cycle-23, or about 440 km/s. Taking the magnetic induction as given by Eq. (4.7) results in

$$\ell_p^{(bulk)} \sim 1.4 \times 10^6 \frac{R^2}{\sqrt{1 + (R\cos\vartheta)^2}} \text{ m} \qquad (4.11)$$

Again, R is expressed in AU; $\ell_p^{(bulk)}$ is high because the solar wind is super-alfvénic.

4.1.6 Collision Mean Free Path

One needs kinetic equations for dealing with collisions in a plasma, e.g. see [4, 19, 28, 36, 41, 45], and [15]. In particular, Fokker-Planck theory is appropriate for describing binary Coulomb collisions. Small-angle (i.e. large impact parameter) electrostatic encounters are the most frequent collisions, and accumulate to eventually give a *large* angular deviation, which means that the *cumulative* scattering angle has become higher than $\pi/2$. If a plasma exhibits a high Λ, then one can speak of binary collisions, and hence of collision frequency, which we may denote by $\nu_{\zeta,\xi}$. This one measures the time rate at which particles of species ζ are scattered by those of species ξ. Considering again only electrons and protons in the solar wind, we are interested in the special interaction case of particles interacting with each other. We refer to the self-collision times, say, $t_{e,c}$ and $t_{p,c}$ for electrons and protons, respectively. They represent the times needed to produce a progressive deviation larger than $90°$. Their reciprocal values determine the electron and proton collision frequency. Adapting symbols and units from [45], one gets

$$\nu_{e,e}^{-1} = t_{e,c} = 1.14 \times 10^7 \sqrt{m_e/m_p} \frac{T_e^{3/2}}{\ln \Lambda n_e} \text{ s} \qquad (4.12a)$$

$$\nu_{p,p}^{-1} = t_{p,c} = 1.14 \times 10^7 \frac{T_p^{3/2}}{\ln \Lambda n_p} \text{ s} \qquad (4.12b)$$

where the various physical and mathematical constants have been embodied in the numerical factors. It is meaningful that, apart from a constant factor, the frequencies in Eqs. (4.12a), (4.12b) can be written as

$$\nu_{\zeta,\zeta} \propto \frac{\ln \Lambda}{\Lambda} \omega_{\zeta,p}, \quad \zeta = e, p \qquad (4.13)$$

4.1 A Sail in the Solar Wind 99

where the proton and electron plasma frequencies have been reported in Sect. 4.1.4. The expression just written overestimates the frequencies calculated in detail by factors ~ 10, e.g. [4, 45], and [36].

One can note that the solar wind electron collision frequency is higher by the factor $\sqrt{m_p/m_e}$. However, in the current framework of solar wind with quasi neutrality and $T_e(R) \sim T_p(R)$, there is no ambiguity in the values of the collision mean free paths. As a point of fact, solar wind collision mean free path is defined as follows:

$$\ell_{e,c} = v_e^{(th)} t_{e,c} \tag{4.14a}$$

$$\ell_{p,c} = v_p^{(th)} t_{p,c} \tag{4.14b}$$

for electrons and protons, respectively. $v^{(th)}$ is a characteristic thermal speed of the particles in an ambient at temperature T. This one may be written as

$$v^{(th)} = \chi \sqrt{\frac{kT}{m}}, \quad \chi \geq 1 \tag{4.15}$$

where, assuming a Maxwellian distribution for speed (not for vector velocity), χ may takes on the following values meaningfully: $\sqrt{2}$ for the most probable speed, $2\sqrt{2/\pi}$ for the mean speed, and $\sqrt{3}$ for the root mean square speed (which corresponds to the energy equipartition). Such values do not change the order of magnitude of the collision mean free path. According to Eqs. (4.12a), (4.12b) and (4.15), electron and proton mean free paths are equal to one another provided both densities and temperatures are equal; though approximately, this is just the case of solar wind. If one considered the basic collisional times as tabulated in [36], electron and proton free paths would differ by only $\sqrt{2}$. Also, the order of magnitude of the current free path is quite in agreement with that calculated in [28]. In all these (and other) references, the kinetic approach to the collision problem is the same, but calculation details are more or less sophisticated, and diversified according to the specific aspects of the collisional problem of interest.

In the current framework, we can summarize the results as follows

$$\ell_{e,c} = \ell_{p,c} \equiv \ell_c \cong \frac{3.59 \times 10^9 T^2}{n \ln{(1.38 \times 10^6 \frac{T^{3/2}}{\sqrt{n}})}}$$

$$= \frac{3.56 \times 10^{12} R^{0.4}}{23.4 - 0.200 \ln R} \text{ m} = \frac{23.8 R^{0.4}}{23.4 - 0.200 \ln R} \text{ AU} \tag{4.16}$$

where again R is expressed in AU. Equation (4.16) should not be used for low R, say, $R < 0.1$ AU. Two important features of solar wind can be read out: (a) collision path changes rather slowly with the distance, and (b) the order of magnitude of such path is 1 AU, namely, collisions (in the sense explained above) in solar wind are very rare events. Therefore, we will omit this mean free path in comparing the sizes of future sailcraft with the here considered solar-wind's characteristic lengths.

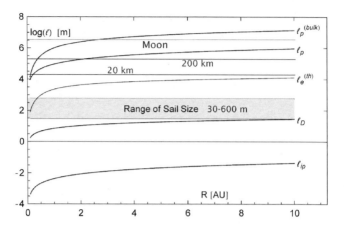

Fig. 4.1 Decimal logarithm of the solar-wind characteristic lengths as function of the distance from the Sun. The reference values (number density, temperature and magnetic induction) at 1 AU are the averages over the cycle-23. The *gray horizontal bar* represents a reasonable range of the solar-sail's linear size (from 30 m to 600 m). The characteristic lengths, denoted generically by ℓ here and in the text, are expressed in meters

4.1.7 Comparing Solar-Wind Lengths to Sail Size

As far as one thought yesterday and may think today, most of the envisaged solar-sail missions exhibit sail's linear size in the (reasonable) range 30–600 m. We have to compare such range with the above five characteristic lengths of the solar wind at various distances from the Sun. This has been done in Fig. 4.1, where the horizontal strip represents the sail range, by plotting the decimal logarithm of the lengths (expressed in meters) versus the distance R given in astronomical units (according to Eqs. (4.1), (4.3), (4.8), (4.9), and (4.11). With regard to the electron's gyro radius, we chose $\vartheta = 0$ because this entails the smallest radius to be compared with the sail size. The graphed functions extend from 0.1 to 10 AU: at larger distance, but lower than 80–100 AU where the termination shock floats (Sect. 3.1.1), lengths do not increase appreciably. At shorter distance, these $\ell(R)$ functions should not be used. The full sail size range is higher than Debye lengths, and lower than the electron gyro radius. Below 1 AU, sail size is comparable with $\ell_e^{(th)}$. Furthermore, sail range is always significantly lower than the proton's inertial length and the proton's gyro radius.

What information we may take out from these features? Recalling the basic plasma properties, one can easily infer that solar sails sense the solar wind as a (quasi) neutral flow. At the sail scales, as considerably smaller than the particle gyro radii, magnetic-field effects on the particles can be neglected. Sailcraft is a *small* object in the solar wind, but large enough to sense collective features of the wind. Such properties should be taken into account in modeling the interaction between a solar-photon sailcraft and the solar wind.

4.2 Ultravioleet Light onto Sail 101

Another useful information comes from comparing the proton and electron thermal speeds, as specified by Eq. (4.15), with the bulk speed of the solar wind. More precisely, we mean the two ratios

$$\upsilon_p^{(th)}/\upsilon^{(bulk)} \cong 0.1186 R^{-0.4} \tag{4.17a}$$

$$\upsilon^{(bulk)}/\upsilon_e^{(th)} \cong 0.1967 R^{-0.4} \tag{4.17b}$$

where the cycle-23 mean $\upsilon^{(bulk)} = 440$ km/s, $\chi = \sqrt{3}$, and the temperature scaling in Eq. (4.3) have been employed. Therefore, as one can easily recognize, the electron flux $n_e \upsilon_e^{(th)} = n_e \sqrt{3kT_e/m_e}$ is notably higher than the bulk flux at any realistic sailcraft distance. Another consequence is that sails are sensitive to electron and proton flows *separately*.

Now, we have to make another observation: the actual effects of the solar wind on solar sails could not be decoupled from the effects stemming from incident photons of wavelength sufficiently shorter than the visible light's. This is the matter of the next section.

4.2 Ultraviolet Light onto Sail

The effects of UV/EUV light on materials may be various, in general, and normally couples with the charged-particle environment the spacecraft moves through. Here, a considerable simplification comes from the fact that the first-generation, or multilayer, sails will not be able to perform *fast* missions—in the sense we will widely develop in Chap. 7. Second-generation sails and the desirable third-generation sails should satisfy the fast-sailing constraints. Second-generation sails might consist of all-metal (either bi-layer or mono-layer in-space processed) membranes of conventional materials improved via nanotechnology. In contrast, sail's subsequent generations may use general nanotechnological materials: probably, carbon nanotubes of metallic type may be valid candidates. Therefore, the high-generation bare sails should be *all metal*. Also, we may not exclude that nanotechnology allows skipping the second generation efficaciously.

Vast literature there exists on the problem of in-orbit spacecraft's *differential* charging, causing electric discharge among different devices onboard, and their ensuing induced malfunctioning. Usually, one is concerned with satellites/spacecraft in the Earth's magnetosphere. However, a sail in the interplanetary space sees a particle environment notably different from that in the Earth-Moon space, as we saw in the previous section and in Chap. 3. The problem of a sail moving inside the solar wind and UV/EUV flows may be simpler—on one hand—than a conventional spacecraft in a planetary magnetosphere, but—on the other hand—may result in some severe optical degradation of the sail material; this affects sailcraft trajectory through temperature increase and thrust decrease. In fast trajectory arcs, the solar-wind proton energy, as sensed by the sail, is increased appreciably. Let us proceed by steps.

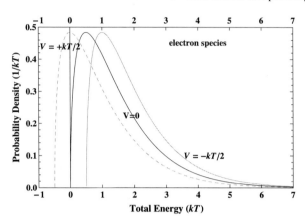

Fig. 4.2 Examples of probability density of the total energy of the electron species under different electrostatic potential. The central zero-potential (i.e. unperturbed) distribution refers to as the Maxwellian distribution of energy at the temperature T. Using kT as general units, the action of (for instance) the positive potential $+kT/2$—at a certain time and distance from the electric source—shifts the population of the electrons with given *total* energy on the left. In contrast, the negative potential of $-kT/2$ causes a right shift of the original distribution with an ensuing energy gap. Reverse situations take place for protons. Thus, if the rightmost distribution represent electrons, the leftmost may be seen as the distribution of protons under the *same* potential

Photon sails so far described are *small* objects in the solar wind, but very *large* with respect to any wavelength practically contributing to TSI. In the current framework, let us further assume the sail flat and non-rotating in an inertial frame (IF). (Modest deviations from these ideal conditions produce negligible effects here). Suppose for a moment that such an object co-moves with the plasma, namely, it moves with the plasma bulk velocity. Such small object is able to sense electron and ion currents separately. If, for simplicity, we repeat the assumption $T \sim T_i \sim T_e$, then the electron current is higher than the ion currents by the factor $\sqrt{m_i/m_e} \geq \sqrt{m_p/m_e} \cong 42.85$. By absorption of the wind electrons, the sail begins being charged negatively, and both electrons and ions will then be immersed in a negative potential. Figure 4.2 shows how a Maxwellian distribution of energy changes under an electrostatic potential. (Note how the essential things can be expressed as function of kT only).

As time goes by, more and more electrons slow down, whereas ions take on higher speeds. If $\mathcal{C}$ is the capacitance of the sail of area A_S, the macroscopic evolution of the sail potential may be described by

$$\mathcal{C}\frac{dV(t)}{dt} - J_{net}(V(t))A_S = 0 \qquad (4.18)$$

where $\mathcal{C} = \varsigma \varepsilon_0 \sqrt{A_S}$, ς being a factor depending on the sail geometry; for a square sail, we may use $\varsigma \cong 3.6$. In Eq. (4.18), J_{net} is the net current density onto *both* sail sides endowed with the uniform potential $V(t)$ as the sail has been assumed to be made of metallic materials. It is plain that the equilibrium charging potential, or the *floating* potential, is achieved when $I_{net} = J_{net}A_S = 0$. Not so obvious are the pieces the total current density consists actually of. In order to understand such point better, let us shorty report some concepts regarding spacecraft charging.

4.2.1 Concepts of Electric Spacecraft Charging

Spacecraft charging is an area of continuing investigation because, essentially, it has a large impact on the spacecraft/satellite lifetime and operations. A compendium of the research in this important area of spaceflight can be found in a very recent book [26].

Whether a spacecraft moves in a type of or another space environment is basic to its design. Electrons, positive ions and high-energy photons all contribute to give the spacecraft an electric potential. Historical perspectives on such effects in spaceflight can be found, for instance, also in [12, 17], and [33]. Here we like to remind the reader a number of concepts for modeling environmental effects that, according to the objectives of this chapter, will tell us whether some degradation of the sail's thermo-optical quantities may affect the trajectories of fast photon sailcraft.

In general, the potential of a macroscopic object (e.g. a spacecraft) in a surrounding or environmental plasma (e.g. planetary magnetospheres or the solar wind) comes from the charge of the whole object with respect to the plasma. This is called the *absolute* charging. A priori, it is not a detrimental effect, especially if all parts the object consists of are equipotential. One can distinguish between surface charging and *internal* charging, namely the charging of those parts of the object not directly exposed to the environment, but reachable by many-keV (or more) charged particles. As a complicated object is made of different materials—typically conducting/dielectric and/or dielectric/dielectric substances—these pieces achieve different levels of charge, namely, there takes place some *differential* charging. Sunlight, eclipse, auto-shadowing, and object geometry affect the various potentials strongly. Ensuing effects can be detrimental.

Models of spacecraft charging in space environment contain several current densities. In principle, one can write the net current as follows:

$$I_{net} = \left(J_e + J_e^{(b)} + J_e^{(s)} \right) A_e + \sum_{ions} \left(J_i + J_i^{(b)} + J_i^{(s)} \right) A_i + \left(J_{phe}^{(esc)} + J_{phe}^{(ba)} \right) A_{phe}$$
$$+ J_{con} A_{con} \qquad (4.19)$$

with the following meaning of the symbols:

1. A_x is the area of the spacecraft surface collecting/emitting charges of species x;
2. J_e denotes the electron current density coming from the external plasma;
3. $J_e^{(b)}$ is the current density of electrons *backscattered* from the object surface. In general, surface backscattering is not an elastic process.

4. $J_e^{(s)}$ represents the current density of *secondary* electrons, namely, those ones produced by energetic particles (usually assumed to be electrons and/or protons generally not belonging to the environmental plasma), which hit the object;

5. J_i denotes the ion current density from the external plasma;

6. $J_i^{(b)}$ is the current density of ions *backscattered* from the object surface;

7. $J_i^{(s)}$ denotes the current density of *secondary* ions emitted as the result of collisions of the object's material with very energetic particles. In our current framework, it is quite negligible;

8. $J_{phe}^{(esc)}$ is the current density of electrons emitted via the solar-ultraviolet irradiance photoelectric effect, or the photoelectrons, which escape from the object definitively;

9. $J_{phe}^{(ba)}$ represents the current density of those photoelectrons back-attracted by the object (if positively charged) because emitted with sufficiently low energy;

10. finally, J_{con} denotes the current density produced onboard by some device of area A_{con} for controlling/mitigating charging dangerous effects.

11. Normally, spacecraft temperature (controlled either actively or passively) is somewhat low so that no appreciable thermionic effect from its surface takes place. Things might change for fast metal sails designed for close Sun flybys, e.g. as low as 0.2–0.3 AU. The electron current density by thermionic emission can be expressed by[6]

$$J_e^{(thn)} = f^{(bands)} \left[f^{(0)} (kT)^2 \exp \left(-\frac{\phi}{kT} \right) \right] f^{(efield)}$$

$$f^{(0)} \equiv \frac{4\pi m_e |q_e|}{h^3} = 1.618311 \times 10^{14} \, \mathrm{A\,m^{-2}\,eV^{-2}} \qquad (4.20)$$

$$f^{(efield)} \equiv \exp \left(\zeta \frac{\sqrt{\mathcal{E}}}{kT} \right), \qquad \zeta \equiv \frac{|q_e|^{3/2}}{(4\pi\varepsilon_0)^{1/2}} = 3.794686 \times 10^{-5} \frac{\mathrm{eV}}{(\mathrm{V/m})^{1/2}}$$

where the brackets enclose the part of $J_e^{(thn)}$ due to the temperature, ϕ denoting the work function (usually measured in eV) of the metallic solid. $f^{(bands)}$ is a factor that accounts mainly for the energy band structure of the emitting solid, whereas $\mathcal{E}$ is the electric field applied to the metal, if any, and not so high to cause tunneling emission. Thus, the overall electron emission responds to an *effective*, or *reduced*, work function given by

$$\phi_{eff} = \phi - \frac{|q_e|^{3/2}}{(4\pi\varepsilon_0)^{1/2}} \sqrt{\mathcal{E}} \qquad (4.21)$$

which therefore enhances electron emission at a given temperature or, equivalently, lessens the emission temperature for a given current density. Depending on the solar

---

[6]In literature, Eq. (4.20), and its factors, may be found under different author names. Here, we preferred to focus on the physics rather than on not so a clear history.

4.2 Ultraviolet Light onto Sail

flyby mission and the materials used for sails, $J_e^{(thn)}$ might be added to the r.h.s. of Eq. (4.19). In any case, strong electric fields are not expected around the sail, so that $f^{(efield)} = 1$ may be assumed.

Note 4.1 With regard to the sign to be assigned to each current density (for internal consistency); the right sign is determined by two factors: (1) the species charge (plainly), and (2) the in/out feature of the particles current. Here, we give positive sign to a positive charge flux coming *in* the spacecraft. Thus, a current of escaping photoelectrons or thermionic electrons will be given the positive sign, i.e. it is equivalent to a positive current entering the object.

Equation (4.19) includes the different species of ions a general plasma consists of. For instance, there are about 30 ions species in the solar wind (Sect. 3.1). Nevertheless, protons are by far the most abundant species. Modeling the above current densities is a difficult task for spacecraft charging, especially if it is composed of many insulating/conducting materials. It is under continuous investigation for its high importance in assuring spacecraft of (at least) the nominal operational lifetime.

The above relationships are not sufficient to solve the problem of determining potential(s) and current densities at the equilibrium. Other two basic equations in such problems are (1) the *Vlasov* equation, (2) the *Poisson* equation. The *collisionless Boltzmann* equation, known also as the Vlasov equation, is applied to every particle species (considered in the problem) via the respective distribution functions (assumed stationary); denoting them by $f_\alpha(\mathbf{r}, \mathbf{v}) \equiv f_\alpha$ (for short), one can write:

$$0 = \partial_t f_\alpha + \mathbf{v} \cdot \partial_\mathbf{r} f_\alpha + \frac{q_\alpha}{m_\alpha} (\mathbf{E} + \mathbf{v} \times \mathbf{B}) \cdot \partial_\mathbf{v} f_\alpha$$

$$\cong \mathbf{v} \cdot \partial_\mathbf{r} f_\alpha - \frac{q_\alpha}{m_\alpha} \nabla V \cdot \partial_\mathbf{v} f_\alpha \qquad (4.22)$$

with the convention $\partial_x \equiv \partial/\partial x$. In Eq. (4.22), the electric field $\mathbf{E}(\mathbf{r}, t)$ and the magnetic induction $\mathbf{B}(\mathbf{r}, t)$ consist of internal (averaged) pieces *and* external contributions. The former come from the long-range interactions between particles and are active even though the time rates of the distribution functions vanish. The last side of Eq. (4.22) represents a notable simplification; for instance, at 1 AU, using the electron's thermal speed and the mean value of the magnetic induction given in Sect. 4.1, the term $\mathbf{v} \times \mathbf{B}$ is less than 0.01 the electric field produced by a macroscopic planar object with 1 V potential in the solar wind.

As a charged object originates an *external* potential acting on the ambient plasma itself, the Poisson equation holds:

$$\nabla^2 V(\mathbf{r}) + \varrho_{net}(\mathbf{r})/\varepsilon_0 = 0 \qquad (4.23)$$

where ϱ_{net} denotes the net charge density. It is obvious that the mathematical system composed by Eqs. (4.19), (4.22), and (4.23) is still rather complicated; numerical methods have to be used in general. Here, we like to remind the reader of some

106   4 Solar Sails in Interplanetary Environment

concepts related to cases of particular importance. We highlight the relevance to the solar-sail sizes considered in Sect. 4.1.

■ The first one consists of the *thin-sheath* concept, which may be summarized as follows. Considering that the ambient plasma and the space object affect each other (i.e. the object perturbs the plasma properties,[7] and the plasma contributes to charge the spacecraft, and to screen its potential), if the plasma Debye length is much *lower* than the linear size of the object, then this interaction volume around the object is space-charge limited. In other words, J_{net} is approximatively proportional to $V^{3/2}$; the plasma sheath around the object may be written simply in terms of the Debye length

$$l_{sheath}/\ell_D \cong \frac{2}{3}\left(\frac{q_e V}{kT_e}\right)^{3/4} \tag{4.24}$$

Such equation holds for space probes with zero or small speed with respect to the solar wind. One should note that on the "surface" of such volume, the potential and its derivative with respect to the distance from the sail's plane vanish both; but not so does the potential's second derivative, which is just proportional to the local (space) charge density.

■ On the other extreme, namely, when the plasma Debye length is much *higher* than the linear size of the object—the *thick-sheath* approximation—current densities are not limited by space charge. For instance, for a planar surface immersed in a quasi-Maxwellian plasma and under the generic action of the ultraviolet radiation, one may write (adapted from [17]) the equilibrium current equation as

$$J_e^{(0)} e^{-q_e V/kT_e} + J_i^{(0)}(1 - q_i V/kT_i) + J_{phe}^{(sat)} = J_{net} = 0 \quad \textbf{[thick sheath]} \tag{4.25}$$

provided both secondary and backscattered currents are negligible, and no charge control is active onboard. In Eq. (4.25), superscript (0) stands for the unperturbed plasma value, whereas *(sat)* refers to as the *saturation* regime. According to our convention, the first term of the l.h.s. of Eq. (4.25) is negative, whereas the second and third ones are positive. For example, considering electrons and protons solely with $T_e = T_i$ in eclipse condition, the plasma quasi-neutrality entails that the ratio between proton and electron current densities results in $J_i/J_e = -\sqrt{m_e/m_p}$, and the solution to Eq. (4.25) is then independent of the electron current density:

$$V \cong -2.504 T_e \text{ V} \tag{4.26}$$

where T_e is expressed in eV. Difficult to find in literature is the more general solution corresponding to $J_{phe} \geqslant 0$. Using the principal branch of the transcendental equation $xe^x - y = 0$, we have plotted the current solution to Eq. (4.25) in Fig. 4.3. Solution (4.26) corresponds to $J_{phe} = 0$.

---

[7]Within plasma sheath, the plasma is no-longer quasi-neutral, strictly speaking. However, if V/l_{sheath} is low, one may go on using $n_e \cong n_i$.

4.2 Ultraviolet Light onto Sail

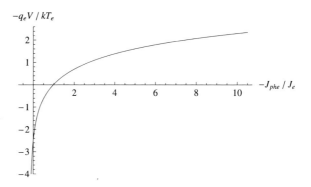

Fig. 4.3 Equilibrium potential achieved by a low-speed planar homogeneous surface in a thick-sheath ambient plasma with emission of photoelectrons. This dimensionless-parameter plot reflects the model assumptions employed here. (Note that $-q_e > 0$ so that $\text{sign}(V) = \text{sign}(-q_e V/kT_e)$ for an immediate readability)

As long as the current density ratio $-J_{phe}/J_e$ is lower than $1 - \sqrt{m_e/m_p}$, the equilibrium potential got by the spacecraft is negative, otherwise it becomes positive.

What described above is a nice approximation only if the artifact speed is much smaller than the ambient-plasma bulk speed. When this is not the case (e.g. fast sailcraft may achieve a non-negligible fraction of the solar-wind bulk speed), things are considerably more complicated because the distribution functions of the species—as observed from the spacecraft—are no longer of Maxwellian type. As shown in Sect. 4.1, sailcraft is a small object with respect to electron and proton gyro radii (except for very large sails at distances lower than 0.2 AU). Even in these cases, part of the electrons and protons of the plasma's current densities are absorbed by the sail *front side* (especially when sailcraft velocity is at large angle with the local radial direction). Behind the sailcraft, arises a *wake* depleted of particles.[8] Out of the two general species in a plasma (electrons and ions), solar-wind's electrons move (thermally) much higher than the bulk speed. This causes a partial refilling of the wake. Neglecting the low-abundance ions more massive than protons, what is the role of the protons? Consider the proton's mean thermal velocity, say, $\mathbf{W}_{thp}$ orthogonal to the solar-wind velocity $\mathbf{V}_{sw}$, both measured in the sailcraft frame; hence, in the time interval $\delta t = \sqrt{A_S}/W_{thp}$, the bulk protons will travel a distance $l_{wake} = V_{sw}\delta t = \sqrt{A_S}V_{sw}/W_{thp}$ behind the sail. This length may be taken as the approximate height of the conical-like wake. For instance, a 400 m-sail fast sailcraft with 14 AU/year crossing perpendicularly the solar wind at 0.25 AU has $l_{wake} \cong 2.3$ km. Of course, this length varies because both speeds change with time.

The picture of a positively charged sailcraft (see Sect. 4.2.2) moving in the solar wind, and endowed with a front-wind thin plasma sheath and a long back-wind wake (with *negative* potential) is confirmed by more sophisticated calculations, e.g. [18], even though more extended analyses are necessary on a mission-by-mission basis.

We end this subsection by mentioning the characteristic times of the main phenomena involved in sailcraft charging. The order of magnitude of the charging time

---

[8] Just for visualization purposes, one might picture a cone-like volume with the planar object as its base.

108   4   Solar Sails in Interplanetary Environment

related to a generic current density J from/to the surface of area A of a body with capacitance $\mathcal{C}$ may be obtained from Eq. (4.18)

$$t_{ch} \sim \mathcal{C}\Delta V/JA_S = \frac{\mathcal{C}(\mathrm{pF})}{J(\mu\mathrm{A\,m}^{-2})A_S(\mathrm{m}^2)} \quad \Delta V\,(\mathrm{V})\,\mu\mathrm{s} \qquad (4.27\mathrm{a})$$

$$t_{ch} \sim \Delta V \varepsilon_r \varepsilon_0/J\delta \qquad\qquad\qquad \text{capacitor with dielectric film} \quad (4.27\mathrm{b})$$

$$t_{ch} \sim 3.6\varepsilon_0 \Delta V/J\sqrt{A_S} \qquad\qquad\qquad \text{large planar sail} \qquad\qquad (4.27\mathrm{c})$$

where ΔV is the potential change caused by the current J. The reader should note the units on the rightmost side of the first equation: for conductors that will achieve a potential of a few tens of Volts, modest current densities cause charging times of the order of few microseconds. In addition, we made explicit two cases: (1) one regarding a simple capacitor having a dielectric film with relative permittivity ε_r and thickness δ, and (2) the charging time of a square sail.

Case-1 regards (approximately) those satellite charging effects that happen because conductor surfaces are separated by some dielectric thin film(s); the second of (4.27a)–(4.27c) tells us that the thinner are such dielectric films the longer are the charging times. Real situations occurred with charging times of minutes or hours up to kilovolts until a destructive arc took place.

Case-2 regards a (bare) squared solar sail; characteristic charging time is given in the third line of Eqs. (4.27a)–(4.27c). It is formally similar to the second equation; however, now the characteristic time may be six orders of magnitude lower (at least), as in the case for large fast sailcraft. Thus, sail charging may be considered an instantaneous process, in practice.

4.2.2 Photoelectrons from a Solar Sail

The photoelectric effect is not a simple thing in its generality, as perhaps one might think of. Here, we will adopt a simple model for taking the main phenomena, occurring in a metallic sail, into account. The purpose of this subsection is to show that the term J_{phe} in Eq. (4.19) is quite important for sailcraft in the interplanetary space.

In what follows, the photon relationship $\lambda[\mathrm{nm}]E[\mathrm{eV}] = hc = 1239.84191$, where λ is the photon wavelength and E denotes its energy, will be useful. In a way equivalent to the practical spectroscopy, one can express energy in terms of length reciprocal, i.e. (nm^{-1}) in our framework. In addition, analogously to the radiometric concept of irradiance, we may define the number or photon spectral irradiance, say, $\mathrm{I}_\lambda^{(ph)}$ as the number of photons, in the wavelength interval $[\lambda, \lambda + d\lambda]$, incident on or crossing a surface of unit area per unit time. Such photons can come from any direction in the semispace containing the positive normal $\mathbf{n}$ to the surface; in particular, they could arrive from one direction $\mathbf{d}$ with incidence $\cos\theta = \mathbf{n} \cdot \mathbf{d}$. The photon broadband irradiance is then the integral of the photon spectral irradiance over $[\lambda_1, \lambda_2]$ with $\lambda_2 > \lambda_1$.

4.2 Ultraviolet Light onto Sail 109

Let us go back to Fig. 2.9 on p. 57. As we noted in Sect. 2.2.2, solar radiation below 380 nm is far from a blackbody spectrum. In particular, in the range 0.004–2.5 keV, SSI exhibits a non-negligible mean line almost parallel to the wavelength-axis. This has consequences even in the present context. As a point of fact, the UV spectrum onto a metallic sail has to be related to the *work function* (precisely defined, e.g., in [1, 13, 21, 30, 40]) values of metals, according to [9, 10]. Apart from proposed sails made of carbon nanotubes, the metals candidate for solar sails of first/second generation have work functions between 4 and 5 eV. Recalling the meaning of work function, this means simply that solar photons with wavelength lower than $1240 \text{ nm} \times \text{eV}/4 \text{ eV} = 310$ nm have a non-zero probability to induce photoelectron emission from the sail. This photoemission regards the valence electrons that, in general, fill some energy band of the metal partially.

A very short reminder is in order. There are many theories that try to explain the various properties that a solid-state structure exhibits. Some approaches have managed to explain some properties very good, but practically have failed to describe others. Generally speaking, (1) the number of valence electrons, say, N_v in a *primitive* (or a minimum-volume) cell[9] of a crystal, (2) the energy band structure, and (3) band overlapping/non-overlapping (here denoted by $\mathcal{O}_b = 1$ or 0, respectively) are key items for distinguishing metal, insulators, semimetals and semiconductors. Normally, for non-overlapped bands, the conduction band is higher than the valence band.

If a crystal is an insulator, then N_v is *even*. But the vice versa is not true: noble and alkali metals exhibit $N_v = 1$ and $\mathcal{O}_b = 0$, and alkaline earth metals (e.g. Be, Mg) have $N_v = 2$ with $\mathcal{O}_b = 1$; as $T \rightarrow 0$, the electrons fill *two* overlapped high-energy bands *partially*; more precisely, one band (the valence band) is most occupied, while the other one (the conduction band) is *almost* empty. Such elements are more appropriately called semimetals or poor-metals: qualitatively, they are not pretty good metallic substances.

For semiconductors, the energy separation between the bottom of an energy band and the top of the just-below band is a very important quantity, shortly named the *band-gap*. In general, an energy gap is a range of energy forbidden to electronic states in solids, namely, no electron exists inside such bands. Band-gap values are small for metals.

With regard to the metals for first-generation sails, Aluminium is normally a face-centered cubic (fcc) solid lattice with $N_v = 3$, and 12 nearest-neighbor (touching) atoms. Thus, one band is certainly full occupied, whereas at least one upper band can be partially filled giving the Aluminium the net character of a metal. The lattice of Chromium, a variable-valence transition metal, exhibits a body-centered cubic (bcc) structure. Its importance in multi-layer sails stems mainly from its high emittance in order to keep sail temperature sufficiently low. Its high density (7.14 g cm^{-3}) constraints us to design only 10–30 nm of it as the sail backside (emissive) layer.

As the incidence photon energy increases, also the core electrons may be involved in photoemission; at the same time, as the sail thickness is very small, the probability of high-energy photon transmission without appreciable energy loss will rise. Let us make such considerations more quantitative. Assuming that the solar

---

[9]Given a lattice, choosing a primitive cell is not unique; however, the number of atoms in any primitive cell is an invariant for the given crystal structure [21]. A primitive cell is a particular type of unit cell.

photons irradiate the sail at incidence θ, and denoting the photoelectron emission polar angle by ϑ, the total number of photoelectrons, per unit time and area, at a distance R from the Sun may be written as

$$J_{phe} = -q_e \int_0^\pi 2\pi \sin\vartheta \, d\vartheta \int_{\lambda_a}^{hc/\Phi} I_\lambda^{(ph)}(R) \cos\theta (1 - \mathcal{R}_\lambda - \mathcal{T}_\lambda) \mathcal{Y}_\lambda(\theta, \vartheta) \, d\lambda$$

(4.28)

where: (1) we have assumed a photoemission symmetric in azimuth, (2) Φ denotes the metal's work function, (3) $\mathcal{R}_\lambda$ and $\mathcal{T}_\lambda$ are the layer's directional-hemispherical spectral reflectance and transmittance respectively (see Sect. 6.4.2), (4) λ_a is the wavelength value below which photon absorption is negligible, in practice, for the reflective layer, and (5) $\mathcal{Y}_\lambda(\theta, \vartheta)$ denotes the bidirectional photoelectron yield at λ, defined as the number of (core and/or valence) electrons, which are brought sufficiently distant from the metal surface, per *absorbed* photon of wavelength λ. Thus, from the viewpoint of the photoelectric effect, no distinction is made between escaping and back-attracted electrons, a phenomenon that pertains to multiple electric-currents competition.

Equation (4.28) is rather complicated to make it explicit someway, in general. However, some notable items regarding solar sailing can be inferred. The photon path inside the sail layer increases with $\cos\theta$

$$\mathcal{Y}_\lambda(\theta, \vartheta) \cos\theta = \mathcal{Y}_\lambda(0, \vartheta) = N_S \delta_S \, d\sigma_{phe}(\vartheta)/d\Omega$$

(4.29)

where the sail's material number density and thickness have been denoted by N_S and δ_S, respectively. $d\Omega$ is the solid angle about ϑ where the photoemission differential cross section $d\sigma_{phe}(\vartheta)/d\Omega$ is considered.[10] What needs attention is that—out of the many phenomena a sufficiently energetic photon can cause inside a solid structure—here one is mainly interested in photoemission (of core and valence electrons) because of its strong contribution to the achievement of the sail equilibrium potential. To such goal, it is also apparent that what matters is the *total* photoemission cross section, say, σ_{phe}. Because the bidirectional photoelectron yield is the only factor dependent on ϑ in the second integral of Eq. (4.28), we can re-write it as follows

$$J_{phe} = -q_e N_S \delta_S \int_{\lambda_a}^{hc/\Phi} I_\lambda^{(ph)}(R)(1 - \mathcal{R}_\lambda - \mathcal{T}_\lambda)\sigma_{phe}(\lambda) \, d\lambda$$

$$\sigma_{phe}(\lambda) \equiv \int_0^\pi \frac{d\sigma_{phe}(\vartheta)}{d\Omega} 2\pi \sin\vartheta \, d\vartheta$$

(4.30)

In solid-state physics, a basic point consists of assuming that the orbitals of the innermost electrons are practically unchanged by the lattice: such electrons "sense" the related nucleus as if the atom were free. In contrast, valence electrons live

---

[10]This cross section includes more-than-one electron emission provided the photon energy is sufficiently high.

4.2 Ultraviolet Light onto Sail

in a periodic potential and undergo electron-electron interactions. Therefore, the contribution of the core electrons to the total photoemission cross section may be taken from the well-developed theory of the photoelectric effect for a free atom. In the current framework, one may consider only non-gamma photons, namely, $E^{(K)} < hc/\lambda \ll m_e c^2$, where $E^{(K)}$ is the ionization energy of the K-shell. More precisely, one should distinguish whether the photon has energy slightly higher or much higher than $E^{(K)}$. The related expressions of the (integrated) cross section $\sigma_{phe}^{(K)}$ can be found in classical texts of atomic physics and astrophysics (e.g. [27]). The cross section of the K-shell can be expressed by[11]

$$\frac{\sigma_{phe}^{(K)}}{\sigma_T} = \frac{128\pi}{\alpha^3 Z^2} \kappa(\lambda) \left(\frac{E^{(K)}\lambda}{hc} \right)^4, \quad E^{(K)} \lesssim hc/\lambda \ll m_e c^2$$

$$\kappa(\lambda) \equiv \frac{e^{-4\varsigma_\lambda \operatorname{arccot} \varsigma_\lambda}}{1 - e^{-2\pi\varsigma_\lambda}}, \quad \varsigma_\lambda \equiv 1/\sqrt{(hc/\lambda)/E^{(K)} - 1}$$

(4.31)

where σ_T denotes the Thomson scattering cross section, α is the fine-structure constant, and Z is the target's atomic number. Inserting the physical constants and expressing the Thomson cross section in barns (1 barn $\equiv 10^{-28}$ m^2), Eq. (4.31) may be recast as follows

$$\sigma_{phe}^{(K)} = 291.327 \times 10^{-6} \kappa(\lambda) \frac{(E^{(K)}\lambda)^4}{Z^2} \text{ barn}$$

$$E^{(K)} \text{ (eV)}, \quad \lambda \text{ (nm)} \lesssim 1239.84/E^{(K)}$$

(4.32)

According to Ref. [27], all other orbitals contribute about 25 percent to the cross section so that $\sigma_{phe} \sim 1.25\sigma_{phe}^{(K)}$ provided that the photon wavelength satisfies the inequality in (4.32). We will take such approximation as the total cross section for photoemission from a solar sail in the XUV range. For instance, an Aluminium atom requires about $(2.3 + 2.1)$ keV for being stripped of both its K-electrons. The L-shell (with 8 electrons) is completely filled; the most energetic electron (*mee*) of this shell has a binding energy of about 120 eV, corresponding to a photon of wavelength $\lambda_{mee} = 10.3$ nm. In this case, one may assume K and L shells exhibiting $1.25\sigma_{phe}^{(K)}$ as their overall photoelectric cross section. Calculations of photo-ionization cross sections can be also found in classical monographs, e.g. [48], and, more recently, can be got online via [11].

Instead, the emitted photoelectrons will have different *initial* kinetic energies according to

$$\frac{1}{2} m_e v_{phe}^2(0) = \begin{cases} hc/\lambda - E^{(band)}, & \min(E^{(band)}) = \Phi \\ hc/\lambda - E^{(n)} \end{cases}$$

(4.33a)

---

[11]The reader, who searches for the photoelectric effect over specialized literature, will also read a different dependence on both Z and λ, specifically a cross section proportional to Z^5 and $\lambda^{7/2}$. Looking carefully, one should note that such function is reported when $E^{(K)} \ll hc/\lambda \ll m_e c^2$, e.g. in phenomena involving hard X-rays.

$$\frac{1}{2}m_e v_{phe}^2(\infty) = q_e V(0) + \frac{1}{2}m_e v_{phe}^2(0) \tag{4.33b}$$

as observed in the sailcraft frame of reference. $E^{(band)}$ is the energy of an electron in the highest occupied energy band of the lattice, whereas $E^{(n)}$ is the atom's nth-stage potential of ionization; even though measured both from the zero-energy level, usually they are taken positive and expressed in eV. Just outside the metal sail, the electrons are immersed in the sail's potential $V(\mathbf{r})$ that comes from the Poisson equation (4.23). If $V(0)$ is positive, then zeroing the l.h.s. of Eq. (4.33b) results in the threshold of electron back-capturing.

Let us now make the photoelectron current density expression easier to calculate by using the SSI means of cycle-23 in the 0.5–310 nm range shown in Fig. 2.9. The relevance of utilizing the SSI's daily means has been pointed out in Chap. 2, also because of the notable difficulty of SSI/TSI prediction. Employing SSI, or $I_\lambda(R)$, in SI units is appropriate as soon as one checks the following equality: $-q_e I_\lambda^{(ph)}\,d\lambda = I_\lambda\,d\lambda\cos\theta \times (1$ ampere 1 second/1 joule), which holds for any R. Therefore, we re-write Eq. (4.28) as follows

$$J_{phe}^{(esc)} \cong J_{phe} = J_{phe}(1)R^{-2}\cos\theta$$

$$J_{phe}(1) \sim N_A \frac{\rho_S \delta_S}{\mathcal{A}_S}$$

$$\times \left(1.25\sigma_{phe}^{(K)} \int_{0.5}^{\lambda_{mee}} (1 - \mathcal{R}_\lambda)I_\lambda(1)\,d\lambda + \int_{\lambda_{mee}}^{310} (1 - \mathcal{R}_\lambda)I_\lambda(1)\sigma_{phe}^{(val)}\,d\lambda\right)$$

$$\tag{4.34}$$

where N_A is the Avogadro number, ρ_S denotes the sail material (mass) density (kg/m^3), and $\mathcal{A}_S$ is its atomic weight (kg/mol).[12] Even though $1 - \mathcal{R}_\lambda$ is a function increasing as wavelength decreases, it may lessen the value of the second integral considerably. $\mathcal{R}_\lambda$ is also a function of the incidence angle; however, in any case, the photoemission decreases at sunlight's large incidence values. Finally, $\sigma_{phe}^{(val)}$ denotes the photoemission cross section related to the valence electrons in the solid structure. Now, one should note that many *concurrent* processes, taking place in a lattice, may release electrons; in other words, *not only* the direct excitation of valence electron contributes to photoemission. To discuss these various phenomena is quite beyond the aims of this chapter. Thus, at this point, a good way to proceed is to use cross section tables (analogously to SSI and absorption) in order to practically compute the second integral in Eq. (4.34).

Remark 4.1 All sailing-oriented considerations done so far reflect—partially—the guidelines of the spectroscopy of the photoemission process, which normally is modeled via a three-step model: (1) photon-absorption/atom-ionization, (2) electron motion *in* the lattice, and (3) electron escape *from* the solid. Quantum formalism is

---

[12] $\rho_S \delta_S$ represents the *sail loading* of the bare reflective layer.

4.2 Ultraviolet Light onto Sail 113

necessary for dealing with the whole process. Utilizing cross sections (and the other quantities previously defined) has allowed us to apply those principles to the current problem. As we shall show below, the low magnitude of the sail's floating potential will allow the analyst to consider only a limited number of known phenomena in order to compute the integrals in J_{phe}.

4.2.3 Evaluation of the Sail's Floating Potential

By modifying Eq. (4.25) a bit, we may set down an analogous equation holding for a planar moving sail with thin-sheath and long wake. First of all, though solar-wind protons have a small thermal speed, however its bulk speed is at least one order of magnitude higher. Second, a fast sailcraft may really have a non-negligible fraction of this speed value in the heliocentric inertial frame. Consequently, in the sailcraft frame of reference, protons will exhibit a somewhat different vector velocity, denoted here by $\mathbb{V}_p \equiv \upsilon_p \mathfrak{p}$ with direction $\mathfrak{p} \neq \mathbf{R}/R$; the proton's current density is the third significant piece entering Eq. (4.19). Such current density is insensitive to *low* potentials (of any sign) a sail may have. In contrast, as we saw above, the electron's thermal speed is rather high with respect to sailcraft and bulk protons. Therefore, we may keep only the ambient-plasma electron shielding factor, say, $\mathfrak{s}_e$ in the balance equation (for the moment) restricted to a three-current model:

$$\mathfrak{s}_e J_e^{(0)} e^{-q_e V/kT_e} + J_p + J_{phe}^{(esc)} = J_{net} = 0 \quad \textbf{[thin sheath]}$$

$$J_e^{(0)} = q_e n_e \upsilon_e, \qquad J_p = q_p n_p \upsilon_p \mathfrak{p} \cdot \mathbf{n}_s, \qquad 0 \leqslant \mathfrak{s}_e \leqslant 1 \tag{4.35}$$

where, again, the superscript (0) denotes an unperturbed quantity. J_{phe} is given by Eq. (4.34); this one has been already put as function of the Sun-sailcraft distance R. Also, in Sect. 4.2.1, we have inferred that sail charging may be regarded as an instantaneous process for all our aims; we can therefore utilize some equations of Sect. 4.1 for making the other terms explicitly dependent on R. Note that, as $\mathbb{V}_p$ is not omnidirectional (like the thermal velocity) and the current density is a vector, J_p depends on the sail orientation, here denoted by the unit vector $\mathbf{n}_s$. Thus, sail orientation affects two current densities out of three, just those ones of the same sign. Note that, unless $\cos\theta = 0$ *and* $\mathfrak{p} \cdot \mathbf{n}_s = 0$ identically during the same time interval,[13] at least one positive current balances the electrons thermal current. The following functions, mean values, and most-probable-values are to be inserted into Eq. (4.35):

$$n_e(R) = n_e(1)R^{-2} \qquad n_p(R) = n_p(1)R^{-2}$$

$$n_p(1) = n_e(1) \qquad \cos\theta = R^{-1}\mathbf{R} \cdot \mathbf{n}_s \equiv \mathbf{r} \cdot \mathbf{n}_s \tag{4.36}$$

$$T_e(R) = T_e(1)R^{-0.8} \qquad \upsilon_e(R) = \sqrt{2kT_e(R)/m_e} \cong 1826R^{-0.4} \text{ km/s}$$

---

[13]This would happen for a sail 90° tilted to sunlight and moving perfectly in solar radial direction for a finite time, an unrealistic situation indeed.

where, again, $f(1)$ means the f-value at 1 AU. Using all values of Eqs. (4.36), the second line of Eq. (4.35) and the most probable speed for the solar-wind electrons, and re-arranging the equilibrium equation results in

$$V = T_e(1)\big(\ln D(1) + 0.4\ln R\big)R^{-0.8} \text{ (V)}$$

$$D(1) \equiv \frac{q_p n_p(1)v_p \mathfrak{p} \cdot \mathbf{n}_s + J_{phe}(1)\cos\theta}{-\mathfrak{s}_e q_e n_e(1)v_e(1)} \tag{4.37}$$

where $T_e(1)$ is expressed in eV, and $J_{phe}(1)$ has been defined in Eq. (4.34). The solution to the sail voltage, as given by Eq. (4.37), depends on the distance R and the two cosines $\cos\theta$ and $\mathfrak{p} \cdot \mathbf{n}_s \equiv \cos\zeta$. Especially when the sailcraft is inside fast streams (discussed Sect. 3.1.1), it is easy to see that $\zeta \cong \theta$. Thus, we will consider the function $V(R, \theta, \zeta = \theta) \equiv V(R, \theta)$ as expressing the sail's floating voltage. Note that Eq. (4.37) is in a form easily generalizable by adding other ion species present in the solar wind (as we saw in Chap. 3), e.g. Helium ions. This may done easily because almost all Helium ions are doubly charged [28] (i.e. α-particles), thus we can replace the previous $D(1)$ by this one

$$D(1) = \frac{q_p n_p(1)v_p + q_\alpha n_\alpha(1)v_\alpha + J_{phe}(1)}{-\mathfrak{s}_e q_e n_e(1)v_e(1)}\cos\theta$$

$$v_\alpha = v_p, \qquad n_\alpha(1) \cong 0.032 n_p(1), \qquad q_\alpha = 2 \tag{4.38}$$

where the Helium's relative abundance averaged over cycle-23, has been carried out from the HMF and plasma data obtained via NASA's OMNIWeb interface. The contours of the function $V(R, \theta)$ have been diagrammed in Fig. 4.4. The central part of this plot refers to the above-mentioned averages related to cycle-23 for both solar-wind and SSI. A mean value of 500 km/s has been considered for the relative speed between solar-wind ions and fast sailcraft. The range of Sun-sailcraft distance has been set to 0.1–5 AU, and the sail orientation pitch angle was varied between 0° and 50°.[14] The screening value for a thin-sheath long-wake moving sail has been taken equal to 1/2. The first, perhaps partially unexpected, result from looking at this first plot in Fig. 4.4 consists of *low and positive* values of the floating potential.[15] If one fixes the incidence value and follows the related horizontal line (from higher to lower R-values), one can see the voltage rising slowly, achieving a maximum and then decaying quickly; this maximum decreases at higher incidence angles. However, we know that solar wind fluctuates significantly (Chap. 3, and Sect. 3.4, in particular). Thus, what may happens in days where the mean conditions are rather different from the above average? We considered the mean values of the solar wind on two days: (1) August 24, 2009,

---

[14] At higher incidence angles, a number of phenomena are expected such that the simple cosine-law, assumed in certain steps of the here-developed sail-voltage model, is no longer valid.

[15] Even adding backscattered and secondary particle currents (Sect. 4.2.1) to the model, this feature does not change qualitatively.

4.2 Ultraviolet Light onto Sail

namely, $(n_e(1) \cong 2.8 \text{ cm}^{-3},\ T_e(1) \cong 0.60 \times 10^5 \text{ K}, v_p = 400 \text{ km/s})$ including 45 km/s of sailcraft speed, and (2) April 1st, 2001 with $(n_e(1) \cong 16 \text{ cm}^{-3}, T_e(1) \cong 2.5 \times 10^5 \text{ K}, v_p = 650 \text{ km/s})$ including 28 km/s of sailcraft speed. The results related to such days are shown in the top and bottom parts of Fig. 4.4, respectively. The first example show a systematic increase of the positive sail voltage as the sailcraft draws closer to Sun. This can be explained by the fact that the electron temperature is not sufficiently high even at $R = 0.1$ for balancing the photoemission current, essentially. In contrast, the V-contours in the second example show opposite curvature and *negative* potential values below about 1 AU; in this case, the electron temperature was considerably high so the electron current prevails not only over the photoemission, but also over a notable increase of the ion bulk speed. Beyond the Earth orbit, the potential returns to positive, but just a few volts.

One could easily apply the previous model of sail charging in whatever day, once the solar wind quantities of interest are measured; thus, plots similar to Fig. 4.4 can be obtained. However, figures of such type are "static" in the sense that they do not take the propagation of particles into account because the scaling equations of Sect. 4.1 are not time dependent. For example, if we measure the wind's proton density at 1 AU, with a bulk speed of 600 km/s around the noon of a certain day $yyyy/ddd/12$, then the density at 2 AU is *not* $n_p(1)/4$ at $yyyy/ddd/12$, but it *will* be so after about 69 hours, or about 3 days. On the other side, 69 hours after, at $yyyy/ddd + 3/09$, density and bulk speed at 1 AU have already changed. Thus, comparing real voltages at different distances from the Sun is a bit more complicated, though the behavior of the floating potential is qualitatively the same.

What are the main features we have learnt hitherto? As the sailcraft moves in the interplanetary space, distance could change and sail orientation could vary, but above all, solar wind rapidly changes especially if the sailcraft moves quickly through heliographic latitudes. Depending on the solar-cycle phase, the sailcraft may experience various particle environments. Thus, along its trajectory the sail's floating potential can vary significantly, and things are not deterministic as well. Nevertheless, for a solar photon sail, which does not rely on the solar wind for getting thrust, these fluctuations appear as perturbation to the sail charging.

The magnitude of these phenomena amounts to a few tens of Volts (at most) with respect to the ambient plasma.

What about the sail supporting structure? In general, one may think of a polyester-type (e.g. Mylar) or polyimide-based (e.g. Kapton, CP1, CP2) support for the very thin reflective-layer, and the sail deployment/sustaining structure (Sect. 3.2). With regard to the bare sail, as mentioned in previous chapters, first-generation sails surely will have a support layer. In principle, one should have two options: (i) sail = reflective-layer + support-layer, (ii) sail = reflective-layer + support-layer + emissive-layer. In the case-i, the arcing problem may arise if the (shadowed) support is not a conductor (like the plastic materials just mentioned), because its negative potential may be sufficiently higher in magnitude than the front-side layer potential. In contrast, the emissive layer of the option-ii should be a metal (e.g. Chromium). If the reflective and emissive layers were not joined electrically, then an interplanetary sail would behave like a capacitor; to connect the two metal

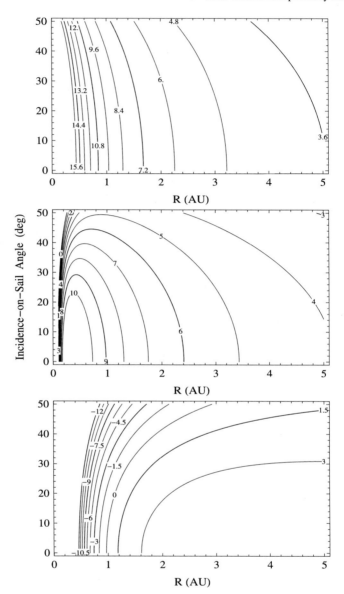

Fig. 4.4 Contours of Al-layer potential *vs* R and TSI/ion incidence angle onto the sail. Values of potential come from Eq. (4.37) with $D(1)$ given by (4.38); *top*: 2009.08.24 *low* electron flux and temperature data, *center*: solar-wind data *averaged* over cycle-23, *bottom*: 2001.04.01 *high* electron flux and temperature data

layers electrically should not be an issue. In so doing, an equipotential surface could be build. Since the wake is mostly depleted of plasma particles, one expects that the sail potential function should be close to that expressed by Eqs. (4.37) and (4.38).

4.3 Solar-Wind Ions Inside the Sail

With regard to the deployment/sustaining sub-system, the main structure of which is the boom,[16] various materials have been utilized by different building companies. For instance, longerons and diagonals can be made of aromatic-rich polyurethane resins, battens may be built via graphite composites; memory polymeric resins and fiberglass have been also selected, just to cite a few. Booms are designed to be resistant to space ionizing radiations. In our framework, the important points are that boom materials could be *non-conductive*, and the surfaces of the boom's elements have one dimension much, much smaller than the characteristic lengths of the solar-wind, which have been plotted in Fig. 4.1; in particular, Debye length is at least a factor 3,000 greater than the diameters of the tubular-like elements in the boom structure. The rod-like elements sense the plasma charge fluctuations while TSI impinges on their front side only. Therefore, the electric and optical properties of the boom's materials, and the detailed geometry of the boom elements are necessary input to the problem that, therefore, is somewhat difficult to deal with in general. Such problem is expected to be highly mission-dependent and should be analyzed on a case-by-case basis.

4.3 Solar-Wind Ions Inside the Sail

In the current framework, one of the notable results from the previous sections is that the equilibrium potential—varying with the Sun-sailcraft distance—is small with respect to the solar-wind proton energies: typically, a few tens of volts (or less) *vs.* few thousands of electron-volts.

We know from Sect. 3.1.1 that there are about thirty ion species in the solar wind; thus, these energy levels are the lower bound of the full particle spectrum. However, because protons are dominant in number, we will first focus on them. The same procedure could be applied to the Helium nuclei (few percent in abundance with respect to protons), but with a bulk energy four times the protons' and a double charge. The theory of passage of ions through solid matter is well developed, and many versatile computer codes there exist today. In particular, we have used the theory summarized in [46] and developed in the many references inside, and its related code named SRIM [47]. We assume that the sail layers are sufficiently homogeneous in the two large dimensions, so that the "amorphous-passage" analysis can be used. Although (as already mentioned) a sail endowed with even few microns of plastic support could not be used for fast sailing, however—as an example—we will show some properties relevant to Al-Kapton-Cr sails. Actually, considered the low energy of the proton/α beams, only the first two layers experience some effects. We have considered the cases of fast-stream protons and α-particles of 4 keV and 16 keV,

---

[16]The truss known as the boom—for spaceflight applications—regards a very-light deployable rigidizable structure, usually periodic in construction. Every "period" consists of some main elements such as the longerons, diagonals, battens, and joints. Each element plays a different role in terms of withstanding/passing loads.

respectively, or about 875 km/s, but apt for a conservative analysis of ion-atom interaction in the lattices of Al-Kapton, where *ion* means either proton or α-particle, and *atom* is referred to as either an atom of Aluminium or one of the Kapton atoms (i.e. Hydrogen, Carbon, Nitrogen, Oxygen).

Physics of damage in a solid target via sufficiently energetic ions is very important from many viewpoints, theoretical and applied (including several medical disciplines). Let us recall few related concepts, very shortly, in the context of the so-called *recoil cascades*. A comprehensive treatment of ion-solid interactions can be found in [34], whereas sputtering via particle bombardment is extensively dealt with in [3].

- displacement (threshold) energy (ε_d): the minimum amount of kinetic energy to be given to a lattice atom in order to be displaced sufficiently far from its lattice site. In so doing, two defects are produced: one vacancy and one interstitial atom.
- a vacancy/interstitial-atom pair is not a stable system, in general; the thermal energy of a lattice may be sufficient to restore the original lattice structure.
- lattice-site binding energy (ε_l): the minimum amount of kinetic energy for removing an atom from its lattice site. $\varepsilon_l < \varepsilon_d$ as displaced atoms have to have additional energy to move far across the lattice, and either collide with another atom or come eventually to a halt.
- surface binding energy (ε_s): the minimum amount of energy for removing a surface atom. The surface of a solid is a *complicated* system (e.g. see [29] for an introduction); this type of binding energy is evaluated via the enthalpy of sublimation. For a chemical compound, one is used to interpolate appropriately among the enthalpy values related to the constituent elements.
- when a target atom is displaced off its site, it may or may not have sufficient energy to induce other displacements; on the other hand, the projectile (that knocked it) may or may not have a residual energy for leaving the lattice site.
- among these cases, two of them initiate or continue the recoil cascade.
- if the atom-projectile has its energy $\varepsilon < \varepsilon_d$ and is of the same target element, then a replacement collision takes place, namely, this atom substitutes the original atom of the vacancy. The number of displacements equals the number of vacancies *plus* the number of replacement collisions.
- depending on the projectile energy, the recoil cascade can be highly branched: one or more of the cascade branches may reverse its/their motion and some atom may leave the lattice back through the ion beam entrance surface. On the other side, other atoms may leave the lattice through the rear surface.

4.3 Solar-Wind Ions Inside the Sail 119

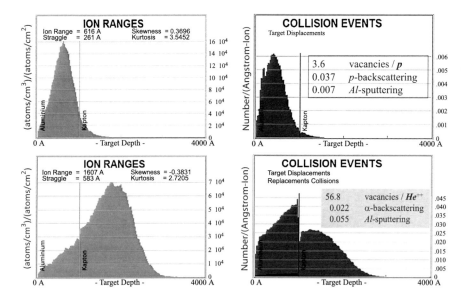

Fig. 4.5 Sail damage by fast-stream protons (*top plots*) and α-particles (*bottom plots*). Conservatively, proton energy has been taken on 4 keV, whereas α-particles have 16 keV. The direction of their momenta is orthogonal to the planar sail. (See text for the explanation of the plots)

- if an atom, sufficiently near the solid surface and at a distance r from it, gets sufficient energy and the direction of its post-collision momentum forms an acute angle, say, ψ with the outward surface normal, then the atom can be *sputtered*, and its residual energy amounts to $\varepsilon - ((d\varepsilon/dx)r/\cos\psi + \varepsilon_s)$. The electron energy loss term can be omitted if the atom belongs to the surface's upper surface.
- the sputtering yield is the mean number of sputtered atoms per bombardment ion. If the target is a chemical compound, then there are as many yields as the compound elements.

In general, surface roughness affects ε_s, and thus the sputtering yield. Roughness may change as a result of the specific ion bombardment, and also via other modifications of the target surface composition.

The following sail has been taken as a reference: Al(100 nm), Kapton(2 μm), Cr(10 nm). Figure 4.5 shows the ion ranges, target displacement and replacement collisions. We are also interested in atom sputtering and ion backscattering. If the former were sufficiently high, then it may alter the sail's optical properties; if the latter were high, then it should be included in the charging analysis of a real sail design. In the current analysis, the many species of heavy ions in the solar wind have not been taken into account: their abundance is much less than that of He-nuclei.

120 4 Solar Sails in Interplanetary Environment

The top part of Fig. 4.5 regards the effects of the protons impinging on the Al-layer, whereas the bottom part refers to as the α-particles. The left sub-figures show the proton/He range distributions as function of the target depth; the right sub-figures regard the displacement and replacement collisions. Depth is expressed in Å; for clearness, only the first 4,000 Å have been shown because none of the 40,000 protons and 40,000 α-particles, used in the Monte Carlo simulations, produced events beyond 300 nm of Kapton. The ordinate units in the ion range distribution $\mathcal{G}(z)$ are (atoms cm^{-3}/atoms cm^{-2}); this means that, if one has a certain proton or α (number) fluence onto the sail, then multiplying the ordinate value (at the considered depth) by such fluence results directly in the number density of ions stopped at that depth.

In the case for protons, $\int_0^{100 \text{ nm}} \mathcal{G}(z) \, dz = 0.897$ (i.e. 89.7 percent of the incident protons stops inside the Al-layer). Of the remaining fraction (0.103), 0.066 penetrates into the Kapton (and crosses 110 nm), whereas the proton backscattering yield amounts to 0.037. There are few displacement events, as expected, and negligible replacement collisions ($\sim$25 times lower, not shown in the figure). On average, 3.6 vacancies per incident proton occur, while Aluminium sputtering yield is 0.007.

With regard to the α-particle impact, about 0.157 halts inside the Al-layer, whereas 0.821 stops in Kapton, with a maximum range around 260 nm from the Al-Kapton interface. The remaining fraction $1 - 0.157 - 0.821 = 0.022$ of the incident α-particles is backscattered (i.e. the replacement collisions are very few). The α-penetration (with 56.8 vacancies per ion) into Kapton is clearly visible from the bottom-left sub-plot. Displacement events are almost equally distributed in Aluminium and Kapton (see the bottom-right subplot). Finally, the Aluminum sputtering yield amounts to 0.055.

In the first part of this chapter, we have shown some criteria for inferring which solar-wind flow properties can be observed from a solar-photon sail. Then, using these features as well, we set up a model for quick computing the floating potential of an interplanetary sail. Subsequently, we have employed theory of damage for evaluating quantitatively the effects induced by the wind's protons and α-particles penetrating the sail layers. Though the abundance of heavier ions is considerably lower, however the analysis could be extended to them by the same code. A key concept should be remarked at this point: the four-current model of sail charging may be embedded in the trajectory design process, whereas the Monte-Carlo code SRIM has been employed for off-line determination of sail layer damages. Of course, the two aspects belong both to the overall mission analysis.

References

1. Ashcroft, N. W., Mermin, N. D. (1976), Solid State Physics. New York: Harcourt College Publishers. ISBN 0-03-083993-9.
2. Balogh, A., Lanzerotti, L. J., Suess, S. T. (2008), The Heliosphere Through the Solar Activity Cycle. Berlin: Springer-Praxis. ISBN 978-3-540-74301-9.
3. Behrisch, R., Eckstein, W. (2007), Sputtering by Particle Bombardment. Berlin: Springer. ISBN 987-3-540-44500-5.

References

4. Bellan, P. M. (2008), Fundamentals of Plasma Physics. Cambridge: Cambridge University Press. ISBN 978-0-521-52800-9.
5. Besse, A. L., Rubin, A. G. (1980), A simple analysis of spacecraft charging involving blocked photo-electron currents. Journal of Geophysical Research, 85(A5), 2324–2328.
6. Callister, W. D. (2007), Materials Science and Engineering—An Introduction. New York: Wiley. ISBN-10: 0-471-73696-1, ISBN-13: 978-0-471-73696-7.
7. Genesis Web Science Document G: Solar Wind Properties (2002), California Institute of Technology, http://www.gps.caltech.edu/genesis/DocumentG.html.
8. Cranmer, S. R., Matthaeus, W. H., Breech, B. A., Kasper, J. C. (2009), Empirical constraints on proton and electron heating in the fast solar wind. The Astrophysical Journal, 702, 1604–1614. doi:10.1088/0004-637X/702/2/1604.
9. Lide, D. R. (Ed.) (2009), CRC Handbook of Chemistry and Physics. Boca Raton: CRC Press.
10. Version 2009 of the Internet Edition of the CRC Handbook of Chemistry and Physics.
11. Calculation of Atomic Photoionization Cross Sections (2009), http://ulisse.elettra.trieste.it/services/elements/WebElements.html.
12. Hastings, D., Garrett, H. B. (1996), Spacecraft-Environment Interactions. Cambridge: Cambridge University Press (first paperback edition, 2004, ISBN 0-521-60756-6).
13. Hofmann, P. (2008), Solid State Physics: An Introduction. New York: Wiley-VCH. ISBN 978-3-527-40861-0.
14. Horedt, G. P. (2004), Polytropes—Applications in Astrophysics and Related Fields. Dordrecht: Kluwer Academic. ISBN 1-4020-2350-2.
15. Fitzpatrick, R. (2011), Plasma Physics. http://farside.ph.utexas.edu/teaching/plasma/lectures/Plasmahtml.html.
16. Friedrich, H. (2005), Theoretical Atomic Physics. Berlin: Springer. ISBN 3-540-25644-X.
17. Garrett, H. B. (1981), The charging of spacecraft surfaces. Reviews of Geophysics and Space Physics, 19(4), 577–616.
18. Garrett, H. B., Minow, J. I. (2004), Charged Particle Effects on Solar Sails: an Overview, NASA/JPL, September 28–29, downloadable from http://hdl.handle.net/2014/40526.
19. Goedbloed, J. P. H., Poedts, S. (2004), Principles of Magnetohydrodynamics: With Applications to Laboratory and Astrophysical Plasmas. Cambridge: Cambridge University Press. ISBN 0-521-62607-2.
20. Groß, A. (2009), Theoretical Surface Science: a Microscopic Perspective (2nd edn.). Berlin: Springer. ISBN 978-3540689669.
21. Kittel, C. (2005), Introduction to Solid State Physics (8th edn.). New York: Wiley. ISBN 0-471-41526-X.
22. Laframboise, J. G., Kamitsuma, M. (1983), The threshold temperature effect in high voltage spacecraft charging. In Proc. Air Force Geophys., Workshop Natural Charg. Large Space Struct. Near Earth Polar Orbit (pp. 293–308). AFRL-TR-83-0046, ADA-134-894.
23. Lai, S. T. (1991), Spacecraft charging thresholds in single and double Maxwellian space environments. IEEE Transactions on Nuclear Science 38(6), 1629–1634.
24. Lai, S. T., Tautz, M. F., Tobiska, K. (2006), Effects of solar UV on spacecraft charging in sunlight. In 44th AIAA Aerospace Sciences Meeting and Exhibit, Reno, NV, 9–12 January 2006. AIAA 2006-407.
25. Lai, S. T., Tautz, M. F. (2006), Aspects of spacecraft charging in sunlight. IEEE Transactions on Plasma Science, 34(5).
26. Lai, S. T. (2012), Fundamentals of Spacecraft Charging: Spacecraft Interactions with Space Plasmas. Princeton: Princeton University Press. ISBN 0691129479, ISBN 978-0691129471.
27. Lang, K. R. (1999), Astrophysical Formulae (3rd edn.). Berlin: Springer. ISBN 3-540-29692-1.
28. Meyer-Vernet, N. (2007), Basics of the Solar Wind, Cambridge Atmospheric and Space Science Series. Cambridge: Cambridge University Press. ISBN 978-0-521-81420-1.
29. Michaelides, A., Sheffler, M. (2006), An Introduction to the Theory of Metal Surfaces, Fritz-Haber-Institut der Max-Planck-Gesellschaft, Berlin, Germany. http://www.fhi-berlin.mpg.de/th/publications

122   4   Solar Sails in Interplanetary Environment

30. Mihaly, L., Martin, M. C. (1996), Solid State Physics (Problems and Solutions). New York: Wiley. ISBN 0-471-15287-0.
31. Mordovskaya, V. G., Oraevsky, V. N. (2003), In-situ measurements of the Phobos magnetic field during the Phobos-2 mission. In Sixth International Conference on Mars, Pasadena, CA, 20–25 July 2003, also http://www.lpi.usra.edu/meetings/sixthmars2003/pdf/3076.pdf.
32. Mordovskaya, V. G., Oraevsky, V. N., Styashkin, V. A. (2003), The peculiarities of the interaction of Phobos with the solar wind are evidence of the Phobos magnetic obstacle (from *Phobos-2* data). arXiv:physics/0212072v2 [physics.space-ph].
33. Minor, J. L. (Ed.) (2004), Proc. 8th Spacecraft Charging Technology Conference, 20–24 October 2003. NASA/CP2004213091.
34. Nastasi, M., Mayer, J., Hirvonen, J. (1996), Ion-Solid Interactions—Fundamentals and Applications. Cambridge: Cambridge University Press. ISBN 0-521-37376-X.
35. Nishi, Y., Doering, R. (2000), Handbook of Semiconductor Manufacturing Technology. New York: Marcel Dekker. ISBN 0-8247-8783-8.
36. NRL Plasma Formulary (2009), http://wwwppd.nrl.navy.mil/nrlformulary/.
37. NASA/GSFC OMNIWeb Interface, http://omniweb.gsfc.nasa.gov/form/dx1.html.
38. Orloff, J. (2009), Handbook of Charged Particle Optics (2nd edn.). Boca Raton: CRC Press. ISBN 978-1-4200-4554-3.
39. Parker, E. N. (2001), A History of the Solar Wind Concept in the Century of Space Science, Vol. 1. Dordrecht: Kluwer Academic. ISBN 978-0-7923-7196-0. Now owned by Springer, Dordrecht, 2002.
40. Patterson, J. D., Bailey, B. C. (2007), Solid-State Physics—Introduction to the Theory. Berlin: Springer. ISBN 103-540-24115-9, ISBN 978-3-540-24115-7.
41. Piel, A. (2010), Plasma Physics, An Introduction to Laboratory, Space, and Fusion Plasmas. Berlin: Springer. ISBN 978-3-642-10490-9, e-ISBN 978-3-642-10491-6.
42. Rozelot, J.-P. (2006), Solar and Heliospheric Origins of Space Weather Phenomena. Berlin: Springer, ISBN 3-540-33758-X, ISBN 978-3-540-33758-4.
43. Nakamura, K., et al. (Particle Data Group) (2010), Journal of Physics G, 37, 075021. http://pdg.lbl.gov/.
44. Soop, M. (1972), Numerical calculations of the perturbation of an electric field around a spacecraft. In R. J. L. Grard (Ed.), Photons and Particle Interactions with Surfaces in Space. Dordrecht: Reidel.
45. Spitzer, L. Jr. (1961), Physics of Fully Ionized Gases. 2nd revised edition by Dover (2006). ISBN 9780486449821, ISBN 978-0486449821.
46. Ziegler, J. F., Biersack, J. P., Ziegler, M. D. (2008), SRIM: The Stopping and Range of Ions in Matter. New York: SRIM. ISBN 978-0-9654207-1-6, ISBN 0-9654207-1-X.
47. Ziegler, J. F. (2008), SRIM-2008.4: The Stopping and Range of Ions in Matter, setup-file downloadable from http://www.SRIM.org.
48. Yeh, J. J. (1993). Atomic Calculation of Photoionization Cross-Sections and Asymmetry Parameters. New York: Gordon and Breach.

Part III
Sailcraft Trajectories

This part of the book deals with sailcraft dynamics in many details, from the various and modern concepts of Time to new classes of trajectories stemming from the non-conservative *and* non-linear features of the solar-sailing equations, where gravity plays an important role as well. Optical diffraction, applied to solar-photon sailing, represents one of the main topics of this part. The concept of lightness vector has a central role in studying non-intuitive properties of special sailcraft trajectories.

Chapter 5
Fundamentals of Sailcraft Trajectory

Motion Under Gravity and Radiation Pressure   Some basic properties of space vehicles endowed with solar sail may be carried out without a detailed model of the thrust due to the solar radiation pressure. This will be next done in a long devoted chapter. Here, we start with establishing the main mathematical formalism that will accompany us in the course of the subsequent chapters. One should realize that such formalism is not merely an option among many others. As a point of fact, as it is known from many branches of science, some methods can favor investigation considerably. The subtitle indicates that two sources of acceleration will be acting on the sailcraft as a whole: one is the solar gravity, the other one is the scalar field of solar radiation pressure in space. Perhaps unexpectedly, the second one generates two local vector fields of which one is conservative. The very large potential of sailcraft missions arise from the simultaneous actions of these three fields, namely, two conservative and one non-conservative. In this and the following chapters, we shall see how trajectories, impracticable to rockets, are possible to sailcraft, thus strongly enlarging the potentials of our space exploration and utilization. In the final part of this chapter, the interested reader can find an introduction to thrust maneuvering from an unusual viewpoint.

Since this book is oriented to graduate students, we will begin with a digression on time and reference frames. We do that for a further reason: much of the scientific software on Astrodynamics that the student can find in her/his university contains a high number of reference frames and time scales; if left unexplained, these may induce confusion or, worse, some misunderstanding, especially when results from different computer codes are to be compared. Such section, we hope, may be considered as a short introduction to the subject, which is vast and complicated indeed.

5.1 Scales of Time and Frames of Reference

In the course of the 20th century, a high number of time scales have been introduced for extending/improving what existed in the previous centuries; new definitions, and

G. Vulpetti, *Fast Solar Sailing*, Space Technology Library 30,
DOI 10.1007/978-94-007-4777-7_5, © Springer Science+Business Media Dordrecht 2013

126   5 Fundamentals of Sailcraft Trajectory

related realizations, reflected the progress of fundamental physics and the increasing requests by the scientific communities. Such needs have been continuing, and already there are proposal to modify/reject some of the time scales we are used to employ today. All that might induce some confusion in beginners to astronomy, geodesy, navigation, and so on. An analogous, but weaker, situation might occur for reference frames. In this section, we will emphasize only some aspects of the latest time scales and frames of reference, and with regard to the formulation and the computation of solar sailing trajectories.

In addition to the usual scientific papers, the topics of this section can be found with full particulars in the publications of the IERS, http://www.iers.org.

In general, time scales may be grouped in two large classes: those ones based on precise artificial clocks, and the other ones relied on the (non-uniform) Earth's rotation. Modern definitions of time and reference frames are based on very accurate and precise measurements via atomic clocks on ground and accurate/precise observations of a number of very far radiative objects in the Universe; most of these sources are quasars. Conceptually speaking, each atomic clock (like any other clock) measures intervals of its proper time (in the sense of Relativity). Atomic clock is a very high standard in both time and frequency. Using the Cesium atom's resonant frequency, it has been possible to define the SI second as just 9,192,631,770 oscillations[1] by notably surpassing—in accuracy and precision—the second's older definition based on the Earth's motion. Although the next generation of time standards (driven by increasing technology demand and scientific research) is under development around the world, the current official standards are stable to within approximately 20–30 ns/year. There is a world-net of over two hundred atomic clocks located in over thirty countries; their measurements of time are processed by the *Bureau International des Poids et Mesures* (BIPM) that then issues the International Atomic Time (TAI).[2] TAI, the epoch of which 1958.01.01 00:00:00 UT2,[3] is almost six decades in evolution; it is a continuous scale that realizes the concept of Terrestrial Time (TT). TT may be thought as what a perfect clock would measure if located anywhere on the geoid. One has

$$TT = 32.184 \text{ s} + TAI \tag{5.1}$$

In contrast, UT1, now linearly related to the ERA,[4] is not a uniform time scale because of the irregularities in the Earth's rotation. The duration of the day is defined to be exactly 86,400 SI seconds. Day is the basic unit for any Julian date system (JD).

---

[1]This is the frequency of the transition between two hyperfine levels of the ^{133}Cs ground state.

[2]As in Chap. 2, here and after, many acronyms come from the corresponding French phrase.

[3]This further version of the Universal Time (UT) was derived from UT1 by adding terms aimed at balancing effects of the annual/seasonal changes in the Earth rotation. The scale of Universal Time named UT1 is based on the Earth's spin rate and takes the planet's polar motion into account.

[4]According to the resolution B1.8 of the XXIV meeting of the International Astronomical Union (2000).

5.1 Scales of Time and Frames of Reference

Because of the conflicting needs to use the highly stable and precise atomic time scale and to retain the consolidated astronomical time scale UT1, a *leap second* has been introduced to be (normally) added to the TAI value at some (but not pre-fixed) calendar date in order to keep its difference to UT1 within ±0.9 s. Such a scale, available worldwide via broadcast signals, is called the Coordinated Universal Time (UTC).[5] UTC is an atomic-clock scale, but with discontinuities inserted artificially for the difference UT1-UTC remains bounded. The difference TAI-UTC can be computed correctly by means of the subroutine iau_DAT, one of 187 Fortran subroutines/functions of SOFA by the International Astronomical Union (IAU). It was decided in the second half of the Seventies to replace the old Greenwich Mean Time by UTC for civil uses. Very recently, a suggestion aiming at replacing UTC by an international uniform scale has been discussed.

Out of the time scales described hitherto, only TT is compliant with General Relativity (GR). As a point of fact, there is a relationship that ties TT to the Geo-centric Coordinate Time (TCG), in the sense of GR, for an inertial frame centered on the barycenter of the Earth, including oceans and atmosphere, also named the *geocenter*:

$$\frac{d(TT)}{d(TCG)} = 1 - L_G \tag{5.2}$$

where $L_G = 6.969290134 \times 10^{-10}$ is a *defining* constant. Perhaps because of this small value, somewhere one may find erroneously written that TT works also as a coordinate time.[6] Actually, a linear transformation has to be applied (see (5.6b)) to get finite values.

Every value of time in a given time scale has its own Julian date, namely the number of days and day fraction from January 1st, 4713 BC, Greenwich noon. Thus, in giving a date, the corresponding time-scale should be specified for avoiding confusion and mistakes in precise calculation. For instance, J2000 stands for January 1, 2000, 12:00:00 TT; this date is equivalent to January 1, 2000, 11:59:27.816 TAI, or January 1, 2000, 11:58:55.816 UTC. In this book, by J2000 we always mean

$$J2000 = 2451545 \text{ TT} \tag{5.3}$$

Jxxxx will denote January 1, Noon, year xxxx, TT.

---

[5]The reader should note that this is a **coordinated** time (distributed over the world), not a *coordinate* time.

[6]Proper time is *invariant*, but not integrable; the converse is true for coordinate time. Therefore, is the second one that can be strictly used as the independent parameter in geocentric motion equations.

In the 20th century, many inertial frames of reference have been introduced and used in theoretical research and numerical investigations and applications, including astrodynamics. In August 1997, at its 23rd General Assembly, IAU decided to adopt, as from January 1, 1998, the International Celestial Reference System (ICRS), which complies with the conditions specified by the IAU Recommendations issued in 1991. Ideally, a reference system in the usual three-dimensional space consists of three axes (considered fixed with respect to something) and a time scale. However, what does *reference system* mean from an operational viewpoint? Essentially, it consists of a set of conventions and prescriptions, and the modeling for defining a triad of spatial axes at any time. In order to scientific disciplines receive real benefits, this concept needs to be carried into effect. This is accomplished by means of a *reference frame*, namely, a set of fiducial points; in this case, such points have to be searched for on the celestial sphere. In contrast to previously adopted (quasi-)inertial systems, ICRS makes use of a number of extra-galactic radio sources, most of which are quasars.

The positions and the uncertainties (on the celestial sphere) of these ones compose the International Celestial Reference Frame (ICRF), which therefore *realizes* the ICRS. ICRF needs to be maintained; there have been two extensions to the first realization or ICRF1, whereas as of January 1st, 2010, a second improved realization, or ICRF2, is the fundamental realization of the ICRS. The total number of considered sources has been 667 in ICRF-Ext.1, 717 in ICRF-Ext.2, and 295 new sources in ICRF2; maintenance of ICRF will be made by such 295 sources, which are the defining objects. ICRF is a high-precision realization of ICRS: the directions of the ICRS pole and the origin of right ascension are maintained fixed relative to the quasars within ± 10 μas. The problems, equations and the evolution of ICRF can be extensively found in [14–20].

Thus, the realized axes of ICRS are those carried out through ICRF. The origin of ICRS is the barycenter of the solar system. Some remarks are in order about the main features of ICRS. *First*, according to IAU recommendations, the reference plane of ICRS is close to the mean equator at J2000, while the origin of right ascensions is close to the dynamical equinox at J2000; this consistency is better than 100 mas. *Second*, ICRS is *kinematically* non-rotating, contrarily to older reference systems that may be dynamically non-rotating. As a point of fact, the proper motions of the (very distant) fiducial points of ICRF are supposed to be negligible for long time and with respect to the precision of their observations. In contrast, previous *dynamically* non-rotating frames were based on modeling the motion of two planes (the Earth's equatorial plane and the ecliptic plane). *Third*, as a further consequence of the need to model the evolution of such planes, due to the motions of solar-system bodies, *dynamical* reference systems always had to have a specific epoch. In this sense, ICRS has no epoch, like B1950 or J2000, attached. In 2000, IAU issued the definition of the Barycentric Celestial Reference System (BCRS), with the origin in the solar-system barycenter (that is the ICRS' origin), within the framework of GR. In the same context, the Geocentric Celestial Reference System (GCRS) was defined with the origin in the geocenter; in particular, "... its spatial coordinates are kinematically non-rotating with respect to those of the BCRS". Although the related metric

5.1 Scales of Time and Frames of Reference

tensors have been detailed, the BCRS definition did not specify the orientation of the spatial axes. Thus, among other things, IAU 2006 Resolution-2 completed the definitions by adding the following recommendation: "For all practical applications, unless otherwise stated, the BCRS is assumed to be oriented according to the ICRS axes. The orientation of the GCRS is derived from the ICRS-oriented BCRS".

What are the time scales of BCRS and GCRS? Of course, one has to resort to the notion of *coordinate time*. With regard to BCRS, the independent time parameter is the TCB, whereas the previously mentioned TCG is the time scale for GCRS. The unit of TCB is the SI second, mentioned previously. These two scales are such that

$$\left\langle \frac{d(TCG)}{d(TCB)} \right\rangle = 1 - L_C \tag{5.4a}$$

$$L_C \equiv 1.48082686741 \times 10^{-8} \pm 2 \times 10^{-17} \tag{5.4b}$$

whereas

$$\left\langle \frac{d(TT)}{d(TCB)} \right\rangle = 1 - L_B \tag{5.5a}$$

$$L_B \equiv 1.550519768 \times 10^{-8} \quad \text{defining constant} \tag{5.5b}$$

In Eqs. (5.4a), (5.4b) and (5.5a), (5.5b) the average process regards a sufficiently long time interval taken at the geocenter. The explicit difference TCB–TCG can be found in [17]. It involves a four-dimensional transformation resulting in $O(c^{-2})$ and $O(c^{-4})$ terms. This difference depends on the masses and positions of the solar-system bodies, and the Earth's velocity; the main part consists of a rate difference, whereas the residual value is periodic (with an amplitude less than 0.0017 s for an observer on the Earth). The former part can be approximated by the constant L_B times the time interval between the TT-date of interest and January 1, 0^h, 1977 (JD = 2443144.5 TT). In the calendar time frame 1600–2200 AD, the overall difference TCB–TCG has been accurately computed. Here, is of interest to know that, at the geocenter on JD = 2455197.5 TT (January 1, Midnight, 2010), it amounted to 15.42099 s, and will increase up to 57.47922 s on January 1, 2100. Such values are not acceptable for precisely computing the coordinates of celestial bodies and spacecraft moving "outside" the Earth-Moon space inasmuch as there is a drift of about 0.467 s/year: TCG could *not* be employed in the place of TCB, which instead is recommended for these purposes. The problem is not closed so simply, though.

For computing space vehicle trajectories and solar-system bodies accurately, one needs to either solve spacecraft and celestial bodies orbits simultaneously or use some source of ephemerides where the large-system evolution (i.e. orbits of the Sun, planets, etc.) has been already carried out and stored in some tabular-form file. This is allowed because the spacecraft mass does not influence the orbits of the other bodies, of course. The most used ephemeris sources are Jet Propulsion Laboratory (JPL) files; in particular the DE405/LE405 (or DE405 for short) file [24] will be employed extensively in this book. This set of planetary/lunar coordinates has an independent parameter, T_{eph}, which has been used in numerically-integrated ephemerides since the Sixties. T_{eph} should not to be confused with the old Ephemeris Time (ET), even though it may be viewed as a modern realization of the ET aims. Some important features of JPL's ephemeris set as a whole are:

- DE405 is aligned with ICRS
- T_{eph} is a linear transformation of TCB
- $\langle \frac{d(T_{eph})}{d(TT)} \rangle = 1$; at the geocenter, $|T_{eph} - TT| < 0.002$ s

Finally, we have to be about the Barycentric Dynamical Time (TDB). The history of TDB started in 1976, when the general assembly of IAU decided to adopt proposals reflecting the need to get a relativistic ephemeris time scale for accurately computing planetary ephemerides. In the subsequent years, although used widely, investigators were aware that the definition of TDB contained a flaw: what defined was not physically possible. In addition, its origin was not defined with a sufficient precision, and there was no transformation for connecting what measured by different observers in the solar system. Thus, fifteen years later, IAU issued the definition of TCB, which is strict from the GR viewpoint. However, a considerable part of the scientific community continued to employ TDB. In 1998, a paper from JPL [38] showed that T_{eph} and TCB differ by an offset and a constant rate. Thus, T_{eph} was realizing what TDB (and the old ET) would have had to do many years ago. In 2006, the IAU resolution-3 reasserted that TCB is the coordinated time for the BCRS; however, (1) there is a strong usefulness for having an unambiguous coordinate scale mapped from TCB via a linear equation *and* such that the difference between this transformed time scale and TT, at the geocenter, is kept small for many centuries; (2) the new TDB is desired to have consistency with T_{eph} and other TDB realizations, especially that in [9].[7] Thus, in 2006 Resolution-3, IAU recommended a TDB definition through a precise linear transformation of TCB. In addition, T_{eph} "is for practical purposes the same as TDB defined in this Resolution".

It is time to synthesize what said above. First of all, let us recall that:

- Observations for getting ephemerides of celestial bodies and artifacts in the solar system or in the Earth-Moon space are time tagged via TT, which, in turn, is realized via TAI;
- The motion equations of natural and artificial bodies are parameterized by means of TCB or TCG, according to the employed frame of reference;
- The observations are to be re-tagged in terms of TCB or TCG in order to be employed in the orbit determination process, a set of algorithms and procedures for carrying out positions and velocities of celestial bodies and spacecraft.

---

[7]The authors carried out an analytical formula for TB–TT valid over a few thousand years around J2000, with an accuracy at 1-ns level. The 127 coefficients in the transformation equation, presented in that paper, produced a formula accurate at the 100-ns level. The transformation could not be fully compliant with the above cited 1976 resolution by IAU, a clear signal that something in the first definition of TDB was erroneous. The implementation of the full equation giving TDB as function of TT has been carried out in SOFA-2006.

5.1 Scales of Time and Frames of Reference 131

In this policy, the following relationships by IAU apply

$$\text{TDB} = \text{TCB} - (\text{JD}_{\text{TCB}} - T_0)86400 L_B + TDB_0 \tag{5.6a}$$

$$\text{TT} = \text{TCG} - (\text{JD}_{\text{TCG}} - T_0)86400 L_G \tag{5.6b}$$

$$T_0 = 2443144.5003725\text{TAI}, \qquad TDB_0 = -6.55 \times 10^{-5}\ \text{s} \tag{5.6c}$$

where L_B and L_G are the defining constants in Eqs. (5.5a), (5.5b) and (5.2), respectively. T_0, which corresponds to 1977.01.01 00:00:32.184, and TDB_0 are defining constants as well. (Of course, Julian dates are expressed in days, whereas the other time values are in seconds.) The above equations add to the above properties regarding T_{eph}; in particular, we may explicitly recast the difference $T_{eph} - \text{TT}$ given in [21] into the form

$$T_{eph} - \text{TT} \cong T_{eph_0} + \frac{1}{(1 - L_C)c^2}\Bigg[(\mathbf{R}_{obs} - \mathbf{R}_\oplus) \cdot \mathbf{V}_\oplus$$

$$+ \int_{T_{eph_0}}^{T_{eph}} \left(\frac{1}{2}V_\oplus^2 + \frac{GM_\odot}{\|\mathbf{R}_\odot - \mathbf{R}_\oplus\|} + \sum_{p \neq earth} \frac{GM_p}{\|\mathbf{R}_p - \mathbf{R}_\oplus\|} \right) dt \Bigg]$$

$$- \frac{L_C}{1 - L_C}(T_{eph} - T_{eph_0}) \tag{5.7}$$

where $\mathbf{R}$ and $\mathbf{V}$ denote position and velocity in BCRS, respectively, of an observer or solar-system bodies. In the relationship (5.7), we have focused on Sun and planets only, and neglected any post-Newtonian effects; also, we put $T_{eph_0} = TDB_0 = -65.5\ \mu\text{s}$. At the geocenter, $\mathbf{R}_{obs} = \mathbf{R}_\oplus$, the integral and the last term of the r.h.s. of (5.7) results in less than 0.002 ms. If the observer (e.g. the spacecraft's clock at a certain position in the solar system) is far from the geocenter, is the dot product[8] (in the squared brackets) that may dominate as the distance from the Earth increases.

In the subroutine PLEPH($\text{JD}_{T_{eph}}$, ι_{target}, $\iota_{central}$, $\mathbf{S}_{target}$), one of the Fortran procedures associated to DE405, the second and third arguments are the numeric identifiers (*input*) of the target and central bodies, whereas the fourth slot denotes the position-velocity array (*output*) of the target-body with respect to the central-body at the desired JD (*input*). The entries of this array are resolved in the ICRF. According to what discussed above, the user can either use JD_{TDB} as $\text{JD}_{T_{eph}}$ or employ JD_{TT} after applying transformation (5.7). Among the other things, once the user's code gets the positions of the major bodies of the solar system at a certain time, it can compute the gravitational perturbations acting on the spacecraft.

---

[8]This term comes from Special Relativity inasmuch as the geocentric frame moves with respect to an observer located in the solar-system barycentric frame.

With regard to TT and TDB in the context of sailcraft trajectories, SOFA and DE405 may be both utilized in a sailcraft trajectory propagation/optimization numeric problem; there is a negative *mean* linear drift between the two from 2000 to 2200 as follows

$$\delta(\text{TT} - \text{TDB})/\delta(T_{eph}) \cong -0.4 \text{ ns/year} \tag{5.8}$$

In particular, TT-TDB starts from a few nanoseconds in 2000.

We end this section by pointing out that—if one aims at computing very high precision trajectories, e.g. a very long-term orbital propagation of some small celestial body—the linear scaling law of time entails two more scalings, namely, those related to length and gravitational mass. This can be expressed as

$$\Theta = \vartheta t + \Theta_0 \quad \Rightarrow$$
$$\mathbf{X} = \vartheta \mathbf{x}, \qquad GM = \vartheta \mu \tag{5.9}$$

where t is a coordinate time scaled by some factor ϑ, $\mathbf{x}$ denotes vector position, and μ is gravitational mass, all in the same reference units. While it is very easy to check the second and third relationships, the basic reason for three simultaneous scalings is that the equations of motion of massive bodies and of propagation of light are left unchanged.

> As anticipated in the chapter summary, this section has been only a short introduction to the matter containing more complicated things; we have finalized the subject to the computation of sailcraft trajectories. Nevertheless, we will not introduce further sophistication because our main aim is to highlight the peculiar features of fast solar sailing. Anyway, the reader should realize that the current status about reference frames and time scales is not *frozen* definitively. The overall situation has been evolving; also new ephemeris constructions have been proposed, e.g. [27]. Independently of what may happen even in the latest months before the manuscript of this book is sent to the editor, in this section we have defined our current (precise) framework of fast solar sailing. Any reference frame(s) we need in the subsequent sections/chapters for dealing with sailcraft trajectories aptly, will be derived from the basic items discussed above.

5.2 The Sun's Orbit in the Solar System

Before proceeding with the basics of sailcraft motion, we need to briefly describe the motion of the Sun in the ICRS endowed with TDB. With an extremely good approximation, one can consider the solar system as an isolated system with constant mass.[9]

---

[9]Over 10 million years, the Sun loses approximately 7×10^{-7} of its mass.

5.2 The Sun's Orbit in the Solar System

Fig. 5.1 Orbit of the solar barycenter in the XY-plane of the Barycentric Invariable-Plane System (BIPS) of reference. Time span is from J1980 to J2100 TDB

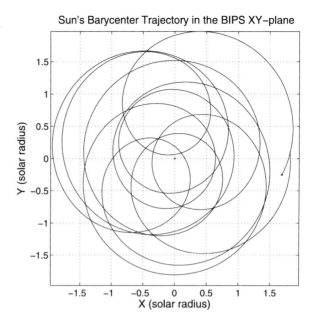

If one considers the solar system as a classical system, where coupling between angular momenta is allowed as well (e.g. spin-orbit coupling, planet-satellite spin-spin coupling, etc.), then its angular momentum is constant. Let $\mathcal{J} \equiv \mathbf{J}/\|\mathbf{J}\|$ denote the direction of this quantity of the solar system. The ICRS angular coordinates (i.e. right ascension and declination) of $\mathcal{J}$ are approximately $\alpha_{\mathcal{J}} = 273.8527°$, $\delta_{\mathcal{J}} = 66.9911°$. The plane orthogonal to $\mathcal{J}$ is named the *invariable plane*, say, $\mathcal{P}$ of the solar system. It intersects the ICRS' XY plane of reference in a line we can direct *positively* with right ascension $\alpha_{\mathcal{P}} = 3.8527° \equiv \mathcal{X}$. Therefore, we can define BIPS of reference with $\mathcal{P}$ as the reference plane where the solar system barycenter, here denoted by $\mathcal{B}$, is the origin and $\mathcal{X}$ represents the X-axis; $\mathcal{Y} = \mathcal{J} \times \mathcal{X}$ completes the triad. Of course, the time scale is TDB again.

We computed—via DE405—the motion of the Sun's barycenter, or $G_\odot$ for short, in the J1980–J2100 time span, and arranged the results of concern here in two figures. Chosen unit for length is the solar radius ($r_\odot = 696{,}000$ km $\cong 214.94^{-1}$ AU) and m/s, respectively. Figures 5.1 and 5.2 show the orbit of $G_\odot$ in $\mathcal{P}$, and the evolution of $Z_\odot$ and $R_\odot$, respectively.

Some important information can be drawn from JPL data file. First, most of the $G_\odot$ motion, essentially due to the giant planets, develops on $\mathcal{P}$, as expected. Second, in the considered time span, $0.064 < R_\odot/r_\odot < 2.1$, while 8.52 m/s $< V_\odot < 16.08$ m/s, with mean values of approximately 1.208 and 12.64 m/s, respectively. Third, the max-to-min ratio amounts to 32.7 for $R_\odot$ whereas the value for $V_\odot$ is almost 1.9. Actually, the angular speed of $G_\odot$ can change considerably over periods of 2–3 years. With regard to the flight classes discussed in Chap. 7, if a sailcraft approaches the Sun, say, with a perihelion as low as 0.2 AU, then it may see $\mathbf{R}_\odot$ under an angle of $2.64°$ at most (during the 21th century); therefore, if one uses

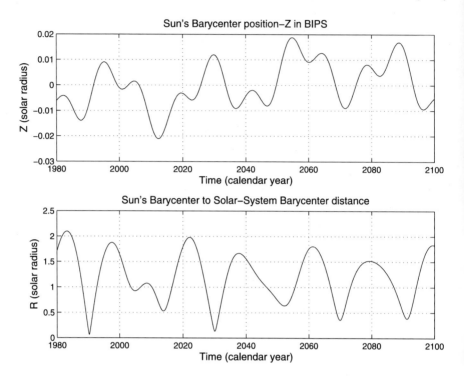

Fig. 5.2 (*Top*) Sun's barycenter motion projected along the angular-momentum direction of the solar system. (*Bottom*) Sun's barycenter distance from the solar-system center of mass during the 1980–2100 time span; it overcame 2 solar radii for a few years in the Eighties. The mean value is 1.208 solar radii

a non-heliocentric reference frame such as the BCRS, the sailcraft's vector radius will be somewhat different from the Sun-to-sailcraft mutual position vector. On the other hand, if one chooses a heliocentric frame, then planetary gravity intervenes with additional terms, as it is known from the general many-body problem of Celestial Mechanics; however, nowadays, propagating spacecraft's orbital state accurately under gravitational perturbations, due to the planets, is not a difficult task.

The analysis shows also that the mean acceleration level is $\sim 2.1 \times 10^{-7}$ m/s^2, namely, close to the acceleration of the Sun subject to the Jupiter gravity (as expected).

In next section, taking what discussed here and in Sect. 5.1 into account, we shall begin studying the motion of a sailcraft in a heliocentric reference frame by means of a general formalism and some concepts directly related to solar sailing.

5.3 Equations of Heliocentric Motion

Let us consider the usual three-dimensional Euclidean space $\mathfrak{R}^3$. The unit vectors of the ICRS provides a reference basis, to which we associate the TDB as the time scale for describing the evolutions of physical quantities represented by scalars, vectors and matrices. Such a space is a vector space, over the field of the real numbers, and endowed with a bilinear symmetric form called the *metric*. In n-dimensional Euclidean spaces, or $\mathfrak{R}^n$, metric is the $n \times n$ identity matrix, i.e. basis is orthonormal.

In mathematics, the term *module* refers to as an additive Abelian group the elements of which can be multiplied by the elements of some *ring* called the *scalars* (of the module). Special modules are represented by *vector spaces*; a vector space is a module over some *field*, e.g. the real field. Vector spaces can be finite or infinite dimensional spaces. (In physics, both types are very useful; an example of infinite-dimensional vector space is the Hilbert space.) In this book, however, we will deal with only finite-dimensional vector spaces.

Any element of a vector space is called a *vector*, which can be quite a general object: from real or complex scalars to the usual vectors, matrices, tensors, spinors, sets of vectors, etc. In general, a vector over a field is an *ordered* list of scalars *from* that field. Any vector can be scaled and/or added to any vector, including itself, of the same space. A vector space is called also *linear* space because two such spaces can be transformed by a linear map.

Any space vector is endowed with sets of free-generators, each element of which is called a *basis*. However, all fundamental concepts regarding space vectors can be stated without resorting to bases. A space vector admits a *dual* space; the two are *isomorphic* to one another. By means of the pair {space, dual-space}, multilinear *tensor* spaces can be built. Tensors are powerful mathematical objects very useful in advanced physics and engineering. Certain physical quantities cannot be described via tensors, however.

A more general kind of mathematical objects can be defined by using two special operators: (1) the exterior (or outer or wedge) product, (2) the interior (or inner) product; their combination gives rise to the Clifford product (that may be defined axiomatically as well). It is then possible to build higher and lower dimensional spaces starting from a vector space of given dimensionality. The new spaces, although of different-dimensionality, can be summed originating a particular vector space, the elements of which are called the *multivectors*, and obeying the Clifford algebra. A Clifford algebra can be defined over a real or complex field.

The *geometric* algebra is a Clifford algebra over a real space where the *geometric* product is defined. This, among many things, allows manipulating subspaces and getting intuitive geometric representations. Geometric algebra is applied to many areas of advanced physics, computer graphics, and (recently) the astrodynamics of solar sailing as well [43].

Introduction to the basic Algebra and structures can be found in many excellent mathematical textbooks, e.g. [31]; a strict treatment of tensors and spinors can be found in the books on differential geometry. Classical textbooks on geometric algebra and geometric calculus are, for instance, [7, 8], and [11]. A large Clifford software package, running under Maple 12™ and later, and under various operating systems, can be downloaded freely [1].

In this and the two subsequent chapters, we suppose that a sailcraft does not exhibit any temporal alterations of the thermo-optical properties, say, $\mathcal{O}$ of its sail. Such properties will depend chiefly on the used sail materials through the radiation spectral bands, and the sail attitude $\mathbb{N}$. In the ICRS/TDB, the equations of motion of such a sailcraft with sail area A can be written formally as

$$\dot{m}_\square = 0 \tag{5.10a}$$

$$\ddot{\mathbb{R}}_\square = \mathbb{G}_\square^{(\odot)} + \mathbb{T}_\square/m_\square + \sum_{\bullet \neq \odot} \mathbb{G}_\square^{(\bullet)} \qquad (5.10b)$$

$$\mathbb{T}_\square \equiv \mathbf{f}(t, \mathcal{O}, A, \mathbb{N}, \mathbb{R}_{\odot\square}, \mathbb{V}_\square) \qquad (5.10c)$$

In (5.10a)–(5.10c), a dot means differentiation with respect to TDB $\equiv t$, the subscript $\square$ denotes the sailcraft, whereas in principle the superscript $(\bullet)$ refers to any celestial body in the solar system: $\mathbb{G}_\square^{(\bullet)}$ represents the gravitational acceleration on the sailcraft caused by the body $(\bullet)$. With regard to relative positions, we use the celestial-mechanics convention

$$\mathbb{R}_{ab} \equiv \mathbb{R}_b - \mathbb{R}_a \qquad (5.11)$$

where a and b denote body labels (symbols or natural numbers), including Sun and sailcraft. Of course, $\mathbb{R}_{ab}$ means the vector from body-a to body-b. In addition, $\mathbb{R}_\bullet$ denotes the position vector of the generic body $(\bullet)$ emanating from $\mathcal{B}$.

Finally, $m_\square$ denotes sailcraft mass and $\mathbb{T}_\square$ is the surface force caused by the solar irradiance and any applicable irradiances from passive sources of light. As such, apart from the environmental impact, Eqs. (5.10a)–(5.10c) are quite general in the current framework, though its formulation is not relativistic.

Why is the above classical formulation sufficiently good in the current astrodynamical context? Essentially for three basic reasons:

1. in the solar system, gravitational fields are weak so that the actual metric is slightly different from that of Minkowski spacetime, namely, $\eta_{\mu\nu} = \mathrm{diag}(+1 - 1 - 1 - 1)$. As usually denoted, $\mu, \nu = 0..3$, with the zero value referred to the time coordinate, i.e. $\eta_{00} = +1$;
2. at a fixed point in the interplanetary space, the time change of the actual metric is notably small;
3. test-body speed is small compared to the speed of light.

As a result, the second coordinate-time derivative of each space coordinate of a test-body is equal to the corresponding space (partial) derivative of the Newtonian potential up to terms $O(c^{-2})$.

Finally, let us note that any tidal gravitational and non-gravitational accelerations acting on the Sun can be neglected. Thus, from the viewpoint of GR, $G_\odot$ follows a *geodesic* and is not accelerated at all. The strict definition of geodesic can be found in many excellent textbooks on Differential Geometry, e.g. [13].

Actually, strictly speaking, Eq. (5.10b) cannot stay alone, namely, the motion of all perturbing bodies should be considered simultaneously in order to get the correct values of their positions through which all terms $\mathbb{G}_\square^{(\bullet)}$ can be computed. Because of an interplanetary sailcraft is sufficiently far from a celestial body to not sense its

5.3 Equations of Heliocentric Motion

aspherical and/or inhomogeneous structure, let us consider $\mathbb{G}_{\square}^{(\bullet)} \to \mathbb{K}_{\square}^{(\bullet)}$, where $\mathbb{K}$ denotes the Keplerian field.

In particular, the Sun's motion in ICRS may be described as follows

$$\ddot{\mathbb{R}}_{\odot} = \sum_{\bullet \neq \odot} \mathbb{K}_{\odot}^{(\bullet)} = \sum_{\bullet \neq \odot} -\mu_{\bullet} \frac{\mathbb{R}_{\bullet \odot}}{\|\mathbb{R}_{\bullet \odot}\|^3} \tag{5.12}$$

where the sum can be restricted to the planets only; $\mu_{\bullet}$ denotes the gravitational constant of the body-$\bullet$.

Let us define a new reference frame, centered on the Sun, with the Y–Z axes obtained from the ICRS' Y-Z axes by a counterclockwise rotation about the ICRS' X-axis; although in principle *unimportant*, let us arbitrarily choose this angle equal to the ecliptic obliquity at J2000, or approximately 23.4392°, which in our case acts like a *defining* constant. Let us call such frame the HIF. It will be particularly useful as the designs of special sailcraft trajectories to interesting targets will develop three-dimensionally about the ecliptic plane. In order to write down the sailcraft motion equations in HIF/TDB, let us first apply the translation[10] via

$$\mathbb{R}_{\odot} = \mathbb{R}_{\square} - \mathbb{R}_{\odot \square} \tag{5.13}$$

Let us call this intermediate frame the ICRS/Sun. From a rigorous viewpoint, the orientation of its axes should be obtained via parallel transportation in a metric curved space. However, considering what said in Sect. 5.2 and in the previous gray-box, we can state that ICRS/Sun has the same spatial orientation of ICRS for any practical purposes.

Inserting (5.13) into (5.10a)–(5.10c) and substituting (5.12) for the $G_{\odot}$ acceleration, and re-arranging terms, we get

$$\ddot{\mathbb{R}}_{\odot \square} = -\mu_{\odot} \frac{\mathbb{R}_{\odot \square}}{\|\mathbb{R}_{\odot \square}\|^3} + \frac{\mathbb{T}_{\square}}{m_{\square}} + \sum_{\bullet \neq \odot} \mu_{\bullet} \left(-\frac{\mathbb{R}_{\bullet \square}}{\|\mathbb{R}_{\bullet \square}\|^3} + \mu_{\bullet} \frac{\mathbb{R}_{\bullet \odot}}{\|\mathbb{R}_{\bullet \odot}\|^3} \right) \tag{5.14}$$

together with the first and third equations (5.10a)–(5.10c). The different form and value of the planetary perturbations are just due to the fact that the origin of the new frame (ICRS/Sun) is (classically) accelerated with respect to the old (inertial) frame (i.e. ICRS).

The final step to HIF is to apply the above mentioned rotation. Because there is no angular velocity between ICRS/Sun and HIF at any arbitrary time, *any* vector and its time derivatives (if they exist), resolved in HIF, can be written as the following linear transformation of the same vector resolved in ICRS/Sun:

$$\mathbf{X} = \begin{pmatrix} 1 & 0 & 0 \\ 0 & \cos \varepsilon_0 & \sin \varepsilon_0 \\ 0 & -\sin \varepsilon_0 & \cos \varepsilon_0 \end{pmatrix} \mathbb{X} \equiv \Gamma \, \mathbb{X} \tag{5.15}$$

---

[10]Let us remind the reader that translation, though very simple formally, is *not* a linear transformation.

where $\varepsilon_0 = 23.4392°$. Left-multiplying both sides of (5.14) by the constant matrix F, and recalling that $\det(F) = 1$, one gets immediately the equation of the motion of a sailcraft in the HIF with TDB as time scale:

$$\ddot{\mathbf{R}}_{\odot\square} = -\mu_\odot \frac{\mathbf{R}_{\odot\square}}{\|\mathbf{R}_{\odot\square}\|^3} + \frac{\mathbf{T}_\square}{m_\square} + \sum_{\bullet\neq\odot} \mu_\bullet \left(-\frac{\mathbf{R}_{\bullet\square}}{\|\mathbf{R}_{\bullet\square}\|^3} + \frac{\mathbf{R}_{\bullet\odot}}{\|\mathbf{R}_{\bullet\odot}\|^3} \right) \qquad (5.16)$$

where $m_\square = constant$ throughout the flight.[11] The planetary position vectors, once extracted from DE405 at TDB $= t$ and multiplied by the rotation matrix F, can be used directly in (5.16). This is allowed by our current framework. Note that the sum on the right-hand side of (5.16) can be written

$$\sum_{k=1}^{N_{ncb}} \mathbf{P}_k \equiv \sum_{\bullet\neq\odot} \mu_\bullet \left(-\frac{\mathbf{R}_{\bullet\square}}{\|\mathbf{R}_{\bullet\square}\|^3} + \frac{\mathbf{R}_{\bullet\odot}}{\|\mathbf{R}_{\bullet\odot}\|^3} \right) = \sum_{\bullet\neq\odot} \left(\mathbf{K}_\square^{(\bullet)} - \mathbf{K}_\odot^{(\bullet)} \right)$$

$$= \nabla_\square \sum_{\bullet\neq\odot} \mu_\bullet \left(\frac{1}{R_{\bullet\square}} - \frac{\mathbf{R}_{\odot\square} \cdot \mathbf{R}_{\odot\bullet}}{\|\mathbf{R}_{\odot\bullet}\|^3} \right) \equiv \nabla_\square \mathcal{D}_\square \qquad (5.17)$$

where $\nabla_\square$ denotes the gradient of with respect to the spatial coordinates of the spacecraft, and N_{ncb} is the number of non-central bodies considered in computing the gravitational perturbation on the sailcraft, which we call "planetary" for simplicity. In words, each planetary gravitational perturbation intervenes as the *difference* of two accelerations: one is the Keplerian acceleration on the spacecraft due to the considered planet, the other one is the acceleration of the Sun free-falling in the field of the planet. The same feature holds for any body, natural or artificial, which orbits close to a planet. This is why the Moon follows a geocentric orbit even though the Keplerian acceleration due to the Sun is higher than that due to the Earth. $\mathcal{D}_\square$ on the rightmost side of (5.17) represents the gravitational *disturbing/perturbing function* or potential, or perturbation function, according to various authors. The first term of this disturbing function sometimes is called the *direct* term, whereas the second one is referred to as the *indirect* term. Mathematically, this latter term is due to the used transformation; physically, its meaning is just that mentioned above, namely, the reference body (in this case, the Sun) *free-falls* in the field of the other bodies. The 'strangeness' of the disturbing function is that it does not vanish, like a usual potential, as the space vehicle (or another body like a comet) recedes more and more from the Sun. It has been extensively studied in Celestial Mechanics (e.g. [5, 36]).

The reader might object that the form of Eqs. (5.16) could be applied even to an advanced rocket with a very, very high exhaust speed so that its continuous thrusting may entail a vehicle mass almost constant. In principle, this is correct. However, as we shall see in Chap. 6, the radiation-pressure thrust scales as the local solar gravity at any (practical) distance from the Sun: this is not a full disadvantage at large distances, as one might think at first glance, standing what we learnt in Sect. 1.5.

---

[11] A constant mass is not strictly necessary, but helps to highlight sailing properties.

5.3 Equations of Heliocentric Motion

At low distances, as we shall see in the following chapters, this is one of the keys to fast solar sailing. In contrast, real throttling of very advanced ion engines might be around a factor 2–3 of the nominal thrust. Furthermore, either high thrust or high jet speed can be got for long time. Any really high-thrust *and* high-exhaust-beam rocket propulsion system may meet some showstopper, in practice, or sometimes after a deep analysis. Radiation-pressure thrust acceleration depends on mechanical and thermo-optical quantities. A good feeling about such points can be inferred from technology and algorithms discussed in Chap. 6.

Now, let us proceed to characterize the thrust acceleration by using the Heliocentric Orbital Frame (HOF) in its *standard* definition; (in attitude control problems, the orientation of the axes can differ from the current one, but the coordinate planes are the same). We have emphasized the adjective 'standard' because, subsequently, we will generalize this reference frame for fast sailing. Therefore, the origin of HOF is the sailcraft barycenter, the Z-axis is the direction $\mathbf{h}_\square$ of the sailcraft's orbital angular momentum per unit mass, say, $\mathbf{H}_\square = \mathbf{R}_\square \times \mathbf{V}_\square$, and the X-axis is the direction $\mathbf{r}_\square$ of the vector position $\mathbf{R}_\square$. We have dropped the symbol of the Sun in the sailcraft-related quantities since there is no ambiguity. The Y-axis completes the counterclockwise triad; therefore, its direction lies on the plane of position and velocity. Such a frame, which plainly is not inertial, is a special case of Spacecraft Orbital Frame (SOF). SOF is not only useful for trajectory maneuvers and attitude control description, but has a conceptual importance; as a point of fact, what matters is the (total) *non*-gravitational acceleration that locally is *sensed onboard* a real spacecraft, and measured via a system of accelerometers, as a consequence of the interaction of some field(s) with the space vehicle. This is allowed by the Principle of Equivalence.

There is a time difference between HOF and SOF, and arises from Relativity. As a point of fact, though both are local and barycenter-centered and may have the (local) axes parallely oriented, nevertheless time intervals are different; if dt is the coordinate-time interval between two (temporally close) events at the same location in HOF, then the proper-time interval for the same events is given by d$\tau = \sqrt{g_{00}}$dt, where g_{00} is the time-time entry of the metric tensor of the gravitational field. However, for the moment, we will not consider such feature here.

The sailcraft thrust acceleration in either SOF or HOF has to be transformed to the inertial reference frame where one has chosen to describe the vehicle motion. This concept does not regards solely time and acceleration, but also observations of external object accomplished onboard for reconstructing the vehicle's trajectory and attitude states. In our case, let us introduce the sailcraft's *lightness vector*, which is the thrust acceleration referred to the solar local gravitational acceleration:

$$\mathbf{L} \equiv (\mathcal{L}_r \ \mathcal{L}_t \ \mathcal{L}_n)^\top, \quad \mathcal{L} \equiv \|\mathbf{L}\|$$
$$\mathbf{L} = \mathbf{f}\big(t, \mathcal{O}, \sigma, \mathbf{n}, \mathbf{R}_{\odot\square}, \mathbf{V}_\square^{(\mathrm{HOF})}\big)$$

(5.18)

where $()^\top$ denotes transposition. $\mathbf{n}$ is the sail orientation resolved in HOF. For reasons that will be clear later, this vector has been chosen to lie in the semispace non-containing the sail's reflective layer, namely, $\mathbf{n}$ is the orientation of the sail backside. σ denotes the sailcraft mass to sail area ratio, namely, the sailcraft sail loading defined in Sect. 3.2. The arguments indicated in the vector function $\mathbf{f}$ are the main ones. Other quantities, on which $\mathbf{L}$ depends, will be discussed in Chap. 6.

Remark 5.1 Two properties from the second line of expressions (5.18) will be used in calculations. *First*, $\mathbf{L}$ could depend explicitly on time; *second*, it may depend explicitly on coordinates after trajectory optimization. However, one could not get $\mathcal{L}_r$ variable, while $\mathcal{L}_t$ and $\mathcal{L}_n$ are non-zero constant. On the other side, $\mathcal{L}_t$ and $\mathcal{L}_n$ could explicitly be time-dependent, while $\mathcal{L}_r$ keeps constant.

Anticipating some result of Chap. 6, it is important to realize that for many purposes $\mathbf{L}$ is independent of the Sun-sailcraft distance. This is due to the total solar irradiance (on which thrust depends linearly) that scales as $1/R^2$ for $R_\square \gtrsim 15 r_\odot \cong 0.07$ AU. We call the magnitude of $\mathbf{L}$ the *lightness number*, and its components in HOF the *radial*, *transversal* and *normal* numbers, respectively . This notion of lightness number represents the (Euclidean) norm of a vector function, which varies with time, in general, according to some mission-peculiar optimization program. It is precisely such *vector* function that determines the sailcraft trajectory evolution.

The sailcraft thrust acceleration, resolved in HIF, can be then expressed via $\mathbf{L}$ as follows:

$$\frac{\mathbf{T}_\square}{m_\square} = \Xi \frac{\mathbf{T}_\square^{(\mathrm{HOF})}}{m_\square} = \frac{\mu_\odot}{R_\square^2} \Xi \mathbf{L} \tag{5.19a}$$

$$\Xi = \left(\mathbf{r}_\square \quad \mathbf{h}_\square \times \mathbf{r}_\square \quad \mathbf{h}_\square \right) \tag{5.19b}$$

where the orthogonal matrix Ξ represents the rotation of the HOF axes onto the HIF axes. The columns of Ξ are the unit vectors (or the direction cosines), resolved in HIF, of the HOF axes according to the previous definition and nomenclature. Of course, $R_\square \equiv \|\mathbf{R}_{\odot\square}\|$. Matrix Ξ has to be computed during the integration of the sailcraft trajectory. (In the case of fast sailing, its computation is not as straightforward as one might think at first glance.) Therefore, the complete equation of the motion of a heliocentric sailcraft can be cast into the form:

$$\ddot{\mathbf{R}}_\square = -\mu_\odot \frac{\mathbf{R}_\square}{\|\mathbf{R}_\square\|^3} + \frac{\mu_\odot}{R_\square^2} \Xi \mathbf{L} + \nabla_\square \mathcal{D}_\square \tag{5.20}$$

Equation (5.20) can be simplified after the following considerations that hold in the current framework:

1. in fast solar sailing, unless the sail is steered such that radiation-pressure thrust is very small, the lightness numbers may be kept at constant values throughout the flight;

5.3 Equations of Heliocentric Motion

2. if some trajectory optimization is performed, the optimal $\mathbf{L}$'s components may be piecewise constant or time changing, but such that the radiation-pressure acceleration is comparable or higher than the local solar gravity, independently of $R_\square$;
3. as we shall see in the following chapters, an interplanetary fast sailcraft does not need planetary flybys to get energy; thus, its distance from planets is rather large;
4. as a consequence of points 1–3, and noting that for Jupiter $\mu_{2_+}/\mu_\odot \cong 9.55 \times 10^{-4}$, one can drop the perturbing planetary accelerations in (5.16) for analytical studies aimed at highlighting the solar sailing features. In computing *numerical* trajectories of interplanetary sailcraft, many planetary perturbations shall be taken into account, as we will see in the next chapters, especially in Chap. 8. This has the additional advantages to check whether the sailcraft, in its launch window for a given mission, flies too close to some planet to receive some *unnecessary* perturbation.

These remarks are to simplify the analytical studies of the heliocentric motion of a sailcraft (especially a fast one), and entail that HIF is considered as an *inertial* reference system; recalling what said about the Sun's motion, such an approximation is well supported. Thus, the following vector equation

$$\ddot{\mathbf{R}}_\square = -\mu_\odot \frac{\mathbf{R}_\square}{\|\mathbf{R}_\square\|^3} + \frac{\mu_\odot}{R_\square^2} \Xi \mathbf{L}$$
$$= \frac{\mu_\odot}{R_\square^2}[-(1-\mathcal{L}_r)\mathbf{r}_\square + \mathcal{L}_t\mathbf{h}_\square \times \mathbf{r}_\square + \mathcal{L}_n\mathbf{h}_\square] \tag{5.21}$$
$$\mathcal{L}_r \geq 0, \qquad \mathcal{L}_r^2 + \mathcal{L}_t^2 + \mathcal{L}_n^2 = \left\| \mathbf{f}\big(t, \mathcal{O}, \sigma, \mathbf{n}, \mathbf{R}_{\odot\square}, \mathbf{V}_\square^{(HOF)}\big) \right\|^2$$

is our current *work* equation of heliocentric sailcraft motion. For its final form, we substituted (5.19b) and (5.18) for Ξ and $\mathbf{L}$, respectively. Of course, any vector in (5.21) is resolved in HIF. The third line in (5.21) points out that the three parameters are not independent of each other for a given sail's material and either attitude or sail loading assigned, but above all that the radial number cannot be negative. Recalling what emphasized after (5.18), this means that, in practice, a conservative repulsive field acts on the sailcraft everywhere in the solar system, unless $\mathbf{n}$ is set by the attitude control system such that $\mathcal{L}_r = 0$; in this case, the whole $\mathbf{L}$ vanishes. Vector equations more general than Eq. (5.21) will be considered in Chap. 8, where the full properties of $\mathbf{L}$ will be explained in the optimization context.

Remark 5.2 One might object that Eq. (5.21) may be written directly from only few assumptions. However, the approach followed so far states *precisely* what are (1) the necessary assumptions, (2) the validity bounds of this motion equation, (3) some of its peculiar points (to be developed later), and (4) how it is related to the planetary ephemerides file from both conceptual and practical viewpoints. Not only, the adopted method to carry out the sailcraft equation of motion lends itself to a further generalization, which is the key for finding new trajectory families of SPS, and how

142   5 Fundamentals of Sailcraft Trajectory

to study them analytically and numerically, even though no general closed-form solution there exists. This is what will be done in the next sub-sections and in the subsequent chapters.

5.3.1 Sailcraft Motion in Radius-Longitude-Latitude Chart

Unless some special situation requires a change of coordinates for analyzing trajectory, we will use Eq. (5.21) in Cartesian coordinates. Nevertheless, in this section we will carry out the motion equations in sailcraft Radius-Longitude-Latitude (RLL) coordinates, which we denote by R, ψ, θ, respectively. These are connected to HIF in the usual way. We will use such equations here to characterize the forces acting on a sailcraft. Dropping the subscript $()_\square$ for simplicity, the local positively-oriented triad related to such coordinates are expressed in HIF as follows:

$$\mathbf{u}_R = \begin{pmatrix} \cos\theta\cos\psi \\ \cos\theta\sin\psi \\ \sin\theta \end{pmatrix}, \qquad \mathbf{u}_\psi = \begin{pmatrix} -\sin\psi \\ \cos\psi \\ 0 \end{pmatrix}, \qquad \mathbf{u}_\theta = \begin{pmatrix} -\sin\theta\cos\psi \\ -\sin\theta\sin\psi \\ \cos\theta \end{pmatrix}$$
$$(5.22)$$

Because $\mathbf{u}_R = \mathbf{r}$, we have to simply find the angle, say, β for rotating the unit vectors $(\mathbf{u}_\psi \mathbf{u}_\theta)$ onto $(\mathbf{h} \times \mathbf{r} \quad \mathbf{h})$ about $\mathbf{r}$. Thus, the transformation between bases can be written as follows:

$$\begin{pmatrix} \mathbf{r} \\ \mathbf{h} \times \mathbf{r} \\ \mathbf{h} \end{pmatrix} = \begin{pmatrix} 1 & 0 & 0 \\ 0 & \cos\beta & \sin\beta \\ 0 & -\sin\beta & \cos\beta \end{pmatrix} \begin{pmatrix} \mathbf{u}_R \\ \mathbf{u}_\psi \\ \mathbf{u}_\theta \end{pmatrix} = \begin{pmatrix} \mathbf{u}_R \\ \cos\beta\mathbf{u}_\psi + \sin\beta\mathbf{u}_\theta \\ \cos\beta\mathbf{u}_\theta - \sin\beta\mathbf{u}_\psi \end{pmatrix} \qquad (5.23)$$

Angle β has to be calculated with the right sign for accounting for the various motion situations, which one can summarize in either inequalities $-\pi \leq \beta < \pi$ or $0 \leq \beta < 2\pi$.[12] It is immediate to recognize that β satisfies the following equations:

$$\mathbf{u}_\theta \times \mathbf{h} = \sin\beta\mathbf{u}_R \quad \Rightarrow \quad \sin\beta = \mathbf{u}_R \cdot \mathbf{u}_\theta \times \mathbf{h}$$
$$\mathbf{u}_\theta \cdot \mathbf{h} = \cos\beta \qquad\qquad\qquad\qquad\qquad (5.24)$$

At this point, we need an independent information about the vector $\mathbf{h}$. In order to achieve this goal *and*, then, to write the sailcraft acceleration by means of the lightness numbers defined previously, we will utilize standard vector algebra. Starting from the vector position expressed via the RLL-basis, we differentiate twice with respect to time, considering that this basis (like the HOF's basis) changes with time. The derivative of each basis vector, orthogonal to the vector itself, are function of the other two unit vectors. Thus, sailcraft's position, velocity, acceleration, and

---

[12]With the right hand side as strict inequality, the triad transformation is bijective.

5.3 Equations of Heliocentric Motion

angular momentum (and any other vector of interest) can be resolved in the RLL coordinates system. One easily gets the following equations:

$$\mathbf{R} = R\mathbf{u}_R$$

$$\mathbf{V} \equiv \dot{\mathbf{R}} = \dot{R}\mathbf{u}_R + R\cos\theta\,\dot{\psi}\mathbf{u}_\psi + R\dot{\theta}\mathbf{u}_\theta$$

$$\mathbf{A} \equiv \ddot{\mathbf{R}} = \left(\ddot{R} - R(\cos\theta\,\dot{\psi})^2 - 2R\dot{\theta}^2\right)\mathbf{u}_R$$

$$+ (R\cos\theta\,\ddot{\psi} + 2\cos\theta\,\dot{R}\dot{\psi} - 2R\sin\theta\,\dot{\theta}\dot{\psi})\mathbf{u}_\psi \qquad (5.25)$$

$$+ \left(R\ddot{\theta} + 2\dot{R}\dot{\theta} + R\sin\theta\cos\theta\,\dot{\psi}^2\right)\mathbf{u}_\theta$$

$$\mathbf{H} = \mathbf{R} \times \mathbf{V} = -R^2\dot{\theta}\mathbf{u}_\psi + R^2\cos\theta\,\dot{\psi}\mathbf{u}_\theta$$

$$\mathbf{h} = (\cos\theta\,\dot{\psi}\mathbf{u}_\theta - \dot{\theta}\mathbf{u}_\psi)/\sqrt{(\cos\theta\,\dot{\psi})^2 + \dot{\theta}^2}$$

Inserting the last equation of (5.25) into (5.24), we obtain

$$\sin\beta = \frac{\dot{\theta}}{\sqrt{(\cos\theta\,\dot{\psi})^2 + \dot{\theta}^2}}, \qquad \cos\beta = \frac{\cos\theta\,\dot{\psi}}{\sqrt{(\cos\theta\,\dot{\psi})^2 + \dot{\theta}^2}} \qquad (5.26)$$

so that Eq. (5.23) is completely solved. Finally, equalling the expression of $\mathbf{A}$ in (5.25) to the rightmost side of (5.21) and using (5.23), solved via (5.26), results in the sailcraft motion equations in the RLL coordinates:

$$\ddot{R} - R(\cos\theta\,\dot{\psi})^2 - 2R\dot{\theta}^2 = -(1 - \mathcal{L}_r)\frac{\mu_\odot}{R^2}$$

$$R\cos\theta\,\ddot{\psi} + 2\cos\theta\,\dot{R}\dot{\psi} - 2R\sin\theta\,\dot{\theta}\dot{\psi} = (\mathcal{L}_t\cos\beta - \mathcal{L}_n\sin\beta)\frac{\mu_\odot}{R^2} \qquad (5.27)$$

$$R\ddot{\theta} + 2\dot{R}\dot{\theta} + R\sin\theta\cos\theta\,\dot{\psi}^2 = (\mathcal{L}_t\sin\beta + \mathcal{L}_n\cos\beta)\frac{\mu_\odot}{R^2}$$

This formulation allows one to highlight trajectory properties oriented to the RLL coordinates, but retaining the physical properties of the lightness vector. We will use these equations in the next section.

5.3.2 Sailcraft Energy and Angular Momentum

The RLL-chart introduced in Sect. 5.3.1 entails calculating the curl of a generic vector $\mathbf{W} = W^{(R)}\mathbf{u}_R + W^{(\psi)}\mathbf{u}_\psi + W^{(\theta)}\mathbf{u}_\theta$. Again, we adopt the convention $\partial_q \equiv \frac{\partial}{\partial q}$ (which will be used interchangeably throughout this book). One can gets the curl in the RLL coordinate system by a straightforward application of the general curl expression in curvilinear coordinates:

$$
\begin{aligned}
R\nabla \times \mathbf{W} = {}& \left(\tan\theta\, W^{(\psi)} - \partial_\theta W^{(\psi)} + \sec\theta\, \partial_\psi W^{(\theta)} \right)\mathbf{u}_R \\
& + \left(-W^{(\theta)} + \partial_\theta W^{(R)} - R\partial_R W^{(\theta)} \right)\mathbf{u}_\psi \\
& + \left(W^{(\psi)} - \sec\theta\, \partial_\psi W^{(R)} + R\partial_R W^{(\psi)} \right)\mathbf{u}_\theta
\end{aligned}
\tag{5.28}
$$

Previously, we emphasized that solar-radiation pressure scales as $1/R^2$ everywhere in the solar system, in practice; therefore, one may suspect that solar-radiation acceleration on sailcraft may be the gradient of some gravity-like potential in terms of the RLL-coordinates. If it were so, the curl of the right-hand sides of Eqs. (5.27) would vanish identically. Applying the curl operator to this vector acceleration, say, $\mathbf{A}^{(\mathrm{RLL})}$ and using the angle β again for compacting the result, we get

$$
\nabla \times \mathbf{A}^{(\mathrm{RLL})} = \frac{\mu_\odot}{R^3}
\begin{pmatrix}
\tan\theta[-\sin^3\beta\mathcal{L}_\mathrm{n} + \cos\beta(1 + \sin^2\beta)\mathcal{L}_\mathrm{t}] \\
\cos\beta\mathcal{L}_\mathrm{n} + \sin\beta\mathcal{L}_\mathrm{t} \\
\sin\beta\mathcal{L}_\mathrm{n} - \cos\beta\mathcal{L}_\mathrm{t}
\end{pmatrix}
\tag{5.29}
$$

provided that $\mathbf{L}$ does not depend explicitly on the coordinates. Because the curl of the solar field is obviously zero, this is a particular case of a more general and basic property of SPS, which can be expressed in the following way:

Theorem 5.1 *The curl of the sailcraft thrust acceleration vector field, observed in HIF, vanishes if and only if both the non-radial components of the lightness vector are zeroed.*

Proof Let us set $\mathbf{L}^{(\mathrm{HIF})} = \Xi\mathbf{L} = \mathcal{L}_\mathrm{r}\mathbf{r} + \mathcal{L}_\mathrm{t}\mathbf{h} \times \mathbf{r} + \mathcal{L}_\mathrm{n}\mathbf{h}$, which is the lightness vector resolved in HIF. We dropped the subscript $\square$ for simplicity here. Let us take the curl of the sailcraft thrust acceleration in HIF:

$$
\begin{aligned}
\nabla \times \mathbf{T}/m = \nabla \times \left(\frac{\mu_\odot}{R^2}\mathbf{L}^{(\mathrm{HIF})} \right) &= \frac{\mu_\odot}{R^2}\nabla \times \mathbf{L}^{(\mathrm{HIF})} + \nabla\left(\frac{\mu_\odot}{R^2} \right) \times \mathbf{L}^{(\mathrm{HIF})} \\
&\equiv \mathbf{M}_1 + \mathbf{M}_2
\end{aligned}
\tag{5.30}
$$

The direction of $\mathbf{M}_1$ is generally different from that of $\mathbf{M}_2$. As a point of fact, $\mathbf{M}_1 \times \mathbf{M}_2 \neq 0$, in general.

(1) Suppose $\nabla \times \mathbf{T}/m = 0$ identically, namely, at every point $\{x, y, z\}$ of an arbitrary sailcraft trajectory. This means that $\mathbf{M}_1 = 0$ and $\mathbf{M}_2 = 0$ have to be satisfied as independent equations. As the gradient of any function of R produces a vector parallel/antiparallel to $\mathbf{r}$, $\mathbf{M}_2 = 0$ results straightforwardly in

$$
\mathbf{M}_2 = \frac{\mathrm{d}}{\mathrm{d}R}\left(\frac{\mu_\odot}{R^2} \right)(-\mathcal{L}_\mathrm{n}\mathbf{h} \times \mathbf{r} + \mathcal{L}_\mathrm{t}\mathbf{h}) = 0
\tag{5.31}
$$

5.3 Equations of Heliocentric Motion 145

This vector equation is satisfied if and only if $\mathcal{L}_t = \mathcal{L}_n = 0$ identically. Inserting them into $\mathbf{M}_1 = 0$, one gets

$$\mathbf{M}_1 = \frac{\mu_\odot}{R^2} \nabla(\mathcal{L}) \times \mathbf{r} = \frac{\mu_\odot}{R^2} \nabla(\mathcal{L}_r) \times \mathbf{r} = 0 \qquad (5.32)$$

because of identity $\nabla \times \mathbf{r} = 0$. From [42], one can see that the most general radial number depends explicitly from the coordinates via $R/r_\odot$, namely, $\mathcal{L}_r = \mathcal{L}_r(R)$ (i.e. when the sail senses the finite size of the Sun). Therefore, $\nabla(\mathcal{L}_r) \times \mathbf{r} = 0$, and the equation is satisfied.

(2) Now, let us assume $\mathcal{L}_t = \mathcal{L}_n = 0$ identically. Then, $\mathbf{M}_2 = 0$ and $\mathbf{M}_1 = (\mu_\odot/R^2)\nabla(\mathcal{L}_r) \times \mathbf{r}$ follow immediately. Since, with regard to coordinates, $\mathcal{L}_r = \mathcal{L}_r(R)$ at most, $\mathbf{M}_1 = 0$, namely, $\nabla \times \mathbf{T}/m = 0$ identically.

This completes the demonstration. □

As a simple but meaningful example, a planar trajectory developing on the reference plane of HIF has $m\nabla \times \mathbf{A}^{(HIF)} = \nabla \times \mathbf{T}$ directed along the normal to this plane and proportional to $\mathcal{L}_t$, as it is very easy to verify. Again, if we wanted a full field-conservative solar sailing on this plane, we should set $\mathcal{L}_t = 0$ all the flight, and viceversa. The generalization to any planar solar-sail trajectory is straightforward.

It appears clear that an arbitrarily-oriented heliocentric sail is an object subject to two conservative (radial) fields and one overall non-conservative (transversal) field. In Chap. 6, we shall analyze the very nature of the interaction between solar radiation and sail in detail.

The geometric meaning of this field splitting is now quantitatively clear, and the physical consequences are profound. As a point of fact, the sailcraft orbital energy per unit mass can then be expressed as follows:

$$E = \frac{1}{2}V^2 - (1 - \mathcal{L}_r)\frac{\mu_\odot}{R} \qquad (5.33)$$

Therefore, only the radial lightness number enters the orbital energy equation. Sailcraft senses the Sun as having a "modified gravity", with repulsive effect if the sailcraft's mass-to-area ratio is sufficiently low. Very interesting astrodynamical features stem from such property, many of them are widely discussed in [32]. Fast solar sailing will use such property as well.

With regard to energy rate, let us differentiate (5.33) with respect to time; recalling that for any vector $\mathbf{w}$ the following identity holds, $w\dot{w} = \mathbf{w} \cdot \dot{\mathbf{w}}$, where $w \equiv \|\mathbf{w}\|$, we get

$$\dot{E} = \mathbf{V} \cdot \left(\dot{\mathbf{V}} + (1 - \mathcal{L}_r)\frac{\mu_\odot}{R^2}\mathbf{r} \right) + \frac{\mu_\odot}{R}\dot{\mathcal{L}}_r$$
$$= \frac{\mu_\odot}{R^2}\mathbf{V} \cdot \mathbf{h} \times \mathbf{r}\mathcal{L}_t + \frac{\mu_\odot}{R}\partial_t\mathcal{L}_r \qquad (5.34)$$

where the r.h.s. of Eq. (5.21) has been substituted for $\dot{\mathbf{V}}$, and the last term holds *if* the radial number depends only and explicitly on time. Some considerations are in order.

146   5 Fundamentals of Sailcraft Trajectory

First, the normal lightness number does not affect the sailcraft energy. Second, the energy change is proportional to the transversal number. Third, the radial number affects the trajectory energy rate only through its time change, if any. To illustrate the different contributions, let us analyze the case for a sailcraft moving on a trajectory arc with the sail at a certain orientation; at a given time t, an attitude maneuver is carried into effect. As a consequence, both the radial and transversal numbers undergo a change that we model here by an *impulse* (for the moment, we ignore the change in the normal number since it has no influence on energy):

$$\mathcal{L}_r(t) = \mathcal{L}_r^- + \left(\mathcal{L}_r^+ - \mathcal{L}_r^-\right)\Theta(t - t) \tag{5.35a}$$

$$\mathcal{L}_t(t) = \mathcal{L}_t^- + \left(\mathcal{L}_t^+ - \mathcal{L}_t^-\right)\Theta(t - t) \tag{5.35b}$$

where here $\Theta(x)$ denotes the Heaviside[13] function equal to 0 for negative x and 1 for positive x. The superscripts in both equations denote values immediately before and after the maneuver time t. The $\mathcal{L}_r$ time derivative results in

$$\partial_t \mathcal{L}_r = \left(\mathcal{L}_r^+ - \mathcal{L}_r^-\right)\delta(t - t) \tag{5.36}$$

where $\delta(x)$ is the Dirac delta distribution. Now, let us calculate the sailcraft energy variation from a time $t_1 < t$ to a time $t_2 > t$; $t_2 - t_1$ is assumed to be sufficiently long to accumulate the effects from both these lightness components, but sufficiently short so the changes over $t_2 - t_1$ of the other quantities in the r.h.s. of Eq. (5.34) may be considered negligible with respect to their values at t. Thus, it is easy to get

$$\Delta E = \int_{t_1}^{t_2} \dot{E}\,dt = \left(\frac{\mu_\odot}{R^2}\mathbf{V}\cdot\mathbf{h}\times\mathbf{r}\right)_t \int_{t_1}^{t_2} \mathcal{L}_t(\xi)\,d\xi + \left(\frac{\mu_\odot}{R}\right)_t \int_{t_1}^{t_2} \partial_t\mathcal{L}_r(\xi)\,d\xi$$

$$= \left(\frac{\mu_\odot}{R^2}\mathbf{V}\cdot\mathbf{h}\times\mathbf{r}\right)_t \left[(t_2 - t)\mathcal{L}_t^+ + (t - t_1)\mathcal{L}_t^-\right] + \left(\frac{\mu_\odot}{R}\right)_t \left(\mathcal{L}_r^+ - \mathcal{L}_r^-\right)$$

$$= \left(\frac{\mu_\odot}{R^2}\mathbf{V}\cdot\mathbf{h}\times\mathbf{r}\right)_t \left[(t_2 - t_1)\mathcal{L}_t\right]_{\Delta\mathcal{L}_t=0} + \left(\frac{\mu_\odot}{R}\right)_t \left(\mathcal{L}_r^+ - \mathcal{L}_r^-\right) \tag{5.37}$$

where the third line corresponds to an attitude maneuver producing no change of $\mathcal{L}_t$. Equation (5.37) shows clearly how the radial and the transversal numbers act: $\mathcal{L}_r$ operates independently of $t_2 - t_1 > 0$ by altering the potential $\mu_\odot/R$; in contrast, the $\mathcal{L}_t$ contribution accumulates over $t_2 - t_1$ by working through the product of the transversal component $(\mathbf{V}\cdot\mathbf{h}\times\mathbf{r})$ of the velocity and the gravity acceleration. In particular, this contribution to change energy needs only $\mathcal{L}_t \neq 0$.

---

[13]Contrarily to the unit step function, Heaviside function is undefined at $x = 0$. The unit step function is not appropriate for defining an impulsive change because its time derivative is indeterminate at the switching instant.

5.3 Equations of Heliocentric Motion 147

Now let us proceed with calculating the time rate of the orbital angular momentum $\mathbf{H}$. We resort again to our work Eqs. (5.21):

$$\dot{\mathbf{H}} = \mathbf{R} \times \ddot{\mathbf{R}} = \frac{\mu_\odot}{R}(\mathcal{L}_t \mathbf{h} - \mathcal{L}_n \mathbf{h} \times \mathbf{r}) \tag{5.38}$$

from which the following basic information can be drawn

$$\mathbf{H} \times \dot{\mathbf{H}} = \frac{\mu_\odot}{R} H \mathcal{L}_n \mathbf{r} \tag{5.39a}$$

$$H\dot{H} = \mathbf{H} \cdot \dot{\mathbf{H}} = \frac{\mu_\odot}{R} H \mathcal{L}_t \quad \Rightarrow \quad \dot{H} = \frac{\mu_\odot}{R} \mathcal{L}_t \tag{5.39b}$$

where $H \equiv \|\mathbf{H}\| = \mathbf{H} \cdot \mathbf{h}$ is normally positive.[14] The radial lightness number does not affect the evolution of the orbital angular momentum. Instead, $\mathbf{H} \times \dot{\mathbf{H}}$ is tilted by $\mathcal{L}_n$, whereas $\mathcal{L}_t$ causes a change of the $\mathbf{H}$'s length. Let us focus again on what regards fast solar sailing. From the rightmost of Eq. (5.39b), one could infer what follows:

Suppose a sailcraft deploys its sail while moving on 1-AU heliocentric circular orbit on XY-plane and keeps the lightness vector fixed. Consider three cases that have an Aluminum-Kapton-Chromium multi-layer flat sail (with the backside normal vector) at constant $-15°$ from the X-axis of HOF. (Strictly speaking, L exhibits small variations during the flight, but they are negligible in the present context of highlighting major features.) According to Eqs. (5.37) and (5.39b), the negative transversal number produces a decrease of energy and angular momentum as time goes by, as shown in Fig. 5.3. We will focus on the H's change. Whereas one expects a downward spiral, somewhat unexpected is the strong deviation from such spiral for a mass-to-area ratio as low as 4 g/m^2. Correspondingly, H undergoes a significant lessening. In the third case, we have $\mathbf{L} = (0.3279, -0.08247, 0)$. Now, some questions arise:

Question 5.1 What happens if $\mathcal{L}_t$ increases in absolute value? One could practically get it by either decrease the mass-to-area ratio or decrease the negative azimuth of the sail axis, or both.

Question 5.2 Is-it reasonable that, after a sufficiently long time interval, (in our previous examples, longer than 1.2 years but significantly shorter for high-$\mathcal{L}$ sailcraft), the angular momentum may be driven to zero?

Question 5.3 What would happen if the propulsion continued with the same sail attitude in HOF?

The answers to such questions will display new sailcraft trajectory families, as we will see in Chap. 7. There, we will need another equation that is a consequence of the above equations. Combining Eqs. (5.34) and (5.39b), and recalling $H = RV \sin\varphi$,

---

[14]This equality will be generalized in the context of fast solar sailing (Chap. 7).

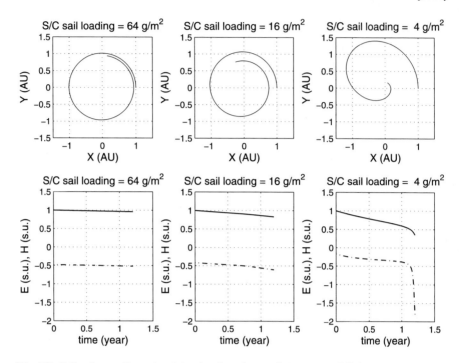

Fig. 5.3 Sailcraft two-dimensional deceleration via negative transversal lightness number. Sailcraft has been assumed to have a multi-layer flat sail of Kapton with Aluminum on the front side and Chromium on the backside. The backside normal is oriented at a prefixed azimuth of $-15°$ and zero elevation in HOF. The three cases differ by the total sailcraft mass on sail area ratio (i.e. the *sailcraft sail loading*): 64, 16, 4 g/m². This could entail different sail size, net payload and mission aims, of course; however, this is irrelevant to the current context. Units are the solar units, namely, $\mu_\odot \equiv 1$, R in AU. *Top row*: sailcraft trajectories for 1.2 years. *Bottom row*: evolutions of the orbital energy (*dot-dashed line*) and angular momentum per unit mass (*solid line*)

results in the following general relationship between energy rate and angular momentum rate

$$\dot{E} = H\dot{H}/R^2 + \mu_\odot \dot{\mathcal{L}}_r/R \qquad (5.40)$$

that is particular expressive when no change in the radial number takes place during the whole flight time, or in each trajectory arc of a piecewise-constant radial-number control.

5.3.3 Sailcraft Trajectory Curvature and Torsion

We will continue to denote the sailcraft-related quantities with no subscript in this section. Let us think of parametrizing a sailcraft trajectory by its arc length s, namely, $\mathbf{R}(s)$ with $s \in [s_0, s_f] \equiv \mathcal{S}$, $s_f > s_0$. According to the theory of curves in

5.3 Equations of Heliocentric Motion

differential geometry, $\mathbf{R}(s)$ describes a *regular* curve if $d\mathbf{R}(s)/ds \neq 0$ for any $s \in \mathcal{S}$. If, in addition, $\mathbf{R}(s)$ is three-times continuously differentiable with $d^2\mathbf{R}/ds^2 \neq 0$ everywhere in $\mathcal{S}$, then this trajectory is a Frenet curve in $\mathfrak{R}^3$. The related Frenet three-dimensional frame is then given by

$$\mathbf{e}_1 = d\mathbf{R}(s)/ds \qquad \text{the tangent vector} \qquad (5.41\text{a})$$

$$\mathbf{e}_2 = \frac{d^2\mathbf{R}(s)/ds^2}{\|d^2\mathbf{R}(s)/ds^2\|} \qquad \text{the principal normal vector} \qquad (5.41\text{b})$$

$$\mathbf{e}_3 = \mathbf{e}_1 \times \mathbf{e}_2 \qquad \text{the binormal vector} \qquad (5.41\text{c})$$

This *orthonormal* frame, co-moving with the sailcraft, evolves according to the following Frenet first-order linear differential equations (also called the Frenet-Serret equations, in literature)

$$\frac{d}{ds}\begin{pmatrix} \mathbf{e}_1 \\ \mathbf{e}_2 \\ \mathbf{e}_3 \end{pmatrix} = \begin{pmatrix} 0 & \kappa & 0 \\ -\kappa & 0 & \tau \\ 0 & -\tau & 0 \end{pmatrix}\begin{pmatrix} \mathbf{e}_1 \\ \mathbf{e}_2 \\ \mathbf{e}_3 \end{pmatrix} \qquad (5.42)$$

where κ and τ denote the trajectory *curvature* and *torsion*, respectively. In $\mathfrak{R}^3$, curvature is defined to be always non-negative, whereas torsion can have any sign. For plane curves, e.g. on the plane determined by $\mathbf{e}_1$ and $\mathbf{e}_2$, one can define an oriented or signed curvature: the positive sign indicates that the tangent vector rotates counterclockwise (as seen from $\mathbf{e}_3$), whereas the negative sign means clockwise rotation of the tangent; at inflection points, if any, $\kappa = 0$, and the tangent is stationary in direction. If the parameter of the curve $(x(t), y(t))$ is time, then the oriented or signed curvature can be expressed as

$$\breve{\kappa} = (\dot{x}\ddot{y} - \dot{y}\ddot{x})/\mathsf{u}^3 \qquad (5.43)$$

where u is the speed of the particle on the curve.

Frenet frame, curvature and torsion are invariant under the Euclidean group of symmetry. The $(\mathbf{e}_1, \mathbf{e}_2)$-plane is named the *osculating* plane. The $(\mathbf{e}_2, \mathbf{e}_3)$-plane is called the *normal* plane, whereas the $(\mathbf{e}_3, \mathbf{e}_1)$-plane is called the *rectifying* plane. A positive torsion means geometrically that, as the body moves along the trajectory, the osculating plane rotates counterclockwise about $\mathbf{e}_1$ (the direction of the body velocity).

A trajectory, fully lying on a plane of $\mathfrak{R}^3$, is characterized by $\tau = 0$ identically in $\mathcal{S}$, and viceversa. The binormal vector is constant and orthogonal to the osculating plane in every point of the trajectory. From the Frenet Eqs. (5.42) with $\tau = 0$, one gets

$$\frac{d^2\mathbf{e}_1(s)}{ds^2} + \kappa(s)^2\mathbf{e}_1(s) = 0 \qquad (5.44)$$

which gives the s-evolution of the unit tangent vector for a plane curve. This equation expresses the key role that the trajectory curvature has in the evolution of the tangent vector (or analogously the normal vector). In particular, for an origin-centered unit radius circle (the unit sphere or S^1), $\kappa(s) = 1$ or, equivalently, $\kappa(t) = \varpi/V$, where ϖ is the angular speed of the circular motion.

To parametrize a sailcraft trajectory in terms of its arc length is not just an easy thing. The natural parameter is time. In this framework, and according to the assumptions made above, the trajectory curvature and torsion, functions of time, are expressed by

$$\kappa(t) = \frac{\|\mathbf{V} \times \dot{\mathbf{V}}\|}{V^3} \tag{5.45a}$$

$$\tau(t) = \frac{\mathbf{V} \times \dot{\mathbf{V}} \cdot \ddot{\mathbf{V}}}{\|\mathbf{V} \times \dot{\mathbf{V}}\|^2} \tag{5.45b}$$

where sailcraft acceleration is given by Eq. (5.21). (The numerator of the r.h.s. of Eq. (5.45b) is one of the many ways to make $\det(\mathbf{V}, \dot{\mathbf{V}}, \ddot{\mathbf{V}})$ explicit.) Curvature and torsion have a common piece of information, namely, $\mathbf{V} \times \dot{\mathbf{V}}$. This can be calculated by means of a bit of vector algebra:

$$\frac{\mathbf{V} \times \dot{\mathbf{V}}}{V \mu_\odot/R^2} = (\cos\varphi \mathbf{r} + \sin\varphi \mathbf{h} \times \mathbf{r}) \times \left(-(1 - \mathcal{L}_r)\mathbf{r} + \mathcal{L}_t \mathbf{h} \times \mathbf{r} + \mathcal{L}_n \mathbf{h} \right)$$

$$= \sin\varphi \mathcal{L}_n \mathbf{r} - \cos\varphi \mathcal{L}_n \mathbf{h} \times \mathbf{r} + \left(\cos\varphi \mathcal{L}_t + \sin\varphi(1 - \mathcal{L}_r) \right) \mathbf{h} \tag{5.46}$$

We have used $\mathbf{V} = V(\cos\varphi \mathbf{r} + \sin\varphi \mathbf{h} \times \mathbf{r})$, φ denoting the counterclockwise angle from $\mathbf{R}$ to $\mathbf{V}$. The importance of such angle in fast solar sailing will be explained in Chap. 7. Inserting expression 5.46 into Eq. (5.45a), one gets

$$\kappa(t) = \frac{\mu_\odot/R^2}{V^2} \sqrt{\left(\cos\varphi \mathcal{L}_t + \sin\varphi(1 - \mathcal{L}_r) \right)^2 + \mathcal{L}_n^2} \tag{5.47}$$

In our framework, sailcraft trajectory is regular so that curvature does not diverge. However, there may be some trajectory limit, not strictly belonging to the class we want to analyze, characterized by $\kappa \to \infty$ as the sailcraft approaches a particular point.

In order to get additional information and analyze torsion as well, we need to calculate the sailcraft acceleration time rate, or the sailcraft *jerk*; to such aim, the time derivative of the HOF basis vectors shall be expressed in the same basis. Calculations are rather straightforward as soon as one uses Eqs. (5.38) and (5.39b). The order to be followed in calculating the time evolution of the HOF basis is: first $\dot{\mathbf{r}}$, then $\dot{\mathbf{h}}$, and $d(\mathbf{h} \times \mathbf{r})/dt$ last. The result can be re-arranged in a way similar to the Frenet equations:

$$\frac{d}{dt} \begin{pmatrix} \mathbf{r} \\ \mathbf{h} \times \mathbf{r} \\ \mathbf{h} \end{pmatrix} = \begin{pmatrix} 0 & \omega_r & 0 \\ -\omega_r & 0 & \omega_h \\ 0 & -\omega_h & 0 \end{pmatrix} \begin{pmatrix} \mathbf{r} \\ \mathbf{h} \times \mathbf{r} \\ \mathbf{h} \end{pmatrix} \tag{5.48}$$

$$\omega_r \equiv \frac{V \sin\varphi}{R} = H/R^2, \qquad \omega_h \equiv \mu_\odot \frac{\mathcal{L}_n}{R^2 V \sin\varphi} = \frac{\mu_\odot/R}{H} \mathcal{L}_n$$

5.3 Equations of Heliocentric Motion

These differential equations depend meaningfully on φ and $\mathcal{L}_n$, but no other component of the lightness vector is involved. Both the quantities ω have the dimensions of an angular speed; they completely determine how the HOF's basis changes with time. They are the analogous to Frenet parameters: in the time parametrization of sailcraft trajectory and HOF, ω_r plays the role of the curvature, and ω_h has the role of torsion; as a point of fact, ω_r is the in-plane angular speed, while ω_h is the out-of-plane angular speed induced by sail thrust. Frenet Eqs. (5.42) and Eqs. (5.48) will help us considerably in carrying out many properties of fast sailcraft trajectories. In particular, we shall recall and discuss Eqs. (5.48) in Sect. 7.1 in detail; for the moment, they allow us to compute the sailcraft jerk that results in:

$$\ddot{\mathbf{V}} \Big/ \left(\frac{\mu_\odot}{R^2} \right) = \left(\dot{\mathcal{L}}_r + \Omega_r(1 - \mathcal{L}_r) - \omega_r \mathcal{L}_t \right) \mathbf{r}$$

$$+ \left(\dot{\mathcal{L}}_t - \Omega_r \mathcal{L}_t - \omega_r(1 - \mathcal{L}_r) - \Omega_h \mathcal{L}_n^2 \right) \mathbf{h} \times \mathbf{r} \qquad (5.49)$$

$$+ \left(\dot{\mathcal{L}}_n - \Omega_r \mathcal{L}_n + \Omega_h \mathcal{L}_t \mathcal{L}_n \right) \mathbf{h}$$

$$\Omega_r \equiv 2\mathbf{R} \cdot \mathbf{V} / R^2, \qquad \Omega_h \equiv (\mu_\odot / R) / H$$

Equation (5.49) has been arranged such way to emphasize physical/geometrical meaning and make implementation on computer sufficiently easy. The sailcraft jerk depends non-linearly on the lightness vector components, and all jerk's components are coupled through frequency-like functions.

At this point, we have the sufficient pieces of information to make the sailcraft torsion explicit. Inserting Eqs. (5.46) and (5.49) into (5.45b), and using the square-root expression of (5.47), we get the result in terms of Ω and ω, the components of $\mathbf{L}$, and the sailcraft position-velocity angle again. Differently from curvature, the torsion depends on the time derivatives of the $\mathbf{L}$ components:

$$\mathcal{W} \equiv \left[\dot{\mathcal{L}}_r + \Omega_r(1 - \mathcal{L}_r) - \omega_r \mathcal{L}_t \right] \sin \varphi \mathcal{L}_n$$

$$- \left[\dot{\mathcal{L}}_t - \omega_r(1 - \mathcal{L}_r) - \Omega_r \mathcal{L}_t - \Omega_h \mathcal{L}_n^2 \right] \cos \varphi \mathcal{L}_n$$

$$+ \left[\dot{\mathcal{L}}_n - \Omega_r \mathcal{L}_n + \Omega_h \mathcal{L}_t \mathcal{L}_n \right] \left[\mathcal{L}_t \cos \varphi + (1 - \mathcal{L}_r) \sin \varphi \right]$$

$$= \mathcal{L}_n \sin \varphi \dot{\mathcal{L}}_r - \mathcal{L}_n \cos \varphi \dot{\mathcal{L}}_t + \left[\mathcal{L}_t \cos \varphi + (1 - \mathcal{L}_r) \sin \varphi \right] \dot{\mathcal{L}}_n \qquad (5.50)$$

$$+ \left[(\mathcal{L}_t^2 + \mathcal{L}_n^2) \cos \varphi + \mathcal{L}_t(1 - \mathcal{L}_r) \sin \varphi \right] \Omega_h \mathcal{L}_n$$

$$+ \left[(1 - \mathcal{L}_r) \cos \varphi - \mathcal{L}_t \sin \varphi \right] \omega_r \mathcal{L}_n$$

$$\mathcal{C} \equiv \left(\mathcal{L}_t \cos \varphi + (1 - \mathcal{L}_r) \sin \varphi \right)^2 + \mathcal{L}_n^2$$

$$\tau(t) = \mathcal{W} / V \mathcal{C}, \qquad \mathcal{C} \neq 0 \qquad (5.51)$$

Equation (5.51), which generalizes the torsion expression given in [41], holds in any interval of a piecewise-continuous control of the lightness vector. In particular, if $\mathcal{L}_n(t) = 0$ and $\mathcal{C}(t) \neq 0$ for the whole flight time, then torsion will vanish everywhere. Let us reason out about such aspect.

Suppose now that, with $\mathcal{L}_n = 0$, the sailcraft achieves a point where $\mathcal{C} = 0$, namely, $\cos\varphi\mathcal{L}_t + \sin\varphi(1 - \mathcal{L}_r) = 0$; this means that torsion is indeterminate, but curvature vanishes there. Let us be more precise. If we have $\mathcal{L}_r(t) = 1$ and $\mathcal{L}_t(t) = \mathcal{L}_n(t) = 0$ for any t in a *finite* time interval $\mathcal{T} \equiv [t_0, t_f]$, then torsion would be indeterminate, but curvature would vanish *identically* throughout $\mathcal{T}$. Regular trajectory would be a straight line, as physically expected. In all cases for which $\mathcal{L}_r \neq 1$ and/or $\mathcal{L}_t \neq 0$, but with $\mathcal{L}_n = 0$ over $\mathcal{T}$, there may be a countable sequence of times, say, $\{t_n\} \subset \mathcal{T}$ at which torsion is indeterminate, but equal to zero at any other time, i.e. $t \in \mathcal{T} \backslash \{t_n\}$. If the sequence is not void, such instants are characterized by particular values of $\varphi(t_n) \equiv \varphi_n$, e.g. $\tan\varphi_n = -\mathcal{L}_t(t_n)/(1 - \mathcal{L}_r(t_n))$. The two trajectory arcs at the left and the right of t_n, namely, in the time intervals $\mathcal{T}_{n-1} := (t_{n-1}, t_n)$ and $\mathcal{T}_n := (t_n, t_{n+1})$, are planar; as a point of fact, *geometrically*, it is possible to join two curves on two different planes at a single point such that the torsion in this one is not zero (e.g. [29]). This would entail a jump discontinuity in $\mathbf{V}(t_n)$; however, this is not *physically* possible because one is dealing with regular heliocentric trajectories driven by an acceleration-*limited* propulsion: low-thrust acceleration may undergo a jump discontinuity, but velocity may not. All this has two consequences: (1) the two trajectory arcs are coplanar, (2) $\lim_{t \to t_n^-} \mathbf{V}(t) = \lim_{t \to t_n^+} \mathbf{V}(t) = \mathbf{V}(t_n)$. (The reasoning can be repeated for any other element of $\{t_n\}$; if there were one element, e.g. $t_0 < t_1 < t_f$, then $\mathcal{T}_0 := [t_0, t_1)$ and $\mathcal{T}_1 := (t_1, t_f]$.) In turn, these properties allow us to assign $\tau(t_n) = 0$, for any n in the sequence, so that torsion vanishes everywhere in $\mathcal{T}$. Differently, aside from the special case of rectilinear arcs, which may be more than one (of course), curvature generally has non-zero values, but at $\{t_n\}$, as it is apparent from $\mathcal{C} = 0$ and Eq. (5.47). What said proves the following statement:

Proposition 5.1 *If SPS is accomplished via $\mathcal{L}_n = 0$ over a finite flight time, then the related trajectory may either: (1) lie on a straight line, or (2) lie on a plane. In the second case, curvature may vanish either over disjoint finite intervals or at the instants of a countable set, which may be empty as well.*

On the other hand, if the normal lightness number is equal to zero only at a single time, say, t, then

$$\tau(t) = \left(\dot{\mathcal{L}}_n / V \left(\cos\varphi\mathcal{L}_t + \sin\varphi(1 - \mathcal{L}_r) \right) \right)_{t=t} \tag{5.52}$$

where now the derivative of $\dot{\mathcal{L}}_n$ is important for $\tau(t)$ in a general 3D trajectory. All such considerations point out the contribution of the normal lightness number to shaping trajectory arcs.

5.4 Equations of Planetocentric Motion 153

The roles of the Frenet-frame and the HOF may seem redundant at first glance, since either can be got from the other one via rotation; as a point of fact, both are co-moving with the spacecraft barycenter and both are orthonormal. However, as the respective evolution equations of the two frames highlight, the Frenet-frame regards the curve's intrinsic properties, while the HOF is kinematical and needs another frame of reference (HIF) because the body's orbital angular momentum is implied. The Frenet-frame evolution equations (5.42) are still geometric in nature, while the HOF evolution equations (5.48) is of dynamical type as they refer explicitly to the acceleration fields generating the trajectory. This is not surprising, of course, since geometry and dynamics are interrelated to one another. Applied to sailcraft, this is just what we need for analyzing special families of trajectories.

We will finish this section by calculating the scalar product of velocity times acceleration as function of the thrust acceleration and position-velocity angle again; It results in:

$$\mathbf{V} \cdot \dot{\mathbf{V}} = V \dot{V} = V \frac{\mu_\odot}{R^2} \left(-\cos\varphi (1 - \mathcal{L}_r) + \sin\varphi \mathcal{L}_t \right) \qquad (5.53)$$

from which the along-track component of the total acceleration, i.e. $\dot{V}$, can be immediately carried out. A simple inspection of the right-hand side of Eq. (5.53) allows stating that:

Proposition 5.2 *For any heliocentric sailcraft, the along-track acceleration does not depend on $\mathcal{L}_n$, namely, the properties implied by (5.53) hold for two and three dimensional trajectories.*

5.4 Equations of Planetocentric Motion

Planetocentric sailcraft trajectories are important on their own. They, furthermore, are necessary for using solar sailing without resorting to any other propulsion system for escaping from planets. On a strict conceptual basis, planetocentric trajectory are different from the general heliocentric motion equations in three main aspects: (i) there is an additional lightness vector definable with respect to the planet's radiance, (ii) the sail attitude control has to be sufficiently fast (for most part of the trajectory) due to the sailcraft's orbital motion, (iii) eclipses may be frequent. We need four reference frames for describing the planetocentric motion of a sailcraft. Because the motion about Earth is the most interesting (at least for long time), we will write the sailcraft motion equations by referring it explicitly to the Earth. The ICRS/TT centered on the Earth barycenter, or ICRS/Earth, is the first frame; HIF and HOF have been already defined in Sect. 5.3. The fourth one is analogous to HOF, but with respect to the Earth; we name it the Geocentric Orbital Frame (GOF), centered on the sailcraft barycenter, and define it as follows: let $\mathbf{R}_{\oplus\square}$ and $\mathbf{U}_\square$ be the sailcrat's vector position and velocity, respectively, in ICRS/Earth. Denoting the angular momentum per unit mass by $\mathbf{J}_\square = \mathbf{R}_{\oplus\square} \times \mathbf{U}_\square$, the matrix of GOF basis, i.e. the rotation matrix from GOF to ICRS/Earth, is given by

$$\Phi = \left(\frac{\mathbf{R}_{\oplus\square}}{R_{\oplus\square}} \quad \frac{\mathbf{J}_\square \times \mathbf{R}_{\oplus\square}}{J_\square R_{\oplus\square}} \quad \frac{\mathbf{J}_\square}{J_\square} \right) \tag{5.54}$$

Rotation matrix Ξ, defined in Eq. (5.19b), can be calculated starting from the following relationships

$$\mathbf{R}^{(HIF)}_{\odot\square} = F\left(\mathbb{R}_{\odot\oplus} + \mathbf{R}_{\oplus\square} \right), \qquad \mathbf{V}^{(HIF)}_\square = F\left(\mathbb{V}_\oplus + \mathbf{U}_\square \right) \tag{5.55}$$

where the matrix F has been defined in Eq. (5.15). $\mathbb{R}_{\odot\oplus}$ and $\mathbb{V}_\oplus$ are the Earth's position and velocity, respectively, with respect to the Sun, and resolved in the ICRS/Sun that has its axes *parallel* to the ICRS/Earth's. Such vectors can be extracted directly from solar-system ephemeris files, e.g. the JPL DE405. We denote the sailcraft's lightness vectors with respect to the Sun and the Earth by $\mathbf{L}$ (again) and $\mathbf{L}^{(\oplus)}$, respectively. What does $\mathbf{L}^{(\oplus)}$ represent? It is the acceleration, normalized to the Earth's Keplerian gravity $K^{(\oplus)}_\square = \mu_\oplus / R^2_{\oplus\square}$, due to the pressure of the radiation emitted by the Earth.

The components of $\mathbf{L}^{(\oplus)}$ are specified in GOF. We will treat the Moon as a point-like body perturbing the sailcraft gravitationally *only*, namely, the lightness vector with respect to the Moon is assumed very small in the current context. In addition, considering that the sailcraft attitude control of a geocentric sailcraft spiralling outward might be sufficiently complicated to require *additional* actuators via (advanced) small rockets, we will add the sailcraft mass rate equation formally. (In principle, a mass change affects both lightness vectors.) Thus, the geocentric trajectory evolution of a sailcraft can be formulated as follows:

$$\dot{\mathbf{R}}_{\oplus\square} = \mathbf{U}_\square$$

$$\dot{\mathbf{U}}_\square = -K^{(\oplus)}_\square \frac{\mathbf{R}_{\oplus\square}}{R_{\oplus\square}} + \mathbf{P}^{(\oplus)}_\square$$

$$+ \left(K^{(\odot)}_\oplus \frac{\mathbf{R}_{\odot\oplus}}{R_{\odot\oplus}} - K^{(\odot)}_\square \frac{\mathbf{R}_{\odot\square}}{R_{\odot\square}} \right) + \left(K^{(\mathbb{C})}_\oplus \frac{\mathbf{R}_{\mathbb{C}\,\oplus}}{R_{\mathbb{C}\,\oplus}} - K^{(\mathbb{C})}_\square \frac{\mathbf{R}_{\mathbb{C}\,\square}}{R_{\mathbb{C}\,\square}} \right) \tag{5.56}$$

$$+ K^{(\odot)}_\square \Xi \mathbf{L} + K^{(\oplus)}_\square \Phi \mathbf{L}^{(\oplus)}$$

$$\dot{m} = -\dot{m}^{(Sail\ ACS)}_p$$

where we have denoted the Keplerian accelerations similarly to $K^{(\oplus)}_\square$:

$$K^{(\odot)}_\square = \frac{\mu_\odot}{R^2_{\odot\square}}, \qquad K^{(\mathbb{C})}_\square = \frac{\mu_{\mathbb{C}}}{R^2_{\mathbb{C}\,\square}}, \qquad K^{(\odot)}_\oplus = \frac{\mu_\odot}{R^2_{\odot\oplus}}, \qquad K^{(\mathbb{C})}_\oplus = \frac{\mu_{\mathbb{C}}}{R^2_{\mathbb{C}\,\oplus}}$$

$$\tag{5.57}$$

5.4 Equations of Planetocentric Motion

In Eq. (5.56), $\mathbf{P}_\square^{(\oplus)}$ represents all gravitational accelerations (in ICRS/Earth), due to the non-spherical and non-homogeneous Earth (including polar motion), assumed necessary to describe the sailcraft's geocentric trajectory with the desired accuracy in the related mission design. If the sailcraft is sufficiently close to the Moon, than analogous terms $\mathbf{P}_\square^{(\leftmoon\,)}$ should be added.

For sailcraft escaping from the Earth-Moon system without the support of launcher's upper stage,[15] a minimum-time escape trajectory should also exhibit a number of properties such as: (**a**) the parking/starting orbit should span an altitude range inside a zone of low space-debris flux (although an interplanetary sail will be designed to "resist" general micro-meteoroids, it should not accumulate punctures in high-flux zones), (**b**) the apogee sequence will be increasing monotonically during the orbit raising, (**c**) the perigee sequence will not violate the given safety minimum altitude, (**d**) the sail should not undergo a high number of large temperature jumps. Of course, properties (**a**) and (**d**) are particularly dependent on the escape time, which could be dramatically decreased by using some technological generation of sails exhibiting a sailcraft sail loading comparable or lower than the critical loading, and a feasible attitude control that produces a high *net* thrust acceleration per orbit. For such a sailcraft, there is no need to flyby the Moon for getting energy; as a result, no Moon-related launch window there will exist.

As a high lightness number sailcraft recedes from the Earth, the quantities $\mathbf{P}_\square^{(\oplus)}$ and $K_\square^{(\oplus)}\Phi\mathbf{L}^{(\oplus)}$ die away, whereas the acceleration $K_\square^{(\odot)}\Xi\mathbf{L}$ eventually becomes the dominant term out of all central and perturbing accelerations in Eq. (5.56). Depending on the terminal attitude control, the sailcraft could reverse its angular momentum $\mathbf{J}_\square$ with respect to the ICRS/Earth, not to be confused with the reversal of angular momentum in heliocentric trajectories (Chap. 7). Some practical aspects and remarks are in order:

1. In attitude dynamics, it is customary to define an orthonormal orbiting frame, also named the Local-Vertical Local-Horizontal Frame (LVLHF), centered on the vehicle barycenter, with the following axes: the third axis is oriented along $-\mathbf{R}_{\oplus\square}$, the first axis is in the plane $(\mathbf{R}_{\oplus\square}, \mathbf{U}_\square)$ and forms an acute angle with the velocity, and (consequently) the second axis is parallel to $-\mathbf{J}_\square$. The rotation matrix from LVLHF to GOF is given by

$$\Omega^{(\mathrm{GOF})}{}_{(\mathrm{LVLHF})} = \begin{pmatrix} 0 & 0 & -1 \\ 1 & 0 & 0 \\ 0 & -1 & 0 \end{pmatrix} \tag{5.58}$$

(The angles of a sailcraft-fixed reference frame with respect to LVLHF, one way of describing the sailcraft attitude, are called the *roll*, *pitch*, and *yaw*, respectively, from the aircraft-dynamics nomenclature.)

---

[15]There are some important reasons that may induce to not launch a sailcraft on an escape trajectory; nevertheless direct escape will be the preferred option for numerical optimization in Chap. 8.

2. In geocentric trajectory computation, to calculate the lightness vector with respect to the Sun may seem only an elegant formal algorithm; actually, the contrary is true in *accurate* numerical integration of the motion Eqs. (5.56). As a matter of fact, $\mathbf{L}$ depends on the sunlight *as actually observed* from the sail at any distance from the Earth; once one gets $\mathbf{L}$ from the thrust model, it is straightforward to perform the multiplication $\Xi\mathbf{L}$ (matrix Ξ at time t has been already stored too).

3. The lightness vector $\mathbf{L}^{(\oplus)}$ is somewhat complicated to calculate because both the Earth irradiance on the sail *and* the relative Earth-sail geometry are highly variable. Especially when a sailcraft is around its perigee, one should take into account the highly variable (and unpredictable during the flight design) global cloud and aerosol coverage, which affect the planetary albedo, and the thermal emission from different zones such as continents and oceans, each of which span large latitude and longitude ranges (i.e. with considerably different temperatures). In each orbit, the sailcraft sees night and day emissions; in addition, depending on the current sail orientation, either sail side can be Earth-irradiated, or *both* sides simultaneously could. One may use simplified models for Earth and Earth atmosphere behaviors; for instance, Earth surface, as a whole, may be modeled as a blackbody emitter at 288 K. Combining radiance and transmittance through the various atmospheric layers, the range 8–$13\ \mu m$ of the infrared (large) band, is the only one that may perturb a geocentric sailcraft trajectory appreciably. At 3 earth radii, a maximum $8\ W/m^2$ of thermal irradiance impinges on sail surface. Analogous simple models applied to the wavelength interval 0.4–$4\ \mu m$ give a maximum of about $17\ W/m^2$ of visible/near-infrared irradiance onto a sail. Such values tell us that, for instance, Earth radiance can perturb a $\mathcal{L} \approx 1$ sailcraft trajectory with accelerations of the same order of the accelerations caused by the oblate Earth, and much more than those ones due to the TSI fluctuations over a solar cycle.

4. When may the geocentric arc of an escaping sailcraft be considered "finished" and the frame switching be applied? This is an old and well-known problem in Celestial Mechanics, which needs an additional piece of comment in the case of a sailcraft autonomously escaping from the Earth-Moon system. Among different switching-frame conditions, that one expressing the non-negativity of the sailcraft energy with respect to the Earth-Moon system entails that one has *not* to include the $\mathcal{L}_r$ term, which is relevant to the orbital energy with respect to the Sun. One may add the potential energy coming from the radial number of $\mathbf{L}^{(\oplus)}$, but this one is very small at large distances from the Earth.

5.5 Thrust Maneuvering: a Few Basic Items

Differently from trajectories, to design an attitude control system for large sails should not present additional conceptual problems other than those ones already extensively dealt with in literature. At the time of this writing, the design of a real

5.5 Thrust Maneuvering: a Few Basic Items

three-axis stabilized deep-space large sailcraft, taking into account the properties related to fast sailcraft trajectories, is still to come.[16] Much probably, some additional care will be required for the control of sailcraft moving around the close perihelia of fast trajectories.

Many excellent books, papers, and journal's special issues deal with this key topic of solar sailing in great detail and generality; we cite (in chronological order) [12, 25, 37, 46], and [47–49], and the many references inside, and very recently many papers in [26]. A particularly interesting topic is sail attitude control via vanes; a recent paper [35] summarizes the results of many numerical studies.

Rather, in this section, we like to change our focus: instead of summarizing the (enormous) results from the just cited sources, we like to carry out some calculations that will allow us to point out the various types of thrust maneuver or, better, of the L-maneuvers admitted in solar-photon sailing. This will enlarge what was discussed in [51] and, recently, in [44]. We will find that the set of admissible thrust changes is larger than one might expect at first glance.

Note 5.1 In the current scope, it will be sufficient to resort to the acceleration model known in literature as the flat-sail model, e.g. [32, 33, 42, 51], and [44]. A significant evolution of such a useful model will be considered in Sect. 6.7.

We wrote L-maneuvers instead of attitude maneuvers because the first set, say, $\mathcal{U}_\mathbf{L}$ and the second set, say, $\mathcal{U}_\mathbf{n}$ satisfy the inequality

$$\mathcal{U}_\mathbf{n} \subseteq \mathcal{U}_\mathbf{L} \tag{5.59}$$

as we will find below. Usually, one associates a change of the sail axis orientation with a variation of the thrust vector due to TSI; however, the converse, may be not true, in general. As an example, if one envisages a sailcraft with two spacecraft as payload, then the first spacecraft can be released on a heliocentric bound orbit, while the second one and the sail proceed to an outbound leg. Mass jettisoning will cause a jump of the lightness number $\mathcal{L}$ with no change of the sail axis $\mathbf{n}$. These simple considerations show that the mission-dependent relationship (5.59) is satisfied. Nevertheless, here below, we will search for an even larger $\mathcal{U}_\mathbf{L}$.

Let us define a sail-attached frame centered on the sail center of mass, say, $O_\square$, and having the sail plane as the reference plane. For a four-boom supported sail, one may choose two nominally-orthogonal booms with directions $\mathbf{b}_1$ and $\mathbf{b}_2$ such that $\mathbf{n} = \mathbf{b}_1 \times \mathbf{b}_2$. For a spinning sail, $\mathbf{b}_1$ may be replaced by the direction of a special-use small object. The transformation between such frame, denoted by $\mathcal{F}^{(\text{sail})}$ here, and HOF may be time-dependent, and including a translation.

Now, we remind the reader that the sail of IKAROS, e.g. [22] and [53], has been endowed with a Reflectance Control Device (RCD) [39, 40], substantially a flexible multi-layer sheet where liquid crystals are able to change their reflectance by applied voltage; this experiment is the first realization of an old, but advanced, concept of

---

[16]The 200-m^2 sail of IKAROS is of spinning type.

158   5 Fundamentals of Sailcraft Trajectory

sail attitude control, first hinted in [51]. Essentially, synchronizing voltage variations
with the IKAROS' sail spinning phase turns into a control torque for changing the
spin direction. We like to generalize this concept in order to enlarge the class of
L-maneuvers.

Consider the overall reflective layer of a sail as consisting of two distinct materials: one covers the fraction $f_b = A_b/A$, where A is the total reflecting area, whereas
the other one spans the fraction $f_c = A_c/A = 1 - f_b$. For reasons that will later
be apparent, subscripts b and c stand for 'bias' and 'control', respectively. Using
polar coordinates on the reference plane of $\mathcal{F}^{(\text{sail})}$, each bias or control elemental area can be written as $dA_k(r_\square, \phi_\square)$, $k \in \{b, c\}$, and is oriented like $\mathbf{n}$, namely,
$dA_k = dA_k\mathbf{n} = (r_\square d\phi_\square dr_\square)\mathbf{n}$, in general. Let us define the moment of the bias
surface and control surface, respectively, with respect to $O_\square$ as

$$\mathbf{Y}_b = \int_{A_b} \mathbf{r}_\square \times d\mathbf{A}_b = \left(\int_{A_b} \mathbf{r}_\square \, dA_b\right) \times \mathbf{n} \qquad (5.60a)$$

$$\mathbf{Y}_c = \int_{A_c} \mathbf{r}_\square \times d\mathbf{A}_c = \left(\int_{A_c} \mathbf{r}_\square \, dA_c\right) \times \mathbf{n} \qquad (5.60b)$$

Vector $\mathbf{Y}$ is a measure of the asymmetry of the distribution of the related reflective
material on the support layer; if it is arranged symmetrically with respect to $O_\square$,
then $\mathbf{Y}_b = \mathbf{Y}_c = 0$. Of course, all such considerations entail that the reflective materials are optically homogeneous as well, otherwise the elemental areas would reflect
differently from each other, and should be weighed with non-unit factors, a situation
that would complicate somewhat the current framework without a real benefit.

Note 5.2 In what follows, we are using the normal, but not orthogonal, basis $\{\mathbf{n}, \mathbf{u}\}$,
where $\mathbf{u}$ is the direction of sunlight as observed in HOF, for expressing the vector
acceleration. Within the current framework, this direction can be considered fixed
in HOF and resolved as $\mathbf{u} = (1, 0, 0)$. We will adapt the simple formalism in [44],
pp. 189–191, to the current context of bias and control surfaces; there, the lightness
vector was made explicit so that, here, the elemental accelerations are carried out
straightforwardly. In Sect. 6.7, we will extend the flat-sail model in the context of
modeling the sunlight-sail interaction by highlighting physics and geometry.

The elemental acceleration, resolved in HOF, due to the sunlight pressure acting
on and interacting with the surface dA_b can be expressed as

$$d\mathbf{a}_b = g_m\mathbf{n}(t) \cdot \mathbf{u}\mathbf{B}(t) \, dA_b \qquad (5.61a)$$

$$g_m = \frac{\mu_\odot}{R^2} \frac{\frac{1}{2}\sigma_{(cr)}}{m(t)}, \quad \sigma_{(cr)} \equiv 2\frac{I_\odot}{cg_\odot} \cong 1.5368 \text{ g/m}^2 \qquad (5.61b)$$

$$\mathbf{B}(t) = P_b(\mathbf{n}(t) \cdot \mathbf{u}, \mathbf{p}_b)\mathbf{n}(t) + Q_b(\mathbf{p}_b)\mathbf{u} \qquad (5.61c)$$

where $I_\odot$ and $g_\odot$ denote the TSI and the solar gravitational acceleration at 1 AU,
respectively, whereas $\sigma_{(cr)}$ is called the critical sailing load; the reported value corresponds to $I_\odot = 1366 \text{ W/m}^2$, which is very close to the mean value of TSI in the

5.5 Thrust Maneuvering: a Few Basic Items

solar cycle 23. In Eqs. (5.61a)–(5.61c), we have emphasized the time-dependent terms of interest here. The term $\mathbf{n} \cdot \mathbf{u}$ comes from the Lambert law applied to the irradiance; it does not depend on the subsequent photon-sail interaction, and for this reason multiplies all terms of $\mathbf{B}$. Functions $P_b()$ and $Q_b()$ are both dependent on the bias layer's optical quantities that have been denoted by a vector here, whereas only $P_b()$ depends explicitly on $\mathbf{n} \cdot \mathbf{u}$. Mainly, $P_b()$ encompasses the photon momenta related to the specular and diffuse reflections, whereas $Q_b()$ accounts for the effect of absorption and the directional irradiance momentum.

With regard to the control surface, the elemental acceleration in HOF due to the sunlight pressure acting on and interacting with the surface $\mathrm{d}A_c$ can be written as

$$\mathrm{d}\mathbf{a}_c = g_m \mathbf{n}(t) \cdot \mathbf{u} \mathbf{C}(t) \, \mathrm{d}A_c \tag{5.62a}$$

$$\mathbf{C}(t) \equiv P_c(\mathbf{n}(t) \cdot \mathbf{u}, \mathbf{p}_c(t)) \mathbf{n}(t) + Q_c(\mathbf{p}_c(t)) \mathbf{u} \tag{5.62b}$$

where functions P_c and Q_c have the same physical meaning of P_b and Q_b, respectively, but the optical quantities $\mathbf{p}_c(t)$ of the control surface are expressed as function of time. This means that, analogously to IKAROS' RCD, the control surface is able to be changed in reflectance/absorptance by some applied electric voltage during any desired time interval, and then to be reset to the nominal values.

Remark 5.3 A feature of the flat-sail model is that each of the vectors $\mathbf{p}_b$ and $\mathbf{p}_c$ consists of optical quantities such as reflectance, absorptance, and diffuse-momentum coefficient; the corresponding net photon momenta are given the direction $\mathbf{n}$ or $\mathbf{u}$. This is a simplified picture with respect to what will be widely discussed in Chap. 6. Nevertheless, it is of meaningful help in modeling radiation-pressure thrust here.

Remark 5.4 Strictly speaking, the set U_q of the optical quantities, which the functions Q_j ($j \in \{b, c\}$) depend on, is contained in the set U_p appearing in P_j (e.g. [32] and [44]). However, we employed the same vectors $\mathbf{p}_j$ because the mathematical results are unchanged.

Now, we suppose that the effects from time variations of R and $I_\odot$ during the L-maneuver intervals are negligible from the trajectory viewpoint. $I_\odot$ fluctuations will be dealt with in the last chapter of this book, but the just mentioned assumption is appropriate to the current context. Therefore, the total elemental jerk is simply given by

$$\begin{aligned}
\mathrm{d}\mathbf{J} = \mathrm{d}\mathbf{J}_b + \mathrm{d}\mathbf{J}_c &= \frac{\mathrm{d}}{\mathrm{d}t}(\mathrm{d}\mathbf{a}_b + \mathrm{d}\mathbf{a}_c) \\
&= g_m \dot{\mathbf{n}}(t) \cdot \mathbf{u}\big(\mathbf{B}(t) \, \mathrm{d}A_b + \mathbf{C}(t) \, \mathrm{d}A_c\big) \\
&\quad + g_m \mathbf{n}(t) \cdot \mathbf{u}\big(\dot{\mathbf{B}}(t) \, \mathrm{d}A_b + \dot{\mathbf{C}}(t) \, \mathrm{d}A_c\big) \\
&\quad + \dot{g}_m \mathbf{n}(t) \cdot \mathbf{u}\big(\mathbf{B}(t) \, \mathrm{d}A_b + \mathbf{C}(t) \, \mathrm{d}A_c\big)
\end{aligned} \tag{5.63}$$

160　　　　　　　　　　　　　　　　　　　　5 Fundamentals of Sailcraft Trajectory

A good consequence of the surface homogeneity assumption is that we may integrate $d\mathbf{J}$ (which is resolved in HOF) with respect to area, employ the above-defined f_b and f_c, and divide the result by $g_m A$. One can obtain easily

$$\mathbf{J} = \int_{A_b} d\mathbf{J}_b + \int_{A_c} d\mathbf{J}_c$$

$$= g_m\left(\dot{\mathbf{n}}(t) \cdot \mathbf{uB}(t) + \mathbf{n}(t) \cdot \mathbf{u\dot{B}}(t)\right)A_b + \dot{g}_m\mathbf{n}(t) \cdot \mathbf{uB}(t)A_b$$

$$+ g_m\left(\dot{\mathbf{n}}(t) \cdot \mathbf{uC}(t) + \mathbf{n}(t) \cdot \mathbf{u\dot{C}}(t)\right)A_c + \dot{g}_m\mathbf{n}(t) \cdot \mathbf{uC}(t)A_c \tag{5.64}$$

$$\tilde{\mathbf{J}} \equiv \mathbf{J}/(g_m A) = \left(\dot{\mathbf{n}}(t) \cdot \mathbf{uB}(t) + \mathbf{n}(t) \cdot \mathbf{u\dot{B}}(t) + \frac{\dot{g}_m}{g_m}\mathbf{n}(t) \cdot \mathbf{uB}(t)\right)f_b$$

$$+ \left(\dot{\mathbf{n}}(t) \cdot \mathbf{uC}(t) + \mathbf{n}(t) \cdot \mathbf{u\dot{C}}(t) + \frac{\dot{g}_m}{g_m}\mathbf{n}(t) \cdot \mathbf{uC}(t)\right)f_c \tag{5.65}$$

Expanding the derivatives in (5.65) by using the definitions in Eqs. (5.61a)–(5.61c) and (5.62a), (5.62b), and re-arranging terms results in the following expression for the normalized jerk (which has the dimensions of a frequency); for higher readability, we will drop time t from the terms already defined clearly:

$$\mathcal{N} \equiv \mathbf{n}(t) \cdot \mathbf{u}, \qquad \dot{\mathcal{N}} = \dot{\mathbf{n}}(t) \cdot \mathbf{u}$$

$$\tilde{\mathbf{J}} = [P_b(\mathcal{N}, \mathbf{p}_b)f_b + P_c(\mathcal{N}, \mathbf{p}_c)f_c]\mathcal{N}\dot{\mathbf{n}}$$

$$+ \left[[P_b(\mathcal{N}, \mathbf{p}_b)f_b + P_c(\mathcal{N}, \mathbf{p}_c)f_c]\mathbf{n} + [Q_b(\mathbf{p}_b)f_b + Q_c(\mathbf{p}_c)f_c]\mathbf{u}\right]\dot{\mathcal{N}}$$

$$+ \left[\frac{\partial P_b(\mathcal{N}, \mathbf{p}_b)}{\partial \mathcal{N}}f_b + \frac{\partial P_c(\mathcal{N}, \mathbf{p}_c)}{\partial \mathcal{N}}f_c\right]\mathcal{N}\dot{\mathcal{N}}\mathbf{n}$$

$$+ \left[[P_b(\mathcal{N}, \mathbf{p}_b)f_b + P_c(\mathcal{N}, \mathbf{p}_c)f_c]\mathbf{n} + [Q_b(\mathbf{p}_b)f_b + Q_c(\mathbf{p}_c)f_c]\mathbf{u}\right]\left(-\frac{\dot{m}}{m}\right)\mathcal{N}$$

$$+ \left[\left(\frac{\partial P_c(\mathcal{N}, \mathbf{p}_c)}{\partial \mathbf{p}_c} \cdot \dot{\mathbf{p}}_c\right)\mathbf{n} + \left(\frac{\partial Q_c(\mathbf{p}_c)}{\partial \mathbf{p}_c} \cdot \dot{\mathbf{p}}_c\right)\mathbf{u}\right]\mathcal{N}f_c \tag{5.66}$$

Normalized jerk (5.66) needs to be explained in detail. We have arranged it in factorized "lines" according to the time-dependent terms; as a point of fact, each time derivative and its factors encompasses one or more modes of varying the resultant of the forces acting on the various elemental surfaces of the sail's reflective layers.

(1) The first line regards a mere change of sail orientation; this set of **L**-maneuvers includes those ones characterized by $\dot{\mathcal{N}} = 0$, for which $\mathbf{n}$ rotates about $\mathbf{u}$ at a constant (incidence) angle. In other words, $\mathcal{L}_r$ may be kept constant while $\mathcal{L}_t$ and $\mathcal{L}_n$ may be varied.

(2) The second line is different from zero if $\dot{\mathcal{N}} \neq 0$, namely, if the sunlight incidence angle is varied; this condition entails that $\dot{\mathbf{n}} \neq 0$. The effect of such a change appears along the directions $\mathbf{n}$ and $\mathbf{u}$.

5.5 Thrust Maneuvering: a Few Basic Items

(3) The third line is due not only to a change of the incidence angle, but also to the derivatives $\partial P_k/\partial \mathcal{N} \neq 0$, if any; the overall contribution is parallel to $\mathbf{n}$.

(4) The fourth line is not zero if there is a change of the sailcraft mass. Since the rocket thrust has not been inserted as a parallel primary propulsion, $\dot{m} \neq 0$ can be ascribed to either (i) the action of micro-rockets developing a pure torque for changing $\mathbf{n}$ in a finite time, (ii) the jettisoning of a mass (e.g. a spacecraft as the second "passenger" in a two-spacecraft sailcraft).[17] In the second case, $m(t)$ may be modeled via Heaviside function, analogously to what done on page 146 for a radial-number change.

(5) The fifth line, perhaps, is a bit unexpected. We can exclude the case $\mathcal{N} = 0$ here with no real loss of generality. This contribution to jerk is not zero if $f_c > 0$ *and* $(\partial P_c/\partial \mathbf{p}_c \cdot \dot{\mathbf{p}}_c, \partial Q_c/\partial \mathbf{p}_c \cdot \dot{\mathbf{p}}_c) \neq (0, 0)$. This one is the only line where solely appear the properties of the control surface. These property variations act parallely to $\mathbf{n}$ and $\mathbf{u}$, respectively. We can assume that such changes—induced by controlled voltage—occur so quickly to be modeled by a Dirac delta relatively to the activation time, say, t_p:

$$\dot{\mathbf{p}}_c(t) = (\mathbf{p}_c{}^+ - \mathbf{p}_c{}^-)\delta(t - t_p) \qquad (5.67)$$

with an obvious meaning of the symbols.

According to the standard flat-sail model, the partial derivatives in the fifth line Eq. (5.66) do not depend of $\mathbf{p}_c$; also, varying the optical quantities of the control surface is independent of any other modification in (5.66). As a result, one can calculate the change of the lightness vector due to controlled alteration of $\mathbf{p}_c$ as given by Eq. (5.67). Therefore, after substituting such function for $\dot{\mathbf{p}}_c$ in the fifth line, integrating from $t_p - \varepsilon$ to $t_p + \varepsilon$ with $\varepsilon > 0$ and arbitrarily small, and employing the definition of lightness vector together with g_m as set in Eqs. (5.61a)–(5.61c), one can carry out

$$(\Delta \mathbf{L})_{\Delta \mathbf{p}_c} = \left(\frac{1}{2} \frac{\sigma_{(cr)}}{\sigma} \right) \mathbf{n} \cdot \mathbf{u}$$

$$\times \left[\frac{\partial P_c}{\partial \mathbf{p}_c} \cdot (\mathbf{p}_c{}^+ - \mathbf{p}_c{}^-)\mathbf{n} + \frac{\partial Q_c}{\partial \mathbf{p}_c} \cdot (\mathbf{p}_c{}^+ - \mathbf{p}_c{}^-)\mathbf{u} \right] f_c \qquad (5.68)$$

where $\sigma = m(t)/A$ denotes the sailcraft sail loading. Equation (5.68) has two significant features:

(α) further supposing that the control sub-layer is built such that $\mathbf{Y}_c = 0$ (defined in $\mathcal{F}^{(\text{sail})}$), and it is activated wholly and simultaneously (a natural operation indeed), then no change of the sail axis occurs. In other words, one could have a sail thrust maneuver *without* re-orienting the sail axis;

---

[17]Such a sailcraft mission configuration has been analyzed in [45] with regard to the motion-reversal fast trajectory (see also Chap. 8). This mission concept will be extended in Chap. 8.

(β) the variation of acceleration takes place in the plane of $(\mathbf{n}, \mathbf{u})$, but not orthogonally, just like the nominal accelerations Eqs. (5.61a)–(5.61c) and (5.62a), (5.62b).

For spinning sails like IKAROS, the control surface has been activated anisotropically in order to produce a torque deviating the spin direction. It would be interesting to analyze the effect of points (α)–(β) for a three-axis stabilized sail; however, that is beyond the aims of this section.

References

1. Ablamowicz, R., Fauser, B. (2009), CLIFFORD—A Maple 12 Package for Clifford Algebra Computations, Version 12, August 8, 2009. http://math.tntech.edu/rafal/cliff12/index.html.
2. Acord, J. D., Nicklas, J. C. (1964), Theoretical and practical aspects of solar pressure attitude control for interplanetary spacecraft, guidance and control II. In Progress in Astronautics and Aeronautics (Vol. 13, pp. 73–101). New York: Academic Press.
3. The Astronomical Almanac for the Year 2009, The US Government Bookstore. http://bookstore.gpo.gov.
4. Angrilli, F., Bortolami, S. (1990), Attitude and orbital modelling of solar sail spacecraft. ESA Journal, 14(4), 431–446.
5. Beutler, G., Mervart, L., Verdun, A. (2005), Physical, Mathematical, and Numerical Principles: Vol. I. Methods of Celestial Mechanics. Berlin: Springer. ISBN 3-540-40749-9, ISSN 0941-7834.
6. Collins, G. W. II (2004), The Foundations of Celestial Mechanics. New York: Pachart Foundation dba Pachart Publishing House, WEB edition.
7. Doran, C., Lasenby, A. (2003), Geometric Algebra for Physicists. Cambridge: Cambridge University Press. ISBN 0-521-48022-1.
8. Dorst, L., Fontijne, D., Mann, S. (2007), Geometric Algebra for Computer Science. Amsterdam: Elsevier/Morgan Kaufmann. ISBN 978-0-12-374942-0.
9. Fairhead, L., Bretagnon, P. (1990), Astronomy & Astrophysics, 229, 240.
10. USNO (2003), http://maia.usno.navy.mil/conv2000/chapter10/software/.
11. Hestenes, D. (1999), New Foundations for Classical Mechanics (2nd edn.). Dordrecht: Kluwer Academic. ISBN 0-7923-5514-8.
12. Hughes, P. C. (2004), Spacecraft Attitude Dynamics. New York: Dover. ISBN 0-486-43925-9.
13. Hurley, D. J., Vandyck, M. A. (2000), Geometry, Spinors and Applications. Berlin: Springer/Praxis. ISBN 1-85233-223-9.
14. IERS (1996), Technical Note No. 21.
15. IERS (2002), Technical Note No. 29.
16. IERS (2004), Technical Note No. 31.
17. IERS (2004), Technical Note No. 32.
18. IERS (2006), Technical Note No. 34.
19. IERS (2009), Technical Note No. 35.
20. IERS (2010), Technical Note No. 36.
21. Irwin, A. W., Fukushima, T. (1999), A numeric time ephemeris of the Earth. Astronomy & Astrophysics, 17, 11.
22. JAXA (2010), IKAROS Project. http://www.jspec.jaxa.jp/e/activity/ikaros.html.
23. Joshi, V. K., Kumar, K. (1980), New solar attitude control approach for satellites in elliptic orbits. Journal of Guidance and Control, 3(1), 42–47.
24. Jet Propulsion Laboratory (2008), host name: ssd.jpl.nasa.gov, remote dir: pub/eph/planets/. Files and scientific documentation on solar-system dynamics can be downloaded via anonymous ftp server.

References

25. Kaplan, M. H. (1976), Modern Spacecraft Dynamics and Control. New York: Wiley. ISBN 0-471-45703-5.
26. Kezerashvili, R. (Ed.) (2011), Solar sailing: concepts, technology, missions. Advances in Space Research, 48(11), 1683–1926.
27. Klioner, S. A. (2007), Relativistic time scales and relativistic time synchronization. In Symposium on Problems of Modern Astrometry, Moscow, 24 October 2007.
28. Klingenberg, W. (1978), A Course in Differential Geometry. New York: Springer.
29. Kühnel, W. (2006), Student Mathematical Library: Vol. 16. Differential Geometry: Curves-Surfaces-Manifolds (2nd edn.). Providence: American Mathematical Society. ISBN 0-8218-3988-8.
30. Lewis, F. L. (1986), Optimal Estimation: with an Introduction to Stochastic Control Theory. New York: Wiley. ISBN 0-471-83741-5.
31. MacLane, S., Birkhoff, G. (1993), Algebra. New York: Chelsea.
32. McInnes, C. R. (2004), Solar Sailing: Technology, Dynamics and Mission Applications (2nd edn.). Berlin: Springer-Praxis. ISBN3540210628, ISBN 978-3540210627.
33. Mengali, G., Quarta, A., Dachwald, B. (2007), Refined solar sail force model with mission application. Journal of Guidance, Control, and Dynamics, 30(2), 512–520.
34. Modi, V. J., Kumar, K. (1972), Attitude control of satellites using the solar radiation pressure. Journal of Spacecraft and Rockets, 9(9), 711–713.
35. Quadrelli, M. B., West, J. (2009), Sensitivity studies of the deployment of a square inflatable solar sail with vanes. Acta Astronautica, 65, 1007–1027.
36. Roy, A. E. (2005), Orbital Motion (4th edn.). Bristol: Institute of Physics Publishing. ISBN 0-7503-10154.
37. Sidi, M. J. (1997), Spacecraft Dynamics and Control, a Practical Engineering Approach. Cambridge: Cambridge University Press. ISBN 0-521-78780-7.
38. Standish, E. M. (1998), Time scales in the JPL and CfA ephemerides. Astronomy & Astrophysics, 336, 381–384.
39. Tsuda, Y., Mori, O., Funase, R., Sawada, H., Yamamoto, T., Saiki, T., Endo, T., Kawaguchi, J. (2011), Flight status of IKAROS deep space solar sail demonstrator. Acta Astronautica, 69(9–10), 833–840.
40. Tsuda, Y., Mori, O., Funase, R., Sawada, H., Yamamoto, T., Saiki, T., Endo, T., Yonekura, K., Hoshino, H., Kawaguchi, J. (2011), Achievement of IKAROS—Japanese deep space solar sail demonstration mission. In The 7th IAA Symposium on Realistic Near-Term Advanced Scientific Space Missions, Aosta, Italy, 11–14 July 2011.
41. Vulpetti, G. (1996), 3D high-speed escape heliocentric trajectories by all-metallic-sail low-mass sailcraft. Acta Astronautica, 39, 161–170.
42. Matloff, G. L., Vulpetti, G., Bangs, C., Haggerty, R. (2002), The Interstellar Probe (ISP): Pre-Perihelion Trajectories and Application of Holography. NASA/CR-2002-211730.
43. Vulpetti, G. (2007), Geocentric-orbit-to-magnetosphere synchronization via solar sailing: about the general astrodynamical solution. In: 1st International Symposium on Solar Sailing (ISSS 2007), Herrsching, Germany, 27–29 June 2007.
44. Vulpetti, G., Johnson, L., Matloff, G. L. (2008), Solar Sails, A Novel Approach to Interplanetary Travel. New York: Springer/Copernicus Books/Praxis. ISBN 978-0-387-34404-1, doi:10.1007/978-0-387-68500-7.
45. Vulpetti, G. (2011), Reaching extra-solar-system targets via large post-perihelion lightness-jumping sailcraft. Acta Astronautica, 68(5–6). doi:10.1016/j.actaastro.2010.02.025.
46. Wertz, J. R. (Ed.) (1978), Spacecraft Attitude Determination and Control, Astrophysics and Space Science Library, New York: Reidel.
47. Wie, B. (2004), Solar-sail attitude, control and dynamics. Journal of Guidance, Control, and Dynamics, 27(4), 526–544.
48. Wie, B., Murphy, D. (2007), Solar-sail attitude control system design for a flight validation mission in sun-synchronous orbit. Journal of Spacecraft and Rockets, 44(4), 809–821.
49. Wie, B. (2008), AIAA Education Series. Space Vehicle Dynamics and Control (2nd edn.). ISBN 978-1-56347-953-3.

50. Wittenburg, J. (2008), Dynamics of Multibody Systems (2nd edn.). Berlin: Springer. ISBN 978-3-540-73913-5.
51. Wright, J. L. (1993), Space Sailing. New York: Gordon and Breach Science. ISBN 2-88124-842-X, ISBN 2-88124-803-9.
52. Yoon, J. H. (2001), Quasi-local conservation equations in general relativity. Physics Letters A, 292(3), 166–172. doi:10.1016/S0375-9601(01)00756-3.
53. Yokota, R., Miyauchi, M., Suzuki, M., Ando, A., Kazama, K., Iwata, M., Ishida, Y., Ishizawa, J. (2011), Heat sealable, novel asymmetric aromatic polyimide having excellent space environmental stability and application for solar sail: IKAROS membrane. In The 28th International Symposium on Space Technology and Science. 2011-o-4-02v.

Chapter 6
Modeling Light-Induced Thrust

Translating the (Scalar) Solar Radiation Pressure Into a (Vector) Acceleration Field   The basic equations of the heliocentric and planetocentric sailcraft motion have been dealt with in Chap. 5 with no explicit equation of the thrust acceleration. The lightness vector time history is necessary and sufficient for numerically integrating trajectories powered by a solar sail. This is a situation quite similar to the rocket-propelled trajectories for which in principle a few parameters suffice for computing trajectories, ignoring which particular engine the thrust comes from. Of course, in a real design of mission trajectories, we have to know how to realize the acceleration fields, and in the early design phases too.

Models of radiation-pressure thrust acceleration could contain many and more small effects, therefore resulting in a very sophisticated algorithm to be verified in flight. First-generation sailcraft exhibit very low thrust accelerations, and small effects result non-measurable: after all, they are beyond the mission's critical aims. On the other side, sailcraft for advanced missions will request the modeling of many relevant effects. Thus, in this chapter, we will set up a detailed model of sail thrust acceleration, which takes into account many factors related to the source of light and the sail, according to a rationale that might work as a general baseline. We shall derive the thrust vector explicitly, namely, chiefly as function of optical parameters, technological parameters, sail structure and sail orientation. Of course, any real mission shall ultimately validate the related acceleration model used for designing the mission.

In the course of about four decades of theoretical and numerical studies, some sail thrust models of increasing complexity have been proposed; in particular, relatively recent papers, such as [89] and [123], have tried to generalize aspects related to the objective difficulty to take a general sail shape into account. Other models (including a past algorithm by the author [103]) normally assume that the thrust may be calculated from hemispherical optical properties (to be determined) and some force directions—supposed to be known; this is not strictly true (though useful) even for the "specularly"-reflected light simply because a realistically specular reflection occurs about a lobe, and this one may be asymmetric with respect to the ideal specular direction. In general, many steps with very different features take place in the thrust

G. Vulpetti, *Fast Solar Sailing*, Space Technology Library 30,
DOI 10.1007/978-94-007-4777-7_6, © Springer Science+Business Media Dordrecht 2013

166   6 Modeling Light-Induced Thrust

algorithm. With no presumption of considering things in their (very wide) generality, however in this chapter we like to face with the various pieces of the same problem with the aim that the reader can realize how complicated this astrodynamical problem is indeed. We will put a special emphasis on the fact that sailcraft thrust, stemming from the interaction between solar photons and sail materials, is driven essentially by the *diffraction of light*. This may be viewed as a sort of necessary condition—the *second* one—to carry out SPS, which adds to the electromagnetic-wave radiation pressure, or, equivalently, the fact that photons carry momenta (in free or curved space-times).

This chapter is arranged in eight sections. Sections 6.1 and 6.2 regard the irradiance and the photon momentum the sail receives. Techniques of real-surface manufacturing are summarized in Sect. 6.3. Mathematical description of the sail's reflective layer and the related optical phenomena are presented in Sect. 6.4. Application of diffraction theories to the calculation of thrust are presented and discussed in Sect. 6.5; the complexity of the topics dealt with in this section have called for many subsections. Section 6.6 derives the thrust acting on a sailcraft from the momentum balance with respect to a control surface enclosing the sail. Section 6.7 is devoted to an important generalization of the conventional flat-sail model. Finally, conclusive remarks on the presented algorithms, and the related conceptual frameworks, can be found in Sect. 6.8.

6.1 Irradiance in the Sailcraft Frame

In this section, we report some results from Relativity for justifying certain expressions that should be applied to trajectories of fast sailcraft. We will use statements well-known in SR and GR. We again use HIF defined in Chap. 5, but we will replace TDB with the TCB we denote by t. With regard to the spacecraft frames, SOF was introduced in Sect. 5.3. Here, SOF is characterized by the use of the sailcraft's proper time $\tau(t)$, which is appropriate for describing local phenomena.

The goal of this section is the relationship between the solar radiance and the irradiance onto the sail $\mathcal{S}$, as sensed in its rest frame. (Any rotational motion of $\mathcal{S}$ with respect to the inertial frame is assumed to be quite negligible for power measurement). Let us detail the general context where we shall operate.

We know from Sect. 2.2 that the Sun is not a solid body and, observing it from a sufficiently distant point in space, one can see different depths at different wavelengths and zenithal angles. One notes that the observed solar radiance is not uniform over the solar *disk*. In particular: (1) the solar brightness decreases from the center to the observed disk edge, or the limb; (2) the radiation tends to be redden as the observer progressively looks toward the limb. The overall phenomenon is referred to as the *limb darkening* and regards the solar photosphere as a whole. As the Sun is not uniform in its properties, (in particular, temperature is height-dependent

6.1 Irradiance in the Sailcraft Frame

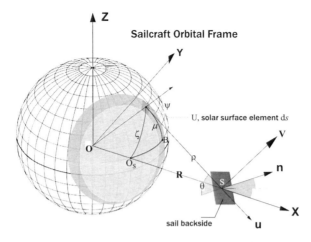

Fig. 6.1 Geometry (not to scale) used for describing the sail receiving photon power from the Sun. The sailcraft orbital frame differs from the heliocentric orbital frame by relativistic time. Adapted from [103]. From the radiance viewpoint, such geometrical picture contrasts the solar images from spacecraft, especially the recent ones from the SDO;they show the wide spatial and temporal variability of the solar radiance. The overall effect of variable TSIon sailcraft trajectories will be discussed in Chap. 9

as shown in Fig. 2.6), when one moves the observation line progressively from the disk center towards the limb, one can observe the opposite of the above effect for a number of spectral emission lines, i.e. a *limb brightening*. The contribution of such lines to the effective TSI is quite negligible for propulsion purposes.

Describing limb darkening quantitatively is not a simple job, as both local and general properties of the outer solar shells are very complicated, as we saw in Chap. 2. However, a simple model is particularly useful in analytic calculations; if one assumes local thermodynamic equilibrium and solar emittance constant over the whole electromagnetic spectrum, i.e. the so-called *gray-atmosphere* approximation, one can get a simple expression of the total radiance as function of the zenithal observation angle, say, ψ as follows

$$L_\odot(\psi) = \bar{L}_\odot(0) \frac{3\cos\psi + 2}{4} = L_\odot(0) \frac{3\cos\psi + 2}{5} \quad (6.1)$$

where $\bar{L}_\odot(0) = (W_\odot/4\pi r_\odot^2)/\pi = 20.08784$ MW m^{-2} sr^{-1} (with $W_\odot$ denoting the mean solar radiant power) is the mean solar radiance, and $L_\odot(0) = (5/4)\bar{L}_\odot$ is the radiance from the sub-sailcraft point. The above value corresponds to a mean TSI of 1366 W m^{-2} at $R = 1$ AU $= 1.4959787 \times 10^{11}$ m. Of course, there are more sophisticated models fitting the observed $L_\odot(\psi)$. Very important is also the spectral solar radiance, here denoted by $L_\odot(v_{em}, \psi)$, which will enter the irradiance calculations below.

Figure 6.1 shows the essential geometry of the current model. There, we have pictured the Sun as having a layer (the photosphere) of negligible thickness, but with a variable radiance, e.g. given by Eq. (6.1), approximating the real mean behavior

168   6 Modeling Light-Induced Thrust

of the solar radiance observed from sailcraft. The photosphere shape is assumed here to be spherical; as a point of fact, in the current context, the solar quadrupole moment $J_2^{(\odot)}$ is rather small, i.e. certainly $<10^{-6}$.

We assume that the sailcraft is immersed only in the Schwarzschild metric (e.g. [2, 107]) of the Sun (i.e. we will ignore planet contributions to the actual metric). As a result, the working metric is fully diagonal, and the related radial and angular coordinates are very close to the conventional spherical coordinates employed in actual observations. An important feature of this model is that any photon moves on a geodesic of null length: the so-called *geometric-optics approximation*.

Note 6.1 Some considerations regarding the assumed metric are in order for the sake of strictness. Strictly put, we should *not* consider the sailcraft in Schwarzschild metric, then add the propulsion from TSI (or the solar wind) and calculate the overall effects on trajectory. As a point of fact, this would mean a sort of mix of classical and relativistic concepts. The strict procedure would be to write the complete energy-momentum tensor by including the solar wind (mass density and pressure) and the TSI pressure, then to solve the Einstein's field equations. In other words, the spacecraft is *exterior* to the Sun, but at the same time it is *interior* to the *heliospheric fluid* including the Sun, the solar wind and the radiative gas. In our current context, solar radiation pressure intervenes in the energy-momentum tensor with a term proportional to $I_{(1\ \text{AU})}/(R^2 c^3)$; even close to the photosphere, the total effect of TSI-pressure is equivalent to $\sim 10^3$ protons per cubic centimeter. The solar-wind's dynamical pressure is negligible with respect to the radiation pressure, whereas its mass density is by far the most important term in this fluid energy density. However, from the photosphere to the termination shock, the order of magnitude of the mean mass of the solar wind may be computed as $\sim 10^{17}$ kg; therefore, the overall effect on the solar-system metric is completely negligible in our framework. Thus, restricting the field up to terms $\mathcal{O}(c^{-2})$ is completely satisfactory for our current aims; in particular, photons propagate according to the metric determined by the solar mass only. Again, in the current framework of sailcraft moving in interplanetary space, analogous considerations can be done for planets the gravity of which is considered (in this book) as established by the classical many-body problem of Celestial Mechanics.

The Schwarzschild line element is expressed meaningfully in spherical coordinates, say, $(R, \grave{\theta}, \grave{\phi})$ (for avoiding confusion with other angle symbols used in this chapter), where $\grave{\theta}$ is the colatitude:

$$c^2 \, \mathrm{d}\tau^2 = (1 - \mathfrak{r}_\odot/R)c^2 \, \mathrm{d}t^2 - (1 - \mathfrak{r}_\odot/R)^{-1} \, \mathrm{d}R^2 - R^2 (\mathrm{d}\grave{\theta}^2 + \sin^2 \grave{\theta} \, \mathrm{d}\grave{\phi}^2) \quad (6.2)$$

where t is the *coordinate* time, and $\mathfrak{r}_\odot \equiv 2GM_\odot/c^2 \cong 2.95$ km is the Schwarzschild radius of the Sun. Dividing both sides of Eq. (6.2) by $c^2 \, \mathrm{d}\tau^2$, recalling the expression of the squared speed in spherical coordinates, and re-arranging results easily in

6.1 Irradiance in the Sailcraft Frame 169

$$\left(\frac{d\tau}{dt}\right)^2 = 1 - \mathfrak{r}_\odot/R - V^2/c^2 - \frac{\mathfrak{r}_\odot/R}{1 - \mathfrak{r}_\odot/R}\frac{(dR/dt)^2}{c^2}$$

$$= 1 - V^2/c^2 - \mathfrak{r}_\odot/R + \mathcal{O}(c^{-2j}), \quad j = 2, 3, \ldots \quad (6.3)$$

Ignoring the terms of order higher than c^{-2}, one gets the relationship between coordinate and proper time appropriate to the current context:

$$dt = d\tau/\sqrt{1 - V^2/c^2 - \mathfrak{r}_\odot/R} \equiv \tilde{\gamma}(R, V)\,d\tau \quad (6.4)$$

$\tilde{\gamma}(R, V)$ may be called the generalized Lorentz factor. In the current framework, sailcraft are endowed with velocity $\mathbf{V}$ at the position $\mathbf{R}$ in the HIF/TCB; the velocity length V is the result of gravity *and* propulsion. Therefore, when Eq. (6.4) is applied to fast sailcraft, the two terms of the order c^{-2} may be comparable.

For our goal in this section, we have to follow the solar-photon energy from the emitting surface to the sail. In Fig. 6.1, $ds = r_\odot^2 \sin \zeta\, d\zeta\, d\mu$ is the surface element emitting the energy dE_{em} during the proper time $d\tau_{em} = \sqrt{1 - \mathfrak{r}_\odot/r_\odot}\,dt_{em}$,[1] where dt_{em} denotes the infinitesimal emission interval of coordinate time, and $r_\odot$ is the solar radius already used in Sect. 5.2.

Let $\mathcal{N}_{em}$ denote the number spectral radiance from the surface ds, namely, the number of photons in the range $[\nu_{em}, (\nu + d\nu)_{em}]$ emitted during $d\tau_{em}$ per unit solid angle and unit *projected* area:

$$d^4\mathcal{N}_{em} = \mathcal{N}_{em}\,d\nu_{em}\,d\omega\,ds_\perp\,d\tau_{em}$$

$$ds_\perp = ds \cos \psi \quad (6.5)$$

One should note that $\mathcal{N}_{em}$ includes the limb-darkening: it depends on the zenithal angle ψ as discussed at Eq. (6.1), and should not be confused with the ψ-dependent factor in the second equation of (6.5). The spectral exitance from ds is defined as

$$dM_{em} = h\nu_{em}\frac{d^4\mathcal{N}_{em}}{d\nu_{em}\,d\tau_{em}\,ds} = h\nu_{em}\frac{ds_\perp}{ds}\frac{d^4\mathcal{N}_{em}}{d\nu_{em}\,d\tau_{em}\,ds_\perp}$$

$$= h\nu_{em}\,\mathcal{N}_{em}\cos\psi\,d\omega \quad (6.6)$$

Now, when a number of electromagnetic waves emitted during dt_{em} travels for impinging on a moving object with velocity $\mathbf{V}_{obs}$, the same number of waves is able to achieve this target in a different coordinate-time interval generally depending on the component of $\mathbf{V}_{obs}$ parallel to the propagation direction $\mathbf{u}$ as measured in the frame of $\mathcal{S}$. This is a configuration of *classical* observation:

$$dt_{obs} = dt_{em}/(1 - \mathbf{u} \cdot \mathbf{V}_{obs}/c) \quad (6.7)$$

Such relationship can be very easily proved by using the world-lines of the signals of light and that of the observer in HIF. How the proper times of the photon emitter

---

[1] The sidereal speed of ds contributes about $(1/2)(4.8 \times 10^{-11})$ to the $\tilde{\gamma}$ value.

and the observer relate to one another can be obtained by combining Eqs. (6.4) and (6.7):

$$d\tau_{obs} = d\tau_{em} \frac{\tilde{\gamma}(r_\odot, 0)}{\tilde{\gamma}(R_{obs}, V_{obs})(1 - \mathbf{u} \cdot \mathbf{V}_{obs}/c)} \tag{6.8}$$

Therefore, one period $T_{em} = 1/\nu_{em}$ of the source corresponds to $T_{obs} \neq T_{em}$ (in general) as measured in the observer's frame of reference. As a consequence, the complete Doppler effect is expressed by

$$\nu_{obs} = \frac{\sqrt{1 - \tau_\odot/r_\odot}(1 - \mathbf{u} \cdot \mathbf{V}_{obs}/c)}{\sqrt{1 - V_{obs}^2/c^2 - \tau_\odot/R_{obs}}} \nu_{em} \equiv \mathcal{D}\nu_{em} \tag{6.9}$$

Now we are able to calculate the infinitesimal spectral irradiance onto the surface S measured in SOF, namely the energy input to an arbitrarily-oriented elementary surface, at a generic position in the solar field and with any velocity satisfying the strict positivity of Eq. (6.3), in its own frame of reference. Using again the solar number spectral radiance, and recalling the definition of spectral irradiance as given by Eq. (2.11), one gets

$$dI_{\nu_{obs}}^{(SOF)} = h\nu_{obs}(\mathcal{N}_{em}\, d\nu_{em}\, d\omega_{obs}\, ds_\perp\, d\tau_{em})/d\nu_{obs}\, dA_{obs}\, d\tau_{obs}$$

$$= (h\nu_{em}\mathcal{N}_{em}) \frac{\nu_{obs}}{\nu_{em}} \frac{d\nu_{em}}{d\nu_{obs}} \frac{d\tau_{em}}{d\tau_{obs}} \frac{ds_\perp}{dA_{obs}} d\omega_{obs} \tag{6.10}$$

where, in the second line, one can easily recognize: (1) the solar spectral radiance $h\nu_{em}\mathcal{N}_{em} = L_\odot(\nu_{em}, \psi)$, (2) the product of the first and second fractions equal to 1 (standing Eq. (6.9)), (3) the third fraction equal to the reciprocal of the fraction in Eq. (6.8) (i.e. the Doppler effect factor $\mathcal{D}$), and (4) the solid angle of observation expressible by $\cos\theta\, dA_{obs}/\rho_{obs}^2$ with ρ_{obs} denoting the distance between ds and dA_{obs}; ρ and θ are shown in Fig. 6.1. Thus, also recalling the second relationship in (6.5), the previous equation can be re-written as

$$dI_{\nu_{obs}}^{(SOF)} = L_\odot(\nu_{em}, \psi)\, \mathcal{D}\frac{\cos\psi \cos\theta}{\rho_{obs}^2}\, ds \tag{6.11}$$

where ν_{em} is the source frequency corresponding to the observational frequency, namely $\nu_{em} = \nu_{obs}/\mathcal{D}$.

The infinitesimal power per unit area onto S due to ds emitting in the range $[\nu_{em}, \nu_{em} + d\nu_{em}]$ follows straightforwardly

$$d^2 I^{(SOF)} = dI_{\nu_{obs}}^{(SOF)}\, d\nu_{obs} = dI_{\nu_{obs}}^{(SOF)}\mathcal{D}\, d\nu_{em}$$

$$= L_\odot(\nu_{em}, \psi)\mathcal{D}^2 \frac{\cos\psi \cos\theta}{\rho_{obs}^2}\, ds\, d\nu_{em} \tag{6.12}$$

which depends on the *square* of the Doppler factor; Eq. (6.11) coincides with Eq. (2.16) if $\mathcal{D} = 1$, namely, without gravity and relative motion between S and

6.1 Irradiance in the Sailcraft Frame

the Sun. $\mathcal{D}$ is a factor, in Eqs. (6.11), (6.12), which may be approximated to the first order in velocity without appreciable error in solar sailing, namely,

$$\mathcal{D} \cong (1 - \mathbf{u} \cdot \mathbf{V}_{obs}/c)(1 - \mathfrak{r}_\odot/2r_\odot)(1 + \mathfrak{r}_\odot/2R_{obs})$$
$$\mathcal{D}^2 \cong (1 - 2\mathbf{u} \cdot \mathbf{V}_{obs}/c)(1 - \mathfrak{r}_\odot/r_\odot)(1 + \mathfrak{r}_\odot/R_{obs}) \tag{6.13}$$

One can note that the r.h.s. of Eq. (6.13) has been arranged according to increasing factors. In fast sailing, $\mathbf{u} \cdot \mathbf{V}_{obs}$ causes the strongest deviation but, at most, in a small interval around the perihelion.

If one likes to express the above equations in terms of wavelength, as we did in Chap. 2 and 4, transformations from ν_{em} to λ_{em} and from $d\nu_{em}$ to $d\lambda_{em}$ are plain. Here, we go on using frequency unless otherwise indicated.

The whole solar disk visible to $\mathcal{S}$ is delimited by the small circle centered on the sub-sailcraft point, denoted by O_S in Fig. 6.1, and with angular radius $\zeta_{limb} = \arccos(r_\odot/R_{obs})$. (Since we are dealing with a surface in interplanetary space, we ignore the irradiance reduction due to a planet passing on the solar disk as seen in SOF). This circle encompasses a spherical area equal to $2\pi(1 - \cos\zeta_{limb})r_\odot^2 \equiv \Delta_\odot$.

Recalling the coordinates $\{\widehat{O_S U}, \widehat{BU}\} \equiv \{\zeta, \mu\}$ of the infinitesimal area ds in Fig. 6.1, the *spectral* irradiance onto $\mathcal{S}$ due to the electro-magnetic radiations stemming from $\Delta_\odot$ is then expressed by

$$I_{\nu_{obs}}^{(SOF)} = \int_{\Delta_\odot} dI_{\nu_{obs}}^{(SOF)}$$
$$= r_\odot^2 \int_0^{\zeta_{limb}} d\zeta \int_0^{2\pi} \sin\zeta \, L_\odot(\mathcal{D}^{-1}\nu_{obs}, \psi)\mathcal{D} \frac{\cos\psi \cos\theta}{\rho_{obs}^2} d\mu \tag{6.14}$$

whereas the *total* solar irradiance brought about by $\Delta_\odot$ results in

$$I^{(SOF)} = \int_0^\infty I_{\nu_{obs}}^{(SOF)} d\nu_{obs} = \int_0^\infty d\nu_{obs} \int_{\Delta_\odot} dI_{\nu_{obs}}^{(SOF)}$$
$$= r_\odot^2 \int_0^{\zeta_{limb}} d\zeta \int_0^{2\pi} \mathcal{D}^2 \sin\zeta \frac{\cos\psi \cos\theta}{\rho_{obs}^2} d\mu \int_0^\infty L_\odot(\nu_{em}, \psi) d\nu_{em} \tag{6.15}$$

Physically, Eqs. (6.14)–(6.15) are sufficiently intuitive; mathematically, their rightmost expressions rely on the well-known Fubini's theorem (in the version regarding the Riemann integrals) that is applicable here.

The following setting will be useful:

$$L_\odot(\psi) \equiv \int_0^\infty L_\odot(\nu_{em}, \psi) d\nu_{em} = \mathcal{D}^{-1} \int_0^\infty L_\odot(\mathcal{D}^{-1}\nu_{obs}, \psi) d\nu_{obs} \tag{6.16}$$

In our case, the sailcraft plays the role of observer, i.e. $\mathbf{R}_{obs} = \mathbf{R}$, $\mathbf{V}_{obs} = \mathbf{V}$, $\rho_{obs} = \rho$, $dA_{obs} = dA$. Integrals in Eqs. (6.14)–(6.15) may seem rather easy to calculate; instead, in the current model, the various quantities depend on the coordinates $\{\zeta, \mu\}$, the sailcraft position and velocity, and the sail orientation in SOF, whose backside normal $\mathbf{n}$ is indicated in Fig. 6.1. As previously mentioned, $\mathbf{u}$ is the photon propagation direction observed in SOF ($\mathbf{u} \neq \boldsymbol{\rho}/\rho$). In the current metric, photons can be deflected by the solar gravitational field; in addition, light's source can appear shifted because the sensor moves with respect to the source. This second effect is much higher than the other one for fast sailcraft, even for photons emitted from the photospheric region. Thus, in computing directions, we will ignore the GR effect and write the properties of the SR-related aberration as

$$\boldsymbol{\rho}/\rho \equiv \overline{\mathbf{u}}, \qquad (\overline{\mathbf{u}} \cdot \mathbf{V})\mathbf{V}/V^2 \equiv \overline{\mathbf{u}}_\parallel, \qquad \overline{\mathbf{u}}_\perp = \overline{\mathbf{u}} - \overline{\mathbf{u}}_\parallel$$

$$\mathbf{u} = [(1 - \overline{\mathbf{u}} \cdot \mathbf{V}/c)\gamma_V]^{-1}(\overline{\mathbf{u}}_\perp + \gamma_V \overline{\mathbf{u}}_\parallel - \gamma_V\, \mathbf{V}/c) \cong \frac{\overline{\mathbf{u}} - \mathbf{V}/c}{1 - \overline{\mathbf{u}} \cdot \mathbf{V}/c}$$

$$\mathbf{u} \times (\overline{\mathbf{u}} \times \mathbf{V}) = 0 \tag{6.17}$$

$$\cos\phi = \frac{\cos\overline{\phi} - V/c}{1 - \cos\overline{\phi}\,V/c} \cong \cos\overline{\phi} - \sin^2\overline{\phi}\,V/c$$

where γ_V is the usual Lorentz factor of SR, also used in Chap. 1. The second equation in (6.17) gives the apparent direction of light in SOF as function of the direction at rest. The third equation tells us that the three directions are coplanar, whereas the fourth one is obtained by scalar multiplication of the second one by the velocity direction $\mathbf{V}/V$; thus, the apparent view angle ϕ is related to the non-aberrated view angle $\overline{\phi}$.

From planar and spherical geometry, we get

$$\rho^2 = R^2 + r_\odot^2 - 2Rr_\odot \cos\zeta \tag{6.18a}$$

$$\sin\widehat{OSU} \equiv \sin\xi = (r_\odot/\rho)\sin\zeta \tag{6.18b}$$

$$\overrightarrow{OU} = r_\odot \begin{pmatrix} \cos\zeta \\ \cos\mu\sin\zeta \\ \sin\mu\sin\zeta \end{pmatrix} \tag{6.18c}$$

$$\boldsymbol{\rho} = \mathbf{R} - \overrightarrow{OU} = \begin{pmatrix} R - r_\odot\cos\zeta \\ -r_\odot\cos\mu\sin\zeta \\ -r_\odot\sin\mu\sin\zeta \end{pmatrix} \tag{6.18d}$$

$$\cos\psi = (R\cos\zeta - r_\odot)/\rho \tag{6.18e}$$

$$\cos\theta = \mathbf{u} \cdot \mathbf{n} \cong \frac{\overline{\mathbf{u}} - \mathbf{V}/c}{1 - \overline{\mathbf{u}} \cdot \mathbf{V}/c} \cdot \mathbf{n} \tag{6.18f}$$

$$\mathbf{u} \cdot \mathbf{V} = \mathbf{u} \cdot \begin{pmatrix} \cos\varphi \\ \sin\varphi \\ 0 \end{pmatrix} V$$

$$\cong \rho^{-1}[(R - r_\odot\cos\zeta)\cos\varphi - r_\odot\cos\mu\sin\zeta\sin\varphi]V \tag{6.18g}$$

6.2 Photon Momentum onto Sail

$$\mathbf{n} = \begin{pmatrix} \cos\delta\cos\alpha \\ \cos\delta\sin\alpha \\ \sin\delta \end{pmatrix}$$

where α and δ denote the azimuth and elevation, respectively, of $\mathbf{n}$ in SOF; $\mathbf{n}$ comes from the calculation of the sailcraft trajectory, either propagated with some prefixed $\mathbf{n}(t)$ or optimized with respect to some index of performance (Chap. 8). Angle φ is measured counterclockwise from $\mathbf{R}$ to $\mathbf{V}$; it will be generalized in Sect. 7.1. The rightmost side of Eqs. (6.18f)–(6.18g) has been obtained by approximating $\mathbf{u}$ by $\bar{\mathbf{u}}$.

When one computes spectral and/or solar irradiances, sailcraft's position and velocity, and sail orientation are given at a time $t(\tau)$; therefore, the various involved quantities can be considered functions as follows:

$$\psi = \psi(\zeta), \qquad \rho = \rho(\zeta) \tag{6.19a}$$

$$\theta = \theta(\zeta, \mu), \qquad \mathbf{u} = \mathbf{u}(\zeta, \mu), \qquad \mathcal{D} = \mathcal{D}(\zeta, \mu) \tag{6.19b}$$

Let us infer some features from the quantities in Eqs. (6.15)–(6.18g).

- A sail $\mathcal{S}$ receives sunlight through a (slightly deformed in SOF) cone of half-angle approximately given by $\xi_{limb} = \arcsin r_{\odot}/R$. Strictly speaking, only for $R \gg r_{\odot}$, the photon beam could be considered really parallel. This inequality may be considered as satisfied at 1 AU; the Sun appears with a radius of about $0.27°$.
- If $\mathbf{R} \cdot \mathbf{n} \cong 0$, there would be some $\mathbf{u} \cdot \mathbf{n} \lesssim 0$ as μ spans the related solar circles, namely, the sail's backside will be irradiated too; if $\mathbf{R} \cdot \mathbf{n}$ were exactly zero, then the two sail sides would be equally irradiated by a non-zero amount because the Sun is finite-size.[2] At very large incidence angles, other phenomena take place as we shall see in the following sections.
- For sailcraft performing a close solar flyby (Chap. 7), Doppler and aberration should not be neglected in highly-accurate trajectory computation because they are mainly linear in sailcraft speed.

Once we have defined the spectral and total irradiance on the sail, as observed in SOF, one may wonder what is the momentum rate, transported by the solar photons impinging onto the sail. This is the object of the next section.

6.2 Photon Momentum onto Sail

The product $I^{(SOF)}A_{sail}/c$ provides only the order of magnitude of the photon force acting onto a sail arbitrarily oriented, even though the sail orientation is included in Eq. (6.15). As we shall see later in this chapter, this is due to several causes. In this section, we begin with calculating the solar-radiation momentum that is the

---

[2]This feature is greatly enhanced when a sail is close to a planet (e.g. the Earth), which perturbs the sailcraft in a complicated manner also because its radiance can be highly time-dependent.

174   6 Modeling Light-Induced Thrust

input momentum to the sail, namely, just *before* the sail-photon interaction. Once we are able (in the next sections) to compute the *output* momentum, i.e. just *after* the interaction, then we could get the vector thrust on the sail.

The sail senses the photon's incidence direction (in SOF) as $\mathbf{u}$, given by the second of Eqs. (6.17), from each respective solar element ds; during the sailcraft proper time $d\tau$ and relatively to the observed frequency range from ν_{obs} to $\nu_{obs} + d\nu_{obs}$, the infinitesimal input momentum onto dA (which has position $\mathbf{R}$ and velocity $\mathbf{V}$ in HIF at $t(\tau)$) is expressed by

$$d^4\mathbf{P}_{(in)}^{(SOF)} = dI_{\nu_{obs}}^{(SOF)} \, d\nu_{obs} \, dA \, d\tau \frac{\mathbf{u}}{c} \tag{6.20}$$

where $dI_{\nu_{obs}}^{(SOF)}$ is given by Eq. (6.11). The spectral input momentum due to the visible solar disk can be defined as

$$d^2\mathbf{P}_{(in),\nu_{obs}}^{(SOF)} = \int_{\Delta_\odot} \frac{d^4\mathbf{P}_{(in)}^{(SOF)}}{d\nu_{obs} \, ds} \, ds \tag{6.21}$$

Using Eq. (6.11), expressions (6.16) and (6.19a), (6.19b), and the Fubini's theorem again, one can easily obtain

$$
\begin{aligned}
d^2\mathbf{P}_{(in)}^{(SOF)} &= \int_0^\infty d^2\mathbf{P}_{(in),\nu_{obs}}^{(SOF)} \, d\nu_{obs} \\
&= dA \, d\tau \frac{r_\odot^2}{c} \int_0^{\zeta_{limb}} L_\odot(\psi) \sin\zeta \cos\psi \rho^{-2} \, d\zeta \int_0^{2\pi} \mathcal{D}^2 \cos\theta \, \mathbf{u} \, d\mu
\end{aligned}
\tag{6.22}
$$

For calculating the photon pressure on dA, then the following dot product

$$\mathcal{P}_{(in)}^{(SOF)} = \frac{d^2\mathbf{P}_{(in)}^{(SOF)} \cdot \mathbf{n}}{dA \, d\tau} \tag{6.23}$$

has to be evaluated. This means that, in the integrand of (6.22), one has to replace $\mathbf{u}$ by $\cos\theta$. In particular, even if $\mathbf{n} = \mathbf{R}/R$, the term $\cos^2\theta$ would become appreciably different from unity as the sailcraft draws sufficiently close to the Sun.

Historically, the first book on solar sailing dealing with the special case consisting of $\mathbf{n} = \mathbf{R}/R$, $\mathbf{V} = 0$ and no relativity was the 1999-edition of [68]. There, it was shown how the solar radiation pressure, though obviously increasing with decreasing R, however increases less than what expected from the $1/R^2$ scaling law. In practice, this effect due to the finite size of the Sun becomes appreciable below 0.1 AU. Equations (6.22) and (6.23) are the generalizations of this special case for an *arbitrarily* oriented sail moving under relativity (as meant in Sect. 6.1).

Very recently, the problem of the Sun as an extended source of radiation has been re-considered in [55] also for potential spherical sail balloons to be utilized in near-Sun orbits for scientific research.

6.2.1 Approximating Irradiance and Input Momentum

The previous sections have regarded the solar irradiation and its momentum time rate onto a sail with any orientation. In the next sections, we have to face with the problem of how the photon pressure produces propulsive effects after the sail *processes* the received power. If the irradiated sail were perfectly flat, applying Eqs. (6.15) and (6.22) would not present particular numerical issues. However, a real sail exhibits a topology highly variable as locally as globally; consequently, the irradiance and momentum integrals, which contain $\cos\theta$ as given by Eq. (6.18f), should be computed for all small or large pieces which the sail is modeled by.

Depending on the sail models, one may wish to decouple the computation of the integrals from the incidence angle. To this aim, we introduce the following approximations:

1. the quantity $\mathcal{D}$ (Eq. (6.9)) changes so very little over $\mu \in [0, 2\pi)$ and $\zeta \in [0, \zeta_{limb}]$ that it may be considered a constant with respect to the solar area visible from the sailcraft. We will consider the value that D takes on for a point-like Sun:

$$\mathcal{D} \cong \mathcal{D}_\sim = \mathcal{D}|_{\mathbf{u}=\mathbf{R}/R} \qquad (6.24)$$

2. in $\cos\theta$, one may approximatively separate the effects coming from speed and finite solar radius:

$$\cos\theta \cong \cos\theta|_{\{V=0,\alpha=0,\delta=0\}} \cos\theta|_{r_\odot=0} \equiv \cos\varsigma \cos\vartheta \qquad (6.25)$$

The right hand-side is the product of the sunlight broadening, caused by the finite-size Sun, on an orthogonal sail with zero speed, and the angle ϑ of incidence on a sail (with velocity $\mathbf{V}$) due to a point-like Sun.

As a result, we can compute the approximated total irradiance as follows

$$I^{(\text{SOF})} \cong r_\odot^2 \mathcal{D}_\sim^2 \cos\vartheta \int_0^{\zeta_{limb}} L_\odot(\psi) \sin\zeta \cos\psi \rho^{-2} \, d\zeta \int_0^{2\pi} \cos\varsigma \, d\mu$$

$$\equiv \left(I^{(\text{SOF})}\right)_\sim \qquad (6.26)$$

176   6 Modeling Light-Induced Thrust

Numerical analysis with the gray-atmosphere Eq. (6.1) has shown a relative difference

$$\left| \left(I^{(\mathrm{SOF})} \right)_{\sim} / I^{(\mathrm{SOF})} - 1 \right| < 4 \, (V/c)^2 \tag{6.27}$$

holding around trajectory perihelion as well.

Similarly, one can employ the expressions (6.24) and (6.25) for calculating the input momentum by keeping the incidence angle (in SOF) separated from the double integral in Eq. (6.22):

$$\mathrm{d}^2 \mathbf{P}_{(in)}^{(\mathrm{SOF})} \cong \mathrm{d}A \, \mathrm{d}\tau \, \frac{r_{\odot}^2}{c} \mathcal{D}_{\sim}^2 \cos \vartheta \int_0^{\zeta_{limb}} L_{\odot}(\psi) \sin \zeta \cos \psi \rho^{-2} \, \mathrm{d}\zeta \int_0^{2\pi} \cos \varsigma \mathbf{u} \, \mathrm{d}\mu$$

$$\equiv \left(\mathrm{d}^2 \mathbf{P}_{(in)}^{(\mathrm{SOF})} \right)_{\sim} \tag{6.28}$$

Numerical analysis has shown the following inequality holding for the same trajectories of (6.27)

$$\frac{\| (\mathrm{d}^2 \mathbf{P}_{(in)}^{(\mathrm{SOF})})_{\sim} - \mathrm{d}^2 \mathbf{P}_{(in)}^{(\mathrm{SOF})} \|}{\| \mathrm{d}^2 \mathbf{P}_{(in)}^{(\mathrm{SOF})} \|} < \frac{V}{c} \tag{6.29}$$

where the numerator is the Euclidean norm of the error vector. The upper limits, which are often conservative, in inequalities (6.27) and (6.29) have been found numerically, namely, they are not the result of analytical expansion. Nevertheless, where appropriate, one may include the speed effect even in the approximated computation of irradiance and photon momentum because the related assumptions entail an error notably smaller than neglecting the effect.

At this point, we suppose that the concept expressed by the approximate relationships (6.24) and (6.25) holds for any realistic model of the actual $L_{\odot}(\nu_{em}, \psi)$, averaged over one or more solar cycles. Consequently, even the (above defined) spectral radiance $I_{\nu_{obs}}^{(\mathrm{SOF})}$ and spectral momentum $\mathrm{d}^2 \mathbf{P}_{(in),\nu_{obs}}^{(\mathrm{SOF})}$ may be approximated in a similar way, namely, by bringing out $\cos \vartheta$. Therefore, one can set:

$$\cos \vartheta \, J^{(\mathrm{SOF})} \equiv \left(I^{(\mathrm{SOF})} \right)_{\sim}, \qquad \cos \vartheta \, \mathrm{d}^2 \mathbf{Q}_{(in)}^{(\mathrm{SOF})} \equiv \left(\mathrm{d}^2 \mathbf{P}_{(in)}^{(\mathrm{SOF})} \right)_{\sim} \tag{6.30a}$$

$$\cos \vartheta \, J_{\nu_{obs}}^{(\mathrm{SOF})} \equiv \left(I_{\nu_{obs}}^{(\mathrm{SOF})} \right)_{\sim}, \qquad \cos \vartheta \, \mathrm{d}^2 \mathbf{Q}_{(in),\nu_{obs}}^{(\mathrm{SOF})} \equiv \left(\mathrm{d}^2 \mathbf{P}_{(in),\nu_{obs}}^{(\mathrm{SOF})} \right)_{\sim} \tag{6.30b}$$

The quantity denoted by $J^{(\mathrm{SOF})}$ may be interpreted as the irradiance at normal-to-surface incidence ($\vartheta = 0$).

6.3 What Does Surface Mean?

The sail surface we are considering is a mathematical abstraction, of course, which acts as an ideal interface between the sail material and its environment. Some effects have been detailed in Chap. 4. Instead, we are concerned progressively with

6.3 What Does Surface Mean?

the "extraction" of thrust from the energy (per unit time and unit area) impinging on the sail "surface". Let us begin this long path with viewing S as a uniform solid with a few outer atomic or molecular layers, which "materialize" the geometric surface, and restrict the considerations to metals. The number of atoms per unit area is approximately given by $n_{surf} \sim N_{AV} \rho_{bulk} \delta_{latt} / W$, where ρ and W denote mass density and atomic weight, respectively, and δ_{latt} is the linear size of the crystal cell. For instance, the mean number density of the three outermost layers of Aluminium is $\sim 7 \times 10^{15}/cm^2$, that is $\sim 10^{-7}$ times the atoms in one cm^3. This order of magnitude is common to many metals. The surface-to-bulk atoms ratio increases as the considered volume decreases. Experimental surface analysis requires tools able to distinguish the surface structure from the bulk for carrying out chemical composition, atomic electronic states, molecular bonds, just to cite a few. Surface probing is carried out into effect by incident beams of radiation of appropriate energy, momentum, charge and spin; after the interaction with the atoms/molecules of the target surface layers, the outgoing radiation is analyzed. Particles used for probing surfaces and collecting information are photons, electrons, ions, and neutrons. Various sensitive techniques are available nowadays (e.g. [102]).

It is fundamental to grasp that, in solids, many physical properties in the material's surface are rather different from their counterparts inside the material's bulk. For instance, the order of magnitude of the *surface* energy density is 1 J/m^2 whereas the *strain* energy density amounts normally to $\sim 10^7$ J/m^3. This indicates that in films ~ 100 nm thick, or less, the surface energy has a non-negligible role in determining the "intensity" of many mechanical phenomena.

Hence a first problem with solar sails, especially if sufficiently *large*; it regards how measuring its thermo-optical, mechanical, chemical, and electric properties. In many mission design concepts envisaged hitherto, sail area ranges from $\sim 10^3$ m^2 to $\sim 10^5$ m^2, or even more; though a four-quadrant square sail, or a square of squares, appears to be appropriate for an efficient support sub-system (i.e. the overall sail is divided into smaller pieces), it is not realistic to think of analyzing the entire surface of each piece (to be flown) for verifying its properties. Once the related observables are built, they can be used in a model describing the sail as a whole. Restricting ourselves to the thermo-optical quantities such as the reflectance, absorptance, and emittance, which are of concern in this chapter, measurements can be effected only on a number of representative samples selected according a combination of various criteria, e.g. importance in the thrust/torque models, cost, rapidity of execution, etc. Not only sampling is important for statistical reasons, but is the only way to carry out the overall sail properties, i.e. through the use of reliable numerical models. Such considerations can be extended to other design properties.

The optical properties of a sail come also from the sail manufacturing process. This also holds for envisaged nanotube sails. Some of the methods of Thin-Film Deposition (TFD) are described below very shortly. General and specific aspects and details regarding thin films can be found in many textbooks, for instance [77, 94, 110], and [102]. In general, a deposition process has to transfer a material, atom by atom, from a source (called also the *target* in the thin-film literature) onto a substrate (i.e. the *destination*) or growth surface because there the film will form and grow.

Source and substrate are put both inside a chamber where the material's transfer is carried out under controlled conditions, including high vacuum. A parameter among the important ones in the TFD technology is the film deposition rate, which may be defined in slightly different ways according to the types of processes of interest; units are $nm\,s^{-1}\,m^{-2}$. Other parameters of particular importance for solar sailing are (1) the film's mass density with respect to the bulk material, (2) the film uniformity, and (3) the grain size.

Many methods have been set up and improved; these are extensively explained in several reference texts, e.g. [30, 110]. Though several variants, the methods share three basic stages: (1) extracting atoms, ions, or molecules from an appropriate holder, (2) transporting such species to the substrate through a controlled environment, and (3) formation and growth of the film on the substrate. We will briefly remind the reader three main methods:

TFD-1 Physical Vapor Deposition (PVD) is the general process utilizing evaporation of, sublimation of, or ion impingement on the source material for moving atoms/ions/molecules to the substrate. The source may be in either solid or molten state. If this 'vapor' is not modified via any chemical reaction, then one has a *physical* deposition. PVD can be mainly categorized as the evaporation method, and the sputtering method.

- In the evaporation technique, the source material is put inside a crucible made typically of refractory metals or carbon (or also Alumina, and Boron nitride); heating is achieved by Joule effect, electron beams with magnetic focusing, or high-frequency induction. Once evaporated, source atoms are driven by pressure difference towards the substrate; a fraction of this flow comes into contact with the substrate surface, i.e. the arriving atoms form chemical bonds with the host ones; the atoms so adhered are named the *adatoms*. A net deposition can be got if the vapor pressure is higher than the equilibrium vapor pressure at the substrate temperature. The higher is such difference the greater is the deposition rate. The final step is the transformation of adatoms into a full condensed layer, which shall grow. An initial cluster of adatoms may either grow or disperse on the substrate surface. Relying on the Gibbs energy, one shows that a critical number of adatoms there exists such that a cluster with a higher number may grow. The actual phenomenon of film grow is a set of complicated processes, which may include a random diffusion of adatoms (with no net macroscopic mass transfer) if the substrate is spatially uniform. If not, a net mass transport occurs.

  A variant of the evaporation process consists of using an electron-beam heating system for high-melting point materials to be evaporated. Also, pulsed-laser deposition employs a laser for ablating the source materials.
- In the sputtering technique,[3] one generally uses Ar ions to bombard the substance to be deposited on the substrate; this substance now acts as the cathode in an electric circuit. When the avalanche conditions are achieved, the gas begins to

---

[3] Also referred to as *back-sputtering* or *spluttering*.

6.3 What Does Surface Mean? 179

glow, and the discharge can become self-sustaining; thus, sputtered atoms will form a vapor, transit through the discharge, and condense onto the substrate to start film growth.

Various sputtering methods are widely used in practical applications: e.g. (a) DC sputtering (even called the cathodic or diode sputtering), (b) radio frequency (RF) sputtering, (c) magnetron sputtering, with the support of magnetic and electric fields, and (d) bias sputtering, where either RF or DC bias-voltage is applied to the substrate, so to change energy and flux of the incident charged particles.

A particular variant of sputtering is suitable for high deposition rates of metal oxides. This is known as the *reactive* sputtering inasmuch as a gas strongly reactive is introduced in the vacuum chamber; it combines with the source material atoms to be deposited.

High-quality sputtering and related testing & diagnostics are widely employed by metallurgical firms specialized also in solar technology and large-area coatings.

TFD-2 Chemical Vapor Deposition (CVD) is the general process regarding the deposition of a substance as the result of some chemical reactions in the vacuum chamber. More precisely, suppose that the vapor (if any) of some material—which contains the substance to be deposited—undergoes a reaction with another substance or thermally dissociates for getting the needed non-volatile substance onto the substrate; in such conditions, the whole process is named chemical vapor deposition. Like the PVD process, high-temperature steps are necessary to end up in film. There are many CVD-process variants among which (1) Plasma-Enhanced, (2) Remote Plasma-Enhanced, (3) Low-Pressure, (4) Laser-Enhanced, (5) metal-organic, and (6) Ultrahigh Vacuum. Each of these ones is employed for specific classes of films (including Carbon nanotubes and synthetic diamonds) and requires different equipments.

TFD-3 In the Thermal Spray Deposition (TSD) technique, a particle stream of molten substance is accelerated onto the growth surface. Particles undergo a strong deceleration at impact; lateral spreading and substance cooling (leading to solidification) occur rapidly. There are many types of TSD: plasma spray and low-pressure plasma techniques are two of them; the second one is used because the first one produces porous films with a density lower than the bulk material.

If the development of a device requires a thin-film, it is not a-priori obvious which method is the best one. The choice can come only from a detailed analysis. In solar sailing, apart from the potential cases entailing one-layer sails, one has to cope with the problem of arrange a thin film on a very, very large substrate, and with sufficient uniformity, adhesion and resistance to the space environment(s) the sailcraft is designed to operate in. Aluminizing plastic substrates (usually polyimides for solar sailing) is a process carried out by PVD. One should realize that the overall quality of a sail requested for a specific mission will *also* depend on the method selected for film deposition referred to both the reflective and the emissive layers.

6.4 Topology of a Solar-Sail Surface

The space ambient is not the sole cause of sail's possible damage and performance degradation. We know that a large sail should/shall be segmented into a number of smaller pieces if nothing else for the sail sustainment during the flight. However, one obvious reason for segmentation is that the substrate of a large sail could not be deployed in *one* (extremely large) vacuum chamber.[4] Therefore, these smaller surfaces (always considerably large with respect to the standard artificial objects in space) shall be on-ground manufactured, side joined, connected to the sail-supporting structure, and folded in the sail packing subsystem for the planned deployment procedure in space. The operations of sail seaming, packing and deployment could affect the sail's actual optical behavior in flight, even if the film production were optically ideal.

Sails are very extended membranes[5] from the viewpoint of the overall mechanical response. Therefore, concepts and features peculiar to membranes can be applied. For instance, a sail to be deployed remains a certain time in the *stowed state* (due to folding) whereupon creases are normally induced.

A *crease* (or fold line) is a keen inelastic discontinuity in a membrane: creasing alters the stress state of a membrane, and can modify its topology.

A *wrinkle* is an elastic response to compressive stress resulting in smooth undulations of the membrane itself. Let us consider a thin film deposited on a much thicker substrate: (1) if the substrate is sufficiently stiff and the film-substrate interface is contaminated, then the film may detach or delaminate, (2) if the two layers are mechanically compatible one another, wrinkles may arise; this is the case for solar sails. Wrinkling entails surface strain, but there is no mass exchange between film and substrate. Membranes cannot support compressive stresses because they have very small bending rigidity. Therefore, when the membrane structure undergoes some compressive stresses, wrinkling occurs. However, this low bending stiffness is important to model if one wants to capture wrinkle details. Wrinkles change the membrane topology, and are virtually unavoidable during the operational life of the membrane. Wrinkling may be triggered also by non-uniform tensioning and temperature. One can meet two types of wrinkles: *material* wrinkles, and *structural* wrinkles. The material ones take place as permanent deformations, e.g. they may be the outcome of membrane manufacturing and packaging. The structural ones, appearing as transitory deformations, stem from membrane boundary conditions and loading.

Even a *seam* is able to induce some anisotropy in the membrane, and to change its topology. A number of seams in a large sail quadrant, or any other geometri-

---

[4]We are not considering sails from in-space manufacturing, and this for a number of reasons related to (big) space infrastructures, although one may not exclude such facilities in the long term.

[5]Normally, a *true* membrane is defined as having a bending stiffness exactly zero, whereas membranes exhibiting very low, but finite, bending rigidity are named elastic sheets. However, in this book, because the term *sheet* is used in a more general context, we use the name membrane for the real thin sheets, i.e. with tiny bending stiffness.

6.4 Topology of a Solar-Sail Surface

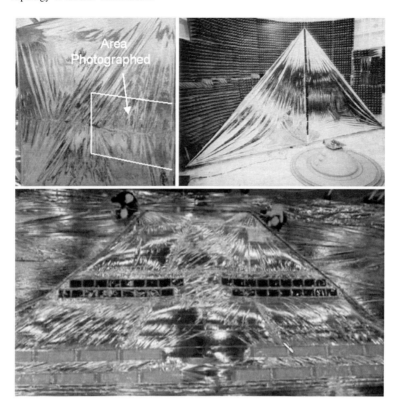

Fig. 6.2 Examples of sail wrinkling: (*top*) on-ground vertical deployment of sail samples at NASA LaRC (from [82] and [60], courtesy of NASA). The photographed area refers to an area under photogrammetric measurements (photogrammetry is a technique for measuring surface profile, static and dynamic, via the determination of 3D coordinates from multiple 2D photographs). (*bottom*) Part of the engineering model of IKAROS, the first full experimental sailcraft (courtesy of JAXA)

cal shape, could not be got around because of the sail manufacturing constraints previously highlighted.

Figure 6.2 shows wrinkles on vertical deployment in gravity. Although gravity enhances 2–3 orders of magnitude the formation of wrinkles through the related tension loads in the unfolding process, one has clear examples of how wrinkling may affect the surface of a sail deployed in space: the topological structure of the sail may be somewhat different from that of an ideal flat surface, even before the sail undergoes some billowing.

Creases and seams may alter wrinkle profiles. In particular, the presence of creases and uneven loading onto the sail sheet brings about wrinkles. An extended analysis of such membrane phenomena can be found, for instance, in [52, 54, 100, 119, 120], and the references inside. Important concepts, relevant to solar sails, are shape, wavelength, amplitude, direction, and distribution of wrinkles on a membrane. A general membrane, depending on its stress and strain state, may exhibit taut (or unwrinkled), wrinkled, or slack regions. A problem, arisen in the computation of wrinkling profiles, consists of selecting an appropriate approach for such a delicate analysis; some numerical methods do not "see" some wrinkle features or any structural wrinkle birth too. A modern efficacious method for computing wrinkle characteristics appears to be the geometrically nonlinear finite-element model based on thin-shell elements. An excellent description of the wrinkling process, from experiments and analytical models to numerical investigations can be found in [117–119] and [54].

With that in mind, we are able to picture the topology of a sail via three levels to be considered in modeling the overall thrust with respect to the sailcraft frame of reference. We will define them in decreasing order of size:

A1: Sail's large-scale undulations and global deviations from the flat surface;
A2: Sail's macroscopic or mean surface behaviors as observed sufficiently far from sail regions significantly smaller than the sail's total area;
A3: Sail's microscopic surface properties that—through the thermo-optical quantities—are able to affect thrust (and torques) induced by the solar irradiance.

In developing a model of thrust, a bottom-up approach will be followed here It starts from the very local geometrical properties of the sail surface, then determines what happens on a macroscopic region, and finally considers the effect of large topological alterations such as wrinkles and sail curvature.

6.4.1 Concepts from Modern Optics

One may wonder whether the amazing advancements in computer graphics/vision are able someway to help us for achieving a satisfactory model of the radiation pressure thrust on a space sail. Computer graphics deals mainly with the reproduction of (static or moving) scenes regarding natural, artificial or phantasy objects, some of which are sources of light, others receive, absorb and re-radiate light, other shade or transmit the irradiated energy, and so forth. The focus is on the history that the beams of light of the scene will follow by interacting with the different surfaces of the various objects. Such objects are considerably heavy with respect to the amounts of radiation pressure on them: no photon momentum transfer is computed because of no concern. Thus, the goal of computing the behavior of the many-body many-beam multiple interaction is centered on the complicated task of *rendering* a scene on some image/video output device.

In contrast, from the point of view of the number of objects, our scene of a sail irradiated by the Sun, or a planet, or both, is very simple. In the interplanetary case, we have one source and one (large) reflecting artifact. Thus, things may be somewhat simple, and hence solar-sail mission analysts (including the author) often adopted a flat sail receiving light from a constant-radiance point-like Sun. To consider (1) the finite-size structure of the Sun with its *variable* output, (2) the thermo-optical changes of the sail, and (3) the non-ideal behavior of the space environment in a sailcraft mission design is relatively recent, or very recent too. In

6.4 Topology of a Solar-Sail Surface   183

a previous chapter, we have seen how much complex is the radiant source of our interplanetary scene; in the above sections, we have mentioned how the processes of manufacturing, stowing, deploying, and sail-deployed keeping contribute all to make the sail having a non-smooth and non-flat surface. Thus, our goal, namely, the computation of the realistic *thrust* acting on the sail structure is not so simple as one might think of from the ideal models of both Sun and sail. In setting up a more realistic model of thrust calculation, we will use some concepts evolved in radiometry, optics, and—more limitedly—computer graphics.

Let us begin with recalling the concept of Bidirectional Scattering Distribution Function (BSDF) shortly, at least in its most common definition (e.g. [98] and the related discussion in its Sect. 1.5). For an optical surface, BSDF is the ratio of the amount of *scattered* radiant power, in the unit *projected* solid angle, on the amount of received radiant power. Such definition is appropriate to the measurement process (where any measured quantity cannot be infinitesimal). BSDF multiplied by the scattering angle is often employed in literature and named the cosine-corrected Bidirectional Scattering Distribution Function (CCBSDF).

In physics, scattering of light is a complicated phenomenon that is highly dependent on the incident photon energy and the scattering-body characteristics. (We will return to this key topic in Sect. 6.5). In our context, one should first realize that (1) even a large sail is sufficiently small to consider the solar power arriving uniformly in the sail's solid angle, and (2) even if sail segmentation appears much probable or necessary, manufacturing sail's large pieces should result in a sufficiently homogeneous product, unless a fraction of the area has to be made non-homogeneous for thrust maneuvering. In any case, apart from specific choices for simplifying the explanation, the algorithm developed below is general. Thus, it is better to split BSDF into two more convenient functions, namely, the Bidirectional Reflectance Distribution Function (BRDF) and the Bidirectional Transmittance Distribution Function (BTDF), which are better addressable to our specific problem. Before discussing and applying them, let us simply inspect Fig. 6.3 for a greater clarity. The surface receiving light is a thin film of non-perfectly smoothed material; a beam of electromagnetic radiation impinges on a small area of such film, and causes a number of phenomena, as shown in the figure, resulting in some radiance profile (arbitrarily indicated there). We like to point out some features that are important in modeling the surface optical response. First, reflected radiation outcomes from the same elemental area receiving radiation (i.e. no sub-surface scattering). Second, the scattered radiance may generally be asymmetric. Third, both reflection and transmission radiances develop three-dimensionally, and depend on the characteristics of the incident waves (wavelength, polarization), some of the thin-film properties (e.g. conductivity, roughness, anisotropy).

Now, the sail's large-scale topology (which we saw to encompass a general mission-dependent non-flat shape with superimposed numerous undulations) may in principle be dealt with general coordinates in the actual curved 2D space of the membrane, according to the intrinsic geometry of surfaces. However, this would lead to a problem very, very difficult with respect to our aims of thrust modeling, even because the details of the sail could change as the flight goes ahead. In addition,

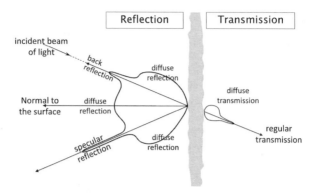

Fig. 6.3 Light reflected from and transmitted through a non-ideal surface. An arbitrary radiance profile has been represented in a 2D plot, with reflected/transmitted radiance proportional to the distance. Specular and back reflection, and diffuse reflection have been pictured on the *left side*; regular (or specular) and diffuse transmission have been sketched on the *right side*

a real surface surely deviates from some ideal geometric shape used for simplifying the calculation of the optical response. Such roughness is not easy to model in general; however, in order to formalize the above points (A2)–(A3), *Gaussian* models will be employed here (Sect. 6.4.3).

6.4.2 BRDF and BTDF

In the *current* framework, the necessary information regarding the incident radiation beam is embedded in the irradiance, whereas the ensuing radiance distribution encloses the information of the radiation after its interaction with the surface. Adopting a "slimmer" symbology, let us set

$$I_\lambda \equiv I^{(\text{SOF})}_{\lambda_{obs}} = I^{(\text{SOF})}_{\nu_{obs}} c/\lambda^2$$

$$J_\lambda \equiv J^{(\text{SOF})}_{\lambda_{obs}} = J^{(\text{SOF})}_{\nu_{obs}} c/\lambda^2 \qquad (6.31)$$

$$dA \equiv dA_{obs}, \qquad \lambda \equiv \lambda_{obs}$$

where the quantity $J^{(\text{SOF})}_{\nu_{obs}}$ is given by Eqs. (6.30a), (6.30b); the concepts established for Eqs. (6.25)–(6.30b) will be used here on. Denoting by $\mathbf{d}_\odot = (\phi_\odot, \vartheta_\odot)$ the direction of the Sun as observed from the sail front side, the incoming and the reflected spectral powers are expressed respectively by

$$\begin{aligned} d\Phi_\lambda &= I_\lambda \, dA \cong J_\lambda \cos\vartheta_\odot \, dA, & 0 \leqslant \vartheta_\odot \leqslant \pi/2 \\ d^2\Psi_{(re),\lambda} &= L_{(re),\lambda} \, dA \cos\vartheta_{(re)} \, d\omega_{(re)}, & 0 \leqslant \vartheta_{(re)} \leqslant \pi/2 \end{aligned} \qquad (6.32)$$

6.4 Topology of a Solar-Sail Surface

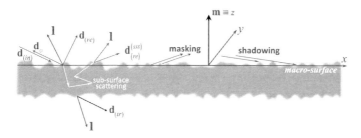

Fig. 6.4 Main phenomena to be considered in models for calculating BRDF and BTDF. **d** is the generic direction vector of the incoming, reflected, or transmitted photon beam. **l** denotes the local normal, and **m** represents the orthogonal direction to the macro-surface plane. Subscripts (*in*), (*re*), and (*tr*) refer to as the incoming, reflected, or transmitted beam, respectively. The local frame of reference has been denoted by $\mathcal{F}_\mathbf{m}$ in the text. Note that $\mathbf{d}_{(in)} = -\mathbf{d}_\odot$; whenever appropriate, we employ the sun direction because, in $\mathcal{F}_\mathbf{m}$, its colatitude is acute. Superscript (*sss*) means sub-surface scattering. In general, random geometry entails that several *valleys* are shadowed partially or totally (i.e. not illuminated by the incident light) and/or masked (i.e. not visible to the far observer). Note also that, due to the subsurface scattering of light, the reflected wave may emerge from a point *different* from the incident one. Possible inter-reflections are not shown

where $L_{(re),\lambda}$ is the reflected spectral radiance, and $d\omega_{(re)}$ is the elemental solid angle about $\mathbf{d}_{(re)}$. In such two equations, every term is known in the current framework (or can be assigned) but $L_{(re),\lambda}$. Now, speaking about reflection in a very general way, the concept of Bidirectional surface-scattering Reflectance Distribution Function (BSSRDF) states that the spectral radiance is proportional to the spectral irradiance:

$$L_{(re),\lambda} = \Upsilon \frac{d\Phi_\lambda}{dA} \Leftrightarrow \Upsilon = \frac{d^2\Psi_{(re),\lambda}/d\omega_{(re)}}{\cos\vartheta_{(re)}\, d\Phi_\lambda} \qquad (6.33)$$

Many are the variables and parameters Υ depends on:

$$\Upsilon = \Upsilon(\lambda, \mathbf{d}_\odot, \mathbf{x}_{(in)}, \mathbf{d}_{(re)}, \mathbf{x}_{(re)}, t \mid \{\wp\}) \qquad (6.34)$$

where $\mathbf{x}_{(in)}$ and $\mathbf{x}_{(re)}$ denote the vector positions (in the reference frame based on the above-mentioned macro-surface) of the points of irradiance and *corresponding* reflected radiance, respectively. $\{\wp\}$ denotes a set of parameters related to the modeling of scattering, refraction, and absorption (e.g. [27]). All this is compliant with Fig. 6.4, where the subsurface scattering has been sketched. The units of BSSRDF are sr^{-1}. The general dependence on time in (6.34) means that the various optical parameters may change with time, e.g. due to chemical reactions of the ambient gas with, particle bombardment on, or temperature-induced physical changes of the object surfaces.

Needless to say, such a function is very complicated to handle, though there exist models for dealing with specific classes of problems, e.g. the "special effects" in cinema. For instance, one needs BSSRDF for rendering many of the objects of every-day life in either static or dynamic configurations. Metallic objects practically exhibit a negligible subsurface scattering provided the incident band has sufficiently long wavelengths. From this point of view, the metallic reflective layer of a sail is expected to behave such a way for most of the incident spectrum. (The major effects of short-wavelength photons have been dealt with in Chap. 4). Therefore, here, we can ignore the position dependence in BSSRDF, and thus *turn* this general concept into the BRDF concept, first introduced by Nicodemus [73]. Here, we can write

$$L_{(re),\lambda} = \Lambda_\lambda \frac{\mathrm{d}\Phi_\lambda}{\mathrm{d}A} \iff \Lambda_\lambda = \frac{\mathrm{d}^2\Psi_{(re),\lambda}/\mathrm{d}\omega_{(re)}}{\cos\vartheta_{(re)}\,\mathrm{d}\Phi_\lambda} \quad \text{or} \tag{6.35a}$$

$$\Lambda_\lambda = \frac{\mathrm{d}F_{(re),\lambda}}{\cos\vartheta_{(re)}\,\mathrm{d}\Phi_\lambda} \cong \frac{\Psi_{(re),\lambda}}{\omega_{(re)}\cos\vartheta_{(re)}\Phi_\lambda} \tag{6.35b}$$

where $F_{(re),\lambda}$ is the reflected intensity (defined in Sect. 2.1), and the distinction between incidence point and exit point disappears in determining the various terms. The rightmost of Eq. (6.35b) is related to finite-area samples irradiated by Φ_λ, and finite-solid-angle sensors. (Some authors, e.g. [17], following the radar research literature, begins with defining the BRDF as it is in Eq. (6.35b), and then calls BSDF that part of BRDF related to the non-specular reflection from non-smooth surfaces). In Eqs. (6.35a), (6.35b) we have set

$$\Lambda(\lambda, \mathbf{d}_\odot, \mathbf{d}_{(re)}, t) \equiv \Lambda_\lambda \tag{6.36}$$

The BRDF is assumed to obey the reversion theorem of Helmholtz (e.g. [11], chapter VIII, and [92]):

$$\Lambda(\lambda, \mathbf{d}_\odot, \mathbf{d}_{(re)}, t) = \Lambda(\lambda, \mathbf{d}_{(re)}, \mathbf{d}_\odot, t) \tag{6.37}$$

With regard to sails, the dependence on time is due to the possible surface modifications in the space environment.

Aside from the symbols used for BRDF, a confusion might arise in reading the BRDF definitions throughout many papers or textbooks on optics, and computer graphics. Often, authors define BRDF as the ratio $\mathrm{d}L_r/\mathrm{d}E_i$, whereupon some perplexity might come from using the ratio between the so-named differential element of reflected radiance, or reflected infinitesimal/elemental radiance $\mathrm{d}L_r$ and the so-named differential element of radiant incidence, or incident radiance, or incident flux $\mathrm{d}E_i$. Another misunderstanding might arise in the explicit writing of the solid angle about the *incidence* direction instead of the solid angle about the *reflection* direction.

As specified in Sect. 2.1, in this book we employ standard nomenclature (even though some symbols have been changed for avoiding confusion with other symbols in different contexts). If one considers infinitesimal powers explicitly, like Eqs. (6.32), (6.33) and (6.35a), (6.35b), then there is no ambiguity or misunderstanding. As a result, it is straightforward to recognize that

6.4 Topology of a Solar-Sail Surface

dL_r appears to be coincident with the (spectral) reflected radiance times the elemental area dA, whereas dE_i is equal to the (spectral) power received by the same area.

Another way to express the scattering function nomenclature used in literature is the following

$$CCBSDF = \cos \vartheta \; BSDF \tag{6.38}$$

to be remembered when sets of results, coming from different theories and/or laboratories, are to be compared.

From the Handbook of Optics [81], it is meaningful that many types of spectral reflectance are related to the concept of BRDF. Perhaps, the concept of BRDF may be better "framed" in that of spectral bidirectional reflectance via Eqs. (6.35a), (6.35b):

$$d\mathcal{R}_\lambda = d^2\Psi_{(re),\lambda}/d\Phi_\lambda = \Lambda_\lambda \cos \vartheta_{(re)} \, d\omega_{(re)} \tag{6.39}$$

Therefore, BRDF can be interpreted as the (spectral) bidirectional reflectance per unit projected solid angle, compliantly with the general concept of BSDF given on p. 183.

From Eq. (6.39), the spectral and total directional-hemispherical reflectance, here denoted by $\mathcal{R}_\lambda$ and $\mathcal{R}$, respectively, can be written as

$$\mathcal{R}_\lambda = \int_{2\pi \, \mathrm{sr}} \Lambda_\lambda \cos \vartheta_{(re)} \, d\omega_{(re)}$$
$$\mathcal{R} = \int_{\lambda_{min}}^{\lambda_{max}} \mathcal{R}_\lambda \Phi_\lambda \, d\lambda \Big/ \int_{\lambda_{min}}^{\lambda_{max}} \Phi_\lambda \, d\lambda \tag{6.40}$$

In the case of perfectly specular reflection on a homogenous medium, one gets from Eq. (6.39)

$$\Lambda_\lambda = \frac{1}{\cos \vartheta_{(re)}} \frac{d\mathcal{R}_\lambda}{d\omega_{(re)}} = \frac{\mathcal{R}^{(F)}(\vartheta_\odot)}{\cos \vartheta_{(re)}} \frac{\delta(\vartheta_{(re)} - \vartheta_\odot)}{2\pi \, \sin \vartheta_{(re)}}, \quad 0 < \vartheta_\odot < \pi/2 \tag{6.41}$$

where $\mathcal{R}^{(F)}(\vartheta_\odot)$ is the Fresnel reflectance, which can be found in many excellent textbooks where the various calculation approaches are described, and shortly reported in Sect. 6.5.3.2. Here, we would like to note that, in the computer graphics literature, the one-dimensional Dirac delta, relative to solid angle, often is employed. The above one-dimensional Dirac delta, which shall be extended to two dimensions, entails the product $2\pi \sin \vartheta_{(re)} = |(d\omega/d\vartheta)_{(re)}|$; it is appropriate to the photon's linear momentum distribution. In any case, one should always be careful to use a Dirac distribution function on account of the associated properties (e.g. [112]). In the current case, the result of the integration over the front-side hemisphere, namely $\mathcal{R}^{(F)}(\vartheta_\odot)$, is a continuous function throughout the colatitude interval $[0, \pi/2]$.

Similar considerations can be done for the BTDF, which might be viewed as the BRDF of the membrane backside. Therefore, referring to Fig. 6.4, one has

$$d^2 \Psi_{(tr),\lambda} = -L_{(tr),\lambda}\, dA_{(b)} \cos \vartheta_{(tr)}\, d\omega_{(tr)}, \quad \pi/2 \leqslant \vartheta_{(tr)} \leqslant \pi$$

$$L_{(tr),\lambda} = \Theta_\lambda \frac{d\Phi_\lambda}{dA}, \quad \Theta_\lambda \equiv \Theta(\lambda, \mathbf{d}_\odot, \mathbf{d}_{(tr)}, t) \tag{6.42a}$$

$$\Theta_\lambda = \frac{dF_{(tr),\lambda}}{-\cos \vartheta_{(tr)}\, d\Phi_\lambda} \simeq \frac{\Psi_{(tr),\lambda}}{-\omega_{(tr)} \cos \vartheta_{(tr)} \Phi_\lambda} \tag{6.42b}$$

where Θ is just the BTDF and usually $dA_{(b)} = dA$. Equation (6.42b) is the form analogous to Eq. (6.35b), and its rightmost expression is a measurable quantity, which can be employed for defining BTDF experimentally by means of finite-area samples (irradiated by Φ_λ), and finite-solid-angle sensors placed in the back hemisphere. In addition, similarly to Eqs. (6.40) and with regard to the backside of the reflective layer, one can define the spectral and total directional-hemispherical transmittance:

$$\mathcal{T}_\lambda = -\int_{2\pi\, \text{sr}} \Theta_\lambda \cos \vartheta_{(tr)}\, d\omega_{(tr)}$$

$$\mathcal{T} = \int_{\lambda_{min}}^{\lambda_{max}} \mathcal{T}_\lambda \Phi_\lambda\, d\lambda \Big/ \int_{\lambda_{min}}^{\lambda_{max}} \Phi_\lambda\, d\lambda \tag{6.43}$$

We now assume that the materials of the reflective layers are such that they emit no appreciable fraction of their internal energy under the action of the solar irradiance at any distance from the Sun; for instance, we will ignore effects such as the anti-Stokes Raman lines at any distance R from the Sun. Thus, one can write

$$\mathcal{A}_\lambda = 1 - (\mathcal{R}_\lambda + \mathcal{T}_\lambda)$$

$$\mathcal{A} = \int_{\lambda_{min}}^{\lambda_{max}} \mathcal{A}_\lambda \Phi_\lambda\, d\lambda \Big/ \int_{\lambda_{min}}^{\lambda_{max}} \Phi_\lambda\, d\lambda = 1 - (\mathcal{R} + \mathcal{T}) \tag{6.44}$$

where $\mathcal{A}_\lambda$ and $\mathcal{A}$ denote the spectral and total absorptance, respectively. Absorptance is of relevance to thrust and sail temperature equilibrium (Sect. 6.5.5).

6.4.3 Describing Sail's Local Roughness

In computer graphics, a microfacet-based model is one of the analytical models for dealing with BRDF/BTDF. It is a middle-way between empirical or semi-empirical models (e.g. [103] for Aluminium-based sails) and the complicated electro-magnetic wave-based models. Computer graphics handles objects in scenes

6.4 Topology of a Solar-Sail Surface 189

irradiated by visible light, i.e. with a max-to-min wavelength ratio of $\sim$2. In contrast, we saw that TSI covers six orders of magnitude, in general, and over two regard solar sailing. In addition, the reflective layer of a solar-photon sail should be of $\sim$100 nm in equivalent thickness (whereas the sail can be a few microns thick or less), and the sail manufacturing process will determine the roughness characteristic lengths, which have to be compared to the incident wavelength ranges.

Let us shortly outline the main points of the microfacet model. The actual surface is assumed to consist of a high number of very small *smooth* surfaces, or microfacets, randomly oriented with respect to each other. The sequential three-dimensional arrangement of such microfacets approximates the actual local surface by portraying its random irregularities. In general, one may assume some pre-assigned shape for every microfacet, e.g. spherical cavities, symmetric V-cavities, holes, symmetric/asymmetric polyhedra. The area, say, da of a generic microfacet is supposed much larger not only than the material's lattice characteristic areas, but also than λ^2. In addition, it is significantly smaller than the maximum surface extension that may be considered as a *flat* piece of sail, on average, i.e. the *macro-surface*. This serves as the mathematical reference plane for the related actual surface, as shown in Fig. 6.4; in other words, on it, one can define a *local* Cartesian reference system, the z-axis of which is along either normals to the macro-surface; in this chapter, we choose the macrosurface's normal in the same hemisphere of the incident radiation. This local Cartesian frame is here denoted by $\mathcal{F}_{\mathbf{m}}$.

The above condition $da \gg \lambda^2$ with the related concept of ray of light, and the tangent plane approximation are *among* the assumptions at the base of various microfacet models. They are well summarized in [122]. Such conditions *should not* apply to the *well*-manufactured reflective layer of a sail, for which the opposite is true for the whole band of the wavelengths related to thrust generation. In other words, we will employ *diffractive* optics instead of geometric optics (Sect. 6.5) in our working out the thrust equation. Later on in this subsection we will describe a random-roughness surface mathematically, a concept used in various diffraction theories. Before that, let us outline the main phenomena occurring when photons impinge on a rough surface.

Usually, an incident beam of photons approaches the thin membrane (only one layer has been sketched in Fig. 6.4) with propagation direction $\mathbf{d}_{(in)}$ and is reflected or transmitted along the direction $\mathbf{d}_{(re)}$ or $\mathbf{d}_{(tr)}$, respectively, or may be absorbed. Parallel incident waves impinge on different surface points and scatter differently: this induces one to resort to statistical methods, as shown below, if the surface is sufficiently irregular. Two important aspects are sketched in Fig. 6.4: one is physical, the other one is geometrical (remembering that also such effects depend on wavelength as well). The former stems from the sub-surface scattering (*sss*), typical of translucent materials, which causes the scattered beam to re-emerge, with a direction $\mathbf{d}_{(re)}^{(sss)}$, from a surface point *different* from the incident one. If the number of scatters is sufficiently high, there remains no memory of $\mathbf{d}_{(in)}$ in $\mathbf{d}_{(re)}^{(sss)}$. The latter aspect consists of many "valleys" that are not illuminated because their neighbor "hills" shade them if the incident angle is sufficiently large. Similarly, the observer is not able to view many "depressions" that are hidden by near "rises". This overall limitation to the reflected radiance is known as the *masking/shadowing* effect.

There may be something induced by the masking/shadowing; at rather large incident and/or observation angles, some details of the surface roughness are concealed so that the "granularity" appears to be decreased from those angles. Thus, qualitatively speaking, the actual roughness is replaced by an effective surface roughness in some scattering models.

Even if nanotechnology were employed for obtaining future sails (e.g. via MWCNT), missions will require layers of metallic characteristics, on which unpolarized very broad bands of electromagnetic radiation will impinge. Therefore, we can ignore subsurface scattering of photons. Very thin sail materials may allow transmission of energetic photons, but most of these ones do not contribute to thrust. Since metals exhibit a complex index of refraction, absorption occurs. The absorbed photon energy transforms ultimately into heat, but its importance is twofold as shown in Sect. 6.5.5. Photons inducing photo-electrons, as widely discussed in Chap. 4, are expected to have a negligible propulsive impact.

Before discussing on how to model the microscopic topography of solid surfaces, looking images obtained via Atomic Force Microscope (AFM) might be enlightening. We have chosen a few images and reported them in Fig. 6.5. One thing appears somewhat impressive: a real surface departs notably from the usual perfect geometric surface as usually envisaged. Such fact has a profound impact on the interaction between the TSI's electromagnetic waves and the sail surface.

There are various types of surface *models*, which can be grouped into two general classes: deterministic and stochastic. A deterministic model makes use of some real-valued injective function dependent on the surface spatial coordinates. A significant example of deterministic model is the 1D sinusoidal surface, which introduces many important features about roughness and the related optical response. A stochastic model describes the surface shape via a stochastic process, either regular or predictable, usually function of the two horizontal coordinates. Let us mention here that a *regular* stochastic process has probability densities (or probability distributions, more generally) *prescribed*. In contrast, a *predictable* stochastic process is *expanded* via functions containing statistical parameters that, in turn, are *described* by probability densities (or distributions). For observing lands from satellites, models are used which rely on the fractal geometry; these ones can ultimately allow the analyst to study the surface either stochastically or deterministically. All such important points are dealt with extensively in literature, e.g. [29].

In the current framework, we shall follow a *regular* stochastic approach to the sail macrosurface, but we will not use the microfacet or similar model due to the above-described reasons. Let us refer to Fig. 6.4 again, and, in particular, the surface side of incident light (the upper one in the figure). Now, the surface is assumed of the type

6.4 Topology of a Solar-Sail Surface

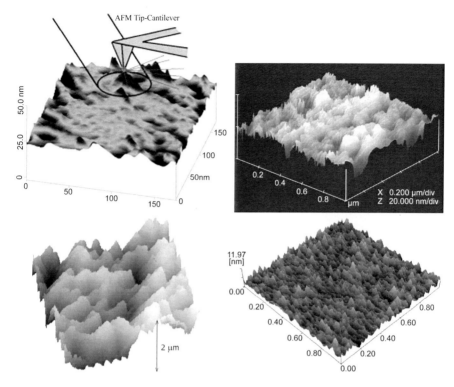

Fig. 6.5 Surface images by Atomic Force Microscope; *top-left*: Schematic of surface probing by AFM (courtesy of Imperial College, London); *top-right*: image of GaN layer grown at near-stoichiometric conditions (courtesy of Materials Research Society, PA, USA); *bottom-left*: 3D image of Pd sample (adapted from [62]), courtesy of R. Li Voti et al.; *bottom-right*: 1 μm × 1 μm Aluminium film sample with peaks less than 12 nm (from [28], courtesy of T. H. Fang and W. J. Chang)

called the *random roughness*, or *random height*, surface for which the local height with respect to the macrosurface level is characterized statistically: let $\mathcal{Z}(z)\,\mathrm{d}z$ be the probability that a generic point has height in the range $[z, z+\mathrm{d}z]$ with mean $\langle z \rangle$ and second central moment $\langle (z - \langle z \rangle)^2 \rangle \equiv \tilde{z}^2$, as usually. $\tilde{z}$ is known as the root mean square (*rms*) roughness, or height, of the surface.

If the sail's reflective layer were ideally uniform, there would be no correlation between $z(x_1, y_1)$ and $z(x_2, y_2)$ at any two different points, say, P_1 and P_2, respectively, on the macrosurface; the horizontal coordinates x and y would play similarly to time in a white Gaussian process. However, in practice, this is not the case.

A *joint* probability density often used in literature is the two-dimensional Gaussian probability density. Its explicit form can be derived from the general multi-dimensional Gaussian probability density (e.g. [51] or [67]); we report it together with its main properties of interest here:

$$P(Z_1, Z_2) = \frac{1}{2\pi \tilde{z}^2 \sqrt{1 - \mathcal{C}(x, y)^2}} \exp\left(-\frac{Z_1^2 + Z_2^2 - 2Z_1 Z_2 \mathcal{C}(x, y)}{2\tilde{z}^2(1 - \mathcal{C}(x, y)^2)}\right)$$

$$\langle Z_1 \rangle = \langle Z_2 \rangle = 0$$

$$\langle Z_1^2 \rangle = \langle Z_2^2 \rangle = \tilde{z}^2 \tag{6.45}$$

$$\langle Z_1 Z_2 \rangle = \tilde{z}^2 \mathcal{C}(x, y) \quad \Leftrightarrow \quad \langle (Z_1 - Z_2)^2 \rangle = 2\tilde{z}^2(1 - \mathcal{C}(x, y))$$

$$0 \leqslant \mathcal{C}(x, y) < 1$$

$$\langle |Z_1| \rangle = \langle |Z_2| \rangle = \tilde{z}\sqrt{2/\pi}$$

where the quantity

$$\tilde{\mathcal{C}}(x, y) = \tilde{z}^2 \mathcal{C}(x, y) \tag{6.46}$$

is the *auto-correlation function* with $x = x_2 - x_1$, $y = y_2 - y_1$. The marginal probability density of either Z_1 or Z_2 is plainly the Gaussian with zero mean and variance $\tilde{z}^2$. The probability density contours of $P(Z_1, Z_2)$ are ellipses; if a level is expressed via $P(0, 0)/l$, with $l \geqslant 1$, then ellipse semiaxes and orientation are given by

$$a = \tilde{z}\sqrt{1 + \mathcal{C}(x, y)}\sqrt{2\ln(l)}$$

$$b = \tilde{z}\sqrt{1 - \mathcal{C}(x, y)}\sqrt{2\ln(l)} \tag{6.47}$$

$$\angle(\mathbf{a}, \mathbf{z}_1) = \pi/4$$

The characteristic function of a Gaussian random vector is well known in its general form (e.g. [67]). Applied to the present case, we may write it as

$$\Psi(\zeta_1, \zeta_2) = \exp\left(-\frac{1}{2}\tilde{z}^2(\zeta_1^2 + \zeta_2^2 + 2\zeta_1\zeta_2 \mathcal{C}(x, y))\right) \tag{6.48}$$

where (ζ_1, ζ_2) is the dummy vector.

As we mentioned above, we are dealing with anisotropy at the sail's global level (the point A1 on p. 182). In contrast, the *micro*-surface could be supposed to be sufficiently isotropic and described via the following auto-correlation coefficient (e.g. [44])

$$\mathcal{C}(r) = \exp(-r^2/\tilde{l}^2), \quad r^2 = x^2 + y^2 \quad \Rightarrow \quad \tilde{\mathcal{C}}(x, y) = \tilde{z}^2 \exp(-r^2/\tilde{l}^2) \tag{6.49}$$

where $\tilde{l}$ is named the *auto-correlation length*. A property of such correlation is interesting:

$$\int_0^\infty r^n \mathcal{C}(r) \, dr = \frac{1}{2}\tilde{l}^{n+1} \Gamma\left(\frac{1+n}{2}\right), \quad \tilde{l} > 0, \, n \geq 0 \tag{6.50}$$

6.4 Topology of a Solar-Sail Surface

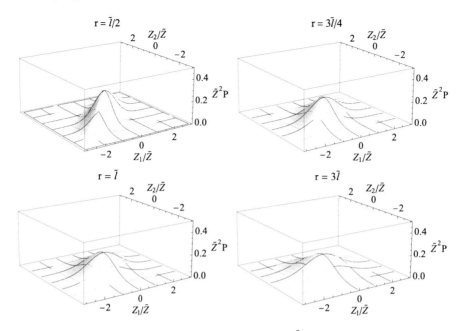

Fig. 6.6 Macrosurface's *joint probability density*: the product $\tilde{z}^2 P$ is plotted *vs* the heights Z_1 and Z_2 expressed in $\tilde{z}$ units. The four r-values are in terms of the auto-correlation length $\tilde{l}$. Note that, at a few correlation lengths, the probability density resembles the one-dimensional Gaussian

where $\Gamma()$ denotes the Euler Gamma function. In principle, Eq. (6.50) does not hold strictly for any real surface; however, at distances of few $\tilde{l}$, it approximates the infinite-extension value very closely.

Then, one can easily build Fig. 6.6. Note that two characteristic lengths are necessary for describing this type of surface feature: the *rms* height $\tilde{z}$ and the auto-correlation length $\tilde{l}$, which represent the vertical spread and the horizontal irregularity spacing, respectively. As shown in Fig. 6.6, the joint probability density of points apart more than $\tilde{l}$ looks like a one-dimensional Gaussian density because $\mathcal{C}(r)^2 \ll 1$.

At this point, the quantity $z(r)$, considered as function of r, can be viewed as a Gaussian scalar stochastic process, which is *mean square* differentiable. The scalar stochastic process $\wp = \mathrm{d}z(r)/\mathrm{d}r$ is Gaussian as well. $\wp(r)$ represents how the surface slope varies with the radial coordinate. It is plain that $\langle \wp \rangle = 0$. As soon as one recognizes that $z(r)$ plays as Z_1 and $z(r + \delta r)$ plays as Z_2, in the limit $\delta r \to 0$ applied to Eq. (6.49), it is not difficult to get the *rms* slope:

$$\tilde{p} \equiv \sqrt{\langle \wp^2 \rangle} = \sqrt{2}\tilde{z}/\tilde{l} \qquad (6.51)$$

Equation (6.51) gives one scalar dimensionless parameter; if there were two autocorrelation lengths because of some anisotropy on the surface, then there would be two different slope scales. *rms* slope helps us to distinguish between different rough surfaces. Of two or more surfaces with the same $\tilde{z}$, the smoothest is that one exhibiting the longest $\tilde{l}$. Quantities $\tilde{z}$, $\tilde{l}$ and $\tilde{p}$ are well-known examples of the so-called topological *finish* parameters of a surface. An important problem consists of determining them univocally via appropriate sets of theoretical tools, instruments, and data processing. In employing various measurement methods, care has to be due in order to not confuse the intrinsic properties of surfaces with instrumental effects.

In Sect. 6.5, we will use the macro and micro surfaces described above. We label the generic element of the set of sail macrosurfaces by ΔS, of area ΔA, to which the micro rough surface $\Delta \tilde{S}$, obeying Eqs. (6.45)–(6.51), can be associated. One should note that such equations represent our model here; as a result of modern techniques of measurement, the sail manufacturing process (outlined on pages from 178) may appear to produce different rough reflective layers, e.g. surfaces characterized by fractal behaviors such as self-affine fractals. Whatever the reflective film may be, it is possible to find an adequate mathematical description, which can be inserted into the current algorithm for thrust calculation.

Actually, using the term *rough surface* is qualitative for a non-ideal surface, either reflecting or transmitting, or both. Although no firm criterion for surface roughness has been established, however often in literature two simple criteria for smoothness are employed. They may be formulated as follows (e.g. [98]): a surface is considered *smooth* if either of the following inequalities is satisfied

$$\left(\frac{4\pi \cos \vartheta_\odot \tilde{z}}{\lambda} \right)^2 \ll 1 \quad \Leftrightarrow \quad \frac{\tilde{z}}{\lambda} \ll 1 \tag{6.52}$$

where the first inequality (Rayleigh criterion) is more stringent than the second one. Either criterion is often used to assess when certain surface-light effects should occur. In our framework, either inequalities (6.52) may be conveniently paired up with what just said about the *rms* slope; thus, at the same incident wavelength, a *smooth and clean* surface exhibits $\tilde{p}^2 \ll 1$ as well. All this has a meaning if the topographic irregularities by far exceed all other defects in diffractive response (Sect. 6.5.2).

Another concept that—on a global scale—tells us how much a surface is rough with respect to the perfectly smooth surface is the Total Integrated Scatter (TIS). If we denote the hemispherical *non*-specularly reflected power by Φ_{diff} and the specularly reflected power by Φ_{spec}, then TIS can be expressed by using the Rayleigh-Rice theory with the polarization factor (Eqs. (6.73a)–(6.73f)) approximated by Fresnel reflectance (e.g. [98], p. 108):

$$\mathrm{TIS} = \frac{\Phi_{diff}}{\Phi_{diff} + \Phi_{spec}} = 1 - \exp\left[-\left(\frac{4\pi \cos \vartheta_\odot \tilde{z}_{eff}}{\lambda} \right)^2 \right]$$

$$\cong \left(\frac{4\pi \cos \vartheta_\odot \tilde{z}_{eff}}{\lambda} \right)^2 \tag{6.53}$$

where the rightmost side agrees with the first smoothness limit in inequalities (6.52). Note that, according to [43], $\tilde{z}_{eff}$ is an effective value of the *rms* height (which can be derived from a band-limited zero-order moment of the quantity given by Eq. (6.54)). A significant feature of the expression (6.53) is that the surface Power

6.4 Topology of a Solar-Sail Surface

Spectral Density (PSD) (see below) is not restricted to be Gaussian. The *rms* slope does not appear in this equation for TIS. This formula has been applied to large incident angles (e.g. [41]). One should remember, though, that at grazing angles, say, $\pi/2 - \vartheta_\odot < \arctan(\bar{p})$ Eq. (6.53) should not be employed. Also, this formula might induce some misinterpretation when one likes to scale the information got (via measurements) at a particular wavelength to a wavelength significantly different, e.g. double, triple, and so on. It assumes that spatial bandwidth is kept fixed. This is not the general case because wavelength variations produce changes in the bandwidth (Eq. (6.55)). Nevertheless, an interesting qualitative behavior of scattering can be inferred from TIS. As the incident wavelength decreases, more energy/momentum is certainly scattered out of the specular direction; in other words, a sufficiently smooth sail is required to get non-negligible thrust from solar short-wavelength photons.

The characteristic lengths of a surface are measurable quantities, or, better, they are observables computed from direct measurements. Even though precise and extended considerations can be only found in specialistic books (e.g. [98, 101]), nevertheless let us focus on what is important for thrust calculation. One of main concepts related to the surface profile measurement and data processing is represented by the PSD. *Formally*, and with regard to an arbitrary macrosurface of area $\Delta A = Ł \cdot Ł$, the *area* PSD can expressed by (e.g. [18, 98])

$$\mathbb{S}_2(\mathbf{f}) = \lim_{Ł \to \infty} \frac{1}{Ł^2} \left| \int_{-Ł/2}^{Ł/2} dy \int_{-Ł/2}^{Ł/2} z(x, y) \exp(i 2\pi \mathbf{f} \cdot \mathbf{r}) dx \right|^2, \quad \mathbf{r} = \begin{pmatrix} x \\ y \end{pmatrix} \quad (6.54)$$

where the symbol of PSD contains dimensionality (2) and argument; this one is the two-dimensional vector $\mathbf{f}$, which is an array of *spatial* frequencies (reciprocal of length) of "propagation" throughout the surface, analogously to the usual temporal frequencies of waves propagating throughout space. This is a generalization of the one-dimensional sinusoidal surface, for which the frequency is the reciprocal of the distance between two consecutive height peaks. Vector $\mathbf{f}$ can be expressed in terms of incidence and scattering angles

$$\mathbf{f} = \begin{pmatrix} \sin \vartheta_{(re)} \cos \phi_{(re)} - \sin \vartheta_\odot \\ \sin \vartheta_{(re)} \sin \phi_{(re)} \end{pmatrix} / \lambda \equiv \begin{pmatrix} f_x \\ f_y \end{pmatrix} \quad (6.55)$$

where f_x and f_y are the first two components of the vector $(\mathbf{k}_{(re)} - \mathbf{k}_{(in)})/2\pi$, where $\mathbf{k}_{()}$ denotes the wave vector of either the incident or reflected wave, and recalling that $\mathbf{k}_\odot = -\mathbf{k}_{(in)}$ with $\phi_\odot = \pi$. The third component (f_z) is equal to $(\cos \vartheta_{(re)} + \cos \vartheta_\odot)/\lambda$. The integral in Eq. (6.54) has the structure of Fourier transform of the surface height profile $z(x, y)$. We will see that, mathematically, the scattered radiance and scattered intensity can be expressed as Fourier spectrograms of the surface deviations from a plane. The units of $\mathbb{S}_2(\mathbf{f})$ are length^4.

A one-dimensional spectrum, or the *profile* PSD, can be found from the area spectrum by integrating over the spatial frequencies having orthogonal propagation direction, namely

$$\mathbb{S}_1(f_x) = \int_{-\infty}^{\infty} \mathbb{S}_2(\mathbf{f}) \, df_y \quad (6.56)$$

The units of $\mathbb{S}_1$ are length^3. Once the PSD of a certain surface is determined (if possible) from some measurable quantities, the finish parameters of that surface can be derived via the even moments of the PSD (where the parameter-related spatial bandwidth has to be used).

The measurement process *probes* various realizations of the stochastic process $z(x)$; these ones are finite in terms of sample length and number of the observation angles. The former fact entails that the information got from measurements is limited in frequency range, whereas the latter tells that some estimator of the observables has to be arranged. The same surface statistics can be extracted from the measurements of a single profile passage. Here, we will report frequency range and formal definition, whereas extended considerations, measurement techniques, and estimation can be found in [98]. Denoting the number of data points of the surface profile by N_p, one has

$$-\frac{N_p}{4\mathrm{Ł}} \leqslant f_x \leqslant -\frac{1}{2\mathrm{Ł}}, \qquad \frac{1}{2\mathrm{Ł}} \leqslant f_x \leqslant \frac{N_p}{4\mathrm{Ł}} \qquad (6.57a)$$

$$\mathbb{S}_1(f_x) = \lim_{\mathrm{Ł}\to\infty} \left\langle \frac{1}{\mathrm{Ł}} \left| \int_{-\mathrm{Ł}/2}^{\mathrm{Ł}/2} z(x) \exp(i 2\pi f_x x)\, \mathrm{d}x \right|^2 \right\rangle \qquad (6.57b)$$

where, this time, the $\langle\rangle$ operator means an appropriate average over an experimental ensemble of the surface *roughness power*, i.e. the square of the Fourier-transformed height divided by the sample length or, equivalently, the roughness energy times the (spatial) roughness frequency. The analogy between the usual concepts of energy/power and the current ones comes from the Plancherel's theorem (e.g. [111], for a quick introduction).

At first glance, all such definitions and concepts might appear almost unnecessary and/or somewhat strained in analogy for solar sailing; however, we soon shall see that, in many diffraction theories, BSDF/BRDF can be expressed in terms of PSD. In particular, in the incidence plane, i.e. $\phi_{(re)} = 0$, one gets $f_x = (\sin\vartheta_{(re)} - \sin\vartheta_\odot)/\lambda$. For a two-dimensional spectrum, varying the scattering azimuthal angle changes $\mathbf{f}$ in both magnitude and direction. However, if the surface is isotropic, then PSD is function only of $\|\mathbf{f}\|$ and all information on its spectrum can be got via measurements of scattering in the incidence plane. In this *particular* case, the two-dimensional $\mathbb{S}_2(\|\mathbf{f}\|)$ can be obtained from the one-dimensional spectral density $\mathbb{S}_1(f_x)$ via the integral transform known as Abel transform, e.g. [1, 12, 98, 113] and [18]. The viceversa can be done via the inverse Abel transform.

▲ One should realize that PSD is a property of the sampled surface, not a measurement method. However, there is the problem to relate PSD *univocally* to BRDF. For instance, the PSD of rough surfaces with a complicated

6.4 Topology of a Solar-Sail Surface 197

structure (e.g. describable by many Fourier components) could not be determined from BRDF measurements. One may want to resort to a certain diffraction theory (because having desired peculiarities) for linking BRDF and PSD; however, if one were not able to tie up with PSD and BRDF univocally, then PSD could not be determined. If PSD were assigned, the corresponding BRDF could be determined, though not univocally; as a point of fact, different PSDs may be employed to give rise to the same BRDF.

The surface structure we are considering is of a particular type, not only for theoretical simplicity, but also because otherwise the sail would be too "irregular" to be controlled efficaciously. Of course, deviations from the baseline will be unavoidable. Here, we mention three out of the scattering modes from a surface.

Topographic Scattering. This is the mode emphasized in our model. The reflected wave is caused to be rippled or wrinkled by the surface's height irregularities: the wavefront phase is modified, or even may fluctuate.

Material Scattering. Even if the surface were geometrically smoothed, the material the surface is made of would exhibit irregularities or fluctuations in density and/or composition. Because, as we shall see below, the BRDF depends also on the dielectric function, such unevenness induces scattering. Note that optically anisotropic materials, as well as distributed deposition imperfections, lead to non-topographic scatter.

Defect Scattering. Surface imperfections, either designed or manufacturing process dependent, may be localized in non point-like zones of the surface material. For instance, a number of "coves" and/or "bulges" (or other more or less defined shapes) may appear like a sparse matrix on the mostly polished surface. Dust on a large smooth surface is another example of scattering from a collection of non-homogeneous structures.

Assumption about Sail Surface: A general surface may exhibit a BRDF stemming from the overall contribution of the above scattering kinds. How to calculate it depends considerably on the diffraction model(s) (Sect. 6.5.2). The field scattered from a stochastically-modeled surface is itself (spatially) stochastic as well. In particular, the field from a *regular* stochastic surface is regular, and so the corresponding scattered radiance. Now, the TSI's wavelengths (from middle ultraviolet up) and the reflective layer of a solar sail should, satisfactorily enough, be consistent with the smoothness criterion and the low-slope criterion. More precisely, this means that the 2nd-order moment of the diffracted radiance could be considered sufficiently small in order to apply classical diffraction theories, which output the BRDF as function of the parameters employed in the regular stochastic model of the sail surface. The smoothness criteria are also important for getting analytical expressions of the wave's polarization states (Sect. 6.5.3). However, medium-roughness sails may not be excluded.

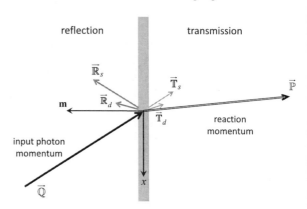

Fig. 6.7 Simple model of sail/photon momentum balance. The reflected and transmitted momenta are supposed to be the respective resultants of the actual momentum distributions related to specular and diffuse reflection, and transmission. All objects are seen from the y-axis of the frame $\mathcal{F}_\mathbf{m}$, but actually they belong to the three-dimensional space, in general

6.5 Calculation of Thrust

In order to realistically calculate the thrust a solar sail gets from its interaction with sunlight, we have to still proceed by a number of steps. In Sect. 6.5.1, we set up a simple model of photon momentum transfer for emphasizing some general behaviors of thrust. In Sect. 6.5.2, we shortly recall some items from optical diffraction; considering the importance of such topic, the subsection is focused on some aspects *relevant* to SPS, but always keeping in mind that the subject is very vast. Finally, in the subsequent subsections, the thrust model will be completed also including the large-scale undulations mentioned in Sect. 6.4.

6.5.1 Simplified Scheme of Sail-Photon Momentum Balance

Only in this subsection, we employ a symbology a bit different from that used hitherto: a vector is denoted by an arrow over a stylized capital letter. Let us recall the reflective thin layer of Fig. 6.4 with its frame $\mathcal{F}_\mathbf{m}$. The layer is here replaced by a thin slab; for clarity, this picture is sketched in Fig. 6.7.

Remark 6.1 Since we do not yet know how to calculate photon momenta stemming from the diffuse part of the total diffraction, the rationale of the current introductory scheme is the following: if we get the *total* reflected/transmitted photon momenta, what is the momentum transfer bringing about thrust? *Momenta* we are speaking about are actually infinitesimal momenta related to the photon energy received (and then diffracted) by an elemental surface in an infinitesimal time interval as meant previously, e.g. in Eq. (6.22). Once divided by $dA\, d\tau$, they represent a vector force on unit area.

Of course, the change of total photon momentum $\vec{M}$, caused by both reflection and transmission (in the same time interval and surface), is sensed onboard as the

6.5 Calculation of Thrust   199

reaction, say, $\overrightarrow{\mathbb{P}}$ to the interaction between electromagnetic waves and the layer; thus, one can write straightforward:

$$\overrightarrow{\mathbb{P}} = -(\overrightarrow{\mathbb{M}}_{out} - \overrightarrow{\mathbb{M}}_{in}) = \overrightarrow{\mathbb{Q}} - (\overrightarrow{\mathbb{R}} + \overrightarrow{\mathbb{T}})$$
$$= \overrightarrow{\mathbb{Q}} - \left[(\overrightarrow{\mathbb{R}}_s + \overrightarrow{\mathbb{R}}_d) + (\overrightarrow{\mathbb{T}}_s + \overrightarrow{\mathbb{T}}_d) \right] \quad (6.58)$$

where the input momentum $\overrightarrow{\mathbb{Q}}$ turns into the momenta reflected $\overrightarrow{\mathbb{R}}$ and transmitted $\overrightarrow{\mathbb{T}}$, respectively. Equation (6.58) is independent of the reference frame; in our case, we will specify the vector components in $\mathcal{F}_{\mathbf{m}}$. Subscripts s and d denote the *specular* and *diffuse* parts, respectively, of reflection/transmission according to Figs. 6.3 and 6.7. Here, the back reflection has been included in the diffuse reflection. We are neglecting temperature effects that will be considered later. Now, let us use the following definitions

$$\mathcal{R}_s = \int_{\omega_{(re,s)}} \cos \vartheta_{(re)} \, d\omega_{(re)} \int_{\lambda_{min}}^{\lambda_{max}} \Lambda_{(\lambda,s)} \Phi_\lambda \, d\lambda \Big/ \int_{\lambda_{min}}^{\lambda_{max}} \Phi_\lambda \, d\lambda$$
$$\mathcal{R}_d \equiv \mathcal{R} - \mathcal{R}_s \quad (6.59)$$

where the solid angle $\omega_{(re,s)}$ refers to the region where the light may be considered as reflected specularly. $\mathcal{R}$ was defined in Eqs. (6.40). Therefore, the two quantities in Eq. (6.59) can represent the specular reflectance and the hemispherical diffuse reflectance, respectively. Similar definitions can be set for transmittance:

$$\mathcal{T}_s = \int_{\omega_{(tr,s)}} -\cos \vartheta_{(tr)} \, d\omega_{(tr)} \int_{\lambda_{min}}^{\lambda_{max}} \Theta_{(\lambda,s)} \Phi_\lambda \, d\lambda \Big/ \int_{\lambda_{min}}^{\lambda_{max}} \Phi_\lambda \, d\lambda$$
$$\mathcal{T}_d \equiv \mathcal{T} - \mathcal{T}_s \quad (6.60)$$

where the solid angle $\omega_{(tr,s)}$ refers to the region where the light may be considered as transmitted specularly. The minus sign comes from the greater-than-$\pi/2$ colatitude of the transmitted beams. $\mathcal{T}$ was defined in Eq. (6.43). Thus, the two quantities in Eq. (6.60) represent the specular transmittance and the hemispherical diffuse transmittance, respectively. Inserting the quantities defined in Eqs. (6.59)–(6.60) into the second of Eqs. (6.44), plainly one has

$$\mathcal{R}_s + \mathcal{R}_d + \mathcal{T}_s + \mathcal{T}_d + \mathcal{A} = 1 \quad (6.61)$$

We will use the above definitions in Eq. (6.58), and make the momenta a bit more explicit by splitting each vector into the parallel-to-ΔS component ($\parallel$) and into the orthogonal one ($\perp$). Thus, the above vectors can be written as

$$\overrightarrow{\mathbb{Q}} = \overrightarrow{\mathbb{Q}}_\parallel + \overrightarrow{\mathbb{Q}}_\perp = \| \overrightarrow{\mathbb{Q}} \| (\overrightarrow{q}_\parallel + \overrightarrow{q}_\perp), \qquad \| \overrightarrow{q}_\parallel + \overrightarrow{q}_\perp \| = 1 \quad (6.62a)$$
$$\overrightarrow{\mathbb{R}}_s = \mathcal{R}_s (\overrightarrow{\mathbb{Q}}_\parallel - \overrightarrow{\mathbb{Q}}_\perp) = \mathcal{R}_s \| \overrightarrow{\mathbb{Q}} \| (\overrightarrow{q}_\parallel - \overrightarrow{q}_\perp) \quad (6.62b)$$

$$\vec{\mathbb{T}}_s = \mathcal{T}_s(\vec{\mathbb{Q}}_\parallel + \vec{\mathbb{Q}}_\perp) = \mathcal{T}_s\|\vec{\mathbb{Q}}\|(\vec{q}_\parallel + \vec{q}_\perp) \tag{6.62c}$$

$$\vec{\mathbb{R}}_d = \vec{\mathbb{R}}_{d,\parallel} + \vec{\mathbb{R}}_{d,\perp} = \chi_{(re)}\mathcal{R}_d\|\vec{\mathbb{Q}}\|(\vec{r}_\parallel + \vec{r}_\perp), \quad \|\vec{r}_\parallel + \vec{r}_\perp\| = 1 \tag{6.62d}$$

$$\vec{\mathbb{T}}_d = \vec{\mathbb{T}}_{d,\parallel} + \vec{\mathbb{T}}_{d,\perp} = \chi_{(tr)}\mathcal{T}_d\|\vec{\mathbb{Q}}\|(\vec{t}_\parallel + \vec{t}_\perp), \quad \|\vec{t}_\parallel + \vec{t}_\perp\| = 1 \tag{6.62e}$$

where the diffuse-momentum coefficients χ have been defined via Eqs. (2.19)–(2.20) (p. 43). They come in the above equations because, contrarily to the specular photons, diffuse-photon momenta are not directional. For simplicity, let us introduce the following vectors

$$\vec{r}_\parallel^{\,\blacksquare} = \chi_{(re)}\,\vec{r}_\parallel, \qquad \vec{r}_\perp^{\,\blacksquare} = \chi_{(re)}\,\vec{r}_\perp, \qquad \|\vec{r}_\parallel^{\,\blacksquare} + \vec{r}_\perp^{\,\blacksquare}\| \leqslant 1 \tag{6.63a}$$

$$\vec{t}_\parallel^{\,\blacksquare} = \chi_{(tr)}\,\vec{t}_\parallel, \qquad \vec{t}_\perp^{\,\blacksquare} = \chi_{(tr)}\,\vec{t}_\perp, \qquad \|\vec{t}_\parallel^{\,\blacksquare} + \vec{t}_\perp^{\,\blacksquare}\| \leqslant 1 \tag{6.63b}$$

Inserting Eqs. (6.61)–(6.62e) into Eq. (6.58), and using definitions (6.63a), (6.63b), after some simple vector re-arrangements we get

$$\vec{\mathbb{P}}_\parallel = (1 - \mathcal{R}_s - \mathcal{T}_s)\vec{\mathbb{Q}}_\parallel - \|\vec{\mathbb{Q}}\|(\mathcal{R}_d\,\vec{r}_\parallel^{\,\blacksquare} + \mathcal{T}_d\,\vec{t}_\parallel^{\,\blacksquare}) \tag{6.64a}$$

$$1 - \mathcal{R}_s - \mathcal{T}_s = \mathcal{R}_d + \mathcal{T}_d + \mathcal{A} \tag{6.64b}$$

$$\vec{\mathbb{P}}_\perp = (1 + \mathcal{R}_s - \mathcal{T}_s)\vec{\mathbb{Q}}_\perp - \|\vec{\mathbb{Q}}\|(\mathcal{R}_d\,\vec{r}_\perp^{\,\blacksquare} + \mathcal{T}_d\,\vec{t}_\perp^{\,\blacksquare}) \tag{6.64c}$$

$$1 + \mathcal{R}_s - \mathcal{T}_s = 2\mathcal{R}_s + \mathcal{R}_d + \mathcal{T}_d + \mathcal{A} \tag{6.64d}$$

where the second and the fourth lines, which we know to come from the energy conservation, are useful to highlight the different effects in the thrust generation. Let us discuss some important behaviors of Eqs. (6.64a)–(6.64d) by considering some particular cases:

Dominant transmission, namely, $\mathcal{R}_s \ll \mathcal{T}_s$ and $\mathcal{R}_d + \mathcal{A} \ll \mathcal{T}_d$. The total momentum transferred to the layer, obtained by summing $\vec{\mathbb{P}}_\parallel$ and $\vec{\mathbb{P}}_\perp$, results in

$$\vec{\mathbb{P}}_{(tr)} \cong \mathcal{T}_d\big[\vec{\mathbb{Q}} - \|\vec{\mathbb{Q}}\|(\vec{t}_\parallel^{\,\blacksquare} + \vec{t}_\perp^{\,\blacksquare})\big] \tag{6.65}$$

which is different from zero, in general. One can note that the right-hand side of the equation is zero if either $\mathcal{T}_d = 0$ or $\vec{t}_\parallel^{\,\blacksquare} + \vec{t}_\perp^{\,\blacksquare} = \vec{\mathbb{Q}}/\|\vec{\mathbb{Q}}\|$. However, one recognizes easily that the second condition is equivalent to say that all transmission is specular, or, equivalently, just what the first condition means. This may be expressed in the following proposition:

Proposition 6.1 *If the sail transmittance does not vanish, only the specular component entails a strict loss of thrust.*

6.5 Calculation of Thrust

No transmission. For instance, if the metallic reflective layer is not too thin, then one has $\mathcal{T}_s = \mathcal{T}_d = 0$. From Eqs. (6.64a)–(6.64d), one carries out

$$\vec{\mathbb{P}}_{\parallel (no\ tr)} = \|\vec{\mathbb{Q}}\|\big[(1 - \mathcal{R}_s)\vec{q}_{\parallel} - \mathcal{R}_d\,\vec{r}^{\,\blacksquare}_{\parallel}\big]$$
$$\vec{\mathbb{P}}_{\perp (no\ tr)} = \|\vec{\mathbb{Q}}\|\big[(1 + \mathcal{R}_s)\vec{q}_{\perp} - \mathcal{R}_d\,\vec{r}^{\,\blacksquare}_{\perp}\big] \tag{6.66}$$

Adding side to side and using the second and fourth relationships of Eqs. (6.64a)–(6.64d) results in

$$\vec{\mathbb{P}}_{(no\ tr)} = 2\mathcal{R}_s\,\vec{\mathbb{Q}}_{\perp} + (\mathcal{A} + \mathcal{R}_d)\,\vec{\mathbb{Q}} - \mathcal{R}_d\|\vec{\mathbb{Q}}\|(\vec{r}^{\,\blacksquare}_{\parallel} + \vec{r}^{\,\blacksquare}_{\perp}) \tag{6.67}$$

where the contribution from absorption appears clearly. Expression (6.67) tells that, even though $\vec{r}^{\,\blacksquare}_{\parallel}$ may vanish (i.e. there is a rotationally symmetric scattering), absorption and diffuse reflection cause the total thrust to deviate from the orthogonal-to-sail direction *provided* the incident sunlight is not paraxial.

Lambertian surface. A non-absorbing sail scattering in Lambertian way is a particular case out of those ones encompassed by Eq. (6.67). It can be obtained by setting $\mathcal{R}_d = 1$, $\vec{r}^{\,\blacksquare}_{\parallel} = 0$ and, after recalling what described on pp. 42–43, $\vec{r}^{\,\blacksquare}_{\perp} = (2/3)\,\mathbf{m}$. Then, the momentum transferred to the sail can be simply put as

$$\vec{\mathbb{P}}_{(\mathfrak{L})} = \vec{\mathbb{Q}} - \frac{2}{3}\|\vec{\mathbb{Q}}\|\mathbf{m} \tag{6.68}$$

from which is apparent that even a Lambertian sail can be controlled via the (vector) control variable $\mathbf{m}$.

Blackbody. This (ideal) particular case might appears a bit surprising. The sail momenta can simply be obtained from Eqs. (6.66)–(6.67) by inserting $\mathcal{A} = 1$, $\mathcal{R}_s = \mathcal{R}_d = 0$, namely,

$$\vec{\mathbb{P}}_{\parallel (bb)} = \vec{\mathbb{Q}}_{\parallel}, \quad \vec{\mathbb{P}}_{\perp (bb)} = \vec{\mathbb{Q}}_{\perp} \quad \Rightarrow \quad \vec{\mathbb{P}}_{(bb)} = \vec{\mathbb{Q}} \tag{6.69}$$

Differently from the previous cases, the total momentum gained by the sail is always equal to the incident one; in other words, thrust on a blackbody sail could not be modulated in direction inasmuch as—in the heliocentric inertial frame—it would always be *radial* (apart from small deviations) regardless one may orient the sail. This would restrict the set of admissible heliocentric (and even planetocentric) sailcraft trajectories enormously.

Therefore, one has the following proposition

Proposition 6.2 *No thrust control in a direction different from the radial one would be possible if the sail were a blackbody.*

The above (very simple) model of sail thrust, based conceptually on quantum physics, may also be used as a test of any electromagnetic-wave-based (complicated) thrust model by suggesting general behaviors which such model has to sat-

202 6 Modeling Light-Induced Thrust

isfy to; after all, it is not possible to keep apart quantum and wave theories completely.

6.5.2 A Reminder of Optical Diffraction

If plane waves impinged on a *perfectly* smooth and reflecting surface at some angle, coherent waves would come about only at specular reflection. Such an ideal condition of smoothness would be met closely if the surface were corrugated very weakly in height with respect to the incident wavelength(s), and a little bit in slope. In practice though, the reflected waves may exhibit considerable phase discrepancy due to many causes, including the surface manufacturing/handling process.

Let us consider two different points $P_1 \equiv (x_1, y_1, z_1)$ and $P_2 \equiv (x_2, y_2, z_2)$ of a corrugated surface receiving a plane wave train with an incidence angle $\vartheta_{(in)}$. With no loss of generality, one may take the coordinate plane $(x{-}z)$ parallel to the incidence plane (with the z-axis orientation as $\mathcal{F}_{\mathbf{m}}$). Denoting the surface vector from P_1 to P_2 by $\Delta \mathbf{r} = (x_2 - x_1, y_2 - y_1)$, then the phase difference, say, $\Delta \hat{\phi}$ between the waves scattered at the two points with spherical angles $(\phi_{(in)}, \vartheta_{(in)})$ and $(\phi_{(out)}, \vartheta_{(out)})$ can be written as follows:

$$\Delta \hat{\phi}_{2,1} = (z_1 - z_2)(\cos \vartheta_{(in)} + \cos \vartheta_{(out)})k - 2\pi \mathbf{f} \cdot \Delta \mathbf{r} \qquad (6.70)$$

where $k = 2\pi / \lambda$ is the wave number, and the vector $\mathbf{f}$ is provided by Eq. (6.55) (with $\odot \to (in)$ and $(re) \to (out)$, in general). Equation (6.70) visualizes the effect of the spatial frequencies as P_2 moves from P_1. The subscript of $\Delta \hat{\phi}$ indicates that the phase of the wave scattered at P_2 is referred to that of the wave scattered at P_1.

When P_2 approaches P_1, the phase difference goes to zero. One may then want to compare the phase of the wave actually scattered at P_1 with the phase of the wave that *would* impinge on the ideal surface, i.e. without height variations. This results in

$$\Delta \hat{\phi}_1 = z_1 (\cos \vartheta_{(in)} + \cos \vartheta_{(out)})k \qquad (6.71)$$

Equation (6.71) may be formally derived from Eq. (6.70) by putting $\Delta \mathbf{r} = 0$ and $z_2 = 0$, but now at P_2 there is no wave with which the actual wave may interfere; thus, one should be careful with the real meanings of Eqs. (6.70)–(6.71).

Note 6.2 From a basic viewpoint, scattering from any surface entails diffraction and no diffraction phenomenon takes place without interference. Some authors distinguish diffraction from scattering in their specific research areas; in the current

6.5 Calculation of Thrust 203

context, we will use the terms *scattering* and *diffraction*, and related quantities, interchangeably. The waves emerging from the surface will interfere optically, thus modulating the field magnitude as well. Any process aiming at measuring the scattered power (in the radiometric sense) will result in some average of the wavefront's very rapid fluctuations over a time much longer than the period(s) of the incident waves.

The oldest known reference to and understanding of diffraction can be traced back to Leonardo da Vinci around 1500 AD, he observed the scattering of sunlight by a wood smoke. However, this phenomenon was studied systematically by Italian Jesuit priest, mathematician, and physicist Francesco Maria Grimaldi (1618–1663). His detailed experiments were published posthumous (1665, a two-tome book) [35]. The word *diffraction* derives from the Latin verb "diffringĕre", which means "to break apart". In the first chapter of his first volume, Grimaldi communicated that light can spread even via a fourth mode (diffraction), in addition to direct transmission, refraction, and reflection.

Thus, waves after the wave-surface interaction exhibit a pattern of diffracted radiance in the scattering volume. Such pattern can be sampled by putting variable-position instruments in such volume, so capturing the overall diffraction signal in "points" of suitable observation planes/surfaces. In particular, the resulting waves can scatter in lobes about the ideal specular direction, diffuse more or less uniformly in the hemisphere, and even backscatter, depending on the metallic or dielectric nature of the surface materials, their anisotropy, imperfections, and so forth. The optical physics, its appropriate mathematical formalism, and its evolution up to including radiometry can be found in many excellent reference books, e.g. [5, 11, 36, 50, 56, 97, 101], and [80].

Physics of scattering encompasses many, many phenomena where the wave-matter interaction plays a big role. Often, the response of the material particle ensemble to waves (not limited to the electromagnetic ones) is linear, and quite satisfactory with respect to the taken modeling goals. Such a response can be described in terms of macroscopic quantities, e.g. the refractive index, and the BSDF. Also, macroscopically speaking, normally the response is independent of time, namely, one is facing with a *static* scattering.

In the classical scattering problem, there are three typical scales: (1) the wavelength of the scattered light, (2) the (linear) size d of the diffracting object C, and (3) the distance D from C to the field observation point P. We have two regimes of diffraction: (a) the Fraunhofer diffraction, i.e. the pattern observed at P when $D \gg d^2/\lambda$, and (b) the Fresnel diffraction, namely, the near-field diffraction pattern observed at P when $D \lesssim d^2/\lambda$.

Historically, the main purpose of a diffraction theory was to calculate the electric disturbance at some observation point achievable by the waves spread by a diffracting element (e.g. a small aperture in a screen, a grating slit, a transparent material with imperfections, rough surfaces, molecules and aerosols, and so forth). The observed field is the sum of the incident field (usually supposed as monochromatic plane wave) and the scattered field. In general, under static scattering and no source of field internally to the scattering medium, the space part of the complex electric field satisfies a less complicated partial differential equation where, nevertheless, the field components are *coupled*. If one assumes that the dielectric function of the medium does not change appreciably over distances comparable to the incident wavelength, then the original equation simplifies: the field components result in being uncoupled. Thus, the scattering phenomenon can be studied by analyzing a single component of the total electric field, which is the solution to a *scalar* equation. Essentially, this is the core of the so-called Scalar Scattering Theory (SST). The simplified partial differential equation can be converted into an integral equation (via the Green's theorem), where the medium is represented by its *potential* scattering (i.e. a function of the refraction index and the incident wave number). This equation is called the integral equation of potential scattering (IEPS) that contains the *total* field as the unknown function, and the *scattering* potential of the medium, which is a known function ([11], Chap. 13).

If the scattering phenomenon includes aspects that entail no field's components decoupling, e.g. *polarized* waves, then one has to switch to Vector Scattering Theory (VST). Theories of this class are significantly complicated in the related equations to be solved; as a point of fact, the system of solving equations is an integro-differential system for the total electric and magnetic fields, and where polarization and magnetization of the scattering medium play a key role through the corresponding electric and magnetic *vector Hertz potentials*. A first-order approximation can be set up for the far field of VST [11]. In such framework, the Hertz potentials turn into the respective Fourier transforms of the medium's polarization and magnetization. In either SST or VST, the scattering cross section and the absorption cross section of electromagnetic waves can be defined.

The Kirchhoff theory of scattering [11] is generally of scalar type. The integral formula, due to Fresnel and Kirchhoff, is applicable to an aperture on an opaque screen. Many shapes of this aperture (screen hole) can be dealt with appropriately. Even well-known and widely applied is the Rayleigh-Sommerfeld formulation, which removed the mathematical inconsistency of the original Kirchhoff treatment. The well-known Born approximation (summarized below) is within the scalar context. The classical Beckmann-Kirchhoff theory [5] is scalar as well; its recent modification will be considered below together with some other formulations employed for describing the scattering from a non-ideal surface. The Rayleigh-Rice perturbation formulation is a vector theory, which relates CCBSDF to the surface PSD. Another vector theory, the vector Kirchhoff's, is not able to provide a complete solution relating the measured intensity of scattered light to PSD. The generalized Harvey-Shack surface scatter theory relies on a two-parameter family of surface transfer functions; it is a linear-systems formulation.

The inverse scatter problem consists of getting quantitative information about the scattering object by measuring the distribution of the scattered field or the transported power. Inferring the properties of random rough surfaces by means of light scattering is a relatively recent problem in the optics of materials. The straightforward quantitative way to probe such surfaces is to measure their morphology by utilizing real-space imaging techniques such as the scanning-tunneling microscopy (STM) and/or the atomic-force microscopy (AFM); nevertheless, differently from

6.5 Calculation of Thrust 205

STM and AFM, optical scattering has the advantages of (1) being a non-contact method, and (2) allowing large sampling sizes, notable indeed for the very thin and wide pieces of a sail that should not be scratched. Of course, by using laboratory sources of very different wavelengths, the optical response of a number of sail's representative samples could be obtained under various irradiance geometries.

At this point, we have to add two pieces of information for the reader. We will embed them into two separated gray boxes. The first is a short reminder of the Born approximation.

⋄ Out of the diffraction formulations considering what happens in the *far*-field regions of light scattered from a generic medium, the Born approximation (not to be confused with the Born-Oppenheimer approximation in the problem of interatomic potentials) is physically significant and very useful in practice inasmuch as it allows inferring notable items on the inverse scatter problem. Let us synthesize the main points without using equations. For the appropriate treatment, of course, one should first refer to [11]; cross section calculations can be found in [50].

As mentioned above, IEPS contains the total field to be found. Once the observation direction is chosen, the problem consists of calculating the integral over the volume of the *scatterer*, or the scattering volume. The *first-order* Born approximation, or more simply the Born approximation, considers the *far* field, (i.e. the field in regions distant from the scatterer much more than a wavelength) with the total field replaced by the incident field in the mentioned integral. The first-order solution becomes straightforward: the total field in the *far* zone is the sum of the incident field (normally supposed to be plane waves) and spherical waves modulated by the so-called scattering amplitude from an incident direction to a diffracted direction. This is a complex quantity equal to the Fourier transform of the scattering potential calculated at the vector given by $\mathbf{K} = (\mathbf{u}_f - \mathbf{u}_0)k$, with $k = 2\pi/\lambda$; $\mathbf{u}_0$ denotes the wave incidence direction onto the scattering object, and $\mathbf{u}_f$ is the direction of the diffracted field. In other words, $\mathbf{K}$ is the difference between the incident and the scattered wave vectors. (In the current case, we have $\mathbf{u}_0 = -\mathbf{d}_\odot$ and $\mathbf{u}_f = \mathbf{d}_{(re)}$, according to Fig. 6.4). If one assumes to measure the diffracted far field for *all* its allowed directions and for *all* the allowed incident directions on the scattering object, then it would be possible to get *all* the Fourier components of the scattering potential. Born approximation may be easily generalized in order to include the diffraction modes mentioned on p. 197.

The second box regards considerations about the sail reflective layer and TIS.

♮ When one requires the design of high-performance surface objects like space mirrors, large antennas, telescope dishes, just to cite a few, the operational electromagnetic band is assigned; among many performance indices, the surface has to be kept as smooth as possible in order to reduce the TIS, namely, TIS $\ll 1$, as said above. On p. 188, we have highlighted that (at least for the currently-conceived sails) the metal reflective layer should be about 100 nm in thickness. This is due to the sail lightness requirements (not limited to this particular sail layer, of course). As a result of the deposition methods outlined in Sect. 6.3, a reasonable *rms* height should be of the order of one third (at most) of its thickness. Therefore, in the wavelength spectrum from about 20–30 nm to at least 4 microns, TIS can become small only close to this upper bound, and beyond. There is an additional meaning to be attached to the word *small* in the solar-sailing context. The sail can be sensitive to the TSI fluctuations, which have been emphasized in Chap. 2. Therefore, the uncertainty (due to surface inhomogeneities and measurement errors) of the radiance *scattered outside* the specular lobe should be kept hopefully at a level lower than the natural TSI variability, i.e. ~ 0.1–0.2 percent, in order to reduce the thrust error. In addition to a good sail manufacturing (of course), this could be done by utilizing accurate diffraction models, as we shall see below.

Here, we shortly summarize a few features of the scattering theories we deem relevant to solar sails in the context of the thrust calculation. We shall return to them in Sect. 6.5.4.

- *Born approximation* is the first stage of an iterative process aiming at calculating the scattered field accurately. It holds if the scattering potential is sufficiently small or, equivalently, if the squared modulus of the refraction index of the scattering object departs weakly from unity; the scale to assess such deviation is the square of the incident wave number.
- *Rayleigh-Rice* theory captures behaviors of experimental data including large angles of incidence and scattering, but solely from *smooth* surfaces.
- *Beckmann-Kirchhoff* theory agrees with experiments analyzing the scattering from even *rough* surfaces, but is limited to incidence and scatter angles sufficiently small, i.e. the so-called *paraxial* approximation.
- *Modified Beckmann-Kirchhoff* theory is recent; more precisely, it is an empirical modification of the Beckmann-Kirchhoff theory. It should combine the advantages of Beckmann-Kirchhoff and Rayleigh-Rice theories, and should avoid the drawbacks of them. It has been proposed for very rough surfaces as well. Numerical validation to the results of such algorithm come from the generalized Harvey-Shack theory.
- *Generalized Harvey-Shack* theory is an extended version of the original Harvey-Shack that relates scattering to the surface topography. It uses the concept of diffracted radiance, includes large incidence angles, and deals with rough surfaces.

A diffraction theory could finally result in some BRDF as function essentially of the incident wavelength, the scattering geometry, the polarization types (which depend on the dielectric function of the scattering medium), and the PSD. This quite important concept in optics is relevant to solar sailing. Put in different words, SPS is not solely due to the electromagnetic radiation pressure, but also to the way this radiation interacts with a physical surface, i.e. chiefly *diffraction*; by using an appropriate theory of this interaction mode, one should measure the surface parameters of the sail to be flown. Such information is important to calculate thrust with the accuracy required by advanced missions.

6.5.3 The Importance of Light's Polarization

This subsection points out the main polarization aspects one should consider in the calculation of thrust. Theory of polarization can be found in many textbooks, e.g. [8, 18, 32, 71, 98] are suggested here.

The reflective layer of a sail is of metallic material. When light reflects on a metal surface, the previous state of polarization may be changed in the scattered light. Normally, the reference plane is the wave's propagation plane (formed by the wave vector and the surface normal), which results in the incidence plane for the incident waves, and the scatter plane for the scattered waves. In the current case, the reference frame is $\mathcal{F}_m$ with the x-axis in the incidence plane such that $\phi_\odot = \pi$. Any orthogonal polarization results in having the electric field parallel to the surface's mean plane, namely, the $x-y$ plane. Polarization is related to the material the surface is made of, not to its shape. BRDF is tied to the spectral bidirectional reflectance via Eq. (6.39); this one depends *also* on the dielectric function of the material through the polarization states of the incident and reflected waves.

A very short additional reminder about the optical response of metals. Normally, the motion of the metal nuclei is not considered when the impinging radiation frequency is sufficiently high. In contrast, core and valence electrons, subject to different restoring forces (if one speaks classically), determine the metal's optical properties. At lower frequencies (typically in the infrared regions) the motion of the nuclei contribute mainly to the optical responses. A metal exhibits a complex dielectric function with a non-zero imaginary part, which implies a non-negligible absorption of light. The index of refraction of a metal is complex, normally with the imaginary part notably greater than the real part. A metal medium can have the squared absolute value of its refractive index considerably larger than unity. For Aluminium, this is well-satisfied in the 280–10,000 nm range, (e.g. [79, 86], and [103]), where about 99 percent of TSI falls on average.

Theoretically well-posed calculations of polarization energy fluxes per unit orthogonal area have been developed in the last five decades, with the mathematical framework based wholly on Maxwell equations. In particular, the field components can be connected to the surface's parameters. Employing the Poynting vector, the four possible spectral bidirectional reflectances per unit solid angle can be calculated. They are proportional to as many collections of terms, say, $\mathcal{Q}_{a \to b}$ that depend on the incidence angle, the scattering angles, *and* the dielectric function of the surface's material. Subscript a indicates the given polarized component (parallel or orthogonal) of the incident wave, whereas b regards the prescribed component (parallel or orthogonal) of the reflected wave. In other words, one gets

$$(d\mathcal{R}_\lambda)_{a \to b}/d\omega_{(re)} \propto \mathcal{Q}_{a \to b} \tag{6.72}$$

Therefore, according to Eq. (6.39), there are four cosine-corrected Bidirectional Reflectance Distribution Functions (CCBRDFs). We will report these polarization factors from [64] and [98] after adapting the symbols and noting that analytical expressions are found only in the small-slope regime (Chap. 4 of [29]):

$$a_s \equiv |\varepsilon - 1|^2 \sin^2 \phi_{(re)}, \qquad a_c \equiv |\varepsilon - 1|^2 \cos^2 \phi_{(re)} \tag{6.73a}$$

$$b_\odot \equiv \sqrt{\varepsilon - \sin^2 \vartheta_\odot}, \qquad b_{(re)} \equiv \sqrt{\varepsilon - \sin^2 \vartheta_{(re)}} \tag{6.73b}$$

$$\mathcal{Q}_{\perp \to \parallel} = \frac{a_s |b_{(re)}|^2}{|(\varepsilon \cos \vartheta_{(re)} + b_{(re)})(\cos \vartheta_\odot + b_\odot)|^2} \tag{6.73c}$$

$$\mathcal{Q}_{\perp \to \perp} = \frac{a_c}{|(\cos \vartheta_{(re)} + b_{(re)})(\cos \vartheta_\odot + b_\odot)|^2} \tag{6.73d}$$

$$\mathcal{Q}_{\parallel \to \perp} = \frac{a_s |b_\odot|^2}{|(\varepsilon \cos \vartheta_\odot + b_\odot)(\cos \vartheta_{(re)} + b_{(re)})|^2} \tag{6.73e}$$

$$\mathcal{Q}_{\parallel \to \parallel} = |\varepsilon - 1|^2 \left| \frac{\cos \phi_{(re)} b_{(re)} b_\odot - \sin \vartheta_{(re)} \sin \vartheta_\odot \varepsilon}{(\varepsilon \cos \vartheta_{(re)} + b_{(re)})(\varepsilon \cos \vartheta_\odot + b_\odot)} \right|^2 \tag{6.73f}$$

where ε denotes the dielectric function, and a factorization has been accomplished for showing symmetries and asymmetries, and for reducing the number of irrational/transcendental complex operations. Note that each $(d\mathcal{R}_\lambda)_{a \to b}/d\omega_{(re)}$ is normalized to the incident irradiance *corresponding* to the given polarization a, namely, either $\parallel$ or $\perp$.

Note 6.3 The above expressions of the polarization factors hold for *non-magnetic* materials. The related generalization is provided in [18] in the framework of the small-perturbation approximation. From [18], we note that the convention $\phi_\odot = \pi$ is compliant with the way equations (6.73a)–(6.73f) have been carried out.

Sunlight is unpolarized, but sail's metal material reflects it polarized, even when its roughness is small. If a detector receives unpolarized light and is insensitive to

6.5 Calculation of Thrust

polarization, then the overall polarization factor can be expressed by (e.g. [29]):

$$Q_{\odot, \Delta S} = \frac{1}{2}(Q_{\perp \to \parallel} + Q_{\perp \to \perp} + Q_{\parallel \to \perp} + Q_{\parallel \to \parallel}) \tag{6.74}$$

Note 6.4 SPS thrust comes from each of the polarization modes, so Eq. (6.74) is important in thrust modeling.

With regard to the incidence plane, one has the significant result that $Q_{\perp \to \parallel} = Q_{\parallel \to \perp} = 0$; thus, in this case, we get

$$(Q_{\odot, \Delta S})_{inc} = \frac{1}{2}(Q_{\perp \to \perp} + Q_{\parallel \to \parallel})_{\phi_{(re)}=0,\pi} \tag{6.75}$$

In Optics, there are two general options in formulating a wave theory: one postulates a time dependence of the electro-magnetic fields as $\exp(-i\omega t)$ while a generic medium is given a complex refractive index as $n + ki$; the other one assumes the fields time varying as $\exp(+i\omega t)$ with an ensuing complex index equal to $n - ki$. Both approaches are valid. The reported polarization factors are independent of the chosen way of dealing with waves mathematically, as expected from a physical viewpoint.

6.5.3.1 The High Refraction Index Approximation

The surface of the sail's metallic reflective layer is a good reflector; in more precise terms, the trigonometric terms could be neglected with respect to the values of the dielectric function in Eqs. (6.73a)–(6.73f). It is an easy task to carry out the related expressions of the polarization factors. We have used a prime for denoting such approximation:

$$Q'_{\perp \to \parallel} = \left(\frac{\sin \phi_{(re)}}{\cos \vartheta_{(re)}}\right)^2 \tag{6.76a}$$

$$Q'_{\perp \to \perp} = \cos^2 \phi_{(re)} \tag{6.76b}$$

$$Q'_{\parallel \to \perp} = \left(\frac{\sin \phi_{(re)}}{\cos \vartheta_{\odot}}\right)^2 \tag{6.76c}$$

$$Q'_{\parallel \to \parallel} = \left(\frac{\cos \phi_{(re)} - \sin \vartheta_{\odot} \sin \vartheta_{(re)}}{\cos \vartheta_{\odot} \cos \vartheta_{(re)}}\right)^2 \tag{6.76d}$$

Then it is straightforward to combine Eq. (6.74) and the expressions (6.76a)–(6.76d) in order to get the overall polarization factor due to unpolarized sunlight

Fig. 6.8 Absolute value of the Aluminium's dielectric function vs wavelength. The decimal logarithms of both variables have been plotted. Wavelengths are expressed in nanometers

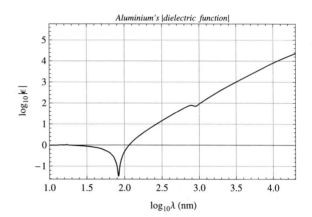

impinging onto high-ε surface sails:

$$Q'_{\odot,\Delta S} = \frac{1}{2}\left(Q'_{\perp\to\parallel} + Q'_{\perp\to\perp} + Q'_{\parallel\to\perp} + Q'_{\parallel\to\parallel}\right) \quad (6.77)$$

Let us note that the product $\sin\vartheta_{(re)}\cos^2\vartheta_{(re)}\,Q'_{\odot,\Delta S}$ does not diverge as $\vartheta_{(re)} \to \pi/2$. This product stems from the product of the polarization factor by the solid angle when the Rayleigh-Rice diffraction theory is used, as we shall see explicitly in Sect. 6.7.2.

The high-ε approximation is good for Aluminium in a broad wavelength band. In Fig. 6.8, we have plotted the decimal logarithm of its dielectric function as function of the decimal logarithm of incident wavelengths expressed in nanometers (data come from [86]). The current approximation may be applied from ~200 nm up to far infrared radiation.

Remark 6.2 With regard to the high-ε approximation, one should note that the four contributing terms are periodic with respect to $\phi_{(re)}$ with period π, whereas the period of $Q'_{\odot,\Delta S}$ is 2π. In a wide interval of $\phi_{(re)}$ around π, the dominant term comes from the parallel-to-parallel polarization. When it is considered in diffuse reflection with $0 < \vartheta_\odot < \pi/2$ and $0 < \vartheta_{(re)} < \pi/2$, light's backscattering is a characteristic result, as it will be shown in Sect. 6.7.2.

6.5.3.2 Fresnel Reflectance

Often, formulas of the scattered power contain the Fresnel reflectances. Therefore, it may be useful to have them within reach, even though restricted to metals with the squared absolute value of the (complex) refraction index much greater than unity. As mentioned several times, such type of metals pertains to solar sailing. Thus, let us denote the index of refraction by $\hat{n} = n - ki$ satisfying $|\hat{n}|^2 \gg 1$. In the current framework, Fresnel polarization reflectances can be got easily by setting $\phi_{(re)} = 0$

6.5 Calculation of Thrust

and $\vartheta_{(re)} = \vartheta_{\odot}$ in Eqs. (6.73a)–(6.73f); afterwards, such expressions are specialized for solar-sail metals, essentially by neglecting $\sin^2 \vartheta_{\odot}$ with respect to $|\varepsilon| = |\hat{n}^2| = n^2 + k^2$.

We write down the exact and the approximated expressions:

$$\mathcal{R}_{\parallel}^{(F)} = \left| \frac{\varepsilon \cos \vartheta_{\odot} - b_{\odot}}{\varepsilon \cos \vartheta_{\odot} + b_{\odot}} \right|^2 \simeq \frac{(n^2 + k^2) \cos^2 \vartheta_{\odot} - 2n \cos \vartheta_{\odot} + 1}{(n^2 + k^2) \cos^2 \vartheta_{\odot} + 2n \cos \vartheta_{\odot} + 1} \tag{6.78a}$$

$$\mathcal{R}_{\perp}^{(F)} = \left| \frac{b_{\odot} - \cos \vartheta_{\odot}}{b_{\odot} + \cos \vartheta_{\odot}} \right|^2 \simeq \frac{(n^2 + k^2) - 2n \cos \vartheta_{\odot} + \cos^2 \vartheta_{\odot}}{(n^2 + k^2) + 2n \cos \vartheta_{\odot} + \cos^2 \vartheta_{\odot}} \tag{6.78b}$$

where, we remind, both the real and the imaginary parts of $\hat{n}$ depend on λ. We employed one subscript per reflectance mode inasmuch as the mixed-polarization modes vanish identically at the specular reflection. Of course, at normal incidence the two reflectances coincide and reduce to the well known expression of reflectance from a perfectly smooth reflecting/absorbing medium.

If the incident light were polarized linearly with an angle φ between its electric field and the incidence plane, the reflectance would amount to

$$\mathcal{R}^{(F)}(\lambda, \vartheta_{\odot}, \varphi) = \mathcal{R}_{\parallel}^{(F)} \cos^2 \varphi + \mathcal{R}_{\perp}^{(F)} \sin^2 \varphi \tag{6.79}$$

For totally incoherent incident light, integrating Eq. (6.79) and averaging over $[0, \pi]$ produces the reflectance:

$$\mathcal{R}^{(F)}(\lambda, \vartheta_{\odot}) = \frac{1}{2} \left(\mathcal{R}_{\parallel}^{(F)} + \mathcal{R}_{\perp}^{(F)} \right) \tag{6.80}$$

In the following subsections, we will drop the symbol λ if there is no ambiguity, and use $\mathcal{R}^{(F)}(\vartheta_{\odot})$ instead.

At this point, we have the ingredients necessary for building a detailed model of the sail thrust, but still limitedly to macrosurfaces, for now.

6.5.4 Diffraction-Based Reflectance Momentum

As we saw extensively, the solar electromagnetic radiation interacting with each elemental area of any macrosurface is scattered or, equivalently, diffracted. Let us begin with considering surface reflection. In each solid angle $d\omega_{(re)}$ about the direction $\mathbf{d}_{(re)}$, solar photons of different energy carry their momenta. In the above defined macrosurface frame, i.e. $\mathcal{F}_{\mathbf{m}}$, we can write the fourth-order momentum associated to the reflected photon spectral radiance:

$$d^4 \mathbf{Q}_{(re)}^{(\mathcal{F}_{\mathbf{m}})} = d^3 \Psi_{(re)} \, \mathbf{d}_{(re)} \frac{d\tau}{c}$$

$$= \left[L_{(re), \lambda} (\cos \vartheta_{(re)} \, dA)(d\omega_{(re)} \, d\lambda) \right] \mathbf{d}_{(re)} \frac{d\tau}{c} \tag{6.81}$$

In order to (subsequently) write the momentum conservation law for the system radiation+sail, and so using the input second-order momenta (Eq. (6.22) or Eq. (6.28), or Eqs. (6.30a), (6.30b) depending on the chosen model), one has to integrate the above momentum over solid angle and wavelength band:

$$d^2\mathbf{Q}_{(re)}^{(\mathcal{F}\mathbf{m})} = dA \frac{d\tau}{c} \int_{\lambda_{min}}^{\lambda_{max}} I_\lambda \, d\lambda \int_{2\pi\,\text{sr}} \cos\vartheta_{(re)} \Lambda_\lambda \, \mathbf{d}_{(re)} \, d\omega_{(re)}$$

$$\mathbf{d}_{(re)} = \begin{pmatrix} \sin\vartheta\cos\phi \\ \sin\vartheta\sin\phi \\ \cos\vartheta \end{pmatrix}_{(re)}$$

(6.82)

having replaced the reflected spectral radiance by the product of the spectral irradiance times the BRDF, as defined via Eqs. (6.35a), (6.35b). Note that we are considering our general model of irradiance calculation. If deemed appropriate, the approximation discussed in Sect. 6.2.1 can be substituted for I_λ in Eq. (6.82). Before going on, some points have to be remarked. *First*, the integrand in the double or more exactly, the triple integral of Eq. (6.82) consists of many pieces; so far, we have been able to calculate or express explicitly all of them but one: BRDF. However, the topics discussed in Sects. 6.3 and 6.4 will allow us to make the BRDF explicit. *Second*, Eq. (6.82) is unifying in the sense that, in principle, we will not worry to separate specular, diffuse, and back-scattered reflection; however, computing such different contributions (where applicable) can be very useful in practice. How a particular sail surface (manufactured in a certain way by design) responds to the environmental spectral irradiance (Sun and/or planets) at the desired incidence directions (according to the sail control law) is enclosed in the BRDF. *Third*, given the (hopefully) broad variety of future sailcraft trajectories and mission aims, one should consider different scattering theories for two reasons: (a) to calculate thrust for any type of admissible sail material irradiated by the real Sun, and (b) to point out BRDF properties that may allow updating trajectory computation during the various phases of mission analysis.

In general, for mathematical description, one resorts to split the actual BRDF from an illuminated surface into the sum of conceptually different contributions. This depends also on the other objects that exchange light with the object of interest. Since, in our framework, only the Sun irradiates the sail, such a complication is not applicable. Suppose that a monochromatic beam of light impinges onto a perfectly-polished planar metal surface of size much larger than the cross section of the beam. It reflects the incident waves specularly with a reflectance given by the Fresnel formulas in Sect. 6.5.3.2. The behaviors of polarized-mode reflectance as function of wavelength and incidence angle is well-known. As the *rms* height and *rms* slope are increased from zero, but such that the above surface smoothness criteria are met, the Fresnel spike enlarges looking like a lobe, whereas energy is scattered out from this small solid angle that, therefore, contains less energy than the Fresnel's. As roughness increases, the specular lobe broadens (and may become asymmetric) as the incidence angle increases; the peak of such lobe (also called the off-specular peak)

6.5 Calculation of Thrust

takes place at a reflection angle greater than the incident angle. More energy is diffused outside this region around the specular direction. If the surface has $\tilde{z}/\lambda$ higher than unity, the scattered light (which develops three dimensionally) is the dominant part of BRDF for any practical incidence angle.

Once again, our sail has a metal reflective layer; thus, sub-surface scattering can be neglected, and one can write:

$$(\Lambda_\lambda)_{a \to b} = \left(\Lambda_\lambda{}^{\varnothing}\right)_{a \to b} + \left(\Lambda_\lambda{}^{\blacksquare}\right)_{a \to b} \tag{6.83}$$

where the superscripts $^{\varnothing}$ and $^{\blacksquare}$ denote specular and diffuse reflection, respectively.

Equation (6.83) means that each diffraction theory has its own regions of quasi-specular and non-specular distribution of power, in general. Considering that different forms of the quasi-specular part can be found in the literature (and sometimes not completely clear in symbology), we have adopted the following factorization:

$$\cos \vartheta_{(re)} \left(\Lambda_\lambda{}^{\varnothing}\right)_{a \to a} = \mathcal{R}_a^{(\mathrm{F})}(\lambda, \vartheta_\odot)$$
$$\times \mathfrak{F}(\lambda, \vartheta_\odot, \vartheta_{(re)}, \{\hat{p}\})$$
$$\times \Omega(\vartheta_\odot, \vartheta_{(re)}, \phi_\odot, \phi_{(re)}, \{\hat{p}\}) \tag{6.84a}$$
$$0 \leqslant \mathfrak{F}(\lambda, \vartheta_\odot, \vartheta_{(re)}, \{\hat{p}\}) \leqslant 1, \quad a \in \{\|, \perp\}$$

$$\mathfrak{F}(\lambda, \vartheta_\odot, \vartheta_{(re)}, \{\tilde{z} = 0\}) = 1$$
$$\lim_{\tilde{z} \to 0} \Omega(\vartheta_\odot, \vartheta_{(re)}, \phi_\odot, \phi_{(re)}, \{\hat{p}\}) = \frac{\delta(\phi_{(re)} - (\phi_\odot + \pi), \vartheta_{(re)} - \vartheta_\odot)}{\sin \vartheta_{(re)}} \tag{6.84b}$$

where the arguments of the functions have been indicated, $\{\hat{p}\}$ denoting the set of surface parameters responsible for any non-specular reflection (even though, in our current problem, topographic scattering is given a dominant importance). Equation (6.84a) means that the quasi-specular region radiates according to Fresnel reflectance *attenuated* by the factor $\mathfrak{F}()$ due essentially to roughness and masking/shadowing. The angular function $\Omega()$ describes the radiance shape in the neighborhood of the specular direction. Relationships (6.84b) have to be satisfied in the limit of no roughness; $\delta(\phi, \vartheta)$ denotes the Dirac's two-dimensional distribution function. In computer graphics, where the full electromagnetic wave approach is not followed essentially for getting a quick rendering, functions $\mathfrak{F}()$ and $\Omega()$ are object of several models, some of which are not 'physical' though, i.e. they do not satisfy the requirements of energy conservation and reciprocality. Thus, a certain care should be used in the choice of such functions for the current problem, also in compliance with the diffraction theory that will be employed to model the diffuse radiance.

We do not insist further on the overall properties of the quasi-specular region. Instead, it is time to see what the diffraction theories mentioned at the end of Sect. 6.5.2 show with regard to either $(\Lambda_\lambda{}^{\blacksquare})_{a \to b}$ or the diffuse radiance, which brings about deviations in photon momentum with respect to the momenta

214   6 Modeling Light-Induced Thrust

from specular reflection. Therefore, again, it is important to know the distribution of the radiance diffracted from a solar sail for accurately computing the total scattered momentum that enters the momentum balance between sail and sunlight.

- **Born Approximation.** An excellent source for the essential points of the Born approximation (B) applied to surface scattering is the Handbook of Optics ([17], Chap. 7). Transferring and adapting such pieces of information to our case (where topographic irregularities have been assumed dominant) result in

$$\left(\Lambda_\lambda^{\;\blacksquare}\right)_{a\to b}^{(B)} = \frac{16\pi^2}{\lambda^4}\cos^2\vartheta_\odot\, Q_{a\to b}\mathbb{S}_2(\mathbf{f}) \tag{6.85a}$$

$$Q_{a\to b} = \begin{cases} \cos^2\phi_{(re)}\mathcal{R}_a^{(F)}(\vartheta_\odot), & a = b \\ \sin^2\phi_{(re)}\mathcal{R}_a^{(F)}(\vartheta_\odot), & a \neq b \end{cases} \tag{6.85b}$$

$$a \in \{\|, \perp\}, \qquad b \in \{\|, \perp\} \tag{6.85c}$$

where the power spectral density is given by Eq. (6.54) with the spatial frequency vector expressed by (6.55). The term $\cos^2\vartheta_\odot$ is also termed the *obliquity* factor; different scattering theories exhibit different obliquity factors.

If each ΔS of the sail has different scattering modes (p. 197) statistically independent, then

$$(\Lambda_\lambda)_{\Delta S} = \sum_{j=1}^{N_{sm}} (\Lambda_\lambda)_{\Delta S,j} \tag{6.86}$$

where N_{sm} is the number of diffraction modes.

One should note that, *in general*, the BRDF does *not* scale like λ^{-4}. As a matter of fact, both polarization and spatial frequencies depend on the incident wavelength: polarization factors do through the material's dielectric function, whereas the reciprocal of wavelength affects the spatial frequency scale.

Comments: the first-order Born approximation is no doubt elegant and simple; however, as explicitly stated in [11] (Chap. 13, pp. 699–700), such approximation is good if the modulus of the index of refraction departs weakly from unity. Now, solar-sailing metals have high refraction indices in the infrared and visible bands; as a result, Born approximation works better if the TSI photons have sufficiently high energy, namely, in the UV regions, in particular in the far ultraviolet band, where the transmittance may be still neglected. At the same time, of course, the above smoothness criteria have to be satisfied even for these low wavelengths.

- **Rayleigh-Rice.** Similarly to the B-related BRDF is that carried out via the Rayleigh-Rice (RR) vector theory. Although the RR results have been extended

6.5 Calculation of Thrust

to high orders in perturbation [75], however in its lowest (and most used) order it can be written as follows

$$\left(\Lambda_\lambda \blacksquare\right)^{(RR)}_{a\to b} = \frac{16\pi^2}{\lambda^4}(\cos\vartheta_\odot\,\cos\vartheta_{(re)})\mathcal{Q}_{a\to b}\mathbb{S}_2(\mathbf{f}) \tag{6.87}$$

where the $\mathcal{Q}_{a\to b}$ are just expressed by Eqs. (6.73a)–(6.73f). It differs from the Born BRDF in two aspects: (1) the obliquity factor, and (2) the more general polarization factors. Often, in literature, the cosine-corrected BRDF is reported (thus, the obliquity factor is given by $\cos\vartheta_\odot\cos^2\vartheta_{(re)}$). Particularly meaningful is the expression of Λ_λ corresponding to the $\perp\to\perp$ mode reflectance, namely,

$$\left(\Lambda_\lambda \blacksquare\right)^{(RR)}_{\perp\to\perp} = \frac{16\pi^2}{\lambda^4}(\cos\vartheta_\odot\,\cos\vartheta_{(re)})$$

$$\times \cos^2\phi_{(re)}\sqrt{\mathcal{R}^{(F)}_\perp(\vartheta_\odot)\mathcal{R}^{(F)}_\perp(\vartheta_{(re)})}$$

$$\times \mathbb{S}_2(\mathbf{f}) \tag{6.88}$$

where the orthogonal-mode reflectance is given by Eq. (6.78b). Equation (6.88) is often called the (scattering) *golden rule*; as a point of fact, it is employed successfully in the data analysis of scattering from high-performance reflectors.

Comments: the Rayleigh-Rice vector theory is widely used in literature because—in the smooth-surface limit—gives satisfactory BRDFs even for large incident and/or scattering angles. However, for UV-waves, it may be difficult to realize large surfaces smooth everywhere in both *rms* height and auto-correlation length (or slope, equivalently). It should be (considerably) less costly for mission analysts to employ theoretical/numerical tools able to predict scattering from the regions of medium roughness in a sail, inasmuch as the contribution to thrust from ultraviolet bands is expected to be non-negligible for accurate trajectory calculation.

- **Beckmann-Kirchhoff**. The Kirchhoff-based theory by Beckmann [5], is of scalar type and has been developed for diffraction from surfaces rougher than those ones satisfying the smoothness criteria, standing the same range of impinging wavelengths. It is customary to define some quantities in order to write the results from the BK theory expressively (e.g. [41, 85]); adapting/changing symbols for this book results in

$$F \equiv \frac{1 + \cos\vartheta_\odot\,\cos\vartheta_{(re)} - \sin\vartheta_\odot\,\sin\vartheta_{(re)}\cos\phi_{(re)}}{(\cos\vartheta_{(re)} + \cos\vartheta_\odot)\cos\vartheta_\odot} \tag{6.89a}$$

$$g \equiv (\cos\vartheta_\odot + \cos\vartheta_{(re)})^2(k\tilde{z})^2 \tag{6.89b}$$

$$\upsilon \equiv (\sin^2\vartheta_\odot + \sin^2\vartheta_{(re)} - 2\sin\vartheta_\odot\,\sin\vartheta_{(re)}\cos\phi_{(re)})(k\tilde{l})^2$$

$$= \|\mathbf{f}\|^2\lambda^2(k\tilde{l})^2 \tag{6.89c}$$

where, again, k is the wave number. The other symbols have been defined in the previous sections, in particular Sect. 6.4.3. Factor F is a pure incidence-scattering geometrical factor, and F^2 becomes the obliquity factor in this theory. In contrast, quantity $\sqrt{g}$ represents the wave phase change caused by $\tilde{z}$ (when sub-surface scattering is negligible as for metals, and according to Eq. (6.70)); the relative amount of specular reflection is a function of g, specifically one can set $\mathfrak{F} = \exp(-g)$ in Eq. (6.84a). Factor υ is tied to the *rms* slope; however, differently from g, it becomes very small for scattering angles sufficiently near the specular reflection.

At the time of [5], there were no radiometric standards, so the results of the Beckmann-Kirchhoff (BK) theory have been differently interpreted in literature. We will report the infinite series expressing the time average of the squared norm of the electric field of the scattered wave, which was named the mean scattered power by the authors. This equation, (which was used in [74] in experimental studies) can be written for a Gaussian height distribution with finite *rms* slope as

$$\mathfrak{D} = \frac{\pi}{\Delta A_s} \tilde{l}^2 F^2 e^{-g} \sum_{j=1}^{\infty} \mathcal{B}_j$$

$$\mathcal{B}_j \equiv \frac{g^j}{j!\,j} e^{-\upsilon/4j}$$

(6.90)

where ΔA_s denotes the irradiated area. One should note that the $\mathfrak{D}$-formula does *not* represent a BRDF. However, as carried out by Stover [98] for $g \gg 1$ (or the *very rough* case), $\mathfrak{D}$ can be reformulated in terms of BRDF. This one, though, shows also one of the limitations of the theory.

If $g \gg \upsilon$ as well, then one can verify by a Computer Algebra System (CAS) that

$$\mathfrak{D}_{(rough)} \cong \frac{\pi}{\Delta A_s} \tilde{l}^2 F^2 g^{-1} \propto \frac{\tilde{l}^2}{\tilde{z}^2}, \qquad g \gg 1,\ g \gg \upsilon$$

(6.91)

In the marked part of the gray box on p. 196, we highlighted a condition that allows PSD to be determined via the inverse diffraction problem. Equation (6.91) tells us clearly that the parameters characterizing a very rough surface cannot be determined independently of each other by the theory: only their ratio can. If the second constraint in (6.91) is not satisfied, then $\mathfrak{D}_{(rough)}$ becomes less simple, but the surface parameters persist solely through their ratio $\tilde{l}^2/\tilde{z}^2$. In addition, if $g \gg 1$, then the specular-reflection contribution appears negligible. Recently in [62], an iterative procedure has been set up in order to get the auto-correlation function related to a surface with *rms* height comparable to the incident wavelength.

In contrast, the PSD of a smooth surface described as above said could be determined via measurements of scattered power per projected solid angle. To such an aim, let us note that the ratio between two consecutive terms of the series is

6.5 Calculation of Thrust

given by

$$\frac{\mathcal{B}_{j+1}}{\mathcal{B}_j} = \frac{jg}{(j+1)^2} e^{v/4j(j+1)} \tag{6.92}$$

from which it is apparent that if $g \ll 1$ and $v < 16 \ln 2 \cong 11$, only the first term of the series will suffice:

$$\mathcal{D}_{(smooth)} \cong \frac{\pi}{\Delta A_s} F^2 \tilde{l}^2 g e^{-(g+v/4)} \tag{6.93}$$

Comments: relatively to solar sailing, the main drawback of the Beckmann-Kirchhoff theory is its underlying paraxial assumption, which would limit its applicability every time sail attitude entails sunlight's incidence angles sufficiently large; nevertheless, scattering from non-smooth sails may be dealt with.

- **Modified Beckmann-Kirchhoff.** A recent view states that the power distribution of the scattered light (from a rough surface), as described by Beckmann-Kirchhoff theory, is actually the diffracted radiance, e.g. [40, 41, 98], and [85]. Other authors, e.g. [74] and [76] utilized the scattered intensity in collecting experimental data which the theory had to be compared to. According to the rightmost side of Eq. (2.9), once the diffracted spectral radiance is calculated, spectral intensity is obtained simply by integrating over the projected area ($\mathrm{d}A_\perp = \cos \vartheta_{(re)} \, \mathrm{d}A$). The modified Beckmann-Kirchhoff (mBK) theory considered here[6] relies mainly on phenomenological arguments. Its purpose is to arrive at a formulation of the above $\mathcal{D}$ equations that can extend the validity to large incidence/scattering angles without giving up the advantage of getting to rough surfaces. The ensuing formulation is well explained in [41, 85]; here, we report the salient points. *First*, it has been noted by several authors that the obliquity F^2 is a source of disagreement with the experimental data; therefore, it is omitted. *Second*, a new renormalization factor, denoted by K, is introduced by considering the effect from the evanescent waves.[7] If no evanescent wave is produced, then $K = 1$. *Third*, the so-modified $\mathcal{D}$ is interpreted in literature as normalized scattered radiance. In formulas, the original BK formulation turns into

$$L(\phi_{(re)}, \vartheta_{(re)})^{(\mathrm{mBK})} = \frac{\pi}{\Delta A_s} \tilde{l}^2 K e^{-g} \sum_{j=1}^{\infty} \mathcal{B}_j \tag{6.94}$$

---

[6]This one is not the only modification, proposed and tested, to the original theory of Beckmann-Kirchhoff; for instance, see [84].

[7]Very shortly, at incidence angles greater than the critical angle of total internal reflection, transmitted electromagnetic waves describable by wave vectors with an imaginary component are observed. These waves are called *evanescent*, and are characterized by an exponential damping in the less dense medium. The decay distance is of the order of a wavelength. Important experiments and theoretical consequences stem from evanescent waves. A good short introduction can be found in [45], whereas the topic is dealt with extensively in devoted monographs, e.g. [23].

where the other symbols have the same meanings as in the original BK theory. Recalling (6.92), only few terms are usually necessary to evaluate the series in Eq. (6.94) with a relative error one order of magnitude lower than the fluctuations of TSI.[8]

Comments: Extensive testing the radiance/intensity predictions of the modified BK-theory have been carried out, especially in [41, 84], and [85], with accurate experimental data in different contexts. The agreement with experimental data resulted satisfactorily, even for rough surfaces. Also, in [41], a comparison is made between results from the modified BK-theory and the RR-theory in the range of applicability of the second one; an excellent agreement is reported.

- **Generalized Harvey-Shack**. Harvey-Shack (HS) theory was formulated almost four decades ago [37], and originally started as an SST. It is a linear-systems formulation based on the surface transfer function concept, which relates the scattering properties to the surface parameters. The original theory entailed a paraxial limitation like the BK-theory. Subsequently, passing through the realization that diffracted radiance is a basic quantity in describing scattering [38], the theory was extended by removing the paraxial restriction, and including wave polarization [40]. Under the smooth-surface condition, the diffuse BRDF of so generalized Harvey-Shack (gHS) theory can be expressed as follows [57]

$$\left(\Lambda_\lambda{}^\blacksquare\right)_{a\to b}^{(gHS)} = \frac{4\pi^2}{\lambda^4}(\cos\vartheta_\odot + \cos\vartheta_{(re)})^2 \mathcal{Q}_{a\to b}\mathbb{S}_2(\mathbf{f}) \tag{6.95}$$

which differs from the RR-theory only in the obliquity term. The inclusion of polarization, here made explicit for comparison with the B-theory and RR-theory, gives the gHS theory a quasi-vector character that did not appear in the original formulation of the HS-theory; however, this holds only for smooth surfaces.

Comments: According to [40], the gHS-theory puts together the advantages of the RR and BK theories, but without the drawbacks of either of them. For smooth surfaces, this theory allows handling the inverse scattering problem better than the RR-theory, a considerable advantage in the problem of determining the surface PSD of the sail surface pieces. In the case of a rougher surface and large incidence/scattering angles, the diffuse BRDF is not given by Eq. (6.95); the related numerical solution is computationally intensive, but it produces results close to experimental data.

■ The above inspection of five diffraction theories aims at recommending that one or more scattering models should be embedded in an accurate mission analysis of fast sailcraft with regard to the thrust calculation and its implementation in

---

[8]Discussing specific numerical results from Eq. (6.94) is beyond the aim of this chapter. The reader is invited to implement this equation by a CAS in order to visualize how sunlight is diffracted by smooth and medium-smooth surfaces. As an exercise, this can be done easily enough with the BK formulations *without* determining the area PSD.

6.5 Calculation of Thrust

the flight design. Calculation of sailcraft trajectories, during the feasibility phase of a space project, should include thrust evaluation by starting with *assumed* surface's shape and material parameters/functions of the candidate sail sheets. In a subsequent (project-dependent) design phase, the inverse scattering problem should eventually be dealt with in order to measure the *actual* parameters/functions and re-calculate thrust by simulating different sail orientation with respect to sunlight. With the sailcraft mass and sail area updated via the progress of the sailcraft design, a new admissible set of mission trajectories can be carried out. And so on, conceptually speaking, until baseline and backup mission profiles are completed.

6.5.5 The Twofold Effect of Absorption

The first part of this subsection is devoted to the calculation of the momentum induced directly by absorption. Subsequently, the sail temperature is considered. Finally, the net momentum brought about by heat emission is modeled. The temperature, or better, the distribution of the temperatures of the sail support subsystem is not considered here; such a subsystem is not unique in its kind and, moreover, its contribution to thrust via its surface exposed to sunlight is expected negligible.

Although the sail may be conceptually organized in various macrosurfaces (Sect. 6.4.3), nevertheless all of them share some key features:

1. front side and backside are generally different in terms of optical properties, but both have the same area;
2. the sail's materials exhibit large thermal conductivities across the membrane whereupon the transversal temperature gradient is small;
3. the sail thickness is so small that (a) both sail sides can be regarded as having the same temperature in the large (i.e. apart from possible local variations produced by some topological behavior), and (b) the sail adjusts its temperature to the changing environment (along its trajectory) with no practical time lag.

♦ Let us calculate the power per unit area *absorbed* by a sail macrosurface relatively to which, we know, sunlight arrives from the direction $(\pi, \vartheta_{\odot})$; each macrosurface has its own $\vartheta_{\odot}$. Using the superscript (f) for the surface front side and other symbols previously defined in this chapter, one gets

$$\mathcal{W}_{(abs)} = \int_{\lambda_{min}}^{\lambda_{max}} \mathcal{A}_{\lambda}^{(f)}(\pi, \vartheta_{\odot})(J_{\lambda} \cos \vartheta_{\odot}) \, d\lambda$$

$$= \cos \vartheta_{\odot} \int_{\lambda_{min}}^{\lambda_{max}} J_{\lambda,(abs)}^{(f)} \, d\lambda = \cos \vartheta_{\odot} \mathcal{A}^{(f)} J \qquad (6.96)$$

where the approximated irradiance (Sect. 6.2.1) has been used for simplicity because the effect of non-point-like Sun on the sail temperature is negligible; J_{λ} was defined in Eqs. (6.31). In Eq. (6.96), $A_\lambda^{(f)}(\pi, \vartheta_\odot)$ denotes the directional spectral absorptance, whereas $A^{(f)}$ is the total absorptance (of the macrosurface); this one coincides with that one expressed via the second of Eqs. (6.44). Note that absorptance is practically independent of the surface temperature. In contrast, it is affected by the source temperature(s) through the spectral irradiance that, in the current case, is the SSI's band in which the front side material is opaque. This means that the energy of photons with such wavelengths eventually goes into heat; thus, in particular, photons producing escaping photoelectrons (Chap. 4) do not contribute to absorptance. The integrand $J_{\lambda,(abs)}^{(f)}$ represents the part of the irradiance absorbed by the macrosurface.

Absorption does not violate the principle of momentum conservation. Thus, from Eq. (6.96), one can write straightforward the momentum absorbed by the macrosurface material, according to the current convention $\mathbf{d}_{(in)} = -\mathbf{d}_\odot$ with $\phi_\odot = \pi$

$$d^2\mathbf{Q}_{(abs)}^{(\mathcal{F}\mathbf{m})} = -\frac{dA\,d\tau}{c}\,W_{(abs)}\,\mathbf{d}_\odot = \frac{dA\,d\tau}{c}\cos\vartheta_\odot\,J A^{(f)}\begin{pmatrix} \sin\vartheta_\odot \\ 0 \\ -\cos\vartheta_\odot \end{pmatrix} \qquad (6.97)$$

This is the major effect, in terms of momentum transferred to the sail system, caused by the sail front-side material's absorptance of solar irradiance.

♦ Sail re-radiates the received energy from both sides. Since it is not a blackbody, emittance has to come in the power released from each macrosurface side. Under the above four assumptions, one can write the total re-radiated (*rer*) power per unit area as follows:

$$\begin{aligned} W_{(rer)} &= W_{(rer)}^{(f)} + W_{(rer)}^{(b)} \\ &= \int_0^\infty d\lambda \int_{f\text{-}hemis} \mathcal{E}_\lambda^{(f)}(T, \phi, \vartheta) B(\lambda, T) \cos\vartheta\, d\omega \\ &\quad + \int_0^\infty d\lambda \int_{b\text{-}hemis} \mathcal{E}_\lambda^{(b)}(T, \phi, \vartheta) B(\lambda, T) \cos\vartheta\, d\omega \\ &= \int_0^\infty \left(M_\lambda^{(f)}(T) + M_\lambda^{(b)}(T)\right) d\lambda \\ &= \left(\mathcal{E}^{(f)}(T) + \mathcal{E}^{(b)}(T)\right)\mathcal{M}(T) = \left(\mathcal{E}^{(f)}(T) + \mathcal{E}^{(b)}(T)\right)sT^4 \qquad (6.98) \end{aligned}$$

where $\mathcal{M}(T)$ and $B(\lambda, T)$ are the blackbody exitance and spectral radiance given by Eqs. (2.22) and (2.23). $M_\lambda^{(j)}(T)$ is the spectral exitance of the j-side ($j = f, b$) at temperature T, $\mathcal{E}_\lambda^{(j)}(T, \phi, \vartheta)$ denotes the directional spectral emittance of the j-side, and $\mathcal{E}^{(j)}(T)$ is the hemispherical emittance of the same side.

Note that each hemispherical emittance is weighed by the blackbody spectrum and extends over all wavelengths by definition. As a result, though the Kirchhoff law states that the directional spectral emittance matches the directional spectral absorptance in any case, there is a-priori no equality relationship between the hemispherical emittance and the absorptance of a material. There are only a few exceptions

6.5 Calculation of Thrust

that regard ideal situations (conceptually useful); for instance, $\mathcal{E}_\lambda(T, \phi, \vartheta) = \mathcal{E} = \mathcal{A}$ for a uniformly-diffusing gray surface at the temperature T that receives light from a blackbody at the same T.

At the equilibrium, the surface temperature is T_s such that

$$\mathcal{A}^{(f)}\left(\cos\vartheta_\odot J + \mathfrak{s}T_{bg}{}^4\right) = \left(\mathcal{E}^{(f)}(T_s) + \mathcal{E}^{(b)}(T_s)\right)\mathfrak{s}T_s{}^4 \tag{6.99}$$

where T_{bg} is the temperature of the space background (modeled as blackbody radiation) in which the sailcraft is immersed. In the interplanetary space, $\mathfrak{s}T_{bg}{}^4 \ll \cos\vartheta_\odot J$ even at $R \cong 100$ AU and at incidence angles as large as $89°$. Analytically or numerically, depending on the complexity of $\mathcal{E}(T)$, Eq. (6.99) allows the sail-system designer to calculate the sail material temperature for each sail attitude and distance.

Note 6.5 Equation (6.99) should be used iteratively in a real project; not only there are different macrosurfaces, but also the membrane's local height (i.e. deviation from the flatness) is variable for a real sail (Sect. 6.5.7).

♦ There is also a contribution to the thrust through the net momentum caused by the thermal radiation from the two sides of the macrosurface. For the front side at a temperature T, and making the solid angle explicit, one gets

$$L_\lambda^{(f)}(T, \phi, \vartheta) = \mathcal{E}_\lambda^{(f)}(T, \phi, \vartheta) B(\lambda, T)$$

$$\mathrm{d}^2\mathbf{Q}_{(rer)}^{(\mathcal{F}\mathbf{m}),(f)} = \frac{\mathrm{d}\tau}{c} \int_0^\infty \mathrm{d}\lambda$$

$$\times \int_0^{2\pi} \mathrm{d}\phi \int_0^{\pi/2} \begin{pmatrix} \sin\vartheta\cos\phi \\ \sin\vartheta\sin\phi \\ \cos\vartheta \end{pmatrix} \left(\mathrm{d}A\cos\vartheta\, L_\lambda^{(f)}(T, \phi, \vartheta)\right)\sin\vartheta\,\mathrm{d}\vartheta \tag{6.100}$$

where the re-radiated radiance has been employed again for calculating the directional spectral photon momentum through the fundamental relationship between photon energy and momentum. This is a fourth-order infinitesimal momentum that, once integrated over solid angle and wavelength, results in the second of Eqs. (6.100). In general, since the emittance depends on ϕ, there is a momentum component parallel to ΔS. Analogously, for the backside, one can write

$$L_\lambda^{(b)}(T, \phi, \pi - \vartheta) = \mathcal{E}_\lambda^{(b)}(T, \phi, \pi - \vartheta) B(\lambda, T)$$

$$\mathrm{d}^2\mathbf{Q}_{(rer)}^{(\mathcal{F}\mathbf{m}),(b)} = \frac{\mathrm{d}\tau}{c} \int_0^\infty \mathrm{d}\lambda \int_0^{2\pi} \mathrm{d}\phi$$

$$\times \int_{\pi/2}^{\pi} \begin{pmatrix} \sin\vartheta\cos\phi \\ \sin\vartheta\sin\phi \\ \cos\vartheta \end{pmatrix} \left(-\mathrm{d}A\cos\vartheta\, L_\lambda^{(b)}(T, \phi, \pi - \vartheta)\right)\sin\vartheta\,\mathrm{d}\vartheta \tag{6.101}$$

where now $\vartheta \in [\pi/2, \pi]$ and $dA_\perp = -dA \cos \vartheta$ for correctly orienting the momentum from the backside emission in $\mathcal{F}_\mathbf{m}$. (Note that the colatitude in the directional spectral emittance usually refers to the corresponding surface's normal; thus, $\vartheta \to \pi - \vartheta$ for the backside.) Therefore, the total re-radiation momentum is expressed by

$$d^2\mathbf{Q}_{(rer)}^{(\mathcal{F}_\mathbf{m})} = d^2\mathbf{Q}_{(rer)}^{(\mathcal{F}_\mathbf{m}),(f)} + d^2\mathbf{Q}_{(rer)}^{(\mathcal{F}_\mathbf{m}),(b)} \tag{6.102}$$

where the two contributions are given by Eqs. (6.100) and (6.101).

It is useful to define the coefficient of emissive momentum, similarly to that of diffuse momentum (Eq. (2.20) on p. 43), for either side of the macrosurface:

$$\hat{\chi}^{(j)} \equiv \frac{\|d^2\mathbf{Q}_{(rer)}^{(\mathcal{F}_\mathbf{m}),(j)}\|/dA\,d\tau}{\mathcal{E}^{(j)}\mathcal{M}/c}, \qquad j = f, b \tag{6.103}$$

where the explicit dependence on temperature has been omitted. The numerator represents the pressure associated to the vector sum of the momenta of the radiation re-emitted from the j-side, whereas the denominator is the pressure one would get if the radiation (re-emitted from the j-side) where mono-directional. Like the coefficient of diffuse momentum, therefore, $\hat{\chi}$ can be viewed as a temperature-induced momentum efficiency. One has two such efficiencies, of course. How do they combine in the net re-emission momentum? A special case could be illuminating. Equation (6.102) and definition (6.103) are general; however, if the directional spectral emittance of either side is symmetric in azimuth, then $\mathcal{E}_\lambda^{(f)}(T, \phi, \vartheta) = \mathcal{E}_\lambda^{(f)}(T, \vartheta)$ and $\mathcal{E}_\lambda^{(b)}(T, \phi, \pi - \vartheta) = \mathcal{E}_\lambda^{(b)}(T, \pi - \vartheta)$ can be inserted into Eqs. (6.100) and (6.101), respectively; one can recognize that only the z-components do not vanish. Applying the definition (6.103) to both sides and using the z-component, then one obtains easily

$$\mathcal{E}^{(f)}\mathcal{M}\hat{\chi}^{(f)} = 2\pi \int_0^\infty B(\lambda, T)\,d\lambda \int_0^{\pi/2} \mathcal{E}_\lambda^{(f)}(T, \vartheta)\cos^2\vartheta \sin\vartheta\,d\vartheta \tag{6.104a}$$

$$\mathcal{E}^{(b)}\mathcal{M}\,\hat{\chi}^{(b)} = 2\pi \int_0^\infty B(\lambda, T)\,d\lambda \int_{\pi/2}^\pi \mathcal{E}_\lambda^{(b)}(T, \pi - \vartheta)\cos^2\vartheta \sin\vartheta\,d\vartheta \tag{6.104b}$$

where the inner integrals still depend on wavelength; these functions weigh the spectral blackbody radiance $B(\lambda, T)$ to give the norm of the re-emission momentum. Thus, the total momentum related to the re-radiation of the absorbed irradiance can be written at the equilibrium temperature as

$$\begin{aligned}
d^2\mathbf{Q}_{(rer)}^{(\mathcal{F}_\mathbf{m})} = \|d^2\mathbf{Q}_{(rer)}^{(\mathcal{F}_\mathbf{m})}\|\mathbf{m} &= \frac{dA\,d\tau}{c}\left(\hat{\chi}^{(f)}\mathcal{E}^{(f)} - \hat{\chi}^{(b)}\mathcal{E}^{(b)}\right)_{T=T_s}\mathcal{M}(T_s)\mathbf{m} \\
&= \frac{dA\,d\tau}{c}\left(\hat{\chi}^{(f)}\mathcal{E}^{(f)} - \hat{\chi}^{(b)}\mathcal{E}^{(b)}\right)_{T=T_s}\mathfrak{s}T_s^{\,4}\mathbf{m} \tag{6.105}
\end{aligned}$$

6.5 Calculation of Thrust

with the coefficients of emissive momentum given by Eqs. (6.104a), (6.104b) at $T = T_s$, T_s being the positive solution of Eq. (6.99). An equivalent form of Eq. (6.105) is useful by replacing the explicit temperature via Eq. (6.99), with $T_{bg} = 0$ for simplicity, to give

$$d^2\mathbf{Q}_{(rer)}^{(\mathcal{F}\mathbf{m})} = \frac{dA\,d\tau}{c}\cos\vartheta_\odot J A^{(f)}\left[\frac{\hat{\chi}^{(f)}\mathcal{E}^{(f)} - \hat{\chi}^{(b)}\mathcal{E}^{(b)}}{\mathcal{E}^{(f)} + \mathcal{E}^{(b)}}\right]_{T=T_s}\mathbf{m} \qquad (6.106)$$

In the squared brackets of Eq. (6.106), one finds which way a positive absorptance causes a net momentum along the macrosurface normal; the quantity in brackets is always less than unity in absolute value, so that $\|d^2\mathbf{Q}_{(rer)}^{(\mathcal{F}\mathbf{m})}\| < \|d^2\mathbf{Q}_{(abs)}^{(\mathcal{F}\mathbf{m})}\|$, as it is apparent from Eq. (6.97). Note that, in practice, the only way to have $d^2\mathbf{Q}_{(rer)}^{(\mathcal{F}\mathbf{m})} = 0$ identically, i.e. for any temperature, is to employ *one* material, namely, a mono-layer sail with both surface sides equally structured. In general, for any practical use, sails will exhibit net re-radiation momenta.

Some further considerations are in order. The various types of emittance of a material are not easy to measure. Emittance does not depend only on temperature, wavelength, and direction—as we supposed above for simplicity—but also on the surface oxidation, roughness, polishing, and any thermal cycle (e.g. [15]) the surface may undergo. These items are general and may be applicable to solar sails; in addition, one should remember that the metallic layers of a sail come from a process of vapor deposition, a PVD in particular, as described in Sect. 6.3. The emittance(s) of the built layer will be affected by such process. Therefore, the measurement of the emittance(s) of the various sail-representative samples should be included in sailcraft design. Such measurements will be necessary, though not sufficient. As a point of fact, since the sail temperature changes along the trajectory, emittance becomes a dynamical parameter that should progressively be updated by including it in the sailcraft trajectory determination process.

One might wonder whether the coefficients of diffuse momentum (p. 43) and emissive momentum are someway interrelated. Conceptually, they are different: the first scalar comes from the distribution of the scattered photon momenta, whereas the second one refers to the momentum distribution of thermal photons. Since the surface topology will certainly act upon radiation diffraction and emission, the two coefficients are expected to be interconnected, even though setting up an interrelation model is not pursued here. In any case, one could not disregard measurements of directional spectral reflectance and emittance. Only in one case we can assert that the two coefficients give the same result: for pure Lambertian surfaces, one gets $\chi = \hat{\chi}^{(f)} = \hat{\chi}^{(b)} = 2/3$. Nevertheless, even in such ideal case, the net re-emission momentum does not vanish.

6.5.6 Transmittance Momentum

In addition to energetic photons that can travel across the sail by inducing various energy-dependent phenomena inside the volume of the sail layers, sail materials may have a transmittance over some portion of the thrust wavelength band (considered in the previous sections) provided that the reflective layer is very, very thin and the plastic support is semitransparent. On the other hand, one should not exclude that more evolved sail systems, based on future nanotechnological developments, may result in sails with a low (but not negligible) transmittance; in other words, sailcraft design may tolerate some transmittance because the sail lightness enhancement would be strikingly high through that technology. Although we shall not detail the process of transmission here, however we will restrict here to a few aspects ultimately relevant to the calculation of the thrust.

In order to proceed with meaningful cases, let suppose that the sail consists of a very thin reflective layer deposited on a much thicker support layer, but still satisfying the first three assumptions of Sect. 6.5.5. If there were an emissive coating backside, then transmittance should be negligible. Formally analogous to Eq. (6.82), the outgoing momentum due to transmission of light though such bilayer membrane is given by

$$d^2 \mathbf{Q}_{(tr)}^{(\mathcal{F}m)} = -dA \frac{d\tau}{c} \int_{\lambda_{min}}^{\lambda_{max}} I_\lambda \, d\lambda \int_{2\pi \text{ sr}} \cos \vartheta_{(tr)} \Theta_\lambda \, \mathbf{d}_{(tr)} \, d\omega_{(tr)}$$

$$\mathbf{d}_{(tr)} = \begin{pmatrix} \sin \vartheta \cos \phi \\ \sin \vartheta \sin \phi \\ \cos \vartheta \end{pmatrix}_{(tr)}, \quad \pi/2 \leqslant \vartheta_{(tr)} \leqslant \pi$$

(6.107)

where the BTDF refers to the whole membrane. For modeling it, let us mention how radiation propagates through a material medium.

In general, electromagnetic radiation will travel throughout some medium according to the noted Radiative Transfer Equation (RTE). Conceptually speaking, RTE describes how the radiance evolves through a medium by concurrent energy *decrease* and energy *increase* in a generic, arbitrarily oriented, elemental volume of the medium; the decrease occurs via absorption and out-scattering from the observation direction, while the increase is due to thermal emission and in-scattering. There is a vast literature on this topic (and its related areas) because radiation transfer is a fundamental feature occurring in a huge number of diverse phenomena at quite different scales: from large astrophysical objects to every-day life scales, and down to microscopical characteristic lengths. For example, the interested reader is suggested referring to [16, 33, 59, 70], [83, 91, 109], and [114] just to cite a few ones in many different disciplines into which one might want to be introduced. RTE can be obtained by employing approaches conceptually different; if one resorts to the Boltzmann equation of statistical mechanics and the radiative transport is quantum-physics based, namely, a *photon gas* spreads throughout and interacts with the medium particles quantum mechanically, then the transfer equation is

6.5 Calculation of Thrust 225

named the Photon Transport Equation (PTE). Here, we will follow the radiometri-
approach.

Remark 6.3 The general structure of RTE is a partial-derivative integro-differential
equation. The quasi-steady form of it is used frequently. One may tries to get analytical solution in special cases of interest, or to resort to Monte Carlo codes for
obtaining reliable data to be compared with the experiments. The above mentioned
references contain extensive analytical and numerical methods. together with the
derivation of the equation. RTE is employed usually in its scalar form; however,
if polarization is important in the problem to solve, then one has to solve a *vector*
RTE.

Adopting the language of the energy approach to RTE, quite compliant with
Eq. (6.107), we may here model our membrane as an ideal plane layer exhibiting very low scattering, but still having (1) non-negligible spectral absorptance and
emittance (over the band(s) of interest in solar sailing), (2) uniform temperature,
and (3) homogeneity even with with respect to electromagnetic radiation; then, one
can get a meaningful analytical solution (e.g. Chap. 9 of [91] and Chap. 9 of [70]).
Adapting such solution to the current framework, i.e. considering that the membrane backside is not perfectly smooth, we can write the following expression for
the transmitted radiance, i.e. the radiance from the membrane backside:

$$L_{(tr),\lambda}(0) = \Theta_\lambda^{(f)} I_\lambda$$

$$L_{(tr),\lambda}^{(RTE)}(z^{(sl)}) = L_{(tr),\lambda}(0) \exp\left(-\beta_\lambda^{(sl)} z^{(sl)}\right)$$

$$+ B(\lambda, T_s)\left[1 - \exp\left(-\beta_\lambda^{(sl)} z^{(sl)}\right)\right] \tag{6.108a}$$

$$L_{(tr),\lambda} = h_\lambda(\mathcal{P}_{(tr)}) L_{(tr),\lambda}^{(RTE)}(z^{(sl)}) \tag{6.108b}$$

$$\mathcal{P}_{(tr)} \equiv \left\{\tilde{z}^{(sl)}, \tilde{l}^{(sl)} \mathbf{d}_{(tr)}\right\}$$

Let us explain the new symbols and the two equations.

New symbols: the superscript (sl) stands for the (polyimide) support layer, which
has thickness $z^{(sl)}$, surface's *rms* roughness $\tilde{z}^{(sl)}$ and slope $\tilde{l}^{(sl)}$, $\beta_\lambda^{(sl)}$ denoting the
spectral extinction coefficient; the superscript ^{RTE} refers to as the solution from the
transfer equation without considering the surface roughness. *Equations*: the first line
of Eq. (6.108a), (6.108b) expresses how much of the SSI becomes radiance of input
to the support layer, $\Theta_\lambda^{(f)}$ denoting the BTDF of the (reflective) front-side layer. The
second line tells us that the radiance at the sail backside consists of two pieces: one is
the input radiance attenuated according to the Beer-Bouguer-Lambert law applied to
an homogeneous medium of optical thickness $\beta_\lambda^{(sl)} z^{(sl)}$; the other contribution is due
to the fraction of the thermal emission of the sail at uniform temperature T_s, which is
not attenuated by the medium itself. Function $h_\lambda()$ in Eq. (6.108b), which depends

226   6 Modeling Light-Induced Thrust

on *rms* roughness and slope of the rear surface, mere redistributes the transmitted power angularly.

We wrote about the photon momentum associated to the thermal emission of both sail macrosurface sides in Sect. 6.5.5. Here, we may employ Eqs. (6.108a), (6.108b) for modeling the BTDF of the sail membrane with regard to the non-thermal component. Θ_λ is just the quantity appearing in Eq. (6.107); obviously, it is a matter of measurements as well. Therefore, in the current model, we can write

$$\Theta_\lambda^{(\text{non-th})} = L_{(tr),\lambda}^{(\text{non-th})}/I_\lambda = L_{(tr),\lambda}(0)\exp\left(-\beta_\lambda^{(sl)}z^{(sl)}\right)/I_\lambda$$

$$= \Theta_\lambda^{(f)}\exp\left(-\beta_\lambda^{(sl)}z^{(sl)}\right) \tag{6.109a}$$

$$\Theta_\lambda = h_\lambda(\mathcal{P}_{(tr)})\Theta_\lambda^{(\text{non-th})} = h_\lambda(\mathcal{P}_{(tr)})\Theta_\lambda^{(f)}\exp\left(-\beta_\lambda^{(sl)}z^{(sl)}\right) \tag{6.109b}$$

Remark 6.4 A few remarks appear appropriate. *First*, one should note that the temperature appearing in Eq. (6.108a) has to be known in advance, for instance via the overall power balance such as Eq. (6.99). *Second*, the above algorithm for computing BTDF ultimately requires the knowledge of $\Theta_\lambda^{(f)}$ and the dimensionless function $h_\lambda(\ )$, which is tied to the roughness of the sail backside. Even BTDF is a matter of experiments to be performed on sail (small) samples *if* the sail is of the kind we described above. In contrast, if the reflective layer or the emissive layer (generally notably thinner) is opaque, there is no transmission momentum to consider in the thrust equation.

6.5.7 Considering Large-Scale Curvature and Wrinkling

Now that the problem of diffraction has been dealt with in the previous sections, a question arises: what about the highest level of topology? In Sect. 6.4, we did a short description on topological modifications induced essentially by membrane wrinkles and sail curvature in the large, due to the sunlight pressure on the sail as a whole, and the temperature. Analysis of membrane wrinkles, in general, and design considerations for gossamer spacecraft, in particular, can be found in many papers and books, e.g. [34, 52, 53, 60, 72, 116], and [54, 117–119]. The occurrence of wrinkles, with their shape, wavelength, amplitude, direction, and distribution throughout a sail, will depend also on the way a sail is deployed and kept open during the mission. Although the considerations of this subsection are somewhat general, with regard to wrinkles we will restrict our attention to their geometry stemming from the sail deployment/keeping modes known as *separate quadrants* (with five-point suspension) and *striped architecture*, with possible different versions. Sail is supported by *unrigged* booms in any case; this represents a significant evolution of the sail configuration promoted in the first decade of the 21th century with respect to what thought mostly in 20th century (e.g. [121] for a comprehensive view).

Criterion for neglecting a thrust cause. In some references, e.g. [61, 72], and Chap. 5 of [54], authors assert that the propulsive effect of wrinkles (through the modification of the reflectance) can be considered negligible. However, when may a thrust effect be considered unimportant? Negligible, unimportant, or uninfluential with respect to what? Let us discuss this point shortly.

In very recent papers, [105] and [106], there was shown numerically—via a sophisticated computer code for sailcraft mission analysis—that heliocentric low-eccentricity orbits of sailcraft may be sensitive to the fluctuations of TSI. (At that time, the investigation was restricted to low-eccentricity orbits, but non-negligible effects on other types of sailcraft trajectories may not be excluded). In the case of sailcraft-Mars rendezvous, using a constant value of TSI could entail the loss of the mission because the Mars-centric trajectory would hard be recoverable. All examples have been built by employing the actual (measured) realization of the solar irradiance in the 2003–2004 time frame. Trajectory perturbations due to variable TSI have resulted much larger than planetary disturbances. The relative amount of TSI fluctuations can be quantitatively set by the right histogram of Fig. 3.10 where the standard deviation and the mean of daily-TSI in the solar-cycles 21–23 have been reported: $\delta_{TSI} = 0.56/1365.94 \cong 4.1 \times 10^{-4}$.

Suppose now that, by considering a set of causes in the algorithm for thrust calculation, one gets the vector $\mathbf{T}_0$; subsequently, *one* new cause, modeled and *added* to the set, results in the new vector thrust $\mathbf{T}_{+1}$ (e.g. as the maximum effect obtainable). By definition, such new element may be neglected in the algorithm if

$$\frac{\|\mathbf{T}_{+1} - \mathbf{T}_0\|}{\|\mathbf{T}_0\|} \ll 3\delta_{TSI} \tag{6.110}$$

Just a few remarks about the meaning of inequality (6.110). *First*, the reference set of elements or causes may or may not contain the TSI fluctuations. In the former case, one actually compares the new cause of possible thrust with the TSI uncertainty. In the latter case, one could test the introduction of TSI fluctuations in classical model as a special case; this entails the analysis of whole trajectories under constant or variable TSI and comparing them with respect to the mission objectives. *Second*, the criterion can be extended to test $\|\mathbf{T}_{-1}\|$, namely, the thrust obtained by *removing* one cause from the reference model. *Third*, the above inequality is a scalar relationship that may help in those cases where the sailcraft position and/or velocity at the target(s) are not stringent.

As a result of what just said, here we do not take for granted that wrinkles produce negligible changes to thrust as claimed in some papers. It will depend on the mission objectives, the sail type, and the actual TSI fluctuations during the mission time frame.

Large-scale (or global) curvature in a sail is due essentially to (i) radiation pressure (mainly close to the Sun), (ii) manufacturing tolerances, (iii) boom's thermal gradients (when a boom component casts a shadow on another one), and (iv) sail's temperature (which changes with sailcraft distance and attitude—Sect. 6.5.5). Let us first clarify a few points useful for modeling these deviations from flatness from an optical response viewpoint. In multi-layer sails of first generation and probably the first part of the second generation, a plastic support (Kapton, Mylar, CP1, CP2, or IKAROS-type polyimide) shall/should always be employed. On this membrane, the reflective and/or emissive layers have to be deposited.

If the sailcraft is designed to run a non-negligible fraction of AU sunward from the Earth, then temperature increases and so the coefficient of linear thermal expansion, say here, CTE for short. At perihelion, plastic support will have the highest CTE among the sail layers. If the sail is endowed with a load distribution net, then its shape could be determined by the net's CTE, which can be chosen lower than the plastic's. Sailcraft of advanced second generation or higher might be all metal (either ground or space manufactured materials), or based on some variety of Carbon nanotubes.

The model surface, representing the actual globally-curved surface, should take the sail architecture into account. For instance, in the cases of three suspension points, a vault-like triangle (which is also right and isosceles) may be appropriate for picturing the sail quadrant. While angles and sides are known, the curvature radius has to be connected to the tensioning field. If, on the other side, the sail consists of a skin-skeleton structure containing chords supporting the plastic layer, then these ones are attached-to-opposite-booms low-CTE chords (parallel to the quadrant edge) that affect the global sail shape (e.g. [34] and [61]). In this striped configuration, the three-dimensional shape of a quadrant has variable surface curvatures, which depend on the boom loading one wants to achieve. For thrust calculation, one may start with a flat surface assumption because the billow strain should, according to [34], of the order of 5×10^{-4} for a sail of 100-m side and thickness of the order of 1 μm. However, one should be careful to not extrapolate straightforward to sails with somewhat larger sides and notably lower thickness, for which non-linear effects have to be analyzed.

In our framework, the total curvature is equivalent to a change of the height profile with respect to an ideal planar sail, which lies on the common median plane of the ideally-*unbent* supporting booms. On this reference plane, we choose the intersection of the diagonal boom centerlines as the origin O, and the bisector of the angle of two *chosen* facing booms as the x-axis. The unit vector $\mathbf{n}$ (which we may call the sail axis for short) is orthogonal to such plane at O in the sail's *backside* hemisphere (i.e. the opposite one to the Sun). The y-axis $\mathbf{y} = \mathbf{n} \times \mathbf{x}$ completes the triad. Let us denote this frame by $\mathcal{F}^{(\text{booms})}$. This Cartesian frame is attached to the sail support system; thus, for a spinning sailcraft, it is non-inertial (in the classical sense). Each quadrant of the sail, with the deployment/keeping architecture chosen for a certain mission, can be logically divided in oriented macrosurfaces

6.5 Calculation of Thrust 229

or regions.[9] Each region may be wrinkled or unwrinkled, whereas slack pieces of surface are excluded from this analysis. For instance, in the case of a three-point suspension quadrant (or five for the whole sail), a possible segmentation consists of a quadrant-inscribed (non-regular) pentagon as unwrinkled, and three remaining triangular regions filled of (structural) wrinkles, which may be modeled via a radial field of wrinkles. In other sail configurations, a uniform field of wrinkles may be appropriate. The functions here below regard the sail swelling related to two architectures.

Function (6.111) may approximate the sail shape related to the ⎧ striped ⎫ architecture: l_s is the nominal sail side, α_q is the quadrant's half-angle (nominally, $\pi/4$), s_{max} is the sag of the quadrant surface section parallel to $\mathbf{n}$ and containing its external edge (i.e. the sail side line). Each quadrant has one coordinate plane as (vertical) plane of symmetry, say, $\mathfrak{S}$. Quadrant's parallel sections, i.e. those ones formed by planes parallel to $\mathfrak{S}$, are *concave* parabolas. Orthogonal sections, namely those ones shaped by planes perpendicular to $\mathfrak{S}$, are *convex* parabolas. For simplicity, the boundary conditions at the booms are $Z(x_b, y_b) = 0$, namely, boom bending has been neglected. Sections obtained by intersection with planes parallel to the plane of the booms are hyperbolas. The subscript κ denotes the order number of the considered quadrant, i.e. $\kappa \in \{0, 1, 2, 3\}$. Angle $\beta_\kappa = \kappa\pi/2$ is the azimuth of the quadrant-κ symmetry axis. Applying the function to the four quadrants results in the shape example shown in the left part of Fig. 6.9:

$$Z_\kappa^{(\text{str})} \equiv Z_\kappa^{(\text{stripes})}\left(x, y, \beta_\kappa \mid \mathfrak{F}^{(\text{booms})}\right)$$

$$= 4s_{max}\left[\left(\frac{x}{l_s}\sin\beta_\kappa + \frac{y}{l_s}\cos\beta_\kappa\right)^2 \cot^2\alpha_q - \left(\frac{x}{l_s}\cos\beta_\kappa - \frac{y}{l_s}\sin\beta_\kappa\right)^2\right]$$

$$-l_s/2 \leqslant x \leqslant l_s/2, \quad -l_s/2 \leqslant y \leqslant l_s/2$$

$$-(x\cos\beta_\kappa + y\sin\beta_\kappa) \leqslant x\sin\beta_\kappa + y\cos\beta_\kappa \leqslant x\cos\beta_\kappa + y\sin\beta_\kappa \qquad (6.111)$$

With regard to the ⎧ suspension ⎫ configuration, function (6.112), which here has been restricted to the first quadrant, can be carried out by starting with the elliptic paraboloid having equation $Z_0^{(3ps)} \equiv z = ax^2 + by^2$, where a and b denoting (positive) scale factors with dimensions length^{-1}; their values affect the mean curvature of the paraboloid, and here are $\sim l_s^{-2}$ numerically. Then, as the paraboloid expands upwards and its vertex is the origin of $\mathfrak{F}^{(\text{booms})}$, a negative-angle rotation about the y-axis moves the vertices of the quadrant (or the boom tips) on the xy-plane. The general implicit form of the rotated paraboloid is given by the first line in Eqs. (6.112). The second and third lines make the surface height explicit. The fourth line gives the value of the rotation angle, say, $\varphi_{rot}^{(0)}$ that produces $Z(x_b, y_b) = 0$. Again, for simplicity, we are neglecting the boom deviations with respect to the sail membrane

---

[9]The algorithm described here may be applied to any sail configuration provided only that the logical regions are *orientable*, in the sense of the Local Theory of Surfaces.

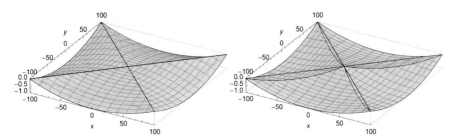

Fig. 6.9 Examples of modeling the shape of membrane swelling for thrust calculation: *left*; approximated shape for the sail's striped architecture; (*right*) approximated shape for a five-point suspension sail. Booms are drawn as *two* (*diagonal*) *lines*. Sail sides and height deviations (both in meters) appear compliant with [34]. Please refer to the next gray box before using the above architecture-dependent models of global sail shape under the action of the solar irradiance, which scales as R^{-2}

deviations from flatness. Finally, the fifth row includes the constraints that the plane coordinates have to satisfy, as it is straightforward to recognize. The other three quadrants are obtained by rotation. An example of what described is shown in the right part of Fig. 6.9:

$$Z_0^{(3ps)} \cos\varphi_{rot} - x \sin\varphi_{rot} = a\left(x \cos\varphi_{rot} + Z_0^{(3ps)} \sin\varphi_{rot}\right)^2 + by^2$$

$$Z_0^{(3ps)} \equiv Z_0^{(3\text{-point suspension})}\left(x, y, 0 \mid \mathcal{F}^{(\text{booms})}\right)$$

$$= \frac{\cos\varphi_{rot} - a\sin(2\varphi_{rot})x - \sqrt{\cos^2\varphi_{rot} - 4a\sin\varphi_{rot}(x + b\sin\varphi_{rot}y^2)}}{2a\sin^2\varphi_{rot}}$$

$$\sin\varphi_{rot}^{(0)} = \frac{1 - \sqrt{1 + (a+b)al_s^2}}{al_s}$$

$$-l_s/2 \leqslant x \leqslant l_s/2, \quad -l_s/2 \leqslant y \leqslant l_s/2, \quad -x \leqslant y \leqslant x$$

$$\text{quadrant}_0 \rightarrow \text{quadrant}_\kappa \, (\kappa = 1, 2, 3) \quad \text{via rotation about } z\text{-axis} \qquad (6.112)$$

A few remarks are in order. *First*, the examples built for Fig. 6.9 regard rotationally symmetric surfaces, i.e. compliant with small incidence angles of sunlight (coming from above). At large incidence angles, the rotational symmetry is expected to be broken. *Second*, the height functions, which may be employed as examples of approximation to quadrant billowing, are introduced here exclusively via geometry for initiating the analysis of the problem of

6.5 Calculation of Thrust

the sail-swollen incidence angle variations. In a real design, one will fit the outcomes from a non-linear analysis of the stress/strain field of the boom-sail system under the action of TSI. *Third*, if one assigns the sail a specific shape-*engineerable* geometry (e.g. one that is able to reduce thrust lessening meaningfully), then this determines the boom loads for a given architecture. Of course, the best solution depends also on the mission aims.

The regions of structural wrinkles are labeled by $\kappa\varepsilon$ here, where κ is the quadrant number again, and ε is the region number. In general, each macrosurface in a quadrant has its normal $\mathbf{m} = \mathbf{m}_\varepsilon$ that determines the local Cartesian frame; therefore, in each region, the considered height function is referred to *its own* $\mathcal{F}_{\mathbf{m}_\varepsilon}$.

Let us describe two approximations to possible wrinkled regions quantitatively. Function (6.113) represents a uniform field of wrinkles of wavelength λ_w[10] and amplitude A_w; ζ_w is the angle that the longitudinal wrinkle direction forms with the x-axis:

$$
\begin{aligned}
Z_{\kappa\varepsilon}^{(\mathrm{wu})} &\equiv Z_{\kappa\varepsilon}^{(\mathrm{wrinkles\ uniform})}(x, y \mid \mathcal{F}_{\mathbf{m}_\varepsilon}) \\
&= A_w \sin\left(\frac{2\pi}{\lambda_w}(y\cos\zeta_w - x\sin\zeta_w) + \phi_w\right)
\end{aligned}
$$

$$
\lambda_w = \text{constant}, \quad -x_{min} \leqslant x \leqslant x_{max}, \quad -y_{min} \leqslant y \leqslant y_{max} \qquad (6.113)
$$

where ϕ_w is the phase shift. Here, the uniform field is defined over a rectangle, but it may be extended to other intervals via a few modifications. An example of uniform wrinkles is shown in the left part of Fig. 6.10.

Finally, function (6.114) describes a field of wrinkles with the following characteristics: they extend *radially* over a fraction of circular corona of inner radius r_0 and outer radius l_w, which is a fraction of the sail side; the parabolic-assumed amplitude begins from zero at r_0, achieves its maximum A_{max}, and decreases to zero at l_w.

$$
\begin{aligned}
Z_{\kappa\varepsilon}^{(\mathrm{wr})} &\equiv Z_{\kappa\varepsilon}^{(\mathrm{wrinkles\ radial})}(x, y \mid \mathcal{F}_{\mathbf{m}_\varepsilon}) \\
&= A_w(r)\sin\left(\pi\frac{r - r_0}{l_w - r_0}\right)\sin(2n_w\phi)
\end{aligned}
$$

$$
A_w(r) \equiv 4\frac{A_{max}}{l_w^2 - r_0^2}r(l_w - r), \quad r \equiv \sqrt{x^2 + y^2}, \quad \phi \equiv \arctan(x, y)
$$

$$
r_0 \leqslant r \leqslant l_w, \quad \phi_{min} \leqslant \phi \leqslant \phi_{max} \qquad (6.114)
$$

An example of radial wrinkles is shown in the right part of Fig. 6.10, where the number n_w of peaks and valleys has been chosen equal to 12.

---

[10]In membrane stress analysis, mechanical engineers are used to calculate the half wavelength denoted by λ.

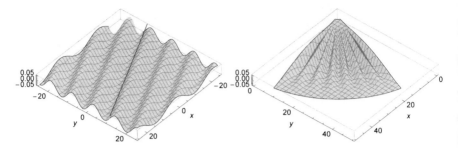

Fig. 6.10 Examples of modeling the shape of structural membrane wrinkles for thrust calculation: (*left*) uniform field of wrinkles characterized by constant amplitude, wavelength, and direction in the *xy*-plane; (*right*) pattern of regional wrinkles (60° wide) exhibiting variable amplitude and emanating radially from the origin of a macrosurface region (e.g. one of the corners of a sail quadrant). Plot box ratios have been adjusted for a better graphic visualization

A remark on the chosen frames of reference. With regard to $\mathcal{F}^{(\text{booms})}$, the orientation of the *x*-axis was based on the boom arrangement for exploiting the (assumed) symmetry of the quadrants. In contrast, an easy choice of the reference direction of $\mathcal{F}_{\mathbf{m}_\varepsilon}$ may depend on the wrinkle pattern in order to simplify the equation of the related surface profile. For instance, both the origin and the *x*, *y*-axes of the corner region of a 3-point suspension quadrant are different from those ones of $\mathcal{F}^{(\text{booms})}$. There is, though, another aspect to be reminded: in the definition of $\mathcal{F}_{\mathbf{m}}$ (where the diffraction calculations are carried out), the *x*-axis was chosen so that the Sun azimuth takes on π; this is very usual in the literature of diffraction theories. Therefore, in order to *superimpose* the wrinkle undulations on the Z-profiles due to the background shape, one needs a translation/rotation bringing $\mathcal{F}^{(\text{booms})}$ to $\mathcal{F}_{\mathbf{m}_\varepsilon}$, then followed by another rotation from $\mathcal{F}_{\mathbf{m}_\varepsilon}$ to $\mathcal{F}_{\mathbf{m}}$. Rotations are simply about the common direction of the 3rd coordinate axis; therefore, they can be summed. Thus, what said is expressed by the following transformations, given separately for clearness:

$$\begin{pmatrix} x \\ y \\ Z \end{pmatrix}^{(\mathcal{F}_{\mathbf{m}_\varepsilon})} = \begin{pmatrix} \cos\alpha_1 & \sin\alpha_1 & 0 \\ -\sin\alpha_1 & \cos\alpha_1 & 0 \\ 0 & 0 & 1 \end{pmatrix} \begin{pmatrix} x - x_O \\ y - y_O \\ Z \end{pmatrix}^{(\mathcal{F}^{(\text{booms})})} \quad (6.115a)$$

$$\begin{pmatrix} x \\ y \\ Z \end{pmatrix}^{(\mathcal{F}_{\mathbf{m}})} = \begin{pmatrix} \cos\alpha_2 & \sin\alpha_2 & 0 \\ -\sin\alpha_2 & \cos\alpha_2 & 0 \\ 0 & 0 & 1 \end{pmatrix} \begin{pmatrix} x \\ y \\ Z \end{pmatrix}^{(\mathcal{F}_{\mathbf{m}_\varepsilon})} \quad (6.115b)$$

where $\alpha_{1,2}$ are the rotation angles and x_O and y_O denote the coordinates of the origin of $\mathcal{F}_{\mathbf{m}_\varepsilon}$ measured in $\mathcal{F}^{(\text{booms})}$. Thus, applying the transformations (6.115a), (6.115b) will result in the twofold aim of having the various out-of-plane profiles referred to the same coordinate system *and* of calculating the local solar colatitude easily.

If we use the concept of the programming language statement *select case* together with the Kronecker delta, we are able to express the general height deviations in a

6.5 Calculation of Thrust

compact form as follows:

$$Z_{\kappa\varepsilon} = \left(\delta^{(str)}_{(n)} Z^{(str)}_{\kappa} + \delta^{(3sp)}_{(n)} Z^{(3sp)}_{\kappa}\right) + \left(\delta^{(wu)}_{(m)} Z^{(wu)}_{\kappa\varepsilon} + \delta^{(wr)}_{(m)} Z^{(wr)}_{\kappa\varepsilon}\right)$$
$$\mathfrak{n} = \text{'str', '3sp', 'flat'} \qquad \mathfrak{m} = \text{'wu', 'wr', 'unw'} \qquad (6.116)$$

where these strings have been employed in the previous configuration functions. Thus, once the region has been selected, one can obtain the total $Z_{\kappa\varepsilon}(x, y)$ to "handle".

In the frame $\mathcal{F}_{\mathbf{m}}$ endowed with the basis $\{\mathbf{i}, \mathbf{j}, \mathbf{k}\}$, a point P of the surface has a vector position $\mathbf{p} = x\mathbf{i} + y\mathbf{j} + Z_{\kappa\varepsilon}(x, y)\mathbf{k}$; for simplicity in the following equations, let set $Z_{\kappa\varepsilon}(x, y) \equiv Z$ and use a subscript for denoting partial derivative. Two independent tangent vectors at P are given by $\mathbf{t}_1 = \partial\mathbf{p}/\partial x = \mathbf{i} + Z_x\mathbf{k}$ and $\mathbf{t}_2 = \partial\mathbf{p}/\partial y = \mathbf{j} + Z_y\mathbf{k}$, as it is known. Out of the two possible normal unit vectors at P, we choose that one satisfying $\mathbf{N} \cdot \mathbf{k} \geq 0$, namely,

$$\mathbf{N} = (-Z_x\mathbf{i} - Z_y\mathbf{j} + \mathbf{k})/\sqrt{1 + Z_x^2 + Z_y^2} \qquad (6.117)$$

Finally, we are able to calculate the angle of solar incidence by considering that $Z(x, y) \neq 0$, as detailed above, in order to replace the angle $\vartheta_\odot$ by the angle $\widehat{\vartheta}_\odot$ in the diffraction-based momentum equations. This is carried out simply via

$$\cos\widehat{\vartheta}_\odot = \mathbf{N} \cdot \mathbf{d}_\odot = \mathbf{N} \cdot (-\sin\vartheta_\odot\mathbf{i} + \cos\vartheta_\odot\mathbf{k})$$
$$= \frac{Z_x \sin\vartheta_\odot + \cos\vartheta_\odot}{\sqrt{1 + Z_x^2 + Z_y^2}} \qquad (6.118)$$

Expanding the last expression with respect to the partial derivatives around the ideal flatness, one obtains to the first order

$$\cos\widehat{\vartheta}_\odot \cong \cos\vartheta_\odot + Z_x \sin\vartheta_\odot \qquad (6.119)$$

Sail background shape shall not be an issue for using Eq. (6.119) in the diffraction equations. In contrast, some difficulty may arise if the sail had regions rather corrugated by wrinkles, e.g. with a non-negligible shadowing/masking. Let us establish some conditions useful in the analysis. Referring to our cases of uniform or radial wrinkle patterns, it is easy to carry out the partial derivatives of the wrinkled-component of Z, according to Eq. (6.116). From inspection and simple numerical analysis, one can gets the following inequalities stating that Eq. (6.119) may be employed directly for the incident angle in the diffraction equations:

$$\text{uniform pattern:} \quad 2\pi A_w/\lambda_w \ll 1$$
$$\text{radial pattern:} \quad 4n_w A_{max}/l_w \ll 1 \qquad (6.120)$$

We can now proceed to the final steps towards the calculation of thrust in the sailcraft orbital frame. The following section regards the momentum balance and the ensuing sailcraft acceleration.

6.6 Total Momentum Balance and Thrust Acceleration

The various 2nd-order momenta calculated in the previous sections depend explicitly on the product $dA\,d\tau$. If one divides such elemental vector momenta by this product, one gets a vector dimensionally equal to a pressure. If one projects this vector normally and tangentially to the local infinitesimal surface and conceptually repeats such operation throughout the sail, then one has the pressure and shear stress distributions (per each physical phenomenon considered) along the sail's whole surface. This is a situation quite similar to the determination of aerodynamic forces and moments on a body: no matter how much complicated the body surface may be, forces and moments on the body depend exclusively on the distributions of pressure and shear stress (e.g. [3]). For reasons that will appear clear as we proceed, we adopt an algorithm a bit different from that used in Aerodynamics. We will continue to speak in terms of photon momentum, but a highlight is put on the 1st-order momenta.

Let us consider the macrosurface (or region) numbered ε of the sail quadrant κ, and integrate the input and the three output 2nd-order photon momenta with respect to the area, namely, the momenta of photons entering and exiting from a control surface enveloping the chosen region. The problem is to express the algebraic sum of vectors in the same meaningful frame of reference.

Let us first recall the input momentum expressed by Eq. (6.28) on page 176 with the flat-sail incidence angle replaced according to Eq. (6.119). One gets

$$
\begin{aligned}
d\mathbf{P}^{(SOF)}_{(in),\kappa\varepsilon} &= \int_{A_{\kappa\varepsilon}} d^2\mathbf{P}^{(SOF)}_{(in),\kappa c} \cong \int_{A_{\kappa\varepsilon}} \left(d^2\mathbf{P}^{(SOF)}_{(in),\kappa\varepsilon} \right)_{\sim} \\
&= \mathbf{U}^{(SOF)}\,d\tau \int_{A_{\kappa\varepsilon}} \cos\widehat{\vartheta}_\odot\,dA \\
&\cong \left(\cos\vartheta_\odot A_{\kappa\varepsilon} + \sin\vartheta_\odot \int_{A_{\kappa\varepsilon}} Z_x\,dA \right) \mathbf{U}^{(SOF)}\,d\tau
\end{aligned}
\tag{6.121}
$$

where $A_{\kappa\varepsilon}$ denotes the area of the chosen region. Since the scalar product is invariant under orthogonal transformations, the integral in the second line is an invariant multiplying the product $\mathbf{U}^{(SOF)}\,d\tau$, which is peculiar to the Sun and SOF, and does not depend on how the sail is segmented, conceptually and practically.

Next, ne will use a compact form for the output momentum:

$$
d\mathbf{Q}^{(\mathcal{F}m)}_{(type),\kappa\varepsilon} = \int_{A_{\kappa\varepsilon}} d^2\mathbf{Q}^{(\mathcal{F}m)}_{(type),\kappa\varepsilon}, \quad type = \text{`}re\text{'}, \text{`}rer\text{'}, \text{`}tr\text{'}
\tag{6.122a}
$$

$$
d\mathbf{Q}^{(\mathcal{F}m)}_{(out),\kappa\varepsilon} = \sum_{type} d\mathbf{Q}^{(\mathcal{F}m)}_{(type),\kappa\varepsilon}
\tag{6.122b}
$$

where the different contributions on the right side are given by Eqs. (6.82), (6.106) and (6.107). The reader will note that 'abs' has been not included in subscripts; in other words, the absorption momentum given by Eq. (6.97) will not appear explicitly in the sail-sunlight momentum balance equation. As a point of fact, a fraction

6.6 Total Momentum Balance and Thrust Acceleration

of the incoming photons may be absorbed by transferring part of the input momentum to the physical body, and give no *output* momentum. The time change of such momentum appears as a body reaction that results in an acceleration measurable in a frame where the body is at rest. In contrast, the absorbed energy (as we saw) is re-emitted totally at the equilibrium temperature, and results in a net output momentum that is explicitly written in Eqs. (6.122a), (6.122b).

One should note that the sum in parentheses in Eq. (6.121) and the right-hand sides of Eqs. (6.122a), (6.122b) no longer depend on the respective xy-coordinates. Therefore, the following transformations will regard rotations, not translations.

In the above equations, the input momentum is resolved in (SOF) whereas the output momenta are expressed in $\mathcal{F}_{\mathbf{m}}$. Frames of reference $\mathcal{F}^{(\text{booms})}$, $\mathcal{F}_{\mathbf{m}}$ and $\mathcal{F}_{\mathbf{m}_\varepsilon}$ are attached to either the sail membrane or the sail support system; they may have different origins and orientations, in general, and may be defined regardless of the sailcraft position and velocity. In particular, the boom frame has advantageous symmetry and could be employed in attitude dynamics. Thus, once the diffraction and heat transfer equations have been utilized as above discussed, one may think of splitting the frame transformation $\mathcal{F}_{\mathbf{m}} \to \text{SOF}$ into $\mathcal{F}_{\mathbf{m}} \to \mathcal{F}^{(\text{booms})}$ and $\mathcal{F}^{(\text{booms})} \to \text{SOF}$. After the various contributions are summed in $\mathcal{F}^{(\text{booms})}$, this *resultant* should be applied to the center of pressure (c.p.).

In the current framework, the center of pressure (c.p.) is defined to be located such a way the resultant of the sail forces produces the same force moment about the *sailcraft*'s barycenter as the actual loads distributed on the (curved and wavy) sail do. In general, c.p. does not coincide with the booms center (even if the sail system payload, i.e. the spacecraft, were absent). In addition, the barycenter-c.p. vector is expected to change with the nominal sunlight angle of incidence $\vartheta_\odot$[11] because the distributions of the diffuse reflected and diffuse transmitted (if any) photon momenta change with this angle. Such features should be taken into account in designing the Attitude and Orbit Control System (AOCS) for a large-sail sailcraft. At the time of this writing, and to the author knowledge, no investigation about the existence of a point that works for sails as the aerodynamic center (i.e. independently of the incidence angle) does for aerofoils (e.g. see [3]) has been carried out.

Let us proceed to the final step of calculating the resultant propulsive force that, together with the gravitational one, certainly will be applied to the barycenter of the sailcraft (sail system+spacecraft) the trajectory of which we are concerned with.

Let us split the $\mathcal{F}_{\mathbf{m}}$-to-SOF transformation as said above, namely,

$$\varXi^{(\text{SOF})}_{(\mathcal{F}_{\mathbf{m}}),\kappa\varepsilon} = \varXi^{(\text{SOF})}_{(\mathcal{F}^{(\text{booms})})} \varXi^{(\mathcal{F}^{(\text{booms})})}_{(\mathcal{F}_{\mathbf{m}}),\kappa\varepsilon} \tag{6.123}$$

where the rotation from $\mathcal{F}_{\mathbf{m}}$ to $\mathcal{F}^{(\text{booms})}$ can be easily obtained by the inverse rotation matrices of Eqs. (6.115a), (6.115b). Matrix $\varXi^{(\text{SOF})}_{(\mathcal{F}^{(\text{booms})})}$ is simply the direction cosine matrix for the $\mathcal{F}^{(\text{booms})}$-to-SOF rotation; it can be parametrized according to the

---

[11]This is like the center of pressure of an aerofoil, which varies on the chord with the angle of attack, in general [3].

parameter set chosen for the attitude dynamics of the considered sailcraft mission; we do not insist further on this well-known item.

If N_κ is the number of the regions in the quadrant κ, then one has to sum the contributions coming from the various regions of the four quadrants. Thus, the 1st-order momentum of input related to the whole sail amounts to

$$dP_{(in)}^{(SOF)} = \sum_{\kappa=0}^{3} \sum_{\varepsilon=1}^{N_\kappa} dP_{(in),\kappa\varepsilon}^{(SOF)} \qquad (6.124)$$

whereas the 1st-order momentum of output is expressed by

$$dQ_{(type),\kappa\varepsilon}^{(\mathcal{F}^{(booms)})} = \Xi_{(\mathcal{F}_\mathbf{m}),\kappa\varepsilon}^{(\mathcal{F}^{(booms)})} dQ_{(type),\kappa\varepsilon}^{(\mathcal{F}_\mathbf{m})} \qquad (6.125a)$$

$$dQ_{(out)}^{\mathcal{F}^{(booms)}} = \sum_{\kappa=0}^{3} \sum_{\varepsilon=1}^{N_\kappa} dQ_{(type),\kappa\varepsilon}^{(\mathcal{F}^{(booms)})} \qquad (6.125b)$$

$$dQ_{(out)}^{(SOF)} = \Xi_{(\mathcal{F}^{(booms)})}^{(SOF)} dQ_{(out)}^{\mathcal{F}^{(booms)}} \qquad (6.125c)$$

Therefore, the total 1st-order reaction stemming from the considered interaction of the solar photons with the surface and volume structures of the sail's material is given by

$$dP_{(sail)}^{(SOF)} = dP_{(in)}^{(SOF)} - dQ_{(out)}^{(SOF)} \qquad (6.126)$$

where we are able to calculate each term on right side *according to* the steps we have previously described. The various momentum contributions are proportional all to the elemental interval $d\tau$ of proper time. Therefore, the following force will act on the sailcraft

$$\mathbf{F}_{(sail)}^{(SOF)} = dP_{(sail)}^{(SOF)}/d\tau \qquad (6.127)$$

This quantity will then be in input to the trajectory design, the results of which will affect the other sailcraft systems, and so forth iteratively until some prefixed halt condition is satisfied.

6.7 The Surface-Isotropic Flat-Sail Model

Three are the purposes of this section. *First,* applying some of the tools developed in the previous sections to a case particularly important in solar sailing, namely, the flat sail, a model of which we have used in Sect. 5.5 under a different mathematical form. *Second,* showing an example procedure with formulas where one can insert particular values of interest for computing thrust. *Third,* adding the isotropic-surface assumption explicitly, and pointing out the related calculation. We will focus on the most complicated aspect, namely, the calculation of the momentum due to the diffuse reflectance. We leave the simpler computation of the other momenta to the interested reader.

6.7 The Surface-Isotropic Flat-Sail Model

Remark 6.5 One may object that the flat-sail model already in literature contains an assumption for azimuthal symmetry, e.g. [104] pp. 188–190, namely, the reflected scattering momentum was taken along the sail normal. Loosely speaking, the diffraction process in the old flat-sail model is "too symmetric". We shall see what the equations of Sects. 6.4–6.5 actually imply.

We will not give examples regarding non-planar shape and wrinkles inasmuch as they do not contain conceptual difficulties; the interested reader could use the equations of Sect. 6.5.7 for setting up examples where deviations from sail flatness may be either negligible or important, depending on the mission aims and employing the criterion (6.110). Equations from (6.111) to (6.120) will require a bit of effort for getting implemented on a computer.

6.7.1 Lightness Vector of Flat Sails

Section 6.5.1 already contains some basic features of a flat sail model. The first extension to be done is to consider Eqs. (6.64a)–(6.64d) applicable directly to the whole sail because it is assumed flat. However, in order to the model be utilizable simply, a number of simplifications regards the Sun and the sailcraft velocity. Thus, it is reasonable to take that Sun as point-like and relativistic effects negligible. It is true that we could ignore most of the relativity and still considering a finite-size Sun with limb-darkening effect for *any* sail orientation; nevertheless, in the framework of this section, such a complication would be unjustified.

We will continue to use the same symbology of Sect. 6.5.1, when applicable. Let us begin with writing a few assumptions for the Sun and sailcraft, which may be taken as a good reference for fast sailcraft with perihelia $R_p \geqslant 0.15$ AU. They are:

1. the Sun is a point-like source of radiation;
2. the Sun is a point-like source of classical gravity;
3. no aberration of light as observed from SOF;
4. proper time is meant to coincide with HIF's time.

Therefore, the momentum, per unit time and area, of input to the sail can be expressed by

$$\vec{\mathbb{Q}} = \left(d^2 \mathbf{P}^{(SOF)}_{(in)} \right)_{\sim} / dA \, d\tau = c^{-1} J^{(SOF)} (\mathbf{R} \cdot \mathbf{n}) \, \mathbf{R}/R^2$$
$$= \left(TSI/cR^2 \right) \cos \vartheta_\odot \mathbf{u} \tag{6.128}$$

where now the unit vector $\mathbf{u}$ reduces to $\mathbf{R}/R$, and R is expressed in AU. Equation (6.128) can be easily checked formally by using the assumptions 1–4 in Eq. (6.28).

Remark 6.6 Before going on, one should note that TSI is the measured value of an irradiance occurring in *nature* "without the assumptions 1–4"; are our equations that

238   6 Modeling Light-Induced Thrust

contain certain assumptions simplifying calculations and results, especially with regard to the irradiance-on-sail scaling.

Let us now sum the two equations in (6.64a)–(6.64d) holding for the full flat sail. One gets

$$\overrightarrow{\mathbb{P}} = \overrightarrow{\mathbb{P}}_\parallel + \overrightarrow{\mathbb{P}}_\perp$$

$$= (2\mathcal{R}_s + \mathcal{R}_d + \mathcal{T}_d + \mathcal{A})\overrightarrow{\mathbb{Q}}_\perp + (\mathcal{R}_d + \mathcal{T}_d + \mathcal{A})\overrightarrow{\mathbb{Q}}_\parallel$$

$$- \|\overrightarrow{\mathbb{Q}}\|\big[\mathcal{R}_d \chi_{(re)}(\overrightarrow{r}_\parallel + \overrightarrow{r}_\perp) + \mathcal{T}_d \chi_{(tr)}(\overrightarrow{t}_\parallel + \overrightarrow{t}_\perp)\big] \quad (6.129)$$

where the substitutions containing the total absorptance and the expressions in (6.63a), (6.63b) have been employed. Equations (6.128)–(6.129) already represent a quasi-general flat-sail model. However, we like to specialize it to a non-transparent sail, and render it more flexible for applications. Thus, let us first recognize that in the current framework the vector thrust per unit area (or thrust density) is given by

$$\overrightarrow{\mathbb{P}} = \mathrm{d}^2\mathbf{P}_{(sail)}^{(SOF)}/\mathrm{d}A\,\mathrm{d}\tau \quad (6.130)$$

with obvious meaning of the symbols. Then, we can write

$$\mathcal{T}_d = 0 \quad (6.131a)$$

$$\overrightarrow{\mathbb{Q}}_\perp = \|\overrightarrow{\mathbb{Q}}\|\cos\vartheta_\odot\mathbf{n}, \qquad \overrightarrow{\mathbb{Q}}_\parallel = \|\overrightarrow{\mathbb{Q}}\|(\mathbf{u} - \cos\vartheta_\odot\mathbf{n}) \quad (6.131b)$$

$$\overrightarrow{r}_\perp = \sin\tilde{\beta}\mathbf{m} = -\sin\tilde{\beta}\mathbf{n}, \qquad \overrightarrow{r}_\parallel = \cos\tilde{\beta}\mathbf{x}_s, \qquad \|\mathbf{x}_s\| = 1 \quad (6.131c)$$

where, we remind, the unit vector $\mathbf{n}$, orthogonal to the sail mean surface, lies on the semispace opposite to that of the sunlight incidence, i.e. $\mathbf{n} = -\mathbf{m}$ in Fig. 6.7. Inequality $0 \leqslant \tilde{\beta} \leqslant \pi$ represents the range of the angle between the diffuse-reflection momentum vector $\overrightarrow{\mathbb{R}}_d$ and the x-axis of the sail frame, say, $\mathcal{F}^{(sail)} \equiv (\mathbf{x}_s, \mathbf{y}_s, \mathbf{m})$, e.g. the boom frame introduced in Sect. 6.6. We remind that in such frame $\phi_\odot = \pi$. If the total diffuse momentum were orthogonal to the sail, then $\tilde{\beta} = \pi/2$. The introduction of such angle pinpoints the fact that, in general, a surface scatters light with no azimuthal symmetry (as we will se later). In contrast, in literature, a further (hidden) assumption is often made: $\overrightarrow{\mathbb{R}}_d$ orthogonal to the sail plane, i.e. just $\tilde{\beta} = \pi/2$; such an assumption results in lower computation cost.

We have to put another ingredient into Eq. (6.129): the effect coming from the re-radiation of the absorbed light, which has been dealt with in Sect. 6.5.5, in particular expressed by Eq. (6.106). This one represents an additional term to the output momentum in the balance Eq. (6.58). One has for the whole sail

$$\mathrm{d}^2\mathbf{Q}_{(rer)}^{(SOF)}/\mathrm{d}A\,\mathrm{d}\tau = -\big(\mathrm{TSI}/cR^2\big)\cos\vartheta_\odot\mathcal{A}^{(f)}\left[\frac{\hat{\chi}^{(f)}\mathcal{E}^{(f)} - \hat{\chi}^{(b)}\mathcal{E}^{(b)}}{\mathcal{E}^{(f)} + \mathcal{E}^{(b)}}\right]_{T=T_s}\mathbf{n}$$

$$= -\|\overrightarrow{\mathbb{Q}}\|\mathcal{A}^{(f)}\kappa_{(sail)}\mathbf{n} \quad (6.132)$$

6.7 The Surface-Isotropic Flat-Sail Model

where the quantity in square brackets has been set to $\kappa_{(sail)}$. Inserting Eqs. (6.131a)–(6.132) into Eq. (6.129), and carrying out some algebraic arrangements, one gets

$$\vec{P} = \|\vec{Q}\|\big[\big(2\mathcal{R}_s \cos\vartheta_\odot + \chi_{(re)}\mathcal{R}_d \sin\tilde{\beta} + \mathcal{A}^{(f)}\kappa_{(sail)}\big)\mathbf{n}$$
$$+ \big(\mathcal{A}^{(f)} + \mathcal{R}_d\big)\mathbf{u} - \big(\chi_{(re)}\mathcal{R}_d \cos\tilde{\beta}\big)\mathbf{x}_s\big] \tag{6.133}$$

This equation provides the vector thrust density under the assumptions 1–4, and the hypothesis of making flat, or sufficiently flat, a sail. The lightness vector can be obtained by

$$\mathbf{L} = \vec{P}\,\frac{A}{m}\Big/ g(R), \quad g(R) \equiv \mu_\odot/\|\mathbf{R}\|^2 \tag{6.134}$$

where $g(R)$ is the magnitude of the solar acceleration at $\mathbf{R}$. Now, in using Eq. (6.128), one should remember that there R is expressed in AU, whereas the R-value in $g(R)$ is in other units, e.g. SI. Therefore, the scalar factor in Eq. (6.134) can be calculated as follows

$$\frac{\|\vec{Q}\|}{m/A}\Big/\frac{\mu_\odot}{R^2} = \frac{\text{TSI}\cos\vartheta_\odot}{c(\mu_\odot/\text{AU}^2)(m/A)} = \frac{1}{2}\frac{\sigma_{(cr)}}{\sigma}\cos\vartheta_\odot$$
$$\sigma_{(cr)} \equiv 2\frac{\text{TSI}}{c(\mu_\odot/\text{AU}^2)}, \quad \sigma \equiv m/A \tag{6.135}$$

where in the second line we repeated the setting made in Eqs. (5.61), which now receive explanation. Finally, by combining Eqs. (6.134) and (6.135), we are able to make the lightness vector explicit:

$$\mathbf{L} = \frac{1}{2}\frac{\sigma_{(cr)}}{\sigma}\cos\vartheta_\odot\big[\big(2\mathcal{R}_s \cos\vartheta_\odot + \chi_{(re)}\mathcal{R}_d \sin\tilde{\beta} + \mathcal{A}^{(f)}\kappa_{(sail)}\big)\mathbf{n}$$
$$+ \big(\mathcal{A}^{(f)} + \mathcal{R}_d\big)\mathbf{u} - \big(\chi_{(re)}\mathcal{R}_d \cos\tilde{\beta}\big)\mathbf{x}_s\big] \tag{6.136}$$

The simplicity of the flat-sail model decouples the various sources of thrust acceleration; the first fraction summarizes the Sun and the sailcraft's main parameter, the second fraction represents the Lambert law, and the bracket is the effect of the sunlight-sail interaction: diffraction, absorption, and re-emission. The maximum scalar value of the bracket terms is 2. This explains why the factor $\frac{1}{2}$ has been pointed out in Eq. (6.136); thus, the highest (theoretical) $\mathcal{L}$-value is given by $\sigma_{(cr)}/\sigma$. Such a sail is often referred to as the ideal or perfect sail. Two remarks about the obtained results are in order.

First, Eqs. (6.133)–(6.136) contain the momentum coefficients and hemispherical quantities (all defined in the previous sections) that one has to determine someway, namely, via experiments and (less simple) interaction models.

Second, if one further assumes that the reflective layer of a sail is homogenous and isotropic, then one might expect than the coefficient of $\mathbf{x}_s$ vanishes! However, this is not so in general because, as we shall see below, isotropic surface means

240   6 Modeling Light-Induced Thrust

that its power spectral density does not depend on the azimuth of the reflected wave direction; but this does not entail that the surface scatters light isotropically. For example, if the surface were probed by a stylus instrument, then the same roughness average would be measured throughout.

Remark 6.7 The model expressed by Eq. (6.136) can be viewed as a relatively simple case of transformation $\mathbf{L}(\sigma, \mathbf{n})$ from the variables σ and $\mathbf{n}$ (only three are independent) to the lightness vector $\mathbf{L}$. It would be a non-linear transformation even if the optical quantities were constant; as a point of fact, the first component of $\mathbf{n}$ in SOF is equal to $\cos \vartheta_{\odot}$. Consequently, the sailcraft acceleration, controlled linearly via the lightness vector, becomes controlled *non-linearly* via $(\sigma, \mathbf{n})$ but the very special case of a radial sail, i.e. $\cos \vartheta_{\odot} = 0$ for a finite time interval. Thus, in general, there is no way to avoid such non-linearity.

Even in Chap. 7, we will continue to use the advantage of the linear control; however, in addition to technological considerations, Chap. 8 will describe what implies for trajectory optimization.

6.7.2 Isotropic-Surface Scattering

Now, we will apply the theory in Sect. 6.5.4 for calculating the total photon momentum diffused by a flat sail to which we add the hypothesis of a reflective surface with isotropic PSD.

Equation (6.82) (p. 212) is applied to the whole sail by substituting $\mathcal{F}^{(sail)}$ for $\mathcal{F}_{\mathbf{m}}$

$$\mathbf{D} \equiv \frac{\mathrm{d}^2 \mathbf{Q}_{(re)}^{(sail)}}{\mathrm{d}A \, \mathrm{d}\tau} = \frac{1}{c} \int_{\lambda_{min}}^{\lambda_{max}} I_\lambda \, \mathrm{d}\lambda \int_{2\pi \text{ sr}} \cos \vartheta_{(re)} \Lambda_\lambda \, \mathbf{d}_{(re)} \, \mathrm{d}\omega_{(re)}$$

$$\mathbf{d}_{(re)} = \begin{pmatrix} \sin \vartheta \cos \phi \\ \sin \vartheta \sin \phi \\ \cos \vartheta \end{pmatrix}_{(re)} \tag{6.137}$$

with all quantities related to the flat sail. Therefore, one can write the total diffuse momentum in SOF as follows

$$\mathbf{D}^{(SOF)} \equiv \frac{\mathrm{d}^2 \mathbf{Q}_{(re)}^{(SOF)}}{\mathrm{d}A \, \mathrm{d}\tau} = (\mathbf{x}_s \ \mathbf{y}_s \ \mathbf{m}) \frac{\mathrm{d}^2 \mathbf{Q}_{(re)}^{(sail)}}{\mathrm{d}A \, \mathrm{d}\tau} = (\mathbf{x}_s \ \mathbf{y}_s \ \mathbf{m})\mathbf{D} = (\mathbf{x}_s \ \mathbf{y}_s \ -\mathbf{n})\mathbf{D} \tag{6.138}$$

where the entries of the direction cosines matrix are meant in SOF. We will now focus on the calculation of the $\mathcal{F}^{(sail)}$-resolved vector $\mathbf{D}$, where we insert the Rayleigh-Rice expression for the *diffuse* BRDF, i.e. Eq. (6.87), and consider the incident light as unpolarized with all contributions to the photon momenta scattered from a high-ε material. To this purpose, we use the sail's polarization factor $\mathcal{Q}'$ in Eq. (6.77) where

6.7 The Surface-Isotropic Flat-Sail Model

the various factors are given by Eqs. (6.76a)–(6.76d). Dropping some unambiguous subscript/superscript for simplicity, the current BRDF results in

$$\Lambda_\lambda^{\blacksquare} = \frac{16\pi^2}{\lambda^4} \cos \vartheta_\odot \cos \vartheta_{(re)} Q' \mathbb{S}_2(\mathbf{f}) \tag{6.139}$$

An important step is to make the surface's power spectral density explicit. As above said, we add the assumptions that the underlying probability density is Gaussian (Eq. (6.45)), and the surface roughness is isotropic, i.e. it does not depend on $\phi_{(re)}$. The related PSD expression can be simply derived by carrying out the Fourier transform of the auto-correlation function, here expressed by Eq. (6.49), and setting $\phi_{(re)} = 0$ in the spatial frequencies given by Eqs. (6.55):

$$\begin{aligned}
\mathbb{S}_2^{(iso)}(\vartheta_{(re)}, \lambda) &= \mathcal{F}\left[\tilde{z}^2 \exp\left(-\frac{x^2 + y^2}{\tilde{l}^2}\right)\right](\mathbf{f})_{\phi_{(re)}=0} \\
&= \pi(\tilde{z}\,\tilde{l})^2 \exp\left[-(\pi\tilde{l})^2\left(f_x^2 + f_y^2\right)_{\phi_{(re)}=0}\right] \\
&= \pi(\tilde{z}\,\tilde{l})^2 \exp\left[-\left(\frac{\pi\tilde{l}}{\lambda}\right)^2 (\cos\vartheta_{(re)} - \cos\vartheta_\odot)^2\right]
\end{aligned} \tag{6.140}$$

where the integration variables have been pointed out. Note that, aside the plain $\tilde{z} \to 0$, the only way to zeroing $\mathbb{S}_2^{(iso)}$ in all non-specular cases is $\tilde{l} \to \infty$; both mean perfectly smooth surfaces. In any real case, the wavelength value is important, namely, smoothness is with respect to the "probe size".

As a first result, after noting that Q' and $\mathbf{d}_{(re)}$ are the sole terms that depend on the azimuthal angle $\phi_{(re)}$, we can re-write Eq. (6.137) as follows

$$\mathbf{D} = \frac{1}{c} \int_{\lambda_{min}}^{\lambda_{max}} \frac{\text{SSI}}{R^2} \mathbf{G}\, d\lambda, \quad \mathbf{G} \equiv 16\pi^3 \frac{(\tilde{z}\,\tilde{l})^2}{\lambda^4} \mathbf{G}_1 \tag{6.141a}$$

$$\mathbf{G}_1 \equiv \int_0^{\pi/2} \sin\vartheta_{(re)} \cos^2\vartheta_{(re)} \exp\left[-\left(\frac{\pi\tilde{l}}{\lambda}\right)^2 (\cos\vartheta_{(re)} - \cos\vartheta_\odot)^2\right] \mathbf{G}_2\, d\vartheta_{(re)} \tag{6.141b}$$

$$\mathbf{G}_2 \equiv \cos\vartheta_\odot^2 \int_0^{2\pi} Q'\, \mathbf{d}_{(re)}\, d\phi_{(re)} \tag{6.141c}$$

One should note that the squared cosines in the vectors $\mathbf{G}_2$ and $\mathbf{G}_1$ eliminate the problem of the incidence and scatter angles very close to $\pi/2$ caused by the incident light with polarization parallel to the incidence plane, according to the expressions (6.76c) and (6.76d) on p. 209. The product of these squared cosines come from the Lambert law, the projected solid angle and the obliquity factor in the RR-BRDF according to Eq. (6.139).

It is easy to calculate the vector $\mathbf{G}_2$ in closed form. What may be surprising (and the author was surprised pleasantly) is that the vector $\mathbf{G}_1$ can be expressed

242   6 Modeling Light-Induced Thrust

analytically in terms of trigonometric and exponential functions, and the usual error function (which is implemented in many CASs). Because nowadays reliable and powerful computer algebra systems are quite available (and some of them are free of charge too), we may view G_1 as resolved in closed form. For this reason, after some careful considerations, the author chose to not report the long expression of the vector G_1. If the reader liked to copy such expressions to her/his computer code/application for checking and/or carrying out other results, the time spent for merely copying would be considerably high; in other words, it is better to recalculate the above integrals by a CAS and leave it store the output to subsequently manipulate.

The fact that G_1 can be carried out analytically is a very favorable circumstance. As a point of fact, this avoids the numerical calculation of the triple integral of Eq. (6.141a)–(6.141c), and many times too. In the examples below, we resort to the blackbody radiance to produce reference results; then, we have used the mean SSI of cycle-23, in which case only one integral has been calculated numerically.

In the current framework, a key property of G_1 is to have the y-component exactly zero for any incidence angle and surface parameters. From Eqs. (6.141a)–(6.141c), it is easy to see that such feature is to be ascribed to the PSD that has been assumed independent of $\phi_{(re)}$ so that the integrand of G_2 is simply the product $Q' \, \mathbf{d}_{(re)}$, the second component of which is antisymmetric with respect to $\phi_{(re)} = \pi$. Thus, vector $\mathbf{D}$ lies wholly on the incident plane, namely, the plane $(\mathbf{x}_s, \mathbf{m})$ according to the current definitions.

The following four figures shows the components of $\mathbf{D}$ for several sunlight angles, i.e. from 0 to $80°$ in step of $5°$, two pairs of $(\tilde{z}, \tilde{l})$, and two different sources of irradiance: the Sun as 5780 K-blackbody, and the mean Sun of cycle-23. Because the Rayleigh-Rice theory is restricted to smooth surfaces, we considered two such examples and a sunlight bandwidth from middle ultraviolet to middle infrared subregions. All input values are reported in each figure. $\mathbf{D}$ has been expressed in µPa; so it can be compared directly with the specular pressure on the ideal sail, or $\mathfrak{P}_{(ideal)} \cong 9.113$ µPa.

Figure 6.11 regards a reference solar blackbody irradiance onto a slightly smooth surface with respect to considered bandwidth. Using the definition of TIS as order of magnitude, i.e. Eq. (6.53), one gets TIS $\cong 0.33$ in the worst case, i.e. $\lambda = 200$ nm and $\vartheta_\odot = 0$. TIS decreases down to about 10 percent at the lower sub-band of the visible spectrum and even for paraxial incidence. However, the total diffuse momentum over the whole bandwidth from 200 nm to 10 µm is not negligible and also exhibits some interesting properties, as apparent from Fig. 6.11 where the sunlight directions and the related scattered momenta are graphed. Aside from the ideal case of the perfect normal incidence, both the distribution of radiance and intensity (not graphed here) have a dominant backscattering; this affects the total scattered momentum that has a negative component in the defined sail frame. The direction of $\mathbf{D}$ lags behind $\vartheta_\odot$ because the forward scattering is sufficiently intense. As $\vartheta_\odot$ increases, $\|\mathbf{D}\|$ first increases, achieves a maximum, and then decreases; for incidence angle close to $\pi/2$, the current formulas do not hold. What matters is the related

6.7 The Surface-Isotropic Flat-Sail Model

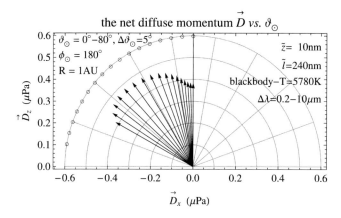

Fig. 6.11 Net diffuse momentum from non-ideal sail computed by Rayleigh-Rice BRDF and isotropic high-ε Gaussian reflective sail. In this case, Sun has been pictured as the 5780 K blackbody. Sail's reflective-layer parameters are reported. See the text for explanation and comments

vector force per unit area in SOF given by

$$\mathbf{F}_d^{(SOF)} = -\mathbf{D}^{(SOF)} = (-\mathbf{x}_s \ -\mathbf{y}_s \ \mathbf{n})\mathbf{D} = -D_x \mathbf{x}_s + D_z \mathbf{n}$$

$$D_x = \frac{\text{TSI}}{cR^2} \chi_{(re)} \mathcal{R}_d \cos \tilde{\beta} \cos \vartheta_\odot, \ D_z = \frac{\text{TSI}}{cR^2} \chi_{(re)} \mathcal{R}_d \sin \tilde{\beta} \cos \vartheta_\odot \quad (6.142)$$

$$\tilde{\beta} - \pi/2 = \tilde{\gamma} \leqslant \vartheta_\odot$$

where the second line has been obtained by the comparison with Eqs. (6.135)–(6.136). The third line expresses the above property of "falling behind", $\tilde{\gamma}$ denoting the angle of **D** from $-\mathbf{n} = \mathbf{m}$, which is the vertical axis in Figs. 6.11–6.14. The equality holds only for the incidence-zero case. One should be careful in employing the **D**-TSI relationships of Eqs. (6.142) with literature data coming from different diffraction theories.

One will note that a diffuse-reflection contribution of 0.1 µPa is equivalent to about 1.1 percent of $\mathfrak{P}_{(ideal)}$, or approximately 18 times the variation of radiation pressure caused by three standard deviations of the TSI distribution in Fig. 3.10 on p. 89, i.e. $18 \times 3 \times 0.56 \ \text{W}/\text{m}^2/c$.

Most of the previous considerations hold for the other three cases shown here. Figure 6.12 regards again the blackbody source at 5780 K, but the *rms* height and autocorrelation length have been changed to 20 nm and 120 nm, which entails a *rms* slope equal to $\sqrt{2}/6$. Note that the product $\tilde{z}\tilde{l} = 2400 \ \text{nm}^2$ is unchanged. The properties noted previously keep holding, but with $\|\mathbf{D}\|$ higher, as expected; the distribution of $\mathbf{D}/\|\mathbf{D}\|$ is slightly narrower.

The next cases have been devised by replacing the blackbody irradiance by the SSI measured in cycle-23. Of course, in order to be utilized in the current framework, we averaged the daily means of all days of cycle-23, as we did with regard to

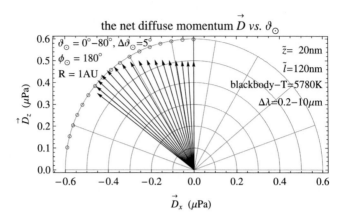

Fig. 6.12 Net diffuse momentum from non-ideal sail computed by Rayleigh-Rice BRDF and isotropic high-ε Gaussian reflective sail. In this case, Sun has been pictured as the 5780 K blackbody. A different set of the sail's reflective-layer parameters has been used, as reported. See the text for explanation and comments

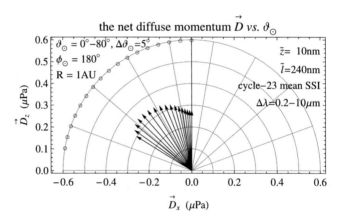

Fig. 6.13 Net diffuse momentum from non-ideal sail computed by Rayleigh-Rice BRDF and isotropic high-ε Gaussian reflective sail. In this case, the spectral irradiance obtained averaging the daily profiles of cycles-23 has been employed (Fig. 2.9). The pair of parameters of the sail's reflective-layer is the same of Fig. 6.11. See the text for explanation and comments

Fig. 2.9 on p. 57. In the current cases, we extracted the data related to the 200 nm–10 μm range. The results for the first set of surface parameters are shown in Fig. 6.13.

Comparing Fig. 6.13 with Fig. 6.11, one notes a reduction of $\|\mathbf{D}\|$ (about 28 percent on average), and only small small decrease (2°) of the "arrows fan". Such reduction is due to the medium-near UV regions, where the cycle-23 SSI is noticeably lower than the 5780 K-blackbody induced irradiance (Fig. 2.9).

Figure 6.14 corresponds to Fig. 6.12 with the cycle-23 SSI again. $\|\mathbf{D}\|$ attenuation is ∼30 percent on average, whereas the fan is slightly decreased (about 1°).

6.7 The Surface-Isotropic Flat-Sail Model

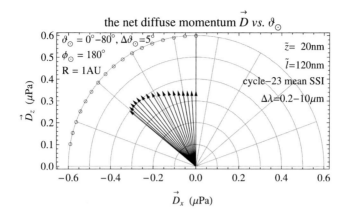

Fig. 6.14 Net diffuse momentum from non-ideal sail computed by Rayleigh-Rice BRDF and isotropic high-ε Gaussian reflective sail. In this case, the spectral irradiance obtained averaging the daily profiles of cycles-23 has been employed (Fig. 2.9). The pair of parameters of the sail's reflective-layer is the same of Fig. 6.12. See the text for explanation and comments

Instead, comparing Figs. 6.13, 6.14, the reduced smoothness gets $\|\mathbf{D}\|$ increased, but **D** changes slowly at large $\vartheta_\odot$.

The cases dealt with here could be viewed as quantitative confirmation of Remark 6.2 on p. 210. Without the parallel-to-parallel polarization, D_x would vanish at any incidence angle, and the current flat-sail model would coincide with the standard flat-sail thrust model, many times employed in literature (and by the author as well).

The chosen values of the sail's surface parameters may be somewhat different from what could be realizable for a large sail. The cases discussed above are only to emphasize basic points in realistic analysis of sailcraft acceleration; it is very difficult to find in literature the surface parameters of the reflective layer, and the method actually employed by companies for making the sail membrane. In Sect. 6.5.4, many comments have been issued about some scattering theories. One of them regards the inverse scattering problem for measuring the surface roughness parameters, though not the only one. Once the sail materials have been selected in a real sailcraft mission project, it is a key point to perform trajectory design by relying on data measured on representative samples of the sail under the various conditions under which the sailcraft will actually fly.

Example 6.1 Let us consider a 2D trajectory arc with $\mathcal{L}_t < 0$ (see Chap. 7) obtainable by orienting **n** in the fourth quadrant of Heliocentric Orbital Frame (HOF); one gets easily

$$\mathbf{n} = \begin{pmatrix} \cos(-\vartheta_\odot) \\ \sin(-\vartheta_\odot) \\ 0 \end{pmatrix}, \quad \mathbf{x}_s = \begin{pmatrix} \sin \vartheta_\odot \\ \cos \vartheta_\odot \\ 0 \end{pmatrix} \quad (6.143)$$

From Fig. 6.14, let us take on $\vartheta_\odot = 30°$; we get $D_x = -0.13281$, $D_z = 0.34774\,\mu\text{Pa}$ at $R = 1$ AU. In addition, $\tilde{\beta} = 110.90°$ and $\tilde{\gamma} = 20.90°$. Thus, using Eqs. (6.142) and (6.143), the reaction force per unit area of the sail amounts to

$$\mathbf{F}_d^{(\text{SOF})} = \begin{pmatrix} 0.36755 \\ -0.05885 \\ 0 \end{pmatrix} \mu\text{Pa} \tag{6.144}$$

Since the signs in $\mathbf{F}_d^{(\text{SOF})}$ do not depend on the value of the incidence angle (as one can verify very easily), we can assert that, in two-dimensional *decelerating* arcs, the diffuse reflection contributes positively to the radial lightness number, and negatively to the transversal lightness number (but when $\vartheta_\odot = 0$).

Note 6.6 One may wonder whether the $D_x(t)$ component may be important in the computation of trajectory profiles. In the case of the flat-model, one might deem this contribution rather small. However, one should consider that not only the magnitude of the lightness vector is altered, but mainly its direction changes with respect to the simpler $\tilde{\beta} = \pi/2$ (standard) situation. This is a very useful exercise for the interested reader who could compute many pairs of trajectories. Here, we report only that H-reversal trajectories are rather sensitive to diffuse reflection (with or without polarization) especially because the reverse-motion time changes in a non-negligible way too. We will take the RR-theory into account in Chap. 8 for mission trajectory optimization.

Remark 6.8 In completing such subsection, we have to recall that in the framework of this book in general, and of this chapter in particular, the main difference between VST and SST consists of including or not including, respectively, the polarization of light in the calculation of the diffracted waves. In this chapter, we have discussed the various aspects of different theories of diffraction with regard to the surface scattering, which is of major concern in SPS. To the author knowledge, no vector diffraction theory has been applied numerically to solar-sail mission concepts, at least in the sense quantitatively explained in this chapter. In any case, differently from the SST, no updates and/or data re-generation have been performed so far for computing the specular and hemispherical diffuse reflectances, and the absorptance of sail metals. Such an open investigation area is not among the aims of this book, though the content of Figs. 6.11–6.14 might represent a first step. In order to carry out the numerical results of Chap. 8, literature data will be employed; Aluminium's optical quantities had been obtained via SST in [103] and re-considered in [69], and employed also in many trajectory cases of the NASA Interstellar Probe mission concept [66]. This author further extended and modified his computer code also for adding the transversal component of the vector $\mathbf{D}$—as shown in the model given by Eq. (6.136)—in order to ascertain the practical effect of polarization in SPS trajectory computation.

6.8 Conclusions

The aim driving such chapter has been to show how many disciplines and technologies intervene in an *accurate* calculation of the thrust generated by the solar photons interacting with a material surface. The accuracy to be achieved depends on the mission objectives, and the type and size of the sail. Once a sophisticated model, embedded in a multi-input multi-output structured logic, has been set up, the mission analyst has the necessary tools for computing and optimizing sailcraft trajectories, and can provide the system engineer with pieces of information directly utilizable by the subsystem designers; these ones, in turn, can give mission analysts physical and technological quantities that affect sailcraft trajectory in non-negligible way. This is a typical logic in conventional space projects; however, there is the risk—in solar photon sailing—to simplify thrust evaluation too much so that such "work package" might seem to be disjoined from the other ones of sailcraft mission project. Thus, the main items of this chapter may be summarized as follows:

- Section 6.1 has been devoted to the calculation of the solar irradiance as measured in the sailcraft frame. We have explained the theoretical framework where the irradiance on the sail has been carried out. We have employed some concepts from GR which affects the spectral flux of photons emitted by the Sun and impinging on the sail.
- Section 6.2 has regarded the "translation" of the irradiance into momentum, namely, the input one. Such momentum is proportional to the irradiated elemental area and the infinitesimal interval of proper time. Both irradiance and related momentum are approximated by useful expressions derived numerically.
- Section 6.3 has shortly explained what a body surface looks like in practice. Since a space sail is a large membrane coated with highly reflective metal layer, various methods of surface manufacturing are discussed shortly. The manufacturing process determines most of the physical properties of interest in SPS.
- Section 6.4 has aimed at modeling an actual piece of sail surface, describing which physical processes will happen on surface, and introducing additional radiometric concepts and quantities appropriate for characterizing and measuring the surface properties.
- Section 6.5 has focused on the main phenomenon responsible for the interaction between electromagnetic waves and real surfaces, namely, optical diffraction. Advantages and drawbacks of five theories are discussed in order to calculate the bidirectional reflectance distribution function; as a matter of fact, the related approaches employed in computer graphics for rendering are expected to not hold in solar sailing. Therefore, the approach we chose is based on using either scalar or vector theories on scattering. The photon momenta emerging from the sail after reflection, re-emission and transmission (i.e. the output momenta) are shown formally, but with all terms coming from the algorithms previously explained.
- Section 6.6 has dealt with the overall force acting on the sailcraft barycenter. Once one calculates the photon-momentum contributions from each selected region of the sail's four quadrants, the various vectors have to be resolved in the same frame of reference.

248   6 Modeling Light-Induced Thrust

- Finally, Sect. 6.7 has carried out the simple, but useful, model of flat sail, so often employed in literature. We added a feature neglected in the past years. Moreover, we built some simple examples showing the calculation of the effect of diffuse reflection to thrust, and pointed out again some of the related specific problems of thrust modeling.

In completing the chapter, we have to remark the following points regarding the algorithms we set up for arriving to a thrust equation with the different terms explained and made explicit.

First, the variation of momentum in Eq. (6.127) is the overall reaction of the sail caused by changes of the photon momenta undergoing reflection, absorption, and transmission during the interval from τ to $\tau + d\tau$. Though the interaction of sunlight and sail membrane has been described chiefly via electromagnetic-wave diffraction and radiative transfer, Eqs. (6.82), (6.106), and (6.107) rely on the basic relationship between photon energy and momentum in quantum physics.

Second, one can note that the Doppler factor $\mathcal{D}$, defined in Eq. (6.9) and approximated by Eqs. (6.13), has been factorized such that one can see the causes of the red shift of the observed photon energy with respect to the solar emission energy. The linear term in the observer velocity (i.e. that regarding the aberration of light) is by far the most disturbing, even grater than $\tau_{\odot}/r_{\odot} \sim 4 \times 10^{-6}$. The factor $\mathcal{D}$ intervenes in quantities such as the photon energy and its associated momentum. Once the irradiance on the sail has been calculated this way, the subsequent interaction with sail's materials has been dealt with the classical electromagnetic theory.

Third, from the definition of the extended Lorentz factor $\tilde{\gamma}(R, V)$, or Eq. (6.4), the quantity $c^2 \tau_{\odot}/R$ is equal to the square of the local escape speed, say, $V(R)_{(esc)}$. For most part of its trajectory, fast sailcraft move with V^2 higher or much higher than $V(R)_{(esc)}^2$; thus, one may turn to simplifying the problem of motion description by using Special Relativity (SR)-based analytical mechanics with the Newtonian gravity acting as a four-force external to sailcraft; in this framework, General Relativity (where the concepts of momentum and force are not given a high importance) intervenes only in calculating the irradiance. This "mixed approach" may be (and has been) employed in designing a (large) computer code able to deal with SPS realistic missions.

Fourth, in the next chapters, we are concerned primarily with astrodynamic and scientific aspects that *could* be realized through engineerable space systems. Thus, in order to quantitatively explain the *numerous* and essential aspects of the fast-sailing theory, we will go on describing it classically, namely

$$\mathbf{L} = \frac{\mathbf{F}_{(sail)}^{(HOF)}/m}{\mu_{\odot}/R^2} \cong \frac{\mathbf{F}_{(sail)}^{(SOF)}/m}{\mu_{\odot}/R^2} \tag{6.145}$$

Many advanced aspects of the employed theoretical tools could find application in advanced and/or specific missions, and efficient computer codes.

We described quantitatively how difficult is to calculate the force acting actually on a sail. Nevertheless, in the next chapter, we will re-consider the heliocentric orbital frame of reference and its *extension* for covering all families of sailcraft trajectories and, hopefully, finding meaningful properties to be applied to a real SPS mission.

References

1. Abel, N. H. (1826), Auflosung einer mechanischen Aufgabe. Journal für die Reine und Angewandte Mathematik, 1, 153–157.
2. Adler, R., Bazin, M., & Schiffer, M. (1975), Introduction to General Relativity (2nd edn.). New York: McGraw-Hill. ISBN 0-07-000423-4.
3. Anderson, J. D. Jr. (2001), Fundamentals of Aerodynamics (3rd edn.). New York: McGraw-Hill. ISBN 0-07-237335-0.
4. Barrick, D. E. (1970), Radar Cross Section Handbook. New York: Plenum (Chap. 9).
5. Beckmann, P., Spizzichino, A. (1963/1987), The Scattering of Electromagnetic Waves from Rough Surfaces. New York: Pergamon. 1963; new edition by Artech House Publishers, March 1987, ISBN 0890062382, ISBN 978-0890062388.
6. Bennett, J. M. (1976), Measurement of the *rms* roughness, autocovariance function and other statistical properties of optical surfaces using a FECO scanning interferometer. Applied Optics 15, 2705–2721.
7. Bennett, J. M., Mattsson, L. (Eds.) (1989), Introduction to Surface Roughness and Scattering. Washington: Optical Society of America.
8. Bennett, J. M. (2010), Polarization. In Handbook of Optics: Vol. I. Geometrical and Physical Optics, Polarized Light, Components and Instruments (3rd edn.). New York: McGraw-Hill. ISBN 978-0-07-163598-1, ISBN 978-0-07-162925-6, MHID 0-07-162925-4.
9. Bergmann, P. G. (1976), Introduction to the Theory of Relativity. New York: Dover. ISBN 0-486-63282-2.
10. Bohren, C. F. (2007), Scattering by particles. In Handbook of Optics: Vol. I. Geometrical and Physical Optics, Polarized Light, Components and Instruments (3rd edn.). New York: McGraw-Hill. ISBN 978-0-07-163598-1, ISBN 978-0-07-162925-6, MHID 0-07-162925-4.
11. Born, M., Wolf, E. (2005), Principles of Optics (7th edn.). Cambridge: Cambridge University Press. ISBN 0-521-642221.
12. Bracewell, R. N. (2000), The Fourier Transform and Its Applications. New York: McGraw-Hill. ISBN 007-303-938-1.
13. Callister, W. D. (2010), Materials Science and Engineering: An Introduction (8th edn.). New York: Wiley. ISBN 978-0-470-41997-7.
14. Carniglia, C. K. (1979), Scalar scattering theory for multilayer optical coatings. Optical Engineering, 18, 104115.
15. Çengel, Y. A. (2007), Introduction to Thermodynamics and Heat Transfer. New York: McGraw-Hill.
16. Chandrasekhar, S. (1950), Radiative Transfer. Oxford: Clarendon.
17. Church, E. L., Takacs, P. Z. (1994), Surface scattering. In Handbook of Optics (2nd edn.). New York: McGraw-Hill. ISBN 007047740X, ISBN 978-0070477407.
18. Church, E. L., Takacs, P. Z. (2010), Surface scattering. In Handbook of Optics: Vol. I. Geometrical and Physical Optics, Polarized Light, Components and Instruments (3rd edn.). New York: McGraw-Hill. ISBN 978-0-07-163598-1, ISBN 978-0-07-162925-6, MHID 0-07-162925-4.
19. Church, E. L., Takacs, P. Z., Leonard, T. A. (1989), The prediction of BRDF's from surface profile measurements. Proceedings of SPIE, 1165, 136–150.

250   6 Modeling Light-Induced Thrust

20. Church, E. L., Jenkinson, H. A., Zavada, J. M. (1979), Relationship between surface scattering and microtopographic features. Optical Engineering, 18, 125–136.
21. Chenmoganadam, T. K. (1919), On the specular reflection from rough surfaces. Physical Review, 13, 96–101.
22. Davies, H. (1954), The reflection of electromagnetic waves from a rough surface. Proceedings of IEEE Part IV, 101, 209–214.
23. de Fornel, F. (2010), Evanescent Waves: from Newtonian Optics to Atomic Optics. Berlin: Springer. ISBN 364208513X, ISBN 978-3642085130.
24. Elson, J. M., Bennett, J. M. (1979), Vector scattering theory. Optical Engineering, 18, 116–124.
25. Fano, U. (1941), The theory of anomalous diffraction gratings and of quasistationary waves on metallic surfaces (Sommerfeld's waves). Journal of the Optical Society of America, 31, 213–222.
26. Fornasini, P. (2005), Basic crystallography. In African School and Workshop on X-Rays in Materials, Dakar, Senegal, December 12–17, 2005.
27. Donner, C., Lawrence, J., Ramamoorthi, R., Hachisuka, T., Jensen, H. W., Nayar, S. (2009), An empirical BSSRDF model. ACM Transactions on Graphics, 28(3). ISSN 0730-0301.
28. Fang, T.-H., Chang, W.-J. (2003), Effects of AFM-based nanomachining process on aluminum surface. Journal of Physics and Chemistry of Solids, 64, 913–918.
29. Franceschetti, G., Riccio, D. (2007), Scattering, Natural Surfaces, and Fractals. New York: Academic Press. ISBN 978-0-12-265655-2, ISBN 0-12-265655-5.
30. Freund, L. B., Suresh, S. (2009), Thin Film Materials—Stress, Defect Formation and Surface Evolution, Cambridge: Cambridge University Press. ISBN 978-0-521-52977-8.
31. van Ginneken, B., Stavridi, M., Koenderink, J. J. (1998), Diffuse and specular reflectance from rough surfaces. Applied Optics, 37(1).
32. Goldstein, D. (2007), Polarized Light (2nd edn.). Revised and expanded. New York: Marcel Dekker. ISBN 0-8247-4053-X.
33. Graziani, F. (Ed.) (2006), Computational Methods in Transport, Granlibakken 2004. Berlin: Springer. ISBN 3-540-28122-3, ISBN 978-3-540-28122-1.
34. Greschik, G., Mikulas, M. M. (2002), Design study of a square solar sail architecture. Journal of Spacecraft and Rockets, 39(5).
35. Grimaldi, F. M. (1665), Physico Mathesis de Lumine, Coloribus, et Iride, aliisque Annexis Libri Duo, Ed. V. Bonati, Bologna, Italy; http://fermi.imss.fi.it/rd/bdv?/bdviewer/bid=300682#.
36. Guenther, R. D. (1990), Modern Optics. New York: Wiley. ISBN 0-471-60538-7.
37. Harvey, J. E. (1976), Light-scattering characteristics of optical surfaces. Ph.D. Dissertation, University of Arizona.
38. Harvey, J. E., Vernold, C. L., Krywonos, A., Thompson, P. L. (1999), Diffracted radiance: a fundamental quantity in non-paraxial scalar diffraction theory. Applied Optics, 38, 6469–6481.
39. Harvey, J. E., Krywonos, A. (2004), A global view of diffraction: revisited. Proceedings of SPIE, AM100-26.
40. Harvey, J. E., Krywonos, A., Stover, J. C. (2007), Unified scatter model for rough surfaces at large incident and scatter angles. Proceedings of SPIE, 6672, 66720C-1. doi:10.1117/12.739139.
41. Harvey, J. E., Krywonos, A., Vernold, C. L. (2007), Modified Beckmann-Kirchhoff scattering model for rough surfaces with large incident and scattering angles. Optical Engineering, 46(7), 078002.
42. Harvey, J. E., Choi, N., Krywonos, A., Marcen, J. (2009), Calculating BRDFs from surface PSDs for moderately rough optical surfaces. Proceedings of SPIE, 7426, 74260I. doi:10.1117/12.831302.
43. Harvey, J. E. Choi, N., Krywonos, A., Schröder, S., Penalver, D. H. (2010), Scattering from moderately rough interfaces between two arbitrary media. Proceedings of SPIE 7794, 77940V. doi:10.1117/12.863995.

References   251

44. He, X. D., Torrance, K. E., Sillion, F. X., Greenberg, D. P. (1991), A comprehensive physical model for light reflection. Computer Graphics, 25(4), 175–186.
45. Hecht, E. (2002), Optics (4th edn.). Reading: Addison Wesley. ISBN 0-321-18878-0.
46. Heavens, O. S. (1965), Optical Properties of Thin Solid Films. Constable and Company, Ltd., Dover edition (1991), ISBN 0-486-66924-6.
47. Helrich, C. S. (2009), Modern Thermodynamics with Statistical Mechanics. Berlin: Springer. ISBN 978-3-540-85417-3, e-ISBN 978-3-540-85418-0. doi:10.1007/978-3-540-85418-0.
48. Huang, R., Stafford, C. M., Vogt, B. D. (2007), Effect of surfaces properties on wrinkling of ultrathin films. Journal of Aerospace Engineering, 20(1).
49. IUPAC (2006–2009), Compendium of Chemical Terminology, 2nd edn. Compiled by A. D. McNaught and A. Wilkinson. Blackwell Scientific, Oxford (1997). XML on-line corrected version: http://goldbook.iupac.org (2006) created by M. Nic, J. Jirat, B. Kosata; updates compiled by A. Jenkins. ISBN 0-9678550-9-8. Last update: 2009-09-07; version: 2.1.5. doi:10.1351/goldbook.
50. Jackson, J. D. (1999), Classical Electrodynamics (3rd edn.). New York: Wiley. ISBN 0-471-30932-X.
51. Jazwinski, A. H. (1970), In Mathematics in Science and Engineering: Vol. 64. Stochastic Processes and Filtering Theory. New York: Academic Press.
52. Jenkins, C. H. M. (2001), In AIAA Progress in Astronautics and Aeronautics: Vol. 191. Gossamer Spacecraft: Membrane and Inflatable Structures Technology for Space Applications. ISBN 1-56347-403-4.
53. Jenkins, C. H. M., Gough, A. R., Pappa, R. S., Carroll, J., Blandino, J. R., Miles, J. J., Racoczy, J. (2004), Design considerations for an integrated solar sail diagnostics system. In 45th AIAA/ASME/ASCE/AHS/ASC Structures, Structural Dynamics & Materials Conference, Palm Springs, CA, 19–22 April 2004. AIAA-2004-1510.
54. Jenkins, C. H. M. (2006), In AIAA Progress in Astronautics and Aeronautics: Vol. 212. Recent Advances in Gossamer Spacecraft. ISBN 1-56347-777-7.
55. Koblik, V., Polyakhova, E., Sokolov, L. (2011), Solar sail near the Sun: point-like and extended models of radiation source. Advances in Space Research, 48(11), 1717–1739.
56. Kong, J. A. (2008), Electromagnetic Wave Theory. New York: EMW. Last edition ISBN 0-9668143-9-8.
57. Krywonos, A. (2006), Predicting surface scatter using a linear systems formulation of nonparaxial scalar diffraction. Ph.D. Dissertation, College of Optics and Photonics, University of Central Florida.
58. Lawrie, I. D. (2001), A Unified Grand Tour of Theoretical Physics (2nd edn.). Bristol: Institute of Physics (IOP) Publishing. ISBN 0-7503-0604-1.
59. Li, Q. (2008), Light scattering of semitransparent media. M.S. Thesis, Georgia Institute of Technology, May 2008. http://smartech.gatech.edu/handle/1853/22686.
60. Lichodziejewski, D., Derbès, B., West, J., Reinert, R. (2003), Bringing an Effective Solar Sail Design toward TRL 6. AIAA 2003-4659.
61. Lichodziejewski, D., Derbès, B., West, J., Reinert, R., Belvin, K., Slade, K., Troy, M. (2004), Development and Ground Testing of a Compactly Stowed Scalable Inflatably Deployed Solar Sail. AIAA 2004-1507.
62. Li Voti, R., Lehau, G. L., Gaetani, S., Sibilia, C., Violante, V., Castagna, E., Bertolotti, M. (2009), Light scattering from a rough metal surface: theory and experiment. Journal of the Optical Society of America B, 26(8).
63. Lord, E. A. (1976), Tensors, Relativity and Cosmology. New Delhi: Tata McGraw-Hill.
64. Maradudin, A. A., Mills, D. L. (1975), Scattering and absorption of electromagnetic radiation by a semi-infinite medium in the presence of surface roughness. Physical Review B, 11(4), 1392–1415.
65. Maradudin, A. A. (2007), Light Scattering and Nanoscale Surface Roughness. Berlin: Springer. ISBN 0-387-25580-X, ISBN 978-0387-25580-4, e-ISBN 0-387-35659-2, e-ISBN 978-0387-35659-4.

66. Matloff, G. L., Vulpetti, G., Bangs, C., Haggerty, R. (2002), The Interstellar Probe (ISP): Pre-Perihelion Trajectories and Application of Holography. NASA/CR-2002-211730.
67. Maybeck, P. S. (1979), Stochastic Models, Estimation and Control (Vol. 1). New York: Academic Press. ISBN 0-12-480701-1.
68. McInnes, C. R. (2004), Solar Sailing: Technology, Dynamics and Mission Applications (2nd edn.). Berlin: Springer-Praxis. ISBN 3540210628, ISBN 978-3540210627.
69. Mengali, G., Quarta, A., Dachwald, B. (2007), Refined solar sail force model with mission application. Journal of Guidance, Control, and Dynamics, 30(2), 512–520.
70. Modest, M. F. (2003), Radiative Heat Transfer (2nd edn.). New York: Academic Press. ISBN 0-12-503163-7.
71. Möller, K. D. (2007), Optics, Learning by Computing with Examples Using Mathcad, Matlab, Mathematica, and Maple (with CD-ROM) (2nd edn.). New York: Springer. ISBN 978-0-387-26168-3, e-ISBN 978-0-387-69492-4.
72. Murphy, D. M., Murphey, T. W., Gierow, P. A. (2003), Scalable solar-sail subsystem design concept. Journal of Spacecraft and Rockets, 40(4).
73. Nicodemus, F. E. (1970), Reflectance nomenclature and directional reflectance and emissivity. Applied Optics, 9, 1474–1475.
74. O'Donnell, K. A., Mendez, E. R. (1987), Experimental study of scattering from characterized random surfaces. Journal of the Optical Society of America A, 4(7), 1194–1205.
75. O'Donnell, K. A. (2001), High order perturbation theory for light scattering from a rough metal surface. Journal of the Optical Society of America A, 18, 1507–1518.
76. Ogilvy, J. A. (1991), Theory of Wave Scattering from Random Rough Surfaces. Bristol: Adam Hilger.
77. Ohring, M. (2002), Materials Science of Thin Films (2nd edn.). New York: Academic Press. ISBN 0-12-524975-6.
78. Oren, M., Nayar, S. K. (1994), Generalization of Lambert's reflectance model. In SIGGRAPH'94, Orlando, FL, 24–29 July 1994. New York: ACM. ISBN 0-89791-667-0.
79. Palik, E. A. (Ed.) (1998), Handbook of Optical Constants of Solids. New York: Academic Press. ISBN 0-12-544420-6.
80. Palmer, J. M., Grant, B. G. (2009), SPIE Press Monograph. The Art of Radiometry. PM184, ISBN 9780819472458, e-ISBN 9780819479167.
81. Palmer, J. M. (2010), The measurement of transmission, absorption, emission and reflection. In Handbook of Optics: Vol. II. Design, Fabrication, and Testing, Sources and Detectors, Radiometry and Photometry (3rd edn.) New York: McGraw-Hill. ISBN 978-0-07162927-0.
82. Pappa, R. S., Black, J. T., Blandino, J. R. (2003), Photogrammetric Measurement of Gossamer Spacecraft Membrane Wrinkling. doi:10.1.1.6.8337
83. Petty, G. W. (2006), A First Course in Atmospheric Radiation (2nd edn.). New York: Sundog. ISBN 978-0-9729033-1-8, ISBN 0-9729033-1-3.
84. Ragheb, H., Hancock, E. R. (2006), Testing new variants of the Beckmann-Kirchhoff model against radiance data. Computer Vision and Image Understanding, 102(2), 145–168. doi:10.1016/j.cviu.2005.11.004.
85. Ragheb, H., Hancock, E. R. (2007), The modified Beckmann-Kirchhoff scattering theory for rough surface analysis. Pattern Recognition, 40(7), 2004–2020. doi:10.1016/j.patcog.2006.10.007.
86. Rakić, A. D. (1995), Algorithm for the determination of intrinsic optical constants of metal films: application to aluminum. Applied Optics, 34(22).
87. Rastogi, P. K. (Ed.) (1997), Optical Measurement Techniques and Applications. Boston: Artec House.
88. Rice, S. O. (1951), Reflection of electromagnetic waves from slightly rough surfaces. Communications on Pure and Applied Mathematics, 4, 351.
89. Rios-Reyes, L., Scheeres, D. J. (2005), Generalized model for solar sails. Journal of Spacecraft and Rockets, 42, 182–185.
90. Ryder, L. (2009), Introduction to General Relativity. Cambridge: Cambridge University Press, ISBN 978-0-511-58004-8 eBook (EBL), ISBN 978-0-521-84563-2.

References

91. Sharkov, E. A. (2003), Passive Microwave Remote Sensing of the Earth, Physical Foundations. Berlin: Springer Praxis. ISBN 978-3-540-43946-2.
92. Schlick, C. (1994), An inexpensive BRDF model for physically-based rendering. In Eurographics '94, Computer Graphics Forum (Vol. 13, pp. 233–246).
93. Schwarzschild, K. (1916), Über das Gravitationsfeld eines Massenpunktes nach der Einsteinschen Theorie, Sitzungsberichte der Königlich Preußische Akademie der Wissenschaften, Physik-Math Klasse, pp. 189–196.
94. Smith, D. L. (1995), Thin-Films Deposition, Principles and Practice. New York: McGraw-Hill. ISBN 0-07-058502-4.
95. Snell, J. F. (1978), Radiometry and photometry. In Handbook of Optics. New York: McGraw-Hill.
96. Simonsen, I., Maradudin, A. A., Leskova, T. A. (2010), The scattering of electromagnetic waves from two-dimensional randomly rough penetrable surfaces. CERN Document Server, record#1240240. arXiv:1002.2200v1 [physics.optics] 10 Feb 2010.
97. Stenzel, O. (2005), The Physics of Thin Film Optical Spectra—An Introduction. Berlin: Springer. ISBN 3-540-23147-1, ISBN 978-3-540-23147-9, ISSN 0931-5195.
98. Stover, J. C. (1995), Optical Scattering: Measurement and Analysis (2nd edn.). New York: SPIE. ISBN 9780819477767, e-ISBN 9780819478443.
99. Tanaka, K., Soma, E., Yokota, R., Shimazaki, K., Tsuda, Y., Kawaguchi, J. (IKAROS demonstration team) (2010), Develpment of thin film solar array for small solar power demonstrator IKAROS. In 61st International Astronautical Congress. IAC-10.C3.4.3.
100. Tessler, A., Sleight, D. W., Wang, J. T. (2005), Effective modeling and nonlinear shell analysis of thin membranes exhibiting structural wrinkling. AIAA Journal of Spacecraft and Rockets, 42(2), 287–298.
101. Tsang, L., Kong, J. A., Ding, K.-H. (2000), Scattering of Electromagnetic Waves: Theories and Applications. Berlin: Wiley. ISBN 0-471-38799-1, e-ISBN 0-471-22428-6.
102. Vickerman, J. C., Gilmore, I. S. (Eds.) (2009), Surface Analysis, the Principal Techniques (2nd edn.). New York: Wiley. ISBN 978-0-470-01763-0.
103. Vulpetti, G., Scaglione, S. (1999), The Aurora project: estimation of the optical sail parameters. Acta Astronautica, 44(2–4), 123–132.
104. Vulpetti, G., Johnson, L., Matloff, G. L. (2008), Solar Sails, a Novel Approach to Interplanetary Travel. New York: Springer/Copernicus Books-Praxis. ISBN 978-0-387-34404-1. doi:10.1007/978-0-387-68500-7.
105. Vulpetti, G. (2010), Effect of the total solar irradiance variations on solar-sail low-eccentricity orbits. Acta Astronautica, 67(1–2), 279–283.
106. Vulpetti, G. (2011), Total solar irradiance fluctuation effect on sailcraft-Mars rendezvous. Acta Astronautica, 68, 644–650.
107. Wald, R. M. (1984), General Relativity. Chicago: The University of Chicago Press. ISBN 0-226-87033-2.
108. Walter, B., Marschner, S. R., Li, H., Torrance, K. E. (2007), Microfacet models for refraction through rough surfaces. In Eurographics Symposium on Rendering, Grenoble, France.
109. Wang, L. V. Wu, H. I. (2007), Biomedical Optics: Principles and Imaging. New York: Wiley. ISBN 978-0-471-74304-0.
110. Wasa, K., Kitabatake, M., Adachi, H. (2004), Thin Film Materials Technology—Sputtering of Compound Materials. New York: William Andrew. ISBN 0-8155-1483-2, also Springer, ISBN 3-540-21118-7.
111. Weisstein, E. W. (2010), Plancherel's Theorem, from MathWorld, a Wolfram Web Resource. http://mathworld.wolfram.com/PlancherelsTheorem.html.
112. Weisstein, E. W. (2010), Delta Function, from MathWorld, a Wolfram Web Resource. http://mathworld.wolfram.com/DeltaFunction.html.
113. Weisstein, E. W. (2011), Abel Transform, from MathWorld, a Wolfram Web Resource, http://mathworld.wolfram.com/AbelTransform.html.
114. Wolf, E. (2010), Progress in Optics (Vol. 55). Amsterdam: Elsevier. ISBN 978-0-444-53705-8, ISSN 0079-6638.

115. Wolff, L. B., Nayar, S. K., Oren, M. (1998), Improved diffuse reflection models for computer vision. International Journal of Computer Vision, 30(1), 55–71.
116. Wong, Y. W., Pellegrino, S. (2003), Prediction of Wrinkle Amplitudes in Square Solar Sails. AIAA 2003-1982.
117. Wong, Y. W., Pellegrino, S. (2006), Wrinkled membranes, part I: experiments. Journal of Mechanics of Materials and Structures, 1(1), 3–25.
118. Wong, Y. W., Pellegrino, S. (2006), Wrinkled membranes, part II: analytical models. Journal of Mechanics of Materials and Structures, 1(1), 27–61.
119. Wong, Y. W., Pellegrino, S. (2006), Wrinkled membranes, part III: numerical simulations. Journal of Mechanics of Materials and Structures, 1(1), 63–95.
120. Woo, K., Jenkins, C. H. (2004), Effect of crease orientation on wrinkle-crease interaction in thin sheets. In 50th AIAA/ASME/ASCE/AHS/ASC Structures, Structural Dynamics, and Materials Conference, Palm Springs, CA, 4–7 May 2009. AIAA 2009-2162.
121. Wright, J. L. (1993), Space Sailing. New York: Gordon and Breach, ISBN 2-88124-842-X, ISBN 2-88124-803-9.
122. Xu, H., Sun, Y. (2008), A physically based transmission model of rough surfaces. Journal of Virtual Reality and Broadcasting, 5(9).
123. Yamaguchi, T., Mimasu, Y., Tsuda, Y., Funase, R., Sawada, H., Mori, O., Morimoto, M. Y., Takeuchi, H., Yoshikawa, M. (2010), Trajectory analysis of small solar sail demonstration spacecraft IKAROS considering the uncertainty of solar radiation pressure. Transactions of the Japan Society for Aeronautical and Space Sciences, Aerospace Technology Japan, 8, 37–43.

Chapter 7
The Theory of Fast Solar Sailing

To Combine Solar Gravity and Solar Radiation Pressure Efficiently This chapter is devoted to study the way a sailcraft can achieve so high a cruise speed that the related mission types may allow the exploration of the heliosphere and beyond in a time interval shorter than the mean human job time, including the design phase. Suppose a spacecraft, endowed with a high-acceleration rocket engine, is accomplishing a solar flyby. If its engine is fired for a short time around the perihelion, then the subsequent cruise speed can be considerably higher than the Keplerian receding speed that the spacecraft would have without the burning phase. This is a well-known astrodynamical effect due to the non-linearity of the energy conservation equation. Under a certain viewpoint, the fast solar sailing may be considered a generalization of the mentioned property; however, this general process is much more difficult to explain mathematically, and the sailcraft trajectory exhibits new features that are impractical for a rocket spacecraft. Four main sections and various subsections are devoted to such an aim; we will extend some concepts discussed in Chap. 5 and introduce a different view of sailcraft trajectory: the orbital angular momentum reversal via solar radiation pressure.

In Chap. 6, we have understood how solar irradiance can result in thrust. How may one utilize *both* thrust and solar gravity in order to notably increase sailcraft speed, hopefully well higher than the speed obtainable by spiraling about the Sun? To this aim, in this chapter, we shall not employ optimization techniques in calculating trajectories for two basic reasons. *First*, we have to try to understand the many features the trajectory of a "particle" (the sailcraft) could exhibit because it undergoes *three* acceleration fields (as we later will prove). *Second*, the related particle's motion equations are highly non-linear in the state variables; although not a trivial task, however we have to prove as many propositions as possible in order to carry out the existence, the conditions, and the effect-highlighting examples of fast sailing.

Note 7.1 About symbology: in this and the following chapters, when a vector is explicitly equal to zero, we do not use the $\mathbf{0}$, but we will write 0, more simply, since the zero vector is a unique element in a vector space: $0 \cdot \mathbf{w} = (0 \ldots 0)$ independently of the length and direction of $\mathbf{w}$.

G. Vulpetti, *Fast Solar Sailing*, Space Technology Library 30,
DOI 10.1007/978-94-007-4777-7_7, © Springer Science+Business Media Dordrecht 2013

256   7 The Theory of Fast Solar Sailing

7.1 Extending the Heliocentric Orbital Frame of Sailcraft

Section 5.3.2 was ended with three questions; in particular, the third one contained an explicit mention to the heliocentric orbital frame, which was defined in the usual way. In this section, we will extend that definition in order to deal with complicated sailcraft trajectories consisting of regular arcs with opposite motions, and without any discontinuity of the vector velocity. Let us proceed by steps.

Note 7.2 Suppose to know the time profiles of both vector position and velocity of a spacecraft powered by *finite*-acceleration propulsion, say, $\mathbf{R}(t)$ and $\mathbf{V}(t)$, respectively. Let us recall shortly a few consequences from such an acceleration-limited thrusting mode in a Newtonian gravity field. Neither jump nor essential discontinuity of $\mathbf{V}(t)$ could occur at $\mathbf{R} \neq 0$, even if the acceleration vector were changed finitely in an infinitesimal time. With the additional continuity of $\mathbf{R}(t)$, the cosine of the angle between $\mathbf{R}$ and $\mathbf{V}$ is a continuous function of time in a regular trajectory.

In the usual framework of heliocentric orbits, $\mathbf{V}$ can be viewed as separated from $\mathbf{R}$ by an angle $\varphi \in [0, \pi]$ for either direct or retrograde motion. In other words, a spacecraft trajectory, where no discontinuous velocity change occurs, is either direct or retrograde. However, equations in Sect. 5.3.2 induces us to consider the following possibility: as the power source for solar sailing is external to the spacecraft, and the heliocentric motion duration is not limited by mass, one may expect that a sufficiently high deceleration applied for sufficiently long time may eventually result in achieving some space point where the orbital angular momentum of the sailcraft is precisely zero. This summarizes two of the three ending questions of Sect. 5.3.2. The third question induces to think of orbital momentum *reversal* if thrusting continues past such point. In other words, sailcraft trajectories might contain arcs of direct *and* retrograde motions *without* velocity discontinuity. Is-it possible to deal with all these aspects in a unified way avoiding non-differentiable functions?

Let us re-consider the HOF evolution equations (5.48), in particular the third one that regards the time derivative of the unit vector $\mathbf{h}$ parallel to the orbital angular momentum $\mathbf{H}$:

$$\dot{\mathbf{h}} = -\omega_h \mathbf{h} \times \mathbf{r} = -\mu_\odot \frac{\mathcal{L}_n}{R^2 V \sin\varphi} \mathbf{h} \times \mathbf{r}, \quad \varphi \in \Psi_\frown \equiv (0, \pi) \qquad (7.1)$$

where the values 0 and π have been excluded because $\mathbf{h}$ has not been yet defined there; nevertheless, spacecraft could achieve such values in its trajectory. If $\mathcal{L}_n \neq 0$, no matter how much small it may be, $\dot{\mathbf{h}}$ diverges as $\sin\varphi$ approaches 0. However, if we consider the following thrusting regime

$$\mathcal{L}_n(t) = 0, \quad t \in [t_0, t_f] \equiv \mathfrak{T}, \ t_0 < t_f$$
$$\varphi(t_0) \in \Psi_\frown, \qquad \varphi(t_f) \in \Psi_\frown \qquad (7.2)$$

then, Proposition 5.1 on p. 152 can be applied. As a result, trajectory may be either a straight line or a continuous planar curve. Let us first consider the case of a

7.1 Extending the Heliocentric Orbital Frame of Sailcraft

rectilinear trajectory. Two cases are possible: (a) the line passes through the origin of HIF, and (b) the line does not contain such point. In the case-a, no HOF can be defined, and there is no need to define it because velocity and acceleration are radial throughout the flight. In the case-b, i.e. a straight line with direction $\mathbf{w} \neq \pm \mathbf{r}$, one has $\mathbf{H} \neq 0$ and curvature $\kappa(t)$ vanishing over $\mathcal{T}$. As the angle of $\mathbf{w}$ from $\mathbf{r}$ is just φ, which changes with $t \in \mathcal{T}$, $\kappa(t) = 0$ and $\mathbf{L} = constant$ entail $\mathcal{L}_r = 1$ and $\mathcal{L}_t = 0$ over $\mathcal{T}$, namely, the sailcraft motion is uniform. The (purely ideal) case of variable lightness vector on a strictly-rectilinear path would be of scarce interest because of the control (if any) to try to achieve the condition $\tan \varphi(t) = \mathcal{L}_t(t)/(\mathcal{L}_r(t) - 1)$ at any $t \in \mathcal{T}$. Instead, if $\varphi(t) \cong 0$ in the terminal arc of some non-rectilinear sailcraft trajectory (e.g. if sailcraft recedes from Sun after a significant acceleration phase), then the previous condition may be approximated very well by $\mathbf{L} = constant$; in particular, one has $\mathcal{L}_r > 1$ and $\mathcal{L}_t \gtrsim 0$, which entail a positive orbital energy strictly increasing, though very slowly.

Let us analyze the case of a continuous planar curve, which may include rectilinear arcs, if any. In general, according to Eq. (5.39b), H decreases with $\mathcal{L}_t < 0$; if $|\mathcal{L}_t|$ is sufficiently high, and $\mathcal{T}$ sufficiently long, the vanishing of $\mathbf{H}$ is admissible. If the trajectory is fully regular, then this means that velocity becomes parallel or antiparallel to position. If thrusting continued just after $H(t^*) = 0$ or, equivalently, $\sin \varphi(t^*) = 0$, with the same vector thrust as observed in HIF, then $\|\mathbf{H}\|$ would become negative if $\mathbf{h}(t^{*+}) = \mathbf{h}(t^{*-})$; in contrast, if $\mathbf{h}$ reverses its direction across t^*, then $\mathcal{L}_t(t^{*+}) = -\mathcal{L}_t(t^{*-})$, and $\dot{H}$ would be positive (with no attitude maneuver). We have two options: (i) to conclude that Eq. (5.39b) is limited in its application range, and (ii) to extend the definition of HOF beyond the intrinsic limitation of $\varphi \in \Psi_\frown$.

We adopt the viewpoint (ii), and make a first extension of HOF in order to unify (if possible) the treatment of direct and retrograde trajectory arcs. To such an aim, let us define the Modified Heliocentric Orbital Frame (MHOF) as follows:

MHOF

A. if the motion with respect to HIF is direct, then MHOF coincides with HOF
B. if the motion with respect to HIF is retrograde, then $X_{MHOF} = X_{HOF}$ again, but the direction of $Z_{MHOF} = -\mathbf{H}/\|\mathbf{H}\|$; Y_{MHOF}, which lies on the semiplane $(\mathbf{R}, -\mathbf{V})$ completes the right-handed triad.

Therefore, differently from HOF, the MHOF basis (by hat-accenting the HOF's) satisfies *all* the following relationships:

$$\hat{\mathbf{r}} = \mathbf{r}, \qquad \hat{\mathbf{y}} = \hat{\mathbf{h}} \times \hat{\mathbf{r}}, \qquad \mathbf{k} \cdot \hat{\mathbf{h}} \geq 0 \tag{7.3}$$

where $\mathbf{k}$ denotes the direction of Z_{HIF}. Let us analyze the impact of MHOF on the main sailcraft motion equations written in Sects. 5.3, 5.3.2, and 5.3.3. The concept of lightness vector and its components have been defined for HOF, which is an element of the set of the right-handed orthonormal frames co-moving with the body barycenter: given two arbitrary elements, either can be got from the other one via a rotation matrix. Therefore, the same concept shall hold for any other element such

258   7 The Theory of Fast Solar Sailing

as the co-moving orthonormal frame named MHOF. However, it might seem that Eq. (5.21) depends on the arbitrary-chosen direction of $\mathbf{h}$; this is not so under the transformation $\mathcal{H} : \mathbf{h} \to -\mathbf{h}$, as one recognizes that only the signs of the $\mathbf{L}$'s components perpendicular to $\mathbf{r}$ change under $\mathcal{H}$. In addition, as the plane orientation reverses as well, $\cos\varphi \to \cos\varphi$ and $\sin\varphi \to -\sin\varphi$. As a result, Eqs. (5.33), (5.34) and (5.38) are invariant under $\mathcal{H}$. Similarly, Eq. (5.47) is invariant, as expected, because curvature is an intrinsic property of curves.

In contrast, Eqs. (5.39a) and (5.39b) require a bit of attention. In them, symbol H represents the Euclidean norm of the angular momentum, as meant hitherto. Applying $\mathcal{H}$ to them, their left-hand sides keep unchanged, whereas their right-hand sides change sign, unless we replace H by a more general expression in terms of the angle φ beyond $\Psi_\frown$. Let us define

$$H(\varphi) = RV\sin\varphi, \quad \varphi \in \Psi_\ominus \equiv (0, \pi) \cup (\pi, 2\pi) \tag{7.4}$$

where we excluded $\varphi > 2\pi$ so that H is injective. Equation (7.1) can be re-written

$$\dot{\mathbf{h}} = -\omega_h \hat{\mathbf{h}} \times \hat{\mathbf{r}} = -\mu_\odot \frac{\mathcal{L}_n}{RH} \hat{\mathbf{h}} \times \hat{\mathbf{r}}, \quad \varphi \in \Psi_\ominus \tag{7.5}$$

Again, we will consider the control $\mathcal{L}_n(t) = 0$ identically in $\mathcal{T}$. Therefore, $\hat{\mathbf{h}}$ is piecewise-constant because it is undetermined when H = 0, if any. By Proposition 5.1 and the above considerations about rectilinear arcs, the most general trajectory lies on a plane for which now the direction of Z_{MHOF}, which is the plane's normal direction, is always unambiguous at any instant t *not* belonging to a countable set $\{t_j^*\}$ where $\sin\varphi(t_j^*) = 0$. Therefore, one can write

$$\lim_{t \to t_j^{*-}} \hat{\mathbf{h}}(t) = \lim_{t \to t_j^{*+}} \hat{\mathbf{h}}(t) = \hat{\mathbf{h}}_0, \quad t \in (t_{j-1}^*, t_j^*) \cup (t_j^*, t_{j+1}^*) \tag{7.6}$$

In words, $\hat{\mathbf{h}}(t)$, although not defined at $\sin\varphi(t_j^*) = 0$, exhibits a *unique* limit as $\mathbf{V}$ comes arbitrarily close to either $\mathbf{R}$ or $-\mathbf{R}$ from any side. This suggests us taking this limit as the definition of the Z-axis of the orbital frame when the angular momentum is exactly zero.

Thus, Eqs. (5.39a) and (5.39b) can be re-written more generally after replacing H by H

$$\mathbf{H} \times \dot{\mathbf{H}} = \frac{\mu_\odot}{R} H \mathcal{L}_n \mathbf{r} \tag{7.7a}$$

$$H\dot{H} = \mathbf{H} \cdot \dot{\mathbf{H}} = \frac{\mu_\odot}{R} H \mathcal{L}_t \quad \Rightarrow \quad \dot{H} = \frac{\mu_\odot}{R} \mathcal{L}_t \tag{7.7b}$$

where now $\mathcal{L}_t(t)$ is continuous in a sufficiently small neighborhood of t_j^*. The time evolution of the H-function is given by the rightmost Eq. (7.7b). Such function can be written as

$$H = \mathrm{H}\,\mathrm{sign}(\mathbf{H} \cdot \hat{\mathbf{h}}) = \mathbf{H} \cdot \hat{\mathbf{h}} \quad \Leftrightarrow \quad \mathbf{H} = H\hat{\mathbf{h}} \tag{7.8}$$

7.1 Extending the Heliocentric Orbital Frame of Sailcraft

The H-function, as the scalar product of two vectors, is invariant under orthogonal transformations in $\mathfrak{R}^3$.

Also, it is straightforward that Eq. (5.40) generalizes into

$$\dot{E} = H\dot{H}/R^2 + \mu_\odot \dot{\mathcal{L}}_r/R \tag{7.9}$$

The reasoning about Eqs. (7.5) and (7.6) regards the vanishing of $\mathbf{H}$ at instants where $\sin\varphi = 0$ under the control $\mathcal{L}_n(t) = 0$, $t \in \mathcal{T}$ that produces a planar trajectory in general. Now, we note that such a trajectory may be *piecewise*-regular, namely, having the tangent vectors of non-null length *but*, at most, in a finite number of points; in such non-regular points (we may exclude the starting and final points of the whole trajectory with no loss of generality), one has $\mathbf{H} = 0$ because $\mathbf{V} = 0$ there. The existence of such points does not preclude the velocity from being a continuous function, which is a key consequence of the assumed acceleration-limited thrusting mode. In these points, and only there, neither curvature nor torsion are defined. Nevertheless, each open trajectory arc, bounded by one or two non-regular point(s), is regular. Therefore, equation (7.6) could be applied to the countable set $\{t_k^\circ\}$, $k = 1..p$, of the instants, where the velocities are equal to zero. As a result, one has

$$\lim_{t \to t_j^{\circ-}} \hat{\mathbf{h}}(t) = \lim_{t \to t_j^{\circ+}} \hat{\mathbf{h}}(t) = \hat{\mathbf{h}}_0, \quad t \in (t_{j-1}^\circ, t_j^\circ) \cup (t_j^\circ, t_{j+1}^\circ) \tag{7.10}$$

Note that nothing forbids having a particular $\mathbf{L} = (\mathcal{L}_r, \mathcal{L}_t, 0)$ that entails $t_k^\circ = t_j^*$ for some value(s) of k and j. In the following, we will use the superscript $()^*$ for denoting time(s) and/or other quantities referring to points where $H = 0$, regardless which factor of H vanishes.

The previous results allow us to further generalize HOF, which we may call the Extended Heliocentric Orbital Frame (EHOF), as follows:

EHOF
1. if the motion with respect to HIF has $\mathbf{H} \neq 0$, then EHOF coincides with MHOF
2. if at some time t^*, $H(t^*) = 0$, then the direction of Z_{EHOF} is the limit of Z_{MHOF}'s as $t \to t^*$.

In particular, Eqs. (7.4) and (7.8) hold for any $\varphi \in [0, 2\pi)$. Definition admits a finite number of points where the orbital angular momentum may vanish. Let us investigate more deeply. Suppose a sailcraft trajectory with any $\mathcal{L}_n(t) \neq 0$ for a finite time interval; it drives trajectory torsion to be different from zero so that trajectory is three dimensional. Commenting again on (7.5), and bar-accenting the new basis, any non-zero $\mathcal{L}_n$ causes $\dot{\bar{\mathbf{h}}}$ to diverge as H draws closer and closer to zero. In particular,

$$\lim_{H \to 0^+} \dot{\bar{\mathbf{h}}}|_{\mathcal{L}_n \neq 0} = -\,\text{sign}(\mathcal{L}_n)\infty, \qquad \lim_{H \to 0^-} \dot{\bar{\mathbf{h}}}|_{\mathcal{L}_n \neq 0} = \text{sign}(\mathcal{L}_n)\infty \tag{7.11}$$

As $\hat{\mathbf{h}}$ has been defined unambiguously, and its length is constant, such a diverging can solely be ascribed to a jump discontinuity of the angular momentum, namely, a *finite* change $\Delta\mathbf{H} = \mathbf{R} \times \Delta\mathbf{V}$ in an infinitesimal time. A jump discontinuity in

260   7 The Theory of Fast Solar Sailing

the vector velocity may take place only via an *infinite* acceleration acting for an infinitesimal time; however, we know that SPS is an acceleration-limited propulsion (in practice, even a rough approximation to such an ideal condition is unrealistic). Therefore, one can write

$$\lim_{H \to 0} \dot{\mathbf{h}}|_{\mathcal{L}_n \neq 0} = \dot{\mathbf{h}}_n < \infty \tag{7.12}$$

meaning that the vector limit has all components finite. Therefore, from Eq. (7.5) H is forbidden vanishing, which entails $\mathbf{H} \neq 0$. All this proves the following statement

Lemma 7.1 *In any three-dimensional sailcraft trajectory with the normal lightness number non-vanishing for a finite time interval, the orbital angular momentum cannot evolve to zero in such interval.*

Remark 7.1 Does this mean that a general 3D sailcraft trajectory cannot reverse motion? No. The above statement does not forbid $\mathbf{H}$ to evolve from the $\Re^3$'s positive semispace ($\mathbf{H} \cdot \mathbf{k} > 0$) to the negative one ($\mathbf{H} \cdot \mathbf{k} < 0$) passing through the plane $(XY)_{\text{HIF}}$ *without* vanishing (or vice versa, depending on the initial flight conditions). Therefore, in these cases, motion could be reversed and the direction of $\mathbf{H}$ would be always well determined. This is made possible because a vector (in particular, $\mathbf{H}$) can rotate in 3D space to get reversed without necessarily vanishing; instead, if constrained to be orthogonal to a plane, it may be reversed only via zeroing. Lemma 7.1 translates this geometric property of space into a behavior of low-acceleration thrusting mode such as the solar-photon sailing, and we will employ it in this chapter.

Formally, we can unify the various aspects of sailcraft trajectories accounted for by EHOF. For generality purposes, denoting by t_j^*, $j = 1..n$, the instants at which $\mathbf{H}(t_j^*) = 0$, if any, then

$$\bar{\mathbf{O}} \equiv \text{sailcraft's barycenter}$$

$$\bar{\mathbf{r}} = \mathbf{R}/R$$

$$\bar{\mathbf{h}} = \begin{cases} U(t)\mathbf{H}/\|\mathbf{H}\|, & t \in [t_0, t_f] \setminus \{t_j^*\}, \quad j = 1..n < \infty \\ \lim_{t \to t_j^*} U(t)\mathbf{H}/\|\mathbf{H}\|, & t \in \{t_j^*\} \end{cases}$$

$$\bar{\mathbf{y}} = \bar{\mathbf{h}} \times \bar{\mathbf{r}} \tag{7.13}$$

$$U(t) \equiv \begin{cases} u(\mathbf{H}_0 \cdot \mathbf{k})(\prod_{j=1}^n u(t_j^* - t)), & n > 0 \\ u(\mathbf{H}_0 \cdot \mathbf{k}), & n = 0 \end{cases}$$

$$u(x) = \begin{cases} +1, & x \geq 0 \\ -1, & x < 0 \end{cases}$$

The product appearing in $U(t)$ results in the sequence $\{+1, -1, +1, \dots\}$ of $n+1$ elements where switchings occur just past $\{t_1^*, t_2^*, t_3^*, \dots\}$ if $n > 0$, respectively;

7.1 Extending the Heliocentric Orbital Frame of Sailcraft

$n = 0$ is meant that the set $\{t_j^*\}$ is empty, namely, $\|\mathbf{H}\| \neq 0$ everywhere in $[t_0, t_f]$. (A special trajectory class will stem from $n = 1$, as we shall see in Sect. 7.2.) The term $u(\mathbf{H}_0 \cdot \mathbf{k})$ takes into account how the motion begins at $t = t_0$; if retrograde when $n > 0$, the alternate sequence of signs in $U(t)$ gets reversed. One should note that, setting $t_j^* = \acute{t}$, $j = 1..n$ where $\acute{t} > t_f$, whatever t_f may be, then the product $\prod u()$ in the definition of $U(t)$ is $+1$, namely, the case $n = 0$ is implied.

EHOF basis is unambiguous and continuous with $\bar{\mathbf{h}} \cdot \mathbf{k} \geqslant 0$ throughout the whole flight of any heliocentric sailcraft trajectory. Particularly important in EHOF is the invariant H, previously defined, which can be summarized as

$$H(t) = \mathbf{H} \cdot \bar{\mathbf{h}} = U(t)\|\mathbf{H}\| = RV \sin\varphi, \quad \varphi \in \Psi_{\circlearrowright} = [0, 2\pi) \quad \Leftrightarrow \quad \mathbf{H} = H(t)\bar{\mathbf{h}} \tag{7.14}$$

Note 7.3 Negative values of H indicate that motion is retrograde. The converse is true if motion is two-dimensional. It may be not true if motion is three-dimensional; for example, the third component of $\mathbf{H}$ evolves from positive to negative values, whereas φ varies in $(0, \pi)$. This observation will be extended in Sect. 7.4.3.

Here below, we re-write the sailcraft's main equations in terms of the EHOF basis, the state variables, the lightness vector, and the invariant H. They represent the basic description from which we will infer most of the mathematical properties characterizing the fast sailing concept.

$$\ddot{\mathbf{R}} = \frac{\mu_\odot}{R^2}[-(1 - \mathcal{L}_r)\bar{\mathbf{r}} + \mathcal{L}_t\bar{\mathbf{h}} \times \bar{\mathbf{r}} + \mathcal{L}_n\bar{\mathbf{h}}]$$

$$\mathcal{L}_r \geq 0, \ \mathcal{L}_r^2 + \mathcal{L}_t^2 + \mathcal{L}_n^2 \geq 0 \tag{7.15}$$

$$E = \frac{1}{2}V^2 - (1 - \mathcal{L}_r)\frac{\mu_\odot}{R} \tag{7.16}$$

$$\dot{E} = \frac{\mu_\odot}{R^2}\mathbf{V} \cdot \bar{\mathbf{h}} \times \bar{\mathbf{r}}\mathcal{L}_t + \frac{\mu_\odot}{R}\dot{\mathcal{L}}_r = H\dot{H}/R^2 + \frac{\mu_\odot}{R}\dot{\mathcal{L}}_r \tag{7.17}$$

$$\dot{\mathbf{H}} = \mathbf{R} \times \ddot{\mathbf{R}} = \frac{\mu_\odot}{R}(\mathcal{L}_t\bar{\mathbf{h}} - \mathcal{L}_n\bar{\mathbf{h}} \times \bar{\mathbf{r}}) \tag{7.18}$$

$$\mathbf{H} \times \dot{\mathbf{H}} = \frac{\mu_\odot}{R}H\mathcal{L}_n\bar{\mathbf{r}} \tag{7.19a}$$

$$\dot{H} = \frac{\mu_\odot}{R}\mathcal{L}_t \tag{7.19b}$$

$$\frac{d}{dt}\begin{pmatrix} \bar{\mathbf{r}} \\ \bar{\mathbf{h}} \times \bar{\mathbf{r}} \\ \bar{\mathbf{h}} \end{pmatrix} = \begin{pmatrix} 0 & \omega_r & 0 \\ -\omega_r & 0 & \omega_h \\ 0 & -\omega_h & 0 \end{pmatrix}\begin{pmatrix} \bar{\mathbf{r}} \\ \bar{\mathbf{h}} \times \bar{\mathbf{r}} \\ \bar{\mathbf{h}} \end{pmatrix}$$

$$\omega_r \equiv \frac{V \sin\varphi}{R} = H/R^2, \ \omega_h \equiv \mu_\odot\frac{\mathcal{L}_n}{R^2 V \sin\varphi} = \frac{\mu_\odot/R}{H}\mathcal{L}_n \tag{7.20}$$

$$\ddot{\mathbf{V}} \Big/ \left(\frac{\mu_\odot}{R^2} \right) = \left(\dot{\mathcal{L}}_r + \Omega_r (1 - \mathcal{L}_r) - \omega_r \mathcal{L}_t \right) \bar{\mathbf{r}}$$

$$+ \left(\dot{\mathcal{L}}_t - \Omega_r \mathcal{L}_t - \omega_r (1 - \mathcal{L}_r) - \Omega_h \mathcal{L}_n^2 \right) \bar{\mathbf{h}} \times \bar{\mathbf{r}}$$

$$+ \left(\dot{\mathcal{L}}_n - \Omega_r \mathcal{L}_n + \Omega_h \mathcal{L}_t \mathcal{L}_n \right) \bar{\mathbf{h}}$$

$$\Omega_r \equiv 2\mathbf{R} \cdot \mathbf{V} / R^2, \quad \Omega_h \equiv (\mu_\odot / R) / H \tag{7.21}$$

Even motion equations (5.27) in the radius-longitude-latitude chart remain formally unchanged once the **L**'s components are evaluated with respect to the EHOF. The properties regarding the $H(t)$ function prove the following theorem:

Theorem 7.1 *The system of ODEs given by Eqs. (7.15) admits a scalar invariant, denoted by H and defined by Eq. (7.14), the time evolution of which is controlled through Eq. (7.19b).*

Another important quantity, which will help us to explain sailcraft motion, is the angle, say, ϑ of the total-acceleration direction *from* the velocity acceleration, measured counterclockwise and in the range $[0, 2\pi)$:

$$\cos\vartheta = \frac{\mathbf{V} \cdot \dot{\mathbf{V}}}{V \|\dot{\mathbf{V}}\|} = \frac{-\cos\varphi(1 - \mathcal{L}_r) + \sin\varphi \mathcal{L}_t}{\sqrt{(1 - \mathcal{L}_r)^2 + \mathcal{L}_t^2 + \mathcal{L}_n^2}}$$

$$\sin\vartheta = \frac{\|\mathbf{V} \times \dot{\mathbf{V}}\|}{V \|\dot{\mathbf{V}}\|} \, \text{sign}(\bar{\mathbf{h}} \cdot \mathbf{V} \times \dot{\mathbf{V}}) \tag{7.22}$$

$$\vartheta = \arctan(\cos\vartheta, \sin\vartheta)$$

The cross product in the second equation of (7.22) has been already made explicit in Eq. (5.46). The time rate of ϑ can be expressed as follows

$$\dot{\vartheta} = a\dot{b} - b\dot{a} \tag{7.23}$$

where a and b denote the right hand sides of $\cos\vartheta$ and $\sin\vartheta$, respectively.

Remark 7.2 We already pointed out that, in the case of a radial rectilinear arc, no HOF can be defined. However, an arbitrary direction orthogonal to the rectilinear arc can be substituted for $\bar{\mathbf{h}}$; the positive triad is completed in the usual way. With respect to such sailcraft-centered frame, one has $\mathbf{L} = (\mathcal{L}_r \ 0 \ 0)$. In particular, equation (7.17) holds as well. Unless the sail is equipped with an advanced reflectance control device (of which IKAROS has first experimented some preliminary items) applied to a considerable fraction of the sail surface, one has $\dot{\mathcal{L}}_r = 0$ in general. Therefore, during such arc, orbital energy is constant.

7.2 Two-Dimensional Motion Reversal

Even in solar sailing, a (regular) planar trajectory is an abstraction since $\mathcal{L}_n = 0$, strictly, for a finite time interval is not physically realizable. However, a detailed analysis of 2D fast trajectories will pave the way to understand the more complicated aspects of the three-dimensional fast trajectories. Therefore, let us begin with discussing what properties has a trajectory controlled by a lightness vector the third component of which is identically zero for the whole flight time of interest. The analysis of these trajectory types starts from numeric evidences coming from the integration of the motion of a sailcraft with lightness number notably higher than those ones shown in Fig. 5.3. Some important properties will be carried out from a large collection of numeric trajectory integrations. Then, the two-dimensional theory of fast solar sailing will be explained by noting the key role of the H-function and energy (per unit mass) in the trajectory properties. Subsequently, we will consider the problem of a sailcraft three-dimensional flight under the control of a time-dependent lightness vector. There, curvature and torsion will help us to infer several properties of high interest in sailcraft mission design.

7.2.1 Numeric Evidence

This section has been devised for giving the reader the visual concreteness of what solar sailing could bring forward in the context of high-end space missions. To such an aim, we first perform a few numeric experiments for getting a set of trajectories that could help us to understand why the general motion of a sailcraft—provided sufficiently light—is so peculiar and different from the chemical, ion and nuclear propulsion. Then, we will state three different numerical problems in order to begin with focusing on the problem of fast solar sailing and to prepare us for a deeper theoretical analysis. This second set of numerical examples consists of some hundreds of runs carried out by the author expressly for this book section by using the author's multi-purpose computer codes utilized in the course of a couple of decades for deep-space solar-sailing studies, including the NASA InterStellar Probe (NASA-ISP) at George C. Marshall Space Flight Center (MSFC).

Let us begin with turning into two-dimensional trajectories driven by the simplest control in our context:

$$\mathbf{L} = \begin{pmatrix} \mathcal{L}_r \\ \mathcal{L}_t \\ 0 \end{pmatrix} = constant, \quad [t_0, t_f], \ t_f > t_0 \tag{7.24}$$

As emphasized in Chap. 5 and the previous sections, we know that $\mathcal{L}_n = 0$ identically throughout the flight time interval (above denoted by $\mathcal{T}$) produces a planar trajectory, which can be assumed to lie on the HIF's reference plane with no loss of generality. The EHOF basis, resolved in HIF, is given by

$$\bar{\mathbf{r}} = \begin{pmatrix} X \\ Y \\ 0 \end{pmatrix} / \sqrt{X^2 + Y^2}, \qquad \bar{\mathbf{h}} \times \bar{\mathbf{r}} = \begin{pmatrix} -Y \\ X \\ 0 \end{pmatrix} / \sqrt{X^2 + Y^2}, \qquad \bar{\mathbf{h}} = \begin{pmatrix} 0 \\ 0 \\ 1 \end{pmatrix} \quad (7.25)$$

where X and Y are the coordinates of the sailcraft in HIF; the motion equation (7.15) results in

$$\dot{X} = V_X$$

$$\dot{Y} = V_Y$$

$$\dot{V}_X = -\frac{\mu_\odot}{(X^2 + Y^2)^{3/2}} \left[(1 - \mathcal{L}_r)X + \mathcal{L}_t Y \right] \qquad (7.26)$$

$$\dot{V}_Y = -\frac{\mu_\odot}{(X^2 + Y^2)^{3/2}} \left[(1 - \mathcal{L}_r)Y - \mathcal{L}_t X \right]$$

where the second-order motion equations have been made equivalent to a first-order system (ready to be implemented on a computer). The reader may note the above system of motion equations can be recast into the following form

$$\dot{\mathbf{R}} = -\frac{\mu_\odot}{R^3}(\mathrm{I}_2 - \Lambda)\mathbf{R} \equiv -\frac{\mu_\odot}{R^3}\Theta\mathbf{R}$$

$$\Lambda \equiv \begin{pmatrix} \mathcal{L}_r & -\mathcal{L}_t \\ \mathcal{L}_t & \mathcal{L}_r \end{pmatrix}, \quad \mathcal{L}_r \geq 0, \ \Theta \equiv \mathrm{I}_2 - \Lambda \qquad (7.27)$$

where matrix I_2 denotes the identity matrix in two dimensions. Matrix Λ is the 2D solar-sailing control matrix. The determinant is equal to the square of the lightness vector. Matrix Θ has a determinant always positive for every value of the real set $\{\mathcal{L}_r, \mathcal{L}_t\}$. Matrices Θ and Λ commute.

Some of the high-precision integrators built in a professional CAS can be utilized for numerically integrating the system (7.26), which is not stiff, but is dominated by the $1/R^2$ behavior of gravity and thrust.

Figure 7.1 can be considered as a continuation of Fig. 5.3 to sailcraft sail loading lower than the critical loading ($\sigma_c \equiv 2\mathrm{TSI}/cg_{1\mathrm{AU}} \cong 1.537$ g/m^2), specifically in the range (1.34–1.47) g/m^2. For simplicity, departure orbit is a circular orbit of unit radius; nevertheless, the core of fast solar sailing theory does not depend on the particular initial orbit. This will be clearer as we proceed with the related concepts and numerical analysis. The examples of Fig. 7.1 show trajectories that exhibit a point where the invariant H vanishes, namely, each trajectory consists of two arcs separated by one point where $H = 0$. The first one is direct, the second is retrograde. This is true even for the example-(3) where the sailcraft begins with decelerating, reaches a point, say, Q where $V = 0$, then goes back exactly on the same path, then continues to deep space. This trajectory is not regular in Q. By changing the values of $\{\mathcal{L}_r, \mathcal{L}_t\}$, we get *full* regular direct-retrograde arcs for which the retrograde arc is regular and either "external" or "internal" with respect to the mentioned non-regular

7.2 Two-Dimensional Motion Reversal

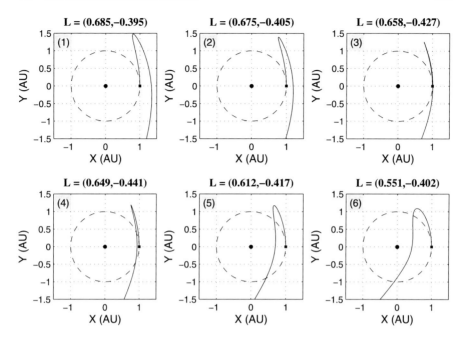

Fig. 7.1 Examples of 2D heliocentric sailcraft trajectories exhibiting motion reversal. Trajectory is driven by the radial and the transversal numbers. A *small square* marks the begin of flight. The *circle of unit radius* is the orbit of a hypothetical test body in the Keplerian solar field only; in these examples, it acts as the departure orbit of the sailcraft

trajectory. More precisely, using the nomenclature of Sect. 7.1, there are sets of direct-reverse motion *regular* trajectories for which the $H(t^*) = 0$ point has either $\varphi^* = 0$ or $\varphi^* = \pi$. We call such curve families the two-dimensional H-reversal trajectories, re-emphasizing the role of the orbital angular momentum through the H-function introduced in Eq. (7.8) and generalized in Eq. (7.14).

Next two figures characterize the six examples further. Figure 7.2 shows the hodographs of the trajectories, the unit speed being the (ideal) Earth Orbital Speed (EOS), or 2π AU/year; in particular, the trajectory speed can be visually compared with the departure orbit hodograph, here coincident with the unit circle like the trajectory plots. The first general behavior is the S-shape, meaning that not only strong deceleration occurs, but also that trajectory recovers speed with the final value higher than the speed of the sailcraft departure planet (the Earth, in this case). Speed re-gaining takes place with no discontinuity.

Figure 7.3 shows the evolution of the orbital energy E and the H-function. All cases share some important features: (1) H monotonically decreases, first rapidly, then slowly; (2) E first decreases, passes through a minimum at t^*, then quickly increases and achieves a long plateau (or cruise phase) where it (strictly speaking) will never lessen; (3) the cruise energy is considerably higher or much higher than the orbital departure energy, even though most of the kinetic energy has been lost during the deceleration phase. In contrast to the energy, which approaches an asymptotic

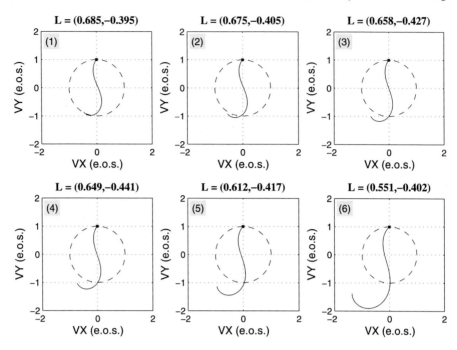

Fig. 7.2 Hodographs of the trajectories shown in Fig. 7.1. A *small square* marks the begin of flight. The *circle of unit radius* is the hodograph of 1-AU circular orbit. This circular speed, which is named the "Earth orbital speed" (e.o.s. or EOS, for short), is equal to 2π AU/year

maximum, speed achieves a maximum for subsequently decreasing to the cruise level. As we shall see in the next sections, this is due to having chosen $\|\mathbf{L}\| < 1$ even in the cruise phase; but this is a matter of optimization, as we shall see in Chap. 8.

Starting from these considerations and observing the six-plot sets of Figs. 7.1–7.3 as a whole, the following questions arise:

1. Why the trajectory shapes are so different with respect to a usual perturbed Keplerian orbit?
2. Since $\mathbf{L}$ is defined in the EHOF and is kept constant there, why does the corresponding time-variable control in HIF induce a motion reversal?
3. How do two sailcraft trajectories, one with a given $\{\mathcal{L}_r, \mathcal{L}_t < 0\} = constant$ and the other driven by the opposite $\{\mathcal{L}_r, -\mathcal{L}_t > 0\}$, compare provided they evolve from the same initial conditions (ICs) for a given flight time? Of course, this question makes sense if both the radial and transversal numbers are sufficiently high.

Qualitative answer to question-1 is very simple: via Theorem 5.1, we have found that the fields acting on sailcraft are three: two are conservative and one is non-conservative. The radial field balances the gravity field to a large extent and increases the orbital energy (with respect to the Keplerian one). The transversal field

7.2 Two-Dimensional Motion Reversal

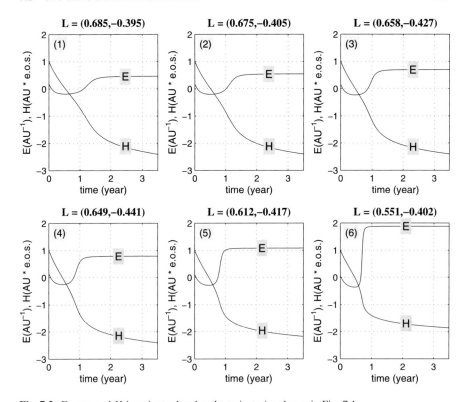

Fig. 7.3 Energy and H-invariant related to the trajectories shown in Fig. 7.1

acts on both energy rate and angular momentum rate. If the fields are all of the same order of magnitude, as they are in the current examples, trajectory and velocity can turn into very unusual profiles.

Less simple is the answer to question-2. For simplicity (but without loss of generality here), let us assume an ideal sail so that the thrust acceleration **a** is directed along the sail axis **n** (defined in Chap. 6). Let also denote the solar gravity acceleration by **g** and set $\mathbf{A} = \mathbf{g} + \mathbf{a}$ for the total acceleration on the sailcraft. For a better conceptual visualization, let us consider the sixth example in Fig. 7.1 for which the sail axis azimuth and sailcraft sail loading are given by $\alpha = -36.1°$ and $\sigma = 1.47$ g/m^2, respectively. In our reasoning, we will use the angle ϑ, defined in Eq. (7.22) and, again, the angle φ. From Eq. (7.22), $\vartheta_0 \cong 131.8°$ (since $\varphi_0 = 90°$); thus, deceleration begins. While the angle between **g** and **a** keeps unchanged (because $L = constant$), **A** bends **V** counterclockwise, and **A** itself rotates counterclockwise as well; it first rotates faster, then slows down considerably with respect to the **V** rotation. This is a key point, which can be verified via the numerical integration of the trajectory. Thus, a point M is eventually reached where $\vartheta_M = 90°$, or $\mathbf{A} \cdot \mathbf{V}|_M = 0$. In M, $\sin \varphi > 0$ and thus the H-function is still positive. Past M, $\mathbf{A} \cdot \mathbf{V}|_M$ becomes positive, this causing V to increase, and **V** can continue to rotate allowing φ to reach 180°. We know that in such a point, denoted by P*, H vanishes, but ϑ is acute (specifically, $\vartheta^* \cong 41.8°$); therefore, $\dot{V}^* > 0$. Past P*, the motion reverses while V continues to increase smoothly.

> Let us note that $\mathbf{L} = \text{constant}$ is not the only way to cause a general motion reversal, as we shall see later. This mode is the simplest to analyze and paves the way to design realistic three-dimensional controls for trajectory optimization. Historically, such an analytic approach was the first to be set up for figuring out the non-simple nature of this non-linear dynamical problem [9].

Once that the existence of the H-reversal mode is evidenced numerically, question-3 is particularly important, because the control one would apply naturally is a lightness vector with both components positive: the radial number balances gravity partially, while the transversal number accelerates the sailcraft. Let us see how the sixth trajectory profile just described ($\mathbf{L} = \{0.551, -0.402\}$) compares with the trajectory induced by $\mathbf{L} = \{0.551, +0.402\}$. The former control mode is termed the H-reversal mode, whereas the latter is named the plain (control) mode. Trajectories and speed evolutions are shown in the top and bottom parts, respectively, of Fig. 7.4. The evolution of sailcraft speed in the plain mode is expected similarly to what happens for rockets. In contrast, unexpected is the evolution of speed for the H-reversal mode. The first part of the profile reflects what we just discussed about question-2; however, the strong rising of V and its subsequent plateau are a notable surprise, which we shall explain in Sect. 7.2.2. In this case, the plain mode exhibits a cruise speed of 7.08 AU/year, or 33.36 km/s; the H-reversal control produces a cruise with 12.28 AU/year, or 58.20 km/s. Another result is that the trajectory curvature, past 1 AU after 280 days from the departure, of the H-reversal mode is very close to zero, as one can qualitatively see in Fig. 7.4. Perihelion is at 0.42 AU reached after almost 251 days from the departure. Let us observe that the two sailcraft could be alike from the realization viewpoint: the big difference in the flight profile comes from *one* scalar control. As we will see in the next sections and chapters, 12 AU/year is in reality a low value for fast solar sailing.

The preliminary numerical results we just discussed induce us to extend the numerical analysis significantly in order to get additional information about the way two-dimensional motion reversal is caused by sets of $\{\mathcal{L}_r, \mathcal{L}_t\}$. This information will further help us in setting up a theory.

Considering again the third example of Fig. 7.1, we first search for a set of H-reversal trajectories—containing one non-regular point, if any. The set of the corresponding controls will act as the delimiter between different regions of the control-plane quadrant defined by $\mathcal{Q} \equiv \{\mathcal{L}_r > 0, \mathcal{L}_t < 0\}$. Many numerical integrations of the motion equations (7.26) show that, independently of the type of osculating orbit at the departure,[1] there exist trajectories exhibiting non-regular $H(t^*) = 0$ for some

---

[1]Here, departure means either heliocentric injection of sailcraft after a planetocentric escape orbit or four directly-specified heliocentric ICs.

7.2 Two-Dimensional Motion Reversal

Fig. 7.4 Plain Acceleration vs H-reversal Mode. $\mathbf{L} = $ constant is applied, as labeled. The strong contrast between the two profiles stems from both the sign of the transversal number and its absolute value, which in turn is sufficiently high to bring about trajectories so different from spirals

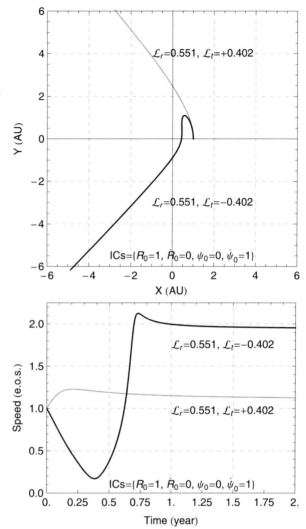

$t^* > t_0$. More precisely, the following properties hold

$$\mathbf{R}(t) = \mathbf{R}(2t^* - t), \qquad \mathbf{V}(t) = -\mathbf{V}(2t^* - t), \qquad t \in [0, 2t^*] \tag{7.28}$$

In order to look into this set of trajectories, one can note that there should be some relationship between the time at which $H = 0$ and the control pair $\{\mathcal{L}_r, \mathcal{L}_t\}$. Therefore, one shall state this problem as a two-boundary value problem with two free parameters. Once the initial conditions are chosen, one has to solve the following augmented motion equations:

$$\dot{R} = \upsilon$$

$$\dot{\psi} = \omega$$

$$\dot{\upsilon} = -(1 - \mathcal{L}_r)\frac{\mu_\odot}{R^2} + R\omega^2$$

$$\dot{\omega} = \mathcal{L}_t \frac{\mu_\odot}{R^3} - 2\frac{\upsilon\omega}{R} \tag{7.29}$$

$$\dot{\mathcal{L}}_r = 0$$

$$\dot{\mathcal{L}}_t = 0$$

where we switched to the polar form of Eq. (7.26), namely, Eqs. (5.27) with $\theta = 0$ identically; ψ denotes the usual polar angle, and ω is its time rate. The following boundary conditions characterize the non-regular H-reversal, including the properties (7.28):

non-regular H-reversal problem:

$$R(0) = R_0, \qquad \psi(0) = \psi_0, \qquad \upsilon(0) = \upsilon_0, \qquad \omega(0) = \omega_0, \tag{7.30}$$

$$\upsilon(t^*) = 0, \qquad \omega(t^*) = 0$$

Conditions at t^* guarantee that velocity vanishes there. Thus, once the H-reversal time is prefixed, the problem consists of finding the values of the radial and transversal numbers such that Eqs. (7.29) are satisfied.[2] The shooting method can solve this numerical problem efficaciously.

Despite its conceptual force and simplicity, an efficient high-accuracy shooting algorithm is not trivial to implement. Nowadays, there exist excellent routines in professional packages of numerical mathematics. Conceptually speaking, a two-point boundary value problem may (in contrast to pure initial-value problems) exhibit either no solution or a finite number of solutions, or infinitely many solutions too. The analyst is requested to provide an initial guess for the expected solution(s), if any. The information contained in such a guess could be critical for the numerical solver (i.e. the ODE integrator). In the current astrodynamical problem, the guess for shooting refers to as the solar-sail trajectory control parameters, which are treated as differentiable functions. We have got a good starting pair by building a low-cardinality random set of trajectories and interpolating in candidate-to-boundary-constraint subsets of **L**. The numerical algorithm is fast and robust, even for high-precision goals.

---

[2]Note that these differential equations lend themselves to a natural generalization of the **L**-control.

7.2 Two-Dimensional Motion Reversal

Thus, by varying t^* progressively with a small step, and also using the previous computed control for next shooting case, one can build the curve, say, $\mathcal{B}^* \equiv \{\mathcal{L}_r(t^*), \mathcal{L}_t(t^*)\}$ belonging to $\mathcal{Q}$. This curve contains all possible pairs of **L**-components inducing a non-regular motion reversal, but *relatively* to the chosen set of ICs, say, $\mathcal{J}_0 \equiv \{R_0, \psi_0, \upsilon_0, \omega_0\}$. Because the procedure is exactly the same for any departure conic of the sailcraft, we chose the simplest one: $\mathcal{J}_0 = \{1, 0, 0, 1\}$.

The last three examples of Fig. 7.1 suggest analyzing H-reversal trajectories having perihelion values lower than that of the non-regular H-reversal mode with the same $\mathcal{J}_0$. This is again a two-boundary value problem, but the second line of the boundary conditions (7.30) has to be replaced. As a matter of fact, in principle one may write two independent conditions characterizing $\mathbf{H}(t^*) = 0$ at some prefixed t^* and apply the previous shooting. However, in real flight designs, it is essential to know the minimum distance from the Sun for aims related to thermal control, radiation dose, trajectory control, and telecommunication problems; even in the current simplified context, solving the two-point value problem by using perihelion characterization does make sense. The following boundary conditions apply:

prefixed perihelion value problem:

$$R(0) = R_0, \qquad \psi(0) = \psi_0, \qquad \upsilon(0) = \upsilon_0, \qquad \omega(0) = \omega_0,$$

$$R(t_p) = R_p, \qquad \sin\varphi(t_p) = \frac{H(t_p)}{R(t_p)V(t_p)} = \left(\frac{R\omega}{\sqrt{\upsilon^2 + (R\omega)^2}}\right)_{t_p} = -1 \tag{7.31}$$

The sixth constraint specifies in a unique way that one wants to get a reversed-motion perihelion: $\sin\varphi = -1$. Once a perihelion value, say, $R_p^{(1)}$ is chosen, one can get the solution $\{\mathcal{L}_r, \mathcal{L}_t\}$ corresponding to a certain value of t_p. Varying this one by a small amount and repeating the solving procedure, one obtains a new pair $\{\mathcal{L}_r, \mathcal{L}_t\}$ for the same value of $R_p^{(1)}$. So the t_p-parametrized curve $(\mathcal{L}_r(R_p^{(1)}, t_p), \mathcal{L}_t(R_p^{(1)}, t_p))$ can be built. By prefixing another perihelion value, say, $R_p^{(2)}$ the algorithm will generate another curve $(\mathcal{L}_r(R_p^{(2)}, t_p), \mathcal{L}_t(R_p^{(2)}, t_p))$, and so forth.

In particular, one can determine the grazing-the-Sun curve by setting $R_p = r_\odot$. Of course, such a curve is for delimiting the $\mathcal{Q}$-region related to the considered initial osculating orbit and the H-reversal mode; no currently-conceived sailcraft would (for a number of reasons, some obvious and some less) be able to flyby the Sun very closely, say, at 2–3 solar radii, even though the ensuing cruise speeds would be somewhat interesting, namely, almost relativistic!

There is one more type of problems we can deal with in order to characterize solar sailing curves in $\mathcal{Q}$; we could find the set of **L**-control that allows a sailcraft to reach different perihelion distances after a prefixed time from the departure. The curves $\{\mathcal{L}_r(R_p), \mathcal{L}_t(R_p)\}$, corresponding to given t_p (and ICs), can be found by solving the two-point boundary problem with the previous $\mathcal{J}_0$ and the new perihelion

272 7 The Theory of Fast Solar Sailing

constraints:

prefixed time-to-perihelion problem:

$$R(0) = R_0, \qquad \psi(0) = \psi_0, \qquad \upsilon(0) = \upsilon_0, \qquad \omega(0) = \omega_0,$$

$$t(R_p) = t_p, \qquad \sin\varphi(R_p) = \frac{R_p\omega(R_p)}{\sqrt{\upsilon(R_p)^2 + (R_p\omega(R_p))^2}} = -1 \qquad (7.32)$$

The solving procedure is similar to that of the prefixed perihelion value problem. One prefixes the perihelion time, say, $t_p^{(1)}$ and gets the solution pair $\{\mathcal{L}_r, \mathcal{L}_t\}$ corresponding to a certain value of R_p. Changing this one by a small amount, a new pair $\{\mathcal{L}_r, \mathcal{L}_t\}$ is found for the same value of $t_p^{(1)}$. So the R_p-parametrized curve $(\mathcal{L}_r(R_p, t_p^{(1)}), \mathcal{L}_t(R_p, t_p^{(1)}))$ can be built. By prefixing another perihelion time, say, $t_p^{(2)}$ the algorithm generates a different curve $(\mathcal{L}_r(R_p, t_p^{(2)}), \mathcal{L}_t(R_p, t_p^{(2)}))$, and so forth. These curves are transversal to those of the problem (7.31).

The author ran his computer codes repeatedly in order to collect over 400 convergent cases and thus build several representative curves in the quadrant $\mathcal{Q}$. This resulted in Fig. 7.5. Let us see how to read this figure appropriately. Quadrant $\mathcal{Q}$ can be divided into three main regions:

$\mathbb{S}$ or the slow H-reversal motion region, where the Sun-sailcraft distance is greater than or equal to the perihelion of the departure orbit; there are small asymptotic gains in speed and energy with respect to the departure values;

$\mathbb{F}$ or the fast H-reversal motion region, where the actual Sun-sailcraft perihelion is lower than the perihelion of the departure orbit; speed and energy gains, depending non-linearly on the reachable perihelion, could be very large;

$\mathbb{P}$ or the *plunging* region, where the sailcraft eventually falls, either quickly or via some slow spiral, onto the Sun (if the sail attitude were not changed, of course).

Region $\mathbb{S}$ is on the right of the non-regular H-reversal curve B^* (i.e. the thicker line in Fig. 7.5), and is characterized by rather high radial lightness numbers.[3] This bound depends on the initial conditions.[4] Region $\mathbb{F}$ is bounded on the right by B^*, and on the left by the *grazing-the-Sun* curve, here denoted by $\mathcal{G}$. In turn, on the left of $\mathcal{G}$, there is the region $\mathbb{P}$, where the sailcraft sail loading is too high to prevent the sailcraft to eventually fall down onto the Sun. Nevertheless, arcs of $\mathbb{P}$-trajectories could be used for various important missions in the solar system; to do this, one has to program a variable control $\mathbf{L}(t)$, either continuous or piecewise-continuous.

---

[3] In the Nineties, the author found numerically only a part of the quadrant, i.e. that one approximately delimited by $0.5 < \mathcal{L}_r < 1$ and $-0.5 < \mathcal{L}_t < 0$. Such subregion, though, allowed him to study many meaningful cases of 2D trajectories.

[4] It is important to remind the reader that—for any $\mathcal{L}_r > 0$—the sailcraft senses Sun as having the reduced gravitational constant $(1 - \mathcal{L}_r)\mu_\odot$. Thus, $\mathcal{L}_r = 1/2$ (and $\mathcal{L}_t = 0$) applied to a classical Keplerian circular orbit means that the orbit is actually a parabola, as one can easily check via the energy equation.

7.2 Two-Dimensional Motion Reversal

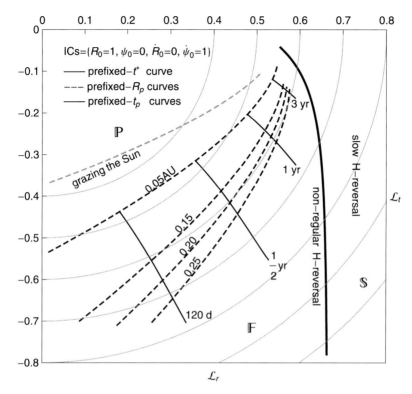

Fig. 7.5 Two-dimensional control regions for H-reversal mode in the quadrant of positive-$\mathcal{L}_r$ and negative $\mathcal{L}_t$. The curves they enclose are the numerical solutions to the system of Eqs. (7.29) subject to the boundary values given by (7.30), (7.31), and (7.32), respectively. The plot region, a portion of the quadrant, has been chosen for graphic readability. A $\mathcal{L}$-constant grid has been added

Let us focus on the curves we plotted in the region $\mathbb{F}$, the fast solar sailing. In order to justify the adjective *fast*, we should wait a few sections, where we show the full characteristics of the reversal mode. For the moment, let us restrict ourselves to describe the curves in $\mathbb{F}$. Those ones, which are concave up, give the control parameters for getting a given perihelion, e.g. $R_p = 0.15$ AU, by varying the time to it (i.e. the problem (7.31)). The concave-down curves provide the control parameters for achieving different perihelia in a given time (i.e. the problem (7.32)). Each pair $(\mathcal{L}_r, \mathcal{L}_t) \in \mathbb{F}$ is the intersection of one curve of the first type and one curve of the second type, and singles out a trajectory that emanates from the given ICs and drives the sailcraft to the aimed H-reversal perihelion in the assigned time. Figure 7.5 allows solving a number of trajectory design problems numerically.

A remarkable point regards time. The time-labeled curves tell us that (i) a higher lightness number $\mathcal{L}$ is required for speeding up trajectories to perihelia, (ii) $\mathcal{L}(t_p)$ is notably non-linear; the upper part of the sets of curves pertain to times-to-perihelion values notably longer than 1 year, and (iii) trajectories particularly fast exhibit $-\mathcal{L}_t > \mathcal{L}_r$.

274   7 The Theory of Fast Solar Sailing

Finally, the $\mathcal{L}$-constant grid allows the analyst to find the intersection of the discussed profiles with a given $\mathcal{L}$; this task becomes very important once one has a model of how the sail attitude changes with the vector $\mathbf{L}$, e.g. Eq. (6.136) with sail elevation set to zero. Once the sail materials are selected, varying the sailcraft sail loading results in curves on the plane $(\mathcal{L}_r, \mathcal{L}_t)$; the superposition of these curves with those ones of Fig. 7.5 enables sailcraft designer to get some key information.

7.2.2 H-Reversal Theory

In this sub-section, we shall develop the two-dimensional theory of fast solar sailing; this is done for the following reasons: (1) to explain the numerical results presented in Sect. 7.2.1, (2) to find a more general context that will lead to the fast three-dimensional theory, (3) to point out some of the features that could be optimized realistically.

▶ In order to get the conditions where the H-reversal mode can take place, how much fast and which the potential applications may be, let us first *assume* that the problem of existence of this sailing mode has been solved; as a result, many properties peculiar of this two-dimensional mode can be carried out; in particular, the existence conditions will be inferred.

Equations (5.47) and (5.53), which hold in the general context of EHOF, may be looked for some value of the angle φ which vanishes the 2D trajectory curvature and simultaneously makes the sailcraft speed a stationary point. In two dimensions, trajectory curvature and speed's time derivative are expressed by

$$\kappa(t) = \frac{\mu_{\odot}}{(RV)^2}(\cos\varphi\,\mathcal{L}_t + \sin\varphi(1 - \mathcal{L}_r)) \tag{7.33a}$$

$$\dot{V} = \frac{\mu_{\odot}}{R^2}(\sin\varphi\,\mathcal{L}_t - \cos\varphi(1 - \mathcal{L}_r)) \tag{7.33b}$$

One immediately recognizes that this statement is equivalent to find the solution to the following matrix equation:

$$\begin{pmatrix} \cos\varphi & \sin\varphi \\ -\sin\varphi & \cos\varphi \end{pmatrix} \begin{pmatrix} \mathcal{L}_t \\ 1 - \mathcal{L}_r \end{pmatrix} = \begin{pmatrix} 0 \\ 0 \end{pmatrix} \tag{7.34}$$

This system admits only one solution: $\mathcal{L}_r = 1, \mathcal{L}_t = 0$, namely, a rectilinear motion in the solar system. Thus, a sailcraft designed to have $\mathcal{L} = \sqrt{\mathcal{L}_r^2 + \mathcal{L}_t^2} = 1$ could, for instance, begin the flight with $\{\mathcal{L}_r < 1, \mathcal{L}_t < 0\}$ and, when appropriate, to switch to faster rectilinear motion via attitude maneuver. These simple considerations are to say that the rectilinear motion may be dealt with as a particular segment of a more complicated fast trajectory. However, for the moment, we will exclude it from the subsequent analysis because such many-arc trajectory entails some optimization process (which we shall deal with in Chap. 8). Let us go back to the single H-reversal mode and begin with searching for the point(s), if any, where sailcraft speed has

7.2 Two-Dimensional Motion Reversal

local extrema. Using Eq. (7.33b), the necessary condition for vanishing the along-track component of the total acceleration is expressed by

$$\sin\varphi\mathcal{L}_t - \cos\varphi(1 - \mathcal{L}_r) = 0 \tag{7.35}$$

This equation admits two solutions

$$\sin\varphi_{(1)} = (1 - \mathcal{L}_r)/\tilde{\mathcal{L}}, \qquad \cos\varphi_{(1)} = \mathcal{L}_t/\tilde{\mathcal{L}}$$
$$\sin\varphi_{(2)} = -(1 - \mathcal{L}_r)/\tilde{\mathcal{L}}, \qquad \cos\varphi_{(2)} = -\mathcal{L}_t/\tilde{\mathcal{L}} \tag{7.36}$$
$$\tilde{\mathcal{L}} \equiv \sqrt{(1 - \mathcal{L}_r)^2 + \mathcal{L}_t^2}$$

Although useful for increasing speed after the perihelion, however here we exclude $\mathcal{L}_r > 1$ from the current analysis, consistently with the numerical results of Fig. 7.5, because in this case the energy would be always positive regardless $\mathcal{L}_t$ and how long the flight may be. Therefore, we define our basic control parameter framework henceforth as expressed by

$$0 < \mathcal{L}_r < 1, \qquad \text{sign}(\mathcal{L}_t) = -\text{sign}(H_0) = -1, \qquad \xi \neq t^*$$
$$\mathcal{L}_r(t) = \mathcal{L}_r^- + \left(\mathcal{L}_r^+ - \mathcal{L}_r^-\right)\Theta(t - \xi) \tag{7.37}$$
$$\mathcal{L}_t(t) = \mathcal{L}_t^- + \left(\mathcal{L}_t^+ - \mathcal{L}_t^-\right)\Theta(t - \xi)$$

compliantly with Eqs. (5.35). The equalities on the first line entail that we will consider sets of trajectories starting *direct* and ending *retrograde*. In addition, sail's impulsive attitude maneuvers are admitted at some time $\xi \neq t^*$, but such that the transversal component keeps negative after maneuvering. Considerations and conclusions drawn from (7.37) will hold exactly for reverse situations such as trajectories starting clockwise and ending counterclockwise.

In this context, solutions to Eq. (7.35) are characterized by

$$\pi/2 < \varphi_{(1)} < \pi, \qquad 3\pi/2 < \varphi_{(2)} < 2\pi \tag{7.38}$$

Since $\text{sign}(H_0) = +1$, $\varphi_{(1)}$ precedes $\varphi_{(2)}$ temporally. As P*, which has $\varphi^* = \pi$, separates the two direct-retrograde arcs, one has

$$t_{(1)} < t^* < t_{(2)} \tag{7.39}$$

Applying (7.37) to Eq. (7.17) at t^*, one gets

$$\dot{E}^* = 0$$
$$\ddot{E}^* = \frac{d}{dt}\left(H\dot{H}/R^2 + \frac{\mu_\odot}{R}\dot{\mathcal{L}}_r\right)_{t=t^*} = (\dot{H}/R)^2_{t=t^*} > 0 \tag{7.40}$$

Therefore, the H-reversal point is a local *minimum* of orbital energy. If there is no **L**-maneuver during the direct arc, the energy time rate is always negative, whereas it is always positive for any $t > t^*$. As a consequence, although in the direct arc of flight a large fraction of the initial energy is lost by deceleration, in the retrograde arc the energy increases monotonically, and at high rate close to the Sun. This is the physical reason of the different behaviors of the plain-acceleration and H-reversal modes shown in Fig. 7.4.

What about $\text{sign}(E^*)$? If the sailcraft starts moving from a bound or parabolic orbit ($E_0 \leqslant 0$), then $E^* < 0$ necessarily. Using the definition of reversal point and the expression of the invariant H, the following additional differential properties of P^* can be got from the polar-form of the motion equations (7.29):

$$\dot{\psi}^* = 0, \qquad \ddot{\psi}^* = \mathcal{L}_t \mu_\odot / R^{*3} < 0$$

$$\dot{R}^* = -V^*, \qquad \ddot{R}^* = -(1 - \mathcal{L}_r)\mu_\odot / R^{*2} < 0 \qquad (7.41)$$

$$\dddot{R}^* = -2(1 - \mathcal{L}_r)\mu_\odot V^*/R^{*3} + \dot{\mathcal{L}}_r \mu_\odot / R^{*2} < 0$$

The first two relationships tell us that the sailcraft's polar angle has a local maximum at t^*, after which ψ begins to lessen, namely, motion is reversed. From the third and fourth relationships, the reversal point cannot be a point of local maximum of distance, in general, except for a trajectory with P^* non-regular. The last relationship regards the radial jerk, whereas the related inequality follows from $\dot{\mathcal{L}}_r = 0$ in the current framework. Therefore, the radial speed at P^* is directed toward the Sun, the radial acceleration enhances this motion, and the radial jerk tells us that such acceleration is enforced. In addition to $H(t^*) = 0$, Eqs. (7.39), (7.40), and (7.41) altogether represent the *differential* properties of the H-reversal point of a 2D trajectory. All such considerations lead to the following propositions:

Proposition 7.1 *If, under the control of type (7.37), a two-dimensional trajectory includes an internal point where orbital angular momentum vanishes, then this point is a local minimum of orbital energy. If the initial osculating orbit is bound or parabolic and the lightness vector undergoes no change until the reversal point, then the energy at such point is necessarily negative. The reversal point is also a local maximum of the polar angle.*

Proposition 7.2 *Under the control of type (7.37), if trajectory is regular at $H(t^*) = 0$, then the reversal point can be neither the aphelion nor the perihelion of the direct-retrograde arcs trajectory.*

We like to briefly analyze a change of lightness vector. To this aim, let us think about a trajectory arc from P_0 to P^* in a given time t^*, where $\dot{R}^* = V^* = 0$, namely, the trajectory is not regular at P^*. There, let us now suppose to perform an impulsive maneuver changing the control variables from $\{\mathcal{L}_r, \mathcal{L}_t\}$ to $\{\mathcal{L}'_r = \mathcal{L}_{max} > \mathcal{L}_r, \mathcal{L}'_t = 0\}$ for $t^* \leq t' < t^{**}$, where t^{**} is some time tag at which a subsequent maneuver may be applied. As this new trajectory arc starts from null speed, path becomes rectilinear and radial, either inwards or outwards according to the radial number $\mathcal{L}'_r < 1$ or $\mathcal{L}'_r > 1$, respectively. Such arc exhibits $\mathbf{H} = 0$ identically. This arc, mentioned on

7.2 Two-Dimensional Motion Reversal

page 257, implies no motion reversal inasmuch as $\dot{\psi}(t') = 0$ and $\ddot{\psi}(t') = 0$ identically. In addition, in this arc, sailcraft energy shows $\dot{E}(t') = 0$, $\ddot{E}(t') = 0$ again. Now, suppose to accomplish a further change of attitude at t^{**} so to reset the control: $\{\mathcal{L}''_r = \mathcal{L}_r, \mathcal{L}''_t = \mathcal{L}_t\}$ kept constant during the rest of the flight $(t^{**}, t_f]$. Because $\mathbf{H}(t^{**}) = 0$, one also gets $\dot{\psi}^{**} = 0$, $\ddot{\psi}^{**} = \mathcal{L}''_t \mu_\odot / R^{**3} < 0$; therefore, motion begins reversing. In addition, from equations (7.29), if $\mathcal{L}'_r < 1$, then $\dot{R}^{**} < 0$ and $\ddot{R}^{**} < 0$ as well. Thus, while the sailcraft goes closer to the Sun, the product of its radial and angular speeds is positive for $t'' > t^{**}$. Furthermore, the angular acceleration takes on stronger negative values; in other words, motion reversal is enhanced.

This two-maneuver example indicates that the (non-independent) conditions

$$R(t^*) > 0, \qquad H(t^*) = 0, \qquad \sin\varphi(t^*) = 0, \qquad \dot{E}(t^*) = 0, \qquad \dot{\psi}(t^*) = 0$$
$$(7.42)$$

are *necessary*, but *not* sufficient conditions, for getting a regular H-reversal motion.

7.2.2.1 Fast Sailing Definition and Existence Conditions

▶ The spontaneous question at this point is whether sufficient conditions there exist. To this aim, we have to unambiguously state the concept of motion reversal for fast solar sailing; we are now able to do this because we have many pieces of information, both numerical and theoretical, previously unavailable. Fast sailing means the achievement of very high speeds compared to the current standards. Not only; cruise speeds comparable or superior to that potentially obtainable by advanced NEP/NIP are quite desirable, and with the basic advantage of using no propellant. However, this goal would be partially offset if the deceleration phase, necessary to go sufficiently close to the Sun, lasted too much; solar-sail flight times to very distant targets should be notably shorter than those ones of flights involving one or more planetary flybys. In fact, we like to remind the reader that fast sailing shall entail no planetary flyby (unless mission involves some special feature(s), which is(are) no way related to energy gain). Thus, combining these requirements, fast sailing can be defined. Without losing in generality, we proceed by considering flights starting counterclockwise.

Definition 7.1 2D Fast Sailing (D2FS). Two-dimensional fast-sailing is a solar-photon sail thrusting mode characterized by:

1. a sailcraft trajectory that is differentiable everywhere and is neither *closed* nor *simply*-closed
2. a pair of direct/retrograde arcs separated by *one* point where $H = 0$;
3. a retrograde arc exhibiting:
   (a) strictly decreasing values of the H-invariant;
   (b) strictly positive time derivative of the orbital energy;
   (c) the $E = 0$ point;

278　　　　　　　　　　　　　　　　　　　　　　7　The Theory of Fast Solar Sailing

(d) a local minimum of $R(t)$, which is also the perihelion, relatively to the time interval from the reversal point to the end of flight.

The enumerated features are compliant with the set of trajectories computed and preliminarily discussed with regard to Fig. 7.5.

We shall now proceed by determining the conditions sufficient for D2FS by noting that the previously mentioned differential features do *not* depend on the initial conditions *explicitly*. In order to characterize the existence of reverse mode trajectories starting from given ICs, one has to *also* resort to relationships involving integral quantities evaluated along the sailcraft path.

▶ Requirement (1) consists of two pieces: (1) differentiability, and (2) nonclosure of the sailcraft trajectory in the sense of the Global Theory of Curves (GTC) (e.g. see [2, 4]). The former avoids strange and unexpected geometrical features, the latter prevents the sailcraft from following a path with periodic properties. Sufficient conditions for function differentiability are well known; in the current case, that $\mathbf{R}(t)$ is differentiable is guaranteed by the motion equations (in various coordinates) we used previously. A non-closed curve acting as sailcraft trajectory can be tested easily by using the total curvature and the total absolute curvature, normalized both to 2π. If sailcraft's regular trajectory is not closed, then the following integral inequalities have to be satisfied:

$$\mathcal{K} \equiv \int_{t_0}^{t_f} \kappa(t) V(t)\, dt, \qquad \mathbb{K} \equiv \int_{t_0}^{t_f} |\kappa(t)| V(t)\, dt$$

$$0 < \frac{\mathcal{K}}{2\pi} \le \frac{\mathbb{K}}{2\pi} < 1 \tag{7.43}$$

In order to understand the meaning of this criterion, let us briefly set out some concepts in the following paragraphs:

C1: In Sect. 5.3.3, we specialized curvature and torsion to solar-sail trajectories. For a two-dimensional curve, torsion is identically zero and curvature may be defined to have a sign (also called the oriented or signed curvature). Observing a planar curve from an *external*-to-the-plane point, a curve has a negative curvature in each point where the tangent vector is rotating clockwise, and positive if rotation is counterclockwise. At inflection points, if any, curvature vanishes, and the tangent-vector direction is stationary. Of course, such definition depends on the chosen parametrization, say, u for describing the curve; for instance, if one changes $u \to -u$, then the curvature sign changes: $\kappa(u) \to -\kappa(-u)$. There are several equivalent ways to state this concept (perhaps, the most expressive and unambiguous is that from Geometric Algebra); however, this is not very important here. What matter is that, along a trajectory, curvature can *vary* in magnitude and sign. One can check it from the 2D-curvature expression (i.e. Eqs. (5.43) and (5.47) with $\mathcal{L}_n = 0$) that can be

7.2 Two-Dimensional Motion Reversal

written as follows:

$$\kappa(t) = \check{\kappa}(t) = \frac{\mu_\odot}{(RV)^2}\left(\frac{H}{RV}(1 - \mathcal{L}_r) + \frac{\dot{R}}{V}\mathcal{L}_t\right) \tag{7.44}$$

from which one sees easily why curvature can have any sign in 2D sailcraft trajectories.

C2: Said that, let us recall that GTC studies the properties of closed curves, with a special emphasis to the total (or integrated) curvature $\mathcal{K}$ and the absolute total curvature $\mathbb{K}$ defined in (7.43). Now, let us note that the GTC's theorem on turning tangents states that—for a simply closed curve (i.e. a curve that does not self-intersect)—the integrated curvature can be equal to either 2π or -2π; in formulas, one gets

$$\oint \kappa(s)\,ds = \pm 2\pi \tag{7.45}$$

where s is the arc length. This may seem a strange result, but it is useful to visualize it by means of the Gauss mapping, which we shall just apply to sailcraft motion. Let us consider the position vector $\overrightarrow{OP} \equiv \mathbf{R}(s)$ of a generic point P on the trajectory, and the unit tangent vector $\overrightarrow{PT} \equiv \mathbf{t} = d\mathbf{R}(s)/ds = \dot{\mathbf{R}}(t)/V(t)$, where O denotes the origin of the coordinates; parallel-transporting $\mathbf{t}$ to O results in $\overrightarrow{O\hat{P}}$, where $\hat{P}$ is the image of P in this injective mapping. As P varies on the trajectory, $\hat{P}$ moves on this 1-sphere: when trajectory curvature entails counterclockwise motion of $\mathbf{t}$, $\hat{P}$ runs counterclockwise about O, and so on. Thus, one can figure out that a closed trajectory induces an integral number of turns of $\hat{P}$ on the circle, no matter how many times the trajectory is convex or concave. (This number is called the *winding number* of the trajectory). In particular, if the trajectory is a simply closed (regular) curve, then $\hat{P}$ covers the circle only once. Gauss mapping can be applied to any regular plane curve.

C3: More precisely, under such mapping, the angle $d\varepsilon = \kappa(s)ds$ of $\mathbf{t}(s + ds)$ with respect to $\mathbf{t}(s)$ is an invariant quantity. Therefore, denoting arc length and curvature of the unit circle by ς and χ, respectively, we can write

$$\varsigma_f - \varsigma_0 = \int_{\varsigma_0}^{\varsigma_f} \chi(\varsigma)\,d\varsigma = \int_{s_0}^{s_f} \kappa(s)\,ds = \int_{t_0}^{t_f} \kappa(t)V(t)\,dt$$

$$\leqslant \int_{t_0}^{t_f} |\kappa(t)|V(t)\,dt \tag{7.46}$$

since $\chi(\varsigma) = 1$. If the particle's trajectory is simply closed (and regular), then its image (i.e. the unit circle) is wholly covered, and once too, namely, $\varsigma_f - \varsigma_0 = 2\pi$.

C4: Furthermore, if the trajectory is also convex (with the usual meaning of set convexity), the last integral in (7.46) amounts to one turn, namely, $\mathbb{K} = 2\pi$. It can be shown in GTC that such relationship is a necessary and sufficient condition for a plane, regular and closed curve to be simple and convex.

280   7 The Theory of Fast Solar Sailing

Therefore, if integrating the product of curvature and speed over the whole solar-sail transfer duration results in satisfying inequality (7.43), then the trajectory is surely not-convex, not-closed, and not simply closed. Such first overall condition guarantees that *fast* sailing does not deal with any strictly periodic features.[5]

▶ Let us now suppose that a solar-sail trajectory shows off the following first set of additional equality/inequality properties of state and control:

$$\int_{t_0}^{\tau} \mathcal{L}_t \frac{\mu_\odot}{R} \, dt = -H_0 < 0, \qquad V(\tau) > 0, \qquad \varphi(\tau) = \pi, \qquad t_0 < \tau < t_f \quad (7.47a)$$

$$\text{constant } \mathcal{L}_t(t) < 0, \quad t \in [t_0, \tau] \tag{7.47b}$$

$$\text{piecewise-constant } \mathcal{L}_t(t) < 0, \quad t > \tau \tag{7.47c}$$

where, again, the subscript $_0$ denotes quantities at the start of flight, whereas the subscript $_f$ refers to as quantities at the considered end of flight. Qualitatively, the first equality relationship is equivalent to assert that if $-\mathcal{L}_t$ is sufficiently high, then the deceleration arc results in zeroing the initial angular momentum. Quantitatively, since the integrand in (7.47a) is just $\dot{H}(t)$, the equality part entails $H(\tau) = 0$. The second inequality of (7.47a) avoids non-regular arcs, whereas the angle equality implies that the sailcraft will go towards the Sun in either control region $\mathbb{F}$ or $\mathbb{P}$, as defined on p. 272. The last inequality of (7.47a) and condition (7.47b) tell us that sailcraft begins with a direct motion that continues up to $t = \tau$. Condition (7.47c) entails that, after τ, motion keeps on exhibiting $\dot{H}(t) < 0$; as a result, the H-function goes on decreasing strictly. Thus, the relationships (7.47a)–(7.47c) imply points (2–3a) of the Definition 7.1.

▶ Finally, let us assume that a sailcraft trajectory, already satisfying the conditions (7.43) and (7.47a)–(7.47c), exhibits the following second set of additional equality/inequality properties of state and control:

$$\dot{\mathcal{L}}_r(t) \geq 0, \quad \tau < t \tag{7.48a}$$

$$\int_{t_0}^{t} (\mathcal{L}_t \dot{\psi} + \dot{\mathcal{L}}_r) \frac{\mu_\odot}{R} \, dt + E_0 = 0, \quad \tau < t < t_f \tag{7.48b}$$

$$\dot{R}(t_p) = 0, \qquad \ddot{R}(t_p) > 0, \qquad \dot{\varphi}(t_p) > 0, \quad \tau < t_p < t_f \tag{7.48c}$$

Times t and t_p are unique. Inequality (7.48a) admits thrust maneuvers past the reversal point, but such that to *increase* sailcraft energy through the radial number. (Note, however, that such maneuver(s) must not violate condition (7.47c).) As the product $H\dot{H} > 0$ in the retrograde arc, the sailcraft energy time rate $\dot{E}$ from Eq. (7.17) is strictly positive. This means that conditions (7.47c) and (7.48a) guarantee point (3) of the definition of D2FS.

---

[5]The H-reversal concept may be utilized for heliocentric periodic orbits provided that the number of reversal points is even (see Sect. 8.5), and sufficiently large even under the (unavoidable) planetary perturbations. However, fast trajectories and periodic orbits would not share some properties, as expected.

7.2 Two-Dimensional Motion Reversal   281

The integrand of the integral (7.48b) is equal to $\dot{E}$, which is to be integrated from epoch to a generic time t after $H(\tau) = 0$. Therefore, the equality in (7.48b) entails $E(t) = 0$. Thus, relationship (7.48b) implies point (3) of Definition 7.1.

Equality and inequalities (7.48c) needs additional explanation. The following description items, labeled by sequential bold capital letter, are devoted to such conditions and what one could infer from it. A fact to be recalled again is that any solar-sail maneuver (including the impulsive one) is acceleration-limited, i.e. no discontinuity of the velocity vector, and in particular of the radial speed, could take place. (Even high-frequency changes of the total solar irradiance are power limited, of course). Therefore, $R(t)$ is always time *differentiable*, and with continuous derivative. Thus, theorems regarding (continuous and) differentiable real functions of one real variable apply.

A. Differentiating the general relationship $\dot{R} - V \cos\varphi = 0$ with respect to time produces

$$\ddot{R} - \dot{V} \cos\varphi + V \sin\varphi\dot{\varphi} = \ddot{R} - \dot{V} \cos\varphi + (H/R)\dot{\varphi} = 0 \qquad (7.49)$$

Let us suppose for a moment that $t_p = t_m$, where $R(t_m) \equiv R_m$ is *any* stationary point of the function $R(t)$. As the equality in (7.48c) means that $\cos\varphi_m = 0$, Eq. (7.49) calculated at t_m can be recast into the form

$$\{R\ddot{R} + H\dot{\varphi}\}_{t_m} = 0 \qquad (7.50)$$

B. Applying the inequalities in (7.48c) to Eq. (7.50) results in $\text{sign}(H_m) = -1$, namely, a local minimum of $R(t)$ occurs in the retrograde-motion arc; this, in turn, means $\varphi_m = 3\pi/2$. Now, suppose one likes to look for a local maximum of $R(t)$ taking place likewise in the retrograde arc: the only change in (7.48c) would be $\ddot{R}_m < 0$. However, Eq. (7.50) would entail $\text{sign}(H_m) = +1$, i.e. R_m would belong to the direct arc, in contrast to the hypothesis. In other words, *no* local maximum of $R(t)$ may occur in the reverse-motion trajectory arc.

C. The same way—via Eq. (7.50) again—one could easily prove that local minima of $R(t)$ are admitted in the direct-motion arc, namely,

$$\dot{R}(t_m) = 0, \qquad \ddot{R}(t_m) > 0, \qquad \dot{\varphi}(t_m) < 0 \qquad (7.51)$$

ICs may admit $\dot{R}_0 = 0$. Therefore, the defined fast sailing exhibits al least one local maximum of $R(t)$, which is found in the direct arc.

D. How many local minima in the retrograde arc? Suppose sailcraft enters a reverse-motion arc characterized by two distinct local minima of $R(t)$. Because $R(t)$ is always differentiable, a local maximum should be located between the two assumed local minima in the clockwise motion phase. However, there is no local maximum of $R(t)$ in the retrograde motion, as we saw in (B). Said differently, only *one* local minimum of the Sun-sailcraft distance can be found in this motion arc.

E. Because the continuity of $R(t)$ and $\dot{R}(t)$, the existence of both one local minimum and no local maximum in the retrograde arc proves that the point of local minimum is also the perihelion, at least throughout the duration of the reverse arc, and t_m can be identified with t_p. In particular, $\varphi_p = 3\pi/2$.

Thus, conditions (7.48c) implies point (3) of D2FS at $t = t_p$.

Such an asymmetry is not to be ascribed to the motion type, i.e. either counterclockwise or clockwise, but to the fact that a retrograde arc *follows* a decelerating direct arc. It is easy to see that, if the sailcraft started clockwise $(\text{sign}(H_0) = -1)$ and were subject to the reverse control, then the *direct* arc following the motion reversal point would exhibit the same asymmetry.

All in one, conditions (7.43), (7.47a)–(7.47c) and (7.48a)–(7.48c) imply D2FS, namely, they represent the *sufficient* conditions for the existence of fast-sailing trajectories with the required properties enumerated in the definition of fast sailing. One can note that such conditions contain two integral equalities with two control parameters that, though constrained to be constant or piecewise-constant along the trajectory, can be assigned in a large region of the quadrant $\mathcal{Q}$ previously defined. Therefore, the sets of two-point boundary value problems dealt with in Sect. 7.2.1 are special subsets of all admissible curves satisfying the above sufficient conditions; all the more, in solving those two-boundary problems numerically, we employed starting values (of the trajectory control) that were compliant with the sufficient conditions so to guarantee convergence to fast-sailing solutions.

7.2.2.2 The Speed Amplification Arc

In this sub-section, we will show why the adjective *fast* is appropriate for this type of solar sailing.

From Eq. (7.33b), it is immediate to carry out

$$\dot{V}_p = -\mathcal{L}_t \mu_\odot / R_p{}^2 > 0 \tag{7.52}$$

as well as applying Eq. (7.17) to perihelion results in

$$\dot{E}_p = V_p \dot{V}_p > 0 \tag{7.53}$$

Differentiating both sides of Eq. (7.33b) with respect to time and using the trajectory curvature $\kappa(t)$ from Eq. (7.33a), one gets

$$\ddot{V} = -2\frac{\dot{R}}{R}\dot{V} + \kappa\dot{\varphi}V^2 \quad \Rightarrow \quad \ddot{V}_p < 0 \tag{7.54}$$

after inserting $\dot{R}_p = 0$, $\dot{\varphi}_p > 0$ and $\kappa_p < 0$.

Note that no conic has perihelion properties like inequalities (7.52) and (7.53). Here, instead, inequalities are satisfied strictly, as it is confirmed by many, many numerical computations of sailcraft trajectories.

These results tell us that, at perihelion, sailcraft is still accelerating, (i.e. the perihelion precedes the point of maximum speed), but such acceleration is decreasing

7.2 Two-Dimensional Motion Reversal

(i.e. the perihelion is reasonably close to the point of maximum speed). This induces us to proceed with the analysis, begun in Sect. 7.2.2, of the local extrema of $V(t)$. Recalling inequalities (7.38) and (7.39), it an easy matter to show that the two point solutions to $\dot{V} = 0$, denoted $\varphi_{(1)}$ and $\varphi_{(2)}$ in Eqs. (7.36), correspond to local extrema of the sailcraft speed $V(t)$. In particular, since point-(2) follows the perihelion where $\dot{V}_p > 0$, $\ddot{V}_{(2)} = 0$ means that $V(t)$ has a local maximum at $t_{(2)}$ provided that no attitude maneuver is performed in $[t_p, t_{(2)}]$. The maximum $V_{(2)}$ is unique because periodic solutions of $\varphi_{(2)}$ cannot take place as fast-sailing trajectories are not closed, as shown above. Analogously, $V_{(1)}$ is a local minimum of speed, and $P_{(1)}$ lies on the direct arc, according to the inequality (7.39). (Attitude maneuvers in $[t_{(1)}, t_p]$ are not permitted by condition (7.47b).)

Time-differentiating equation (7.33b), inserting either Eq. (7.36), and again assuming no attitude maneuver at $t_{(1)}$ or $t_{(2)}$, one can carry out

$$\ddot{V}(t_{(1)}) = \left(\mu_\odot \frac{\tilde{\mathcal{L}}}{R^2}\dot{\varphi}\right)_{t_{(1)}}, \qquad \ddot{V}(t_{(2)}) = -\left(\mu_\odot \frac{\tilde{\mathcal{L}}}{R^2}\dot{\varphi}\right)_{t_{(2)}} \qquad (7.55)$$

where $\tilde{\mathcal{L}}$ was defined in (7.36). Therefore, $\dot{\varphi}_{(1)}$ and $\dot{\varphi}_{(2)}$ are positive both.

Now, we are interested in the sailcraft speed increase that accumulates during $[t_{(1)}, t_{(2)}]$. Let us integrate both sides of Eq. (7.33b):

$$\mathcal{V} \equiv \frac{(V_{max} - V_{min})}{\mu_\odot} = -(1 - \mathcal{L}_r)\int_{t_{min}}^{t_{max}} \frac{\cos\varphi}{R^2}\,dt + \mathcal{L}_t\int_{t_{min}}^{t_{max}} \frac{\sin\varphi}{R^2}\,dt \equiv \mathcal{V}_r + \mathcal{V}_t$$

$$(7.56)$$

where we changed the time subscripts as a result of the related established properties. For simplicity of analysis, we assumed once again that **L** is kept constant throughout the integration interval. However, Eq. (7.56) is sufficiently general to carry out one of the most significant properties of D2FS.

Let us divide the interval from t_{min} to t_{max} in three subintervals:

$$\mathcal{T}^{(1)} \equiv [t_{min}, t^*], \qquad \mathcal{T}^{(2)} \equiv (t^*, t_p], \qquad \mathcal{T}^{(3)} \equiv (t_p, t_{max}]$$

$$\mathcal{V} = \sum_{k=1}^{3}(\mathcal{V}_r^{(k)} + \mathcal{V}_t^{(k)}) \equiv \sum_{k=1}^{3} \mathcal{V}^{(k)} \qquad (7.57)$$

During $\mathcal{T}^{(1)}$, $\cos\varphi(t)$ is negative (and increases in magnitude); therefore, the first integral, $\mathcal{V}_r^{(1)}$, is positive. On the other side, $\sin\varphi(t)$ is positive (but small); thus, $\mathcal{V}_t^{(1)}$ is negative. At P*, the sailcraft is accelerating, even though energy is decreasing in $\mathcal{T}^{(1)}$. If, for some reason, the solar-sail thrusting mode were halted (e.g. by jettisoning the sail) at t^*, then the spacecraft would point directly to the Sun on a rectilinear trajectory. However, Definition 7.1 implies unambiguously that the flight shall proceed via a safe perihelion. During the interval $\mathcal{T}^{(2)}$, not only $\mathcal{V}_r^{(2)}$ and $\mathcal{V}_t^{(2)}$ are positive both, but also considerably larger than the previous one because $1/R^2$ is

Table 7.1 How the speed increase is distributed over the time interval from the minimum to the maximum of speed obtainable under a constant **L**. $\mathcal{V}_r$ and $\mathcal{V}_t$ are defined in Eqs. (7.56)–(7.57)

intervals	$\mathcal{T}^{(1)} \equiv [t_{min}, t^*]$	$\mathcal{T}^{(2)} \equiv (t^*, t_p]$	$\mathcal{T}^{(3)} \equiv (t_p, t_{max}]$
$\mathcal{V}_r^{(k)}$	+	+	−
$\mathcal{V}_t^{(k)}$	−	+	+
$\mathcal{V}^{(k)}$	$\mathcal{V}^{(1)} > 0$	$\mathcal{V}^{(2)} > \mathcal{V}^{(1)}$	$\mathcal{V}^{(3)} < \mathcal{V}^{(2)}$

notably higher. After the perihelion, since $\dot{V}(t_p) > 0$ and $\varphi_p = 3\pi/2$, speed continues to increase, but its increment is smaller because $\cos\varphi > 0$ and $\sin\varphi < 0$; thus, $\mathcal{V}_t^{(3)}$ keeps positive, but $\mathcal{V}_r^{(3)}$ becomes negative. Both contributions are small in absolute value as the sailcraft is rapidly receding from the Sun. Table 7.1 summarizes the properties just described. Extensive numerical analysis shows that the second inequality in Table 7.1 is satisfied typically over a factor 10.

How is the behavior of the speed after its maximum at $t = t_{max}$? From the previous analysis, we known that trajectory is not closed, and the sailcraft will recede from the Sun. No particular item would be noted if we had a trajectory very similar to a parabola or a hyperbola; for conics, one knows that the along-track acceleration is equal to $-\cos\varphi\mu_\odot/R^2$ and the derivatives of energy and angular momentum are identically zero. For the current receding sailcraft, the energy rate is positive, but decreases as $1/R^2$; the derivative of the H-function scales as $1/R$, and (differently from a Keplerian orbit) the derivative of V is proportional to $(\sin\varphi\mathcal{L}_t - \cos\varphi(1 - \mathcal{L}_r))$: for large R, the angle φ approaches 2π. Therefore, with respect to the gravity-only case, here the sailcraft has a waning acceleration plus a reduced deceleration. Therefore, the speed decreasing is slow, and the sailcraft moves under a quasi-cruise regime. This is particular important for missions to the outer bodies of the solar system and beyond, for which the flight time is known to be a key parameter. In [12], approximate formulas for post-perihelion quantities were derived. Here, we like to describe a simpler approach characterizing the cruise arc.

After the local maximum of speed, sailcraft continues to fast recede from the Sun with the H-invariant decreasing swiftly as well. We picture the quasi-cruising beginning at the instant t_c such that

$$H(t_c) = \zeta H(t_p), \quad \zeta > 1 \tag{7.58}$$

and the sailcraft proceeds with a constant speed given by

$$V_{\text{cruise}} = \sqrt{2E(t_c)} \tag{7.59}$$

Since $H(t)$ is strictly decreasing, then (1) it is always possible to assign a number $\zeta > 1$ such that a positive number $t_c > t_p$ satisfies Eq. (7.58), and (2) the function $t_c(\zeta)$ is injective. Since $E(t)$ is strictly increasing in the reversal arc (as already pointed out), if ζ is sufficiently greater than unity then the quantity defined by

7.2 Two-Dimensional Motion Reversal

Eq. (7.59) is sufficiently close to the actual sailcraft speed at t_c and to the terminal speed; in quantitative terms, one has

$$\left|1 - V_{\text{cruise}}/V(t_c)\right| < \varepsilon, \qquad \left|1 - V_{\text{cruise}}/V(t_f)\right| < \varepsilon \qquad (7.60)$$

where ε is a given small positive number. Thus, from t_c to t_f, the sailcraft can be considered as moving rectilinearly with the constant speed V_{cruise}. This may be summarized via the following proposition:

Proposition 7.3 *A fast-sailing trajectory, which satisfies the sufficient conditions given by the equalities and inequalities in (7.43), (7.47a)–(7.47c), and (7.48a)– (7.48c), exhibits a quasi-rectilinear arc with a cruising regime that is eventually achieved after the perihelion; the cruise speed is defined via Eqs. (7.58)–(7.59) and inequalities (7.60).*

Such statement may not used to evaluate the cruise speed without integrating the motion equations, of course. Numerical analysis shows that $\zeta \sim 2$ and $\varepsilon \sim 0.005$ satisfy the above-mentioned equalities and inequalities.

A way to measure the importance of the post-perihelion sailing in terms of speed is to compute the difference

$$\Delta V_{\text{gain}} = V_{\text{cruise}} - \sqrt{2E(t_p)} \qquad (7.61)$$

which measures the increment of speed that one gets with respect to the option of jettisoning the sail at the perihelion. We will to expand this point in the subsection below.

7.2.2.3 Two Alternatives for Comparison

This subsection is devoted to clarify the *H*-reversal mode further through two different strategies for achieving high escape speeds.

One may note that the pre-perihelion deceleration-acceleration profile could be considered as a sort of analogy between solar-sailing and the well-known two-impulse rocket strategy for getting a high-speed hyperbola, and escaping from the solar system. In addition to the enormous advantage of using no propellant, one may wonder whether the post-perihelion acceleration may be of use. In other words, why not jettisoning the sail and continue the flight on a hyperbola? Let us consider two scenarios, where the superscripts (e), (p), (h), (r), or (s) stands for elliptic, parabolic, hyperbolic, rectilinear, or sailing trajectory, respectively.

▶ 1 Let assume an ideal restartable rocket orbiting the Sun on a circular orbit of radius R_0. A first tangential impulse opposite to the velocity at point A causes the vehicle to follow an inner elliptical path to reach the perihelion P at distance R_p, where a second tangential impulse makes the transition from this ellipse to a co-tangential hyperbolic orbit. It is straightforward to calculate the two impulsive delta-V and the hyperbolic excess:

$$\Delta V_1 = \sqrt{\frac{\mu_\odot}{R_0}}\left(1 - \sqrt{\frac{2R_p}{R_0 + R_p}}\right)$$

$$\Delta V_2 = V_p{}^{(h)} - V_p{}^{(e)} = \left(V_p{}^{(p)} + \delta V\right) - V_p{}^{(e)}$$

$$= \sqrt{2\frac{\mu_\odot}{R_p}}\left(1 - \sqrt{\frac{R_0}{R_0 + R_p}}\right) + \delta V \qquad (7.62)$$

$$V_\infty{}^{(h)} = \sqrt{(\delta V)^2 + 2\sqrt{2\frac{\mu_\odot}{R_p}}\,\delta V}$$

where $V_p{}^{(p)}$ denotes the parabolic speed at R_p, and δV is the overplus of impulse to get a hyperbolic orbit; if $\delta V = 0$, then one obtains the co-tangential parabolic orbit. If one assigns $V_\infty{}^{(h)}$, then the ensuing overplus is given by

$$\delta V = \sqrt{\left(V_p{}^{(p)}\right)^2 + \left(V_\infty{}^{(h)}\right)^2} - V_p{}^{(p)}, \quad V_p{}^{(p)} = \sqrt{2\frac{\mu_\odot}{R_p}} \qquad (7.63)$$

▶ 2 There is another interesting alternative, namely, using 1-burn rocket and sail in the following way: the rocket develops one burn as just described above, then it is jettisoned during the semi-ellipse path; when the perihelion is achieved, the sail—with a performance of $\mathcal{L} = 1$ is quickly deployed and kept radially (i.e. $\mathcal{L}_r = 1, \mathcal{L}_t = 0$) for the rest of the flight. The ensuing trajectory is then a straight line with a constant speed, which is obviously equal to

$$V_\infty{}^{(s)} = V_{cruise}{}^{(s)} = V_\infty{}^{(r)} = V_p{}^{(e)} = \sqrt{2\frac{\mu_\odot}{R_p}\frac{R_0}{R_0 + R_p}} \qquad (7.64)$$

These two conceptual (reference) profiles are sketched in Fig. 7.6, where A is the starting point on the circular orbit, where to apply the antiparallel (impulsive) burn, and P is the target perihelion. (Of course, in these two flight concepts the post-perihelion arcs have different directions at infinity; in order to get a prefixed direction, $\psi(A)$ has to be chosen accordingly.) Some considerations are in order:

i. in the rocket/sail option, the quantity in Eq. (7.64) may be considered a *sailing excess*, namely, what could be gained with respect to the parabolic escape by means of a sailcraft balancing the solar gravity exactly.

ii. Such a speed excess is considerably high with respect to the current and near-term rocket performances; it suffices to give the parameters of Eqs. (7.62) and (7.64) some realistic values for an escaping mission, e.g. $R_0 = 1$ AU and $R_p = 0.2$ AU, to get $\Delta V_1 \cong 2.656$ AU/year, $(\Delta V_2)_{\delta V = 0} \cong 1.731$ AU/year,

7.3 Two-Dimensional Direct Motion

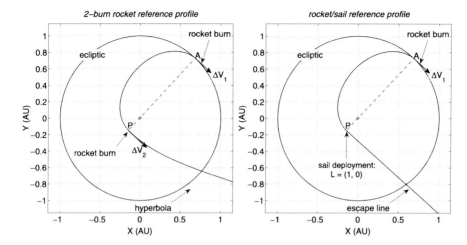

Fig. 7.6 The H-reversal mode can be compared to (1) two-burn restartable ideal rocket for achieving a hyperbola for rapidly escaping from the solar system, (2) one-burn-rocket/sail for a rectilinear escape path. Equations (7.62) and (7.64) summarize the quantities of comparison. Adapted and enlarged from Fig. 17.2 of [16], courtesy of Springer

$V_\infty^{(p)} = 0$, and $V_\infty^{(s)} \cong 18.14$ AU/year. Finally, one should apply an additional $\delta V \cong 7.034$ AU/year for achieving a hyperbolic excess equal to the sailing one. Thus, in this case, the sailing avoids applying the second rocket maneuver amounting to $\Delta V_2 \cong 8.765$ AU/year, or about 41.6 km/s.

iii. These quantitative points answer the question set at the beginning of this subsection. The sailing-mode post-perihelion acceleration is essential to get high cruise speeds, whereas the rocket mode in the rocket/sail scenario can be efficiently substituted by a pre-perihelion sailing.

iv. The reference rocket flights are ideal, namely, no gravity losses occur. In practice, considering the very large amount of total ΔV, maneuvers would be implemented via long thrusting times with further penalization in terms of propellant (as we discussed about in Chap. 1); that makes escaping much larger in mass and considerably longer in flight time, and thus unpractical. In other words, the low-acceleration pre-perihelion and post-perihelion arcs of the H-reversal mode could be considered equivalent to a set of very expensive rocket impulses, and with no mass expenditure.

7.3 Two-Dimensional Direct Motion

In addition to the reversal mode, sailcraft's *full* direct motion is another solution for obtaining very high solar-sailing excess; however, and perhaps strange enough, this mode requires a particular thrust maneuver. Direct-motion may be as much useful as the reversal mode, *provided* that this thrust maneuver (which is not required in the reversal mode) is not accomplished too early. Let us see why.

288   7 The Theory of Fast Solar Sailing

The definition of 2D fast sailing (on p. 277), its related sufficient conditions, and its properties expressed by the above-proved propositions can be viewed as the simplest ones for getting fast sailing. Nevertheless, they do not exclude attitude maneuvers that could change the thrust vector, or better, the lightness vector (that is defined in EHOF). The set, say, $\mathfrak{M}$ of maneuvers contains admissible elements that do not preclude the sailcraft from achieving a reversal point.

Suppose to have obtained a particular pair of $\{\mathcal{L}_r, \mathcal{L}_t\}$ satisfying the D2FS sufficient conditions, for instance by utilizing Fig. 7.5 or solving one of the two-point boundary-conditions problems (7.31), or (7.32). In any case, one has proceeded to integrate numerically the system of equations (7.27) in either its Cartesian form (7.26) or its augmented polar form (7.29). By using a high-precision interpolating function (often representing the numeric solutions to Ordinary Differential Equations (ODEs) in a modern CAS), one may then compute t_{min} and t^*, in particular. At this point, one may asking for what would happen if a thrust maneuver, say, $m \notin \mathfrak{M}$ were inserted sufficiently close to t^*. More precisely, one may assume to accomplish the following impulsive **L**-maneuver:

$$\mathcal{L}_r(t_d^+) = \mathcal{L}_r(t_d^-), \qquad \mathcal{L}_t(t_d^+) = -\mathcal{L}_t(t_d^-), \quad t_{min} < t_d < t^*$$
$$\dot{\mathcal{L}}_r(t) = 0, \quad t \geqslant t_d, \qquad \dot{\mathcal{L}}_t(t) = 0, \quad t > t_d \tag{7.65}$$

where $\mathcal{L}_t(t_d^-) < 0$ is the value of the transversal number before the maneuver,[6] t_{min} is the time of the local minimum of $V(t)$, and t^* is the instant that *would* eventually be achieved if no maneuver were performed. This change of transversal number violates the equality condition $\text{sign}(\mathcal{L}_t) = -1$ on the control parameters in (7.37). Therefore, we have to modify our previous reasoning about the reversal motion arc in order to carry out some of the general properties of the post-maneuver trajectory.

As a result of the assumed control law in (7.65), the left and right derivatives of H at t_d do not match, and $\dot{H}(t) > 0$ for any $t > t_d$. Hence, as $t_d < t^*$, the continuous H-function (which was decreasing strictly before t_d) will never vanish, and $H(t)$ will increase strictly after t_d. This means that, though $H(t)$ is not differentiable at t_d, $H(t_d)$ is a local minimum for it. Motion will continue to be counterclockwise after the maneuver, and one can write consequently

$$H(t) = R(t)V(t)\sin\varphi(t) > 0, \quad t_d < t$$
$$H(t_1) < H(t_2) \ \forall t_d < t_1 < t_2 \tag{7.66}$$
$$\dot{H}(t) < 0, \quad t < t_d, \qquad \dot{H}(t) > 0, \quad t > t_d$$

As the maneuver is carried into effect between t_{min} and t^*, $\cos\varphi(t_d) < 0$, which entails that the radial speed $\dot{R}_d = V_d\cos\varphi(t_d)$ is negative, namely, the sailcraft continues to decrease its distance from the Sun. This item and the inequalities (7.66) mean that the perihelion is not zero and lies in the post-maneuver arc.

---

[6]Using a flat-sail model for visualization, the azimuth of the sail axis changes from α_d to $-\alpha_d$, then it is kept unchanged.

7.3 Two-Dimensional Direct Motion

Control law (7.65) and the just mentioned properties of $H(t)$, applied to Eqs. (7.16)–(7.17), result easily in

$$0 > E(t_d^-) = E(t_d^+)$$

$$\dot{E}(t_d^-) < 0, \qquad \dot{E}(t_d^+) > 0 \qquad (7.67)$$

$$\dot{E}(t) < 0, \quad t < t_d, \qquad \dot{E}(t) > 0, \quad t > t_d$$

Thus, even sailcraft's orbital energy is continuous, but not differentiable at t_d. However, we know that $E(t)$ was negative and strictly decreasing for t sufficiently close to t^*, whereas after the maneuver it results negative, but strictly increasing because of the third inequality in (7.67). Thus, $E(t)$ has a local minimum at t_d.

The time rate of curvature and speed at t_d are also meaningful; from Eqs. (7.33a)–(7.33b), it is straightforward to obtain

$$\kappa_d^+ - \kappa_d^- = \left(\mu_\odot / R_d^2 V_d^2\right) \cos \varphi_d \left(\mathcal{L}_t^+ - \mathcal{L}_t^-\right)$$

$$= \left(\mu_\odot / R_d^2 V_d^2\right) \cos \varphi_d \left(-2\mathcal{L}_t^-\right) < 0 \qquad (7.68a)$$

$$\dot{V}_d^+ - \dot{V}_d^- = \left(\mu_\odot / R_d^2\right) \sin \varphi_d \left(\mathcal{L}_t^+ - \mathcal{L}_t^-\right) = \left(\mu_\odot / R_d^2\right) \sin \varphi_d \left(-2\mathcal{L}_t^-\right) > 0$$

$$\Rightarrow \dot{V}_d^+ > \dot{V}_d^- > 0 \qquad (7.68b)$$

where the properties just before the maneuver are already known, of course. The above equalities/inequalities show quantitatively that trajectory curvature and speed rate both exhibit jump discontinuities at the maneuver time.

Relationships (7.66), (7.67), (7.68a) and (7.68b) represent the main properties that it is possible to infer easily from the application of a maneuver of type (7.65) that alters the original reversal-mode evolution. Other properties should not be known analytically in straightforward way, simply because in the statement of the L-maneuver there is a basic degree of freedom, i.e. the application time, though confined in the interval (t_{min}, t^*); the specific trajectory properties are expected to depend peculiarly on the chosen t_d. Thus, we have to switch to the numerical analysis.

Figure 7.7 shows the profiles of five cases; one H-reversal trajectory and four completely direct trajectories, according to the control (7.65), with four different maneuver times. The reference profile is the H-reversal trajectory with R_p and t_p given both. We chose $R_p = 1/5$ AU achievable after 248 days, whereupon the control $\mathbf{L} \cong (0.493628, -0.330663)$ has been obtained numerically, and with a reversal time $t^* - t_0 \cong 197.743$ days as reported in Table 7.2, which contains the main data of the five trajectories. The four full direct-motion cases stem from applying the L-maneuver at $t_d = t_{min} + (t^* - t_{min})\iota$ with $\iota = 0.99, 0.75, 0.50, 0.25$, chosen arbitrarily; the corresponding time intervals in day units are reported in the table. Maneuver times with respect to three different instants are reported for a better characterization of them.

The top-left panel of Fig. 7.7 shows the trajectory arc inside a square with side 4 AU centered on the origin of EHOF. At each maneuver point, one may recognize

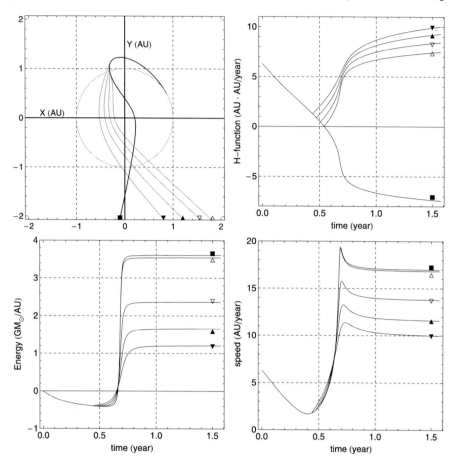

Fig. 7.7 Two-dimensional direct motion according to the control law (7.65). The H-reversal trajectory is taken as the reference flight profile. These plots should be read with Table 7.2 where the chosen markers are explained, and with the related text. *Top-left*: trajectories from the common initial conditions to the rectilinear arcs (which go on indefinitely); *top-right*: evolution of the H-function; *bottom-left*: evolution of the orbital energy; *bottom-right*: evolution of the sailcraft speed. The last three sets of plots are truncated at $t \lesssim 1.6$ years, whereas trajectories have been integrated up to 1240 days

a change of the trajectory curvature, quantified by Eq. (7.68a), which swerves the trajectory to flyby the Sun in the quadrant opposite to that of the H-reversal arc. Perihelion distances and pass times, as well as other quantities as function of the interval $t_d - t_{min} \equiv \bar{t}_d$, are written in Table 7.2. As $\bar{t}_d$ decreases (i.e. reading from left to right in the table), one can see that (1) perihelion pass time increases, (2) the perihelion's polar angle ψ_p decreases (with the obvious exception of the reversal mode), and (3) the trajectory arc from t_d to t_p increases. The overall feature is that, as t_d approaches t^*, the direct-motion arc for $t > t_d$ draws close the reverse-motion profile symmetrically; in other words, considering the line, say, $\mathbb{X}$ passing through

7.3 Two-Dimensional Direct Motion

Table 7.2 Examples of direct-motion trajectories coming from control (7.65) applied to a reference H-reversal motion. The columns containing the various markers show the main quantities related to the curves plotted in Fig. 7.7 and marked the same way. This table is partitioned: the first part specifies the motion mode and the initial conditions; the second one reports the related controls; the third one contains the time intervals related to each maneuver; perihelion properties are written in the fourth section; finally, the fifth one reports some quantities at the common final time, arbitrarily set to five times the perihelion pass time of the reversal mode. See text for the meaning of b

marker		■	△	▽	▲	▼
mode		reverse	direct	direct	direct	direct
ICs		1-AU circ. orbit $\psi_0 = 30°$	idem	idem	idem	idem
$\mathcal{L}_r$		0.493628	0.493628	0.493628	0.493628	0.493628
$\mathcal{L}_t$		−0.330663	0.330663	0.330663	0.330663	0.330663
$t^* - t_0$	day	197.743				
$t^* - t_{min} \equiv \Delta t$	day	50.788				
$t_d - t_{min}$			$0.99\Delta t$	$0.75\Delta t$	$0.50\Delta t$	$0.25\Delta t$
$t_d - t^*$	day		−0.508	−12.697	−25.394	−38.091
$t_d - t_0$	day		197.235	185.046	172.349	159.652
$t_p - t_0$	day	248	248.026	248.704	249.344	249.532
R_p	AU	1/5	0.203	0.292	0.397	0.514
ψ_p	deg	17.424	198.906	197.372	194.245	189.812
$\int_{t_d}^{t_p} V\,dt$	AU	0.955	0.961	1.070	1.165	1.242
$t_f - t_0$	day	1240	1240	1240	1240	1240
b	deg	0	62.537	61.873	59.772	56.519
R_f	AU	46.080	45.652	37.359	31.342	26.839
V_f	AU/yr	16.908	16.750	13.698	11.486	9.832

the reversal point and the Sun's barycenter, the direct arc obtained by $t_d \to t^*$ and the reverse arc are closely symmetric with respect to $\mathbb{X}$. The meaning of such symmetry is clearer by analyzing the other plots in Fig. 7.7.

The top-right plot of Fig. 7.7 shows how the invariant H evolves. In each of the direct-motion cases, its shape looks like that of the square-root sign, and follows clearly the properties expressed by the inequalities (7.66). One can see easily that the increasing $H(t)$ of the case $t_d = t_{min} + 0.99(t^* - t_{min})$ (i.e. that marked by $\triangle$) is almost symmetric of the reverse-motion $H(t) < 0$ arc with respect to the time axis. As $t_d \to t^*$, the ideal symmetry is approached. Note that the other cases exhibit higher H-values, confirming that sailcraft "keeps out" of the Sun.

The bottom-left plot of Fig. 7.7 shows the energy behaviors in the various cases. Aside from the discontinuity at the maneuver times, the curves are quite similar in shape. In particular, direct-motion flights exhibit cruise arcs, but the energy level depends remarkably on when the maneuver is accomplished. Speed profiles, shown in the bottom-right of Fig. 7.7, exhibit each a maximum value after which a cruise arc appears. Like energy, the (pseudo)-cruise speed is higher as $t^* - t_d$ is smaller, and approaches closer and closer the reverse-motion rectilinear arc.

Finally, $\mathfrak{b}$ is the angle between the direct-motion rectilinear arc and the rectilinear arc of the reversal-motion profile. The positive sign means counterclockwise from the reversal-motion direction. What is the practical importance of such quantity? Let us clarify just using the examples marked by $\blacksquare$ and $\triangle$, for which $\mathfrak{b} \cong 62.5°$. Suppose that the flight start is shifted to $\psi_0 - \mathfrak{b} = -32.5°$ (or 63.4 days in advance), and the maneuver (7.65) is applied again at $t_d = t^* - 0.508$ days. By the rotational symmetry of the 2D sailcraft motion equations, it is straightforward to recognize that the new $\mathfrak{b}$, say, $\hat{\mathfrak{b}}$ is equal to zero. Put equivalently, we have *two* launch opportunities in a year, every year. In the launch scenario, the reversal-motion launch date should follow chronologically the direct-motion option; otherwise, the second opportunity would delay by 301.8 days, namely, the sail should be kept folded for 10 months: inadvisable situation indeed. Of course, this scenario assumes that the sailcraft is explicitly designed for including that maneuver.

Remark 7.3 Among the set of trajectories analyzed so far in this chapter, the *H*-reversal solution is characterized by the highest cruise speed. Nevertheless, the full direct-motion option may double the launch opportunities. One should be careful in comparing full direct-motion and direct-reverse motion trajectories. As a point of fact, sometimes in literature, time-varying control minimum-flight-time direct-motion trajectories have been compared with constant-control reverse-motion trajectories. In general, an unconstrained time-varying optimal control will perform better than a fixed control, as it is well known. The problem of which is the best one (with respect to some mission-dependent index of performance) should not be an ill-defined problem.

This book has been focusing on the motion-reversal set of trajectories not for conflicting with the direct-motion set, but for a number of many other reasons that will appear clearer as we proceed. One of these ones is the enlargement of the set of solutions to the SPS motion equations. A critical summary can be found in Sect. 8.6, put just at the end of Chap. 8.

7.4 Three-Dimensional Motion Reversal

In this section, we will face with the problem of getting a set of motion-reversal three-dimensional trajectories, if any, by employing particular $\mathbf{L}(t)$ control. For the moment, we are not interested in optimizing the lightness vector; rather, we aim at proving that such set of trajectory exists, and which are its main mathematical and physical properties. How to optimize some index of performance, e.g. maximizing the cruise speed or the terminal speed at some prefixed time, will be the topic of Chap. 8.

7.4 Three-Dimensional Motion Reversal

7.4.1 Meaning of 3D Motion Reversal

In two dimensions, the concept of motion reversal is plain enough, and the use of the orbital angular momentum is clear. In three dimensions, we have to define what we mean by reversal of sailcraft motion because it is not straightforward. As a point of fact, we remind the reader that, in general, two fields act on the space vehicle under SPS other than gravity; as a result, the shape and the properties of the various families of SPS trajectories can vary greatly.

In general, we have two options for characterizing the sailcraft angular momentum from the orbit viewpoint. The following definitions apply:

$$\mathbf{H}^{(int)} = \kappa^{-1}(-\mathbf{e}_2) \times \mathbf{V} = V^4 \frac{\mathbf{V} \times \dot{\mathbf{V}}}{\|\mathbf{V} \times \dot{\mathbf{V}}\|^2} \quad \text{(intrinsic)} \tag{7.69a}$$

$$\mathbf{H} = \mathbf{R} \times \mathbf{V} \tag{7.69b}$$

where $\mathbf{e}_2$ is the principal normal vector defined via equation (5.41b). Thus, the two definitions differ only by the radius to be cross-multiplied by $\mathbf{V}$. If C is the center of curvature related to a given point P on the trajectory, the radius considered for the intrinsic angular momentum is the vector $\overrightarrow{\text{CP}}$. Point C is not fixed (in general), but moves on the trajectory's evolute. The second, usual, definition of angular momentum is based on the HIF origin, i.e. $\mathbf{R} = \overrightarrow{\text{OP}}$, of course.

The intrinsic $\mathbf{H}^{(int)}$ gives intuitive results for the Sun-centered non-Keplerian circular orbit [5], the orbital plane of which may be orthogonal to some axis, e.g. the X-axis, but with its center displaced by a prefixed quantity from the origin of HIF. In this case, the usual $\mathbf{H}$ says that sailcraft changes motion type twice per orbit. On the contrary, $\mathbf{H}^{(int)} \neq 0$ since $\mathbf{V} \times \dot{\mathbf{V}} \neq 0$ at every point of non-Keplerian circular orbits, as one could check easily. However, our focus is on SPS trajectories very different from circular orbits.

We go on using of the standard concept of orbital angular momentum (per unit mass), always referred to the origin of the inertial frame. The following definition is like to what has been done for two-dimensional trajectories:

Definition 7.2 If the third component of $\mathbf{H}(t)$, relatively to a *regular* sailcraft trajectory, passes through zero in strictly monotonic way in an open neighborhood of t^*, then we say that motion *reverses* at t^*.

7.4.2 Numerical Issues

In order to understand the problem, let us first employ equations (7.29) to propagate the state $\mathfrak{I}_0 = (1, 0, 1, 0)$ via the control $\mathcal{L}_r = 0.5110$, $\mathcal{L}_t = -0.3001$. One gets a simple, strictly planar, trajectory with a reversal time $t^* = 3.935966$, and with a

294   7 The Theory of Fast Solar Sailing

Table 7.3 Inability of a *constant* lightness vector to produce a reversal 3D trajectory starting from a well-established 2D H-reversal trajectory. The overall behavior of the terminal state as the integration accuracy/precision level changes does not depend on the particular $\mathcal{L}_n = 0.0001$ here employed. See text for explanation

input values			
$\mathbf{R}_0$	1	0	0
$\mathbf{V}_0$	0	1	0
$\mathbf{L}$	0.511	-0.3001	0.0001
$t_f - t_0$	$1.5 \cdot 2\pi$		
t^* (2D ref. traj.)	$0.62642844 \cdot 2\pi$		
2D reference state with $\mathcal{L}_n = 0$			
$\mathbf{R}_f$	-7.342682	-9.179460	0
$\mathbf{V}_f$	-1.688921	-1.960965	0
3D output values			
digits of accuracy, precision	$7, 7$		
$\mathbf{R}_f$	-7.327945	-9.181274	0.429476
$\mathbf{V}_f$	-1.685490	-1.961384	0.099963
digits of accuracy, precision	$9, 9$		
$\mathbf{R}_f$	-7.342672	-9.179459	-0.013619
$\mathbf{V}_f$	-1.688919	-1.960965	-0.003164
digits of accuracy, precision	$10, 10$		
$\mathbf{R}_f$	-7.342659	-9.179466	0.015296
$\mathbf{V}_f$	-1.688916	-1.960966	0.003566
digits of accuracy, precision	$11, 11$		
$\mathbf{R}_f$	-7.342677	-9.179459	-0.010249
$\mathbf{V}_f$	-1.688920	-1.960965	-0.002379

perihelion very close to 0.2 reached at $t_p = 4.91502$, all in solar units again. After 1.5 years (or $1.5 \cdot 2\pi$), sailcraft velocity takes on $(-1.688921, -1.960965)$. Now, one may think of using this profile as a reference trajectory, and of giving $\mathcal{L}_n$ a small value, say 0.0001, for obtaining a slightly-perturbed three-dimensional trajectory similar to the H-reversal reference one. In order to discover what happens, we have to switch to the system of Eqs. (7.15), and adding $\dot{\mathbf{L}} = 0$ throughout $[t_0, t_f]$.

Table 7.3 shows the results of the simple attempt of obtaining a motion-reversal 3D trajectory by employing the just mentioned 2D trajectory as the starting profile.

Various runs, differing only in the accuracy/precision levels, have been accomplished. Seemingly, the corresponding trajectories are of motion-reversal type; however, the terminal states (especially the z-components) reported in Table 7.3 change well beyond the requested accuracies. In other words, such profiles vary consider-

7.4 Three-Dimensional Motion Reversal

ably with the number of effective steps employed by the high-precision integrators! This overall behavior can be met again starting from different 2D reference profiles, and also trying other values of the lightness vector.

A question arises spontaneously: is-it a numerical effect or is-there a physical reason in getting these "random" results. In the described problem, we were forcing the orbital momentum to pass through zero by a sufficiently high lightness number; as a point of fact, the same $(0.511, -0.3001)$ control was successful for obtaining $\mathbf{H} = 0$ in the 2D reference trajectory. However, by Lemma 7.1, this goal is not possible inasmuch as $\mathcal{L}_n \neq 0$ throughout the flight. From the viewpoint of the numerical integration of the motion equations, if $t^\diamond$ is the time where one assumes $H = 0$ in a 3D trajectory, then one may define the function $\Delta \bar{\mathbf{h}}_n \equiv \bar{\mathbf{h}}(t_n) - \bar{\mathbf{h}}(t_{n+1}), t_n < t^\diamond < t_{n+1}$, where either n or $n + 1$ is the step number closest to $t^\diamond$; denoting the precision/accuracy goal by γ, it is plain that n can change with it, in general, i.e. $n = n(\gamma)$. One may then note that $\Delta \bar{\mathbf{h}}_n$ can vary notably with $n(\gamma)$ on the very steep sides of $\bar{\mathbf{h}}(t)$, which diverges as either t_n or t_{n+1} approaches $t^\diamond$. Loosely speaking, after the integrator "jumps across" $t^\diamond$, the sailcraft state $(\mathbf{R}_{n+1}, \mathbf{V}_{n+1})$ can be so remarkably different from $(\mathbf{R}_n, \mathbf{V}_n)$ to cause the terminal states at $t_f > t^\diamond$ changing non-smoothly with γ. This is a conceptual difficulty brought about by the evolution of $\bar{\mathbf{h}}(t)$, independently of the integration methods.

This inability of a constant $\mathbf{L}$ to build a smooth trajectory, having $\mathbf{H}(t^\diamond) = 0$ strictly, induces some considerations. One may think qualitatively of using a three-dimensional $\mathbf{L}_1$ for a certain time interval, then switching to $\mathbf{L}_2$ with $\mathcal{L}_n = 0$ for the subsequent interval where the angular momentum is zeroed, then applying another three-dimensional $\mathbf{L}_3$ until the required terminal sailcraft state is achieved. Although conceptually correct, this one (or a similar strategy) entails that the third component of the control thrust has to be exactly zero for a finite time, otherwise any non-zero value of $\mathcal{L}_n$ (no matter how small may be) would cause the divergence of $\dot{\bar{\mathbf{h}}}(t)$, according to the limits in (7.11). In the wake of what discussed in Chap. 6, this $\mathcal{L}_n$-vanishing strategy should not be viable.

In the next subsection, we will analyze a different control policy in order to obtain realistic 3D motion-reversal trajectories.

7.4.3 Motion-Reversal Three-Dimensional Trajectories

This subsection aims at proving that, in $\mathfrak{R}^3$, there exists a set, the elements of which are three-dimensional <u>sailcraft</u> trajectories where for each, and at least in one point, the motion reverses. We first focus on the existence of such set theoretically, then we shall analyze the overall features numerically.

⊙ Note 7.3 on p. 261 mentioned that, in 3D-space, a negative H does not entail a motion reversal necessarily. Let us make things more precise.

Theorem 7.2 *Let the curve $\mathfrak{S}$ be the projection of a regular 3D trajectory $\mathcal{S}$ onto the reference plane of HIF. Let us denote the projected vector position and velocity by $\tilde{\mathbf{R}}(t)$ and $\tilde{\mathbf{V}}(t)$, respectively, and the HIF's basis by $(\mathbf{i}, \mathbf{j}, \mathbf{k})$;*

296   7 The Theory of Fast Solar Sailing

if:

(a) $\mathfrak{S}$ *is regular;*
(b) *there is at least one point* $\tilde{P}^* \equiv \tilde{P}(t^*) \in \mathfrak{S}, t_0 < t^* < t_f$, *where the product* $\tilde{\mathbf{R}}(t^*) \times \tilde{\mathbf{V}}(t^*)$ *vanishes;*
(c) *the continuous function* $\tilde{\mathbf{H}}(t) \cdot \mathbf{k}$ *is either strictly decreasing or strictly increasing in a sufficiently small open neighborhood of* t^*, *say,* $\tilde{T}^*$;
(d) $\mathcal{L}_n(t) \neq 0$, *any* $t \in \tilde{T}^*$;

then the motion along $\mathcal{S}$ *reverses at* t^* *with* $\mathbf{H}(t) \neq 0$ *at any* $t \in \tilde{T}^*$.

Proof Assumption-(a) meets the first part of Definition 7.2. Therefore, let us consider a regular 3D trajectory characterized by the evolution $\mathcal{S}(t) \equiv (\mathbf{R}(t), \mathbf{V}(t))$, with $t_0 \leqslant t \leqslant t_f$, driven by $\mathbf{L}(t) = (\mathcal{L}_r(t), \mathcal{L}_t(t), \mathcal{L}_n(t))$, and project it on the reference plane of HIF, or the XY-plane, with respect to which the inclination of the osculating-orbit is measured by definition. One has

$$\mathfrak{S}(t) \equiv \left(\tilde{\mathbf{R}}(t), \tilde{\mathbf{V}}(t) \right) = \left(\begin{bmatrix} X \\ Y \\ 0 \end{bmatrix}, \begin{bmatrix} V_x \\ V_y \\ 0 \end{bmatrix} \right), \quad t \in [t_0, t_f], \ t_f > t_0 \qquad (7.70)$$

First, let us suppose that $\tilde{\mathbf{H}}(t) = \tilde{\mathbf{R}}(t) \times \tilde{\mathbf{V}}(t) \neq 0$ at any t except for a non-empty countable set of instants for which $\tilde{\mathbf{H}}(t^*_j) = 0$, $j = 1, \ldots, n$, $n \geqslant 1$. For our aims here, it is sufficient to analyze the case $n = 1$. If $\mathfrak{S}$ is regular, $\tilde{\mathbf{H}}(t^*) = 0 \Leftrightarrow \tilde{\mathbf{R}}(t^*)$ parallel to either $\tilde{\mathbf{V}}(t^*)$ or $-\tilde{\mathbf{V}}(t^*) \neq 0$. One can write the relationship between the angular momentum of $\mathcal{S}$ and that of $\mathfrak{S}$ as

$$\mathbf{H}(t) = \mathbf{R} \times \mathbf{V} = (\tilde{\mathbf{R}} + Z\mathbf{k}) \times (\tilde{\mathbf{V}} + V_z\mathbf{k})$$

$$= (YV_z - ZV_y)\mathbf{i} + (ZV_x - XV_z)\mathbf{j} + \tilde{\mathbf{R}} \times \tilde{\mathbf{V}} \qquad (7.71)$$

At $\tilde{P}^*$, one has obviously

$$\mathbf{H}(t^*) = (YV_z - ZV_y)_{t^*}\mathbf{i} + (ZV_x - XV_z)_{t^*}\mathbf{j} \qquad (7.72)$$

Equation (7.72) tells us that the third component of the angular momentum of $\mathcal{S}$ vanishes at t^* or $\mathbf{H}(t^*)$ lies on the XY-plane.

Second, assume that in a sufficiently small, but finite, open neighborhood of $\tilde{P}^*$, say, the arc of $\mathfrak{S}$ from $\tilde{P}(t_a)$ to $\tilde{P}(t_b)$, $t^* \in (t_a, t_b) \equiv \tilde{T}^*$, is finite and regular, and with the continuous real function $\tilde{H}_z(t) \equiv \tilde{\mathbf{H}} \cdot \mathbf{k}$ strictly decreasing, or strictly increasing. As $\tilde{\mathbf{H}}(t^*) = 0$, $\tilde{H}_z(t)$ is strictly monotonic through zero in $\tilde{T}^*$, namely, the projected motion changes from direct to retrograde, or vice versa. Consequently, so behaves the third component of $\mathbf{H}$ because $H_z = \mathbf{H} \cdot \mathbf{k} = \tilde{\mathbf{H}} \cdot \mathbf{k} = \tilde{H}_z$. Thus, the 3D motion changes from direct to retrograde, or vice versa; the second part of Definition 7.2 is met. In addition, because of the assumption $\mathcal{L}_n \neq 0$ in $\tilde{T}^*$ and Lemma 7.1 (p. 260), one has $\mathbf{H} \neq 0$ in $\tilde{T}^*$. $\square$

7.4 Three-Dimensional Motion Reversal

The converse, i.e. the *only if* version of this theorem does not hold in general. As an example, we suggest the interested reader analyzing the following simple case of three-dimensional curve:

$$\mathbf{R} = \left(\sin(t)^2 - \cos(t)^2, \left(1 - \sin(t)\right)\cos(t) + 1/3, \cos(t)\right), \quad 0 \leqslant t \leqslant \pi \quad (7.73)$$

Around $t = \pi/2$, the above 3D curve is regular and exhibits a motion reversal: the component $H_z(t)$ vanishes there, and is strictly increasing around this instant. However, the projected-on-XY-plane curve is not regular just at $t = \pi/2$. Therefore, Definition 7.2 does not apply to the projected curve.

Note 7.4 Suppose that a sailcraft trajectory may be described via a circular Keplerian orbit rotating about an axis contained in its plane, and passing through the origin of HIF. Motion reverses if thrusting time is sufficiently long, and speed is high if the orbit radius is lower than Mercury's semi-major axis. However, this is not a fast sailing because orbital energy is negative all the time (see Sect. 7.4.4).

Remark 7.4 Hypothesis (b) does not exclude that 3D fast sailing mode may have more than one point where motion reverses in the sense made precise by Theorem 7.2. This observation will be apparent in numerical examples regarding 3D trajectories.

One could question whether the above theorem may be of some practical importance. Yes, indeed, because its content and the considerations of the previous section suggest us how building a control in order to show that 3D motion-reversal trajectories can be induced by simple profiles. As a point of fact, we already know special families of two-dimensional curves where the product $\tilde{\mathbf{H}} \cdot \mathbf{k}$ has the properties above described: they are the H-reversal trajectories.

⊙ Suppose that at the time t_0 of injection into the solar gravitational field, the sailcraft is given a lightness vector $\mathbf{L}^{(1)} = (\mathcal{L}_r^{(1)}, \mathcal{L}_t^{(1)}, \mathcal{L}_n^{(1)})$. Without losing in generality, we focus on trajectories starting counterclockwise with the control obeying the following constraints:

$$\dot{\mathbf{L}}^{(1)}(t) = 0, \quad t \in [t_0, t_1], \ 0 < H(t_1) \ll H(t_0) \quad (7.74a)$$

$$\left(\mathcal{L}_r^{(1)}, \mathcal{L}_t^{(1)}\right) \in \mathbb{F}, \quad 0 < \mathcal{L}_n^{(1)}, \ \mathcal{L}^{(1)} < 1 \quad (7.74b)$$

where we have employed the region $\mathbb{F}$ of Fig. 7.5 defined on p. 272. Thus, this control will be kept constant until the H-function is sufficiently smaller than H_0, but positive; at t_1, the derivative of $H(t)$ is still negative. Now, we relax the constraint of constant $\mathbf{L}$ and allow $\mathcal{L}_t$ to increase smoothly, to cross zero and to achieve some positive value. In contrast, $\mathcal{L}_r$ is kept constant; consequently, $\mathcal{L}_n$ increases, achieves a maximum, then decreases. (Were $\mathcal{L}_n^{(1)}$ assumed negative, then we would have a minimum.) For simplicity of calculation, the control $\mathbf{L}^{(2)}$ lasts from t_1 to t_2 such that

$$\dot{\mathcal{L}}_r^{(2)}(t) = 0, \quad \dot{\mathcal{L}}^{(2)}(t) = 0, \quad t \in [t_1, t_2] \quad (7.75a)$$

$$\mathcal{L}_t^{(2)}(t_2) = -\mathcal{L}_t^{(2)}(t_1), \qquad \mathcal{L}_t^{(2)}(t^\square) = 0, \quad t^\square \in (t_1, t_2) \tag{7.75b}$$

$$\mathcal{L}_n^{(2)}(t_2) = \mathcal{L}_n^{(2)}(t_1) \tag{7.75c}$$

The point to note is that, since the radial and the transversal numbers belong to $\mathbb{F}$, the decrease of the action of $\mathcal{L}_t^{(2)}$ on the negative $\dot{H}$ induces a value $H(t^\square)$ small, but positive. Subsequently, $\dot{H}$ becomes positive and hence $H(t)$ increases.

Finally, in the third interval of flight, the control can be frozen to the values at t_2, namely

$$\mathbf{L}^{(3)}(t) = \mathbf{L}^{(2)}(t_2), \quad t \geqslant t_2 \tag{7.76}$$

For simplicity, we call the just defined piecewise lightness vector as the three-$\mathbf{L}$ (smooth) control, though the behavior of $\mathcal{L}_t(t)$ in the relationships (7.75a)–(7.75c) has not been specified.

Note 7.5 From what discussed in Sect. 5.5 and Chap. 6, the control given by Eqs. (7.75a)–(7.75c) could be approximated by keeping the solar incidence angle constant in the $\mathcal{F}^{(\text{sail})}$, defined on page 157, while the Sun moves as observed by onboard sensors.

Let us see the meaning of the time $t^\square$. From the above definition, it is easy to recognize that

$$\dot{\mathcal{L}}_t(t^\square) > 0 \tag{7.77a}$$

$$\dot{H}(t^\square) = 0 \tag{7.77b}$$

$$\ddot{H}(t^\square) = \left[\frac{\mu_\odot}{R} \left(\dot{\mathcal{L}}_t - \mathcal{L}_t \frac{\dot{R}}{R} \right) \right]_{t^\square} = \frac{\mu_\odot}{R(t^\square)} \dot{\mathcal{L}}_t(t^\square) > 0 \tag{7.77c}$$

where we differentiated $\dot{H}(t)$ given by Eq. (7.19b) with respect to time. Therefore, the above qualitatively-inferred minimum of H is actually a local one. Note that, if the sailcraft started clockwise, H would be negative; however, in order to get a deceleration like the counterclockwise case, $\mathcal{L}_t$ would be positive in the arc-1 and, consequently, would decrease to cross zero in the arc-2. Thus, $\dot{\mathcal{L}}_t(t^\square) < 0$, i.e. the H-function would exhibit a local maximum.

Now, let us employ the rightmost of Eq. (7.17) and differentiate it with respect to time. Because the three-$\mathbf{L}$ control entails the constancy of the radial number, we can write

$$\dot{E}(t^\square) = \left[H\dot{H}/R^2 \right]_{t^\square} = 0 \tag{7.78a}$$

$$\ddot{E}(t) = (\dot{H})^2/R^2 + H\ddot{H}/R^2 - 2H\dot{H}\dot{R}/R^3$$
$$\ddot{E}(t^\square) = \left[H\mu_\odot \dot{\mathcal{L}}_t/R^3 \right]_{t^\square} > 0 \tag{7.78b}$$

where Eqs. (7.77a)–(7.77c) have been employed. Thus, orbital energy has a local minimum at $t = t^\square$. Again, if the sailcraft started clockwise, H would be negative

7.4 Three-Dimensional Motion Reversal

and, reasoning as made at Eqs. (7.77a)–(7.77c), one can recognize that the sign of $H\dot{\mathcal{L}}_t$ would keep unchanged at $t^\square$, and energy would has a minimum there, as physically expected from a deceleration followed by an acceleration. The results about H and E can be summarized in the following statement:

Proposition 7.4 *The control given by Eqs. (7.74a), (7.74b), (7.75a)–(7.75c) and (7.76) causes the H-function to achieve a local minimum, at an instant $t^\square$ internal to the second time interval, if the sailcraft motion starts counterclockwise; if it begins clockwise, then $H(t)$ has a local maximum at $t^\square$. At this time, orbital energy $E(t)$ exhibits a local minimum, no matter how the sailcraft starts in.*

Note 7.6 The above three-**L** control has been built as a simple variation of the control $\mathbf{L} = (\mathcal{L}_r^{(1)}, \mathcal{L}_t^{(1)}, 0) \in \mathbb{F}$, namely, a two-dimensional H-reversal motion endowed with a reversal time $t^\circledast$, as we know. We call the related solution of the 2D motion equations the *base profile* of the set of 3D profiles stemming from Eqs. (7.74a), (7.74b), (7.75a)–(7.75c) and (7.76).

There is a further property we can infer from the above defined **L**-control. This is the ratio of curvature to torsion evaluated at $t^\square$. We employ the expressions given by Eqs. (5.47) and (5.51), respectively. (These ones hold also in EHOF, of course.) The procedure consists of two steps: (1) inserting $\mathcal{L}_t(t^\square) = 0$, $\dot{\mathcal{L}}_r(t^\square) = 0$, $\dot{\mathcal{L}}_n(t^\square) = 0$ in the curvature and torsion expressions, and doing the ratio; (2) because $H(t^\square)$ is very small compared to $H(t_0) \sim 1$ (i.e. the orbital angular momenta of inner departure planets), one expects that φ is close to π; therefore, one could expand the ratio $[\kappa/\tau]_{t^\square}$ in power series about $\varphi = \pi$. Up to the second order, this algorithm results in

$$\left[\frac{\kappa}{\tau}\right]_{t^\square} \cong -\operatorname{sign}(\mathcal{L}_n(t^\square))\left[(\pi - \varphi) + \dot{\mathcal{L}}_t \frac{R^2 V}{\mu_\odot \mathcal{L}_n^2}(\pi - \varphi)^2\right]_{t^\square} \tag{7.79}$$

As we will see in what follows, such ratio may be significantly lower than 1 in absolute value even for $|\mathcal{L}_n|/\mathcal{L} \ll 1$, namely, torsion is dominant already before the end of the second interval. Far targets considerably outside the ecliptic may be reached.

In the control $\mathbf{L}^{(2)}$, we left the function $\mathcal{L}_t^{(2)}(t)$ undetermined but at three instants. Next questions are: (i) what about the shape of the transversal number in $[t_1, t_2]$, and (ii) are the location of t_1 and the length of $t_2 - t_1$ with respect to $t^\circledast$ able to affect the actual trajectory? With regard to (i), we can expect that, though the general properties expressed in Proposition 7.4 are independent of the particular $\mathcal{L}_t^{(2)}(t)$, the overall trajectory profile is not so simply because the right-hand sides of Eqs. (7.18), (7.19a) and (7.19b) could be somewhat different in $[t_1, t_2]$. As to question (ii), recalling the direct-motion solutions in two dimensions, one may see them coming again in three dimensions.

Due to the complexity of the structure of 3D motion equations, we deem it is better to switch to numerical integration of sailcraft trajectories in order to show the effects of what said.

Let us first assume that the transversal number varies linearly in the second time interval; from the constraints prefixed in the three-**L** control definition, one gets easily

$$\mathcal{L}_t^{(2)}(t) = \mathcal{L}_t^{(1)}\left(1 - 2\frac{t - t_1}{t_2 - t_1}\right)$$

$$t^{\square} = (t_1 + t_2)/2 \tag{7.80}$$

whereas the function $\mathcal{L}_n^{(2)}(t)$ is general once the law $\mathcal{L}_t^{(2)}(t)$ is assigned, that is

$$\mathcal{L}_n^{(2)}(t) = \text{sign}\big(\mathcal{L}_n^{(1)}\big)\sqrt{\mathcal{L}_t^{(1)2} + \mathcal{L}_n^{(1)2} - \mathcal{L}_t^{(2)}(t)^2} \tag{7.81}$$

Note 7.7 We continue to use solar units in the trajectory computation. In particular, the units time is the year/2π and the speed unit amounts to 2π AU/year, which is called the Earth Orbit Speed (EOS) here. Sometimes, some trajectory quantities expressed in SI-units are of more immediate grasp; it might be useful to remind the reader that the JPL ephemeris files are based on 1 AU = $1.4959787061 \times 10^{11}$ m; then the value 1 year is approximately equal to 365.2569 days, which differs from the standard year. In Chap. 8, we will employ the JPL DE405 for numerically integrate sailcraft trajectories perturbed by the planets.

All the numerical results that we will show from this point on come from integrating the system of Eqs. (7.15) by 8th-order Runge-Kutta with relative accuracy/precision values of 1.0×10^{-11} or 1.0×10^{-12}, depending on the case complexity, and with 24-digit precision in internal computations. The interpolation order ranged from 3 to 5. The fifth-order interpolation produced a difference < 0.1 s between the theoretical time and the computed one of the energy minimum. In the current context, interpolation orders higher than 5 does not produce state-derived quantities meaningfully different.

Case 7.1 $\boxed{\mathcal{L}_t \text{ linear}}$ Table 7.4 summarizes the initial conditions, the control, and some pieces of the output. Main output curves have been plotted in Fig. 7.8. Let us discuss the peculiar points. The first two components of $\mathbf{L}^{(1)}$ have been chosen from the 0.2-AU curve of Fig. 7.5; $\mathbf{L}^{(2)}$ has been chosen of linear type according to Eqs. (7.80), and $\mathbf{L}^{(3)}$ follows from its definition (7.76). A key point is to select the interval $[t_1, t_2]$. The reversal time of the base profile has been computed as shown in Sect. 7.2, and resulted in $t^{\circledast} = 0.711294 \cdot 2\pi$. We selected the time interval such that this instant is in the middle, as reported in Table 7.4, and with t_1 close to $t^{\circledast}$ for getting a trajectory similar to the base profile, if admitted, but allowing sufficient time for the thrust maneuver.

The top-left panel of Fig. 7.8 shows the projections of trajectory on two coordinate planes from which the sailcraft Z-motion is apparent. The reverse-motion character is evident from the XY-projection and the top-right part, where the H-function

7.4 Three-Dimensional Motion Reversal

Table 7.4 Case 7.1. Main data of the 3D sailcraft trajectory driven by the three-**L** control with a *linear* transversal number in the trajectory arc-2. Such data generate the profiles reported in Fig. 7.8 and discussed in the text. θ_f denotes the latitude of the terminal escape direction

$\mathbf{R}_0$	$(0.95, -0.35, 0)$ AU		
$\mathbf{V}_0$	$(2.15, 5.873, 0)/2\pi$ EOS		EOS $= (\text{AU/year}) \cdot (2\pi)$
t_0, t_1	0	$0.7 \cdot 2\pi$	time unit $= \text{year}/2\pi$
$\mathbf{L}^{(1)}$	$(0.52, -0.28, 0.014)$		
t_1, t_2	$0.7 \cdot 2\pi$	$0.7225884 \cdot 2\pi$	
$\mathbf{L}^{(2)}$	$(0.52, \mathcal{L}_t^{(2)}(t), \mathcal{L}_n^{(2)}(t))$		**Eqs.** (7.80) **and** (7.81)
t_2, t_f	$0.7225884 \cdot 2\pi$	$7 \cdot 2\pi$	
$\mathbf{L}^{(3)}$	$(0.52, 0.28, 0.014)$		
$t^{\square}, \varphi(t^{\square})$	$0.7112942 \cdot 2\pi$	$178.921°$	
$\kappa(t^{\square}), \tau(t^{\square})$	1.14737	-30.8263	AU^{-1}
$t^{\square} - t^{\circledast}$		0	
$t^{\square} - t^r$		-0.0028078	(-3.9174 hour)
R_p, t_p	0.199499 AU	5.54983	(322.625 day)
$\mathbf{R}_f$	$(-80.939, -50.011, -15.089)$ AU		
$\mathbf{V}_f$	$(-2.1075, -1.2864, -0.39338)$ EOS		
R_f, V_f	96.332 AU	2.5002 EOS	
θ_f	$-9.01°$		
$\mathbf{R}_f - \mathbf{R}_f^{(2D)}$	$(1.382, -0.03794, -15.089)$ AU		
$\mathbf{V}_f - \mathbf{V}_f^{(2D)}$	$(0.03608, -0.00100, -0.39338)$ EOS		

and the Z-component, say, H_Z of the angular momentum have been plotted. As previously proved, H exhibits a local minimum (because sailcraft motion started counterclockwise), which takes on a value much smaller than H_0, but not zero; while H increases, H_Z passes through zero, i.e. motion is reversed, but $\|\mathbf{H}\| \neq 0$; $\mathbf{H}$ has been decreased and simultaneously bent through the XY-plane without vanishing. The actual reversal time t^r, namely the instant at which $\mathbf{H}$ crosses its xy-plane, is slightly subsequent to the H-minimum time $t^{\square}$, as reported in Table 7.4.

The evolution of $\mathbf{H}$ is shown in the bottom-left panel of Fig. 7.8: after t^r, the length of $\mathbf{H}$, $|H_Z|$ and H all increase differently from the base profile where, we know, the invariant decreases monotonically and strictly. In contrast, the evolutions of energy and speed is quite similar to the base profile, including a quasi-cruise regime, as shown in the bottom-right panel of the figure.

The last three lines of Table 7.4 report the latitude of the terminal escape direction, and the differences between the current terminal position and velocity (at $t_f = 7$ years) and the position and velocity (at the same time) of the base profile. The small increase of cruise speed is due to the slight increase of $\|\mathbf{L}\|$ with respect to

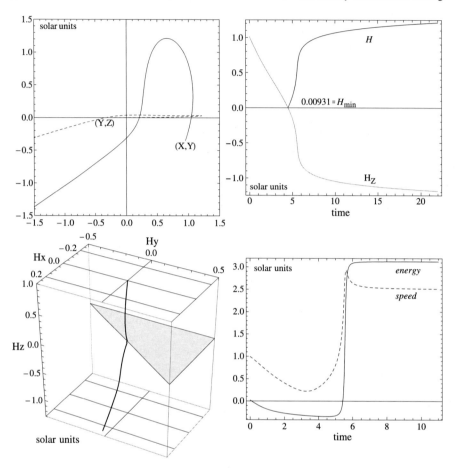

Fig. 7.8 Case 7.1. Sailcraft flight characterized by the three-**L** control with a **linear** transversal number in the trajectory arc-2 as in Table 7.4. Solar units have been used throughout; *top left*: two projections of the 3D trajectory on the XY-plane and on the YZ-plane, respectively, of HIF; *top right*: behaviors of the invariant H and the third component of **H**; *bottom left*: path of the **H**-vector tip; *bottom right*: time profiles of energy and speed

the base profile (because $\mathcal{L}_n$ is slightly positive, on the average). The $|Z_f - Z_f^{(2D)}|$ value increases linearly with time; however, its time rate of about 2.47 AU/year is perhaps unexpected, thus confirming again the high non-linearity of the sailcraft motion equations.

end case 7.1 ∎

Case 7.2 $\boxed{\mathcal{L}_t \text{ sine}}$ One could object that a linear $\mathcal{L}_t^{(2)}(t)$ is not very realistic; yes, it would. Thus, let us raise the complexity a bit by taking the sine function for $\mathcal{L}_t^{(2)}(t)$;

7.4 Three-Dimensional Motion Reversal

Table 7.5 Case 7.2. Main data of the 3D sailcraft trajectory driven by the three-**L** control with a *sinusoidal* transversal number in the trajectory arc-2. Such data generate the profiles reported in Fig. 7.9 and discussed in the text. θ_f denotes the latitude of the terminal escape direction

$\mathbf{R}_0$	$(0.95, -0.35, 0)$ AU		
$\mathbf{V}_0$	$(2.15, 5.873, 0)/2\pi$ EOS		EOS $= $ (AU/year) $\cdot (2\pi)$
t_0, t_1	0	$0.7 \cdot 2\pi$	time unit $=$ year$/2\pi$
$\mathbf{L}^{(1)}$	$(0.52, -0.28, 0.014)$		
t_1, t_2	$0.7 \cdot 2\pi$	$0.7225884 \cdot 2\pi$	
$\mathbf{L}^{(2)}$	$(0.52, \mathcal{L}_t^{(2)}(t), \mathcal{L}_n^{(2)}(t))$		**Eqs. (7.82) and (7.81)**
t_2, t_f	$0.7225884 \cdot 2\pi$	$7 \cdot 2\pi$	
$\mathbf{L}^{(3)}$	$(0.52, 0.28, 0.014)$		
$t^{\square}, \varphi(t^{\square})$	$0.7112942 \cdot 2\pi$	$179.214°$	
$\kappa(t^{\square}), \tau(t^{\square})$	1.14727	-36.3859	AU^{-1}
$t^{\square} - t^{\circledast}$		0	
$t^{\square} - t^r$		-0.0004550	$(-0.63481$ hour$)$
R_p, t_p	0.199549 AU	5.54984	$(322.626$ day$)$
$\mathbf{R}_f$	$(-82.036, -49.773, -8.3832)$ AU		
$\mathbf{V}_f$	$(-2.1361, -1.2802, -0.21804)$ EOS		
R_f, V_f	96.32 AU	2.4999 EOS	
θ_f	$-4.993°$		
$\mathbf{R}_f - \mathbf{R}_f^{(2D)}$	$(0.28573, 0.20041, -8.3832)$ AU		
$\mathbf{V}_f - \mathbf{V}_f^{(2D)}$	$(0.00746, 0.00521, -0.21804)$ EOS		

with it, we can also match the derivatives of the **L**-pieces at t_1 and t_2, and get:

$$\mathcal{L}_t^{(2)}(t) = \mathcal{L}_t^{(1)} \sin\left(\frac{\pi}{2} - \pi\frac{t - t_1}{t_2 - t_1}\right)$$

$$t^{\square} = (t_1 + t_2)/2$$

(7.82)

We computed the trajectory and arranged Table 7.5 and Fig. 7.9. Comparing them with the corresponding Table 7.4 and Fig. 7.8, one can see that the two flight profiles are quite similar. The main differences can be read in the tables or are apparent from the various plots; the general feature is that the $\mathcal{L}_t$-sinusoidal profile is closer than the $\mathcal{L}_t$-linear one to the ideal 2D-reversal solution.

end case 7.2 ■

At this point, one has various options for continuing the analysis of *fast* 3D trajectories. As the next step, one may think of analyzing a more realistic function for

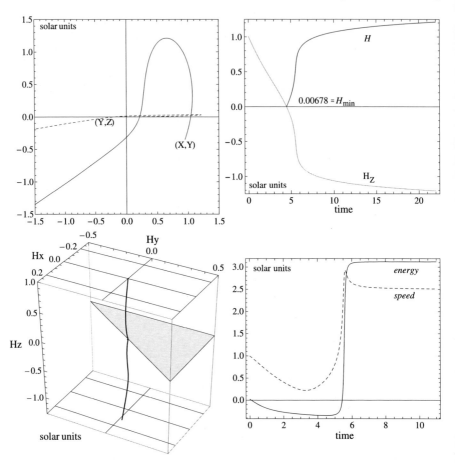

Fig. 7.9 Case 7.2. Sailcraft flight characterized by the three-**L** control with a **sinusoidal** transversal number in the trajectory arc-2, as in Table 7.5. Solar units have been used throughout; *top left:* two projections of the 3D trajectory on the XY-plane and on the YZ-plane, respectively, of HIF; *top right:* behaviors of the invariant H and the third component of **H**; *bottom left:* path of the **H**-vector tip; *bottom right:* time profiles of energy and speed

$\mathcal{L}_t^{(2)}(t)$ such as, for instance, the response coming from a flat-sail model of Aluminum on CP-1 or the new IKAROS plastic support. However, because of the complexity of this response, we prefer to postpone such step to the next chapter. Thus, one could proceed by

option (1) changing the amplitude of the interval $[t_1, t_2]$, but with $t^\circledast$ kept on the middle;

option (2) varying the allocation of $t^\circledast$ in the interval $[t_1, t_2]$. There are two choices: one in the first semi-interval, the other one in the second semi-interval;

option (3) choosing the interval such that $t^\circledast > t_2$;

7.4 Three-Dimensional Motion Reversal

option (4) varying the normal lightness in the first interval such a way to be a significant fraction of the transversal number;

option (5) considering another more realistic trial function for $\mathcal{L}_t^{(2)}(t)$, and using the related sail surface for the whole three-**L** control;

option (6) dividing the interval $[t_2, t_f]$ into $[t_2, t_3)$ and $[t_3, t_f]$ and applying a thrust maneuver at t_3 (modeled as impulsive, for simplicity); one could a-priori opt for

a. maneuvering before the computed three-**L** perihelion, which will then be changed;

b. maneuvering after the three-**L** perihelion;

option (7) inserting planetary gravitational perturbations in the computation;

option (8) considering the TSI time fluctuations.

One could argue that the first six options of the above list belong to the more general problem of flight optimization with respect to some index of performance. They do. Nevertheless, we have preferred to enumerate them (though they do not cover the whole range of features) in order to better understand what an algorithm of optimization should at least include. Here, according to the current line of explanation, we like to keep the sinusoidal option active, and focus on the options (1)–(4). Options (5)–(8) will be considered later; we will discuss options (5)–(7) in Chap. 8 with regard to the optimization of three dimensional trajectories in the sense of NLP. Option-(8) is a stand-alone problem from the viewpoint of a realistic mission, as we shall see in Sect. 9.1. Let us go on with a case related to the *option*-(1) of the above list.

Case 7.3 $\boxed{t_2 - t_1 \text{ amplified}}$ The second time interval has been amplified from 8.25 day of Case 7.2 to about 30 day, but keeping $t^\square = t^\circledast$. Table 7.6 and Fig. 7.10 are to be compared with Table 7.5 and Fig. 7.9, respectively. Few are the meaningful differences: (i) the minimum of the H-function has increased, (ii) the reversal time t^r has delayed further with respect to time $t^\square$ of this minimum from 0.6 to almost 11 hours, (iii) perihelion has decreased by 0.0008 AU, (iv) the escape latitude has increased in absolute value. Energy and speed are practically unchanged ($V_f \cong 74.6$ km/s), whereas **H** shows an undulation around the origin of its space. The progressive tendencies of all such points have been confirmed by a number of numerical simulations with different lengths of the second interval $[t_1, t_2]$.

end case 7.3 ∎

Case 7.4 $\boxed{t^\square < t^\circledast < t_2}$ In the previous cases, the middle instant ($t^\square$) of the interval $[t_1, t_2]$ coincided with the reversal time ($t^\circledast$) of the base profile; in addition, $t^\circledast$ does not appear in the $\mathcal{L}_t^{(2)}$ profiles given by the expressions (7.80) and (7.82). Now, let us move the interval $[t_1, t_2]$ of Case 7.3 *earlier* so that $t^\circledast$ falls in the open semi-interval $(t^\square, t_2)$. We considered a shift equal to $(t_2 - t_1)/10$, or about 3 days. All other inputs from the previous case have been employed. The results from the

306   7 The Theory of Fast Solar Sailing

Table 7.6 Case 7.3. Main data of the 3D sailcraft trajectory driven by the three-L control with a *sinusoidal* transversal number in the trajectory arc-2. In this case, the amplitude of the second interval has been significantly increased with respect to Case 7.2. Such data generate the profiles reported in Fig. 7.10 and discussed in the text. θ_f denotes the latitude of the terminal escape direction

$\mathbf{R}_0$	$(0.95, -0.35, 0)$ AU		
$\mathbf{V}_0$	$(2.15, 5.873, 0)/2\pi$ EOS		EOS $= (\text{AU}/\text{year}) \cdot (2\pi)$
t_0, t_1	0	$0.67 \cdot 2\pi$	time unit $= \text{year}/2\pi$
$\mathbf{L}^{(1)}$	$(0.52, -0.28, 0.014)$		
t_1, t_2	$0.67 \cdot 2\pi$	$0.7525884 \cdot 2\pi$	
$\mathbf{L}^{(2)}$	$(0.52, \mathcal{L}_t^{(2)}(t), \mathcal{L}_n^{(2)}(t))$		**Eqs. (7.82) and (7.81)**
t_2, t_f	$0.7525884 \cdot 2\pi$	$7 \cdot 2\pi$	
$\mathbf{L}^{(3)}$	$(0.52, 0.28, 0.014)$		
$t^{\square}, \varphi(t^{\square})$	$0.7112942 \cdot 2\pi$	$177.184°$	
$\kappa(t^{\square}), \tau(t^{\square})$	1.14869	-10.3911	AU^{-1}
$t^{\square} - t^{\circledast}$		0	
$t^{\square} - t^r$		-0.0076975	$(-10.7394$ hour$)$
R_p, t_p	0.198739 AU	5.54979	$(322.623$ day$)$
$\mathbf{R}_f$	$(-81.539, -49.972, -13.128)$ AU		
$\mathbf{V}_f$	$(-2.1232, -1.2854, -0.34199)$ EOS		
R_f, V_f	96.531 AU	2.5054 EOS	
θ_f	$-7.816°$		
$\mathbf{R}_f - \mathbf{R}_f^{(2D)}$	$(0.78205, 0.001376, -13.128)$ AU		
$\mathbf{V}_f - \mathbf{V}_f^{(2D)}$	$(0.02043, -0.00001608, -0.34199)$ EOS		

numerical integration have been reported on Table 7.7 and Fig. 7.11. Let us discuss the most significant points with respect to Case 7.3.

The first striking result comes from the top-left panel of Fig. 7.11: the out-of-plane part of the trajectory has changed significantly, even though the various ranges of the normal lightness number has been kept unchanged; in particular, sailcraft rises over the XY-plane before diving into negative latitudes. The **H**-crossing time takes place about 4.68 day after $t^{\square}$ (against 10.7 hour of Case 7.3), and the minimum value of $H(t)$ increases by over fifty percent.

Perihelion is achieved almost 323 days after departure, and is higher than the reference case by about 0.017 AU. As a result, the sail temperature via radiative emission (see Sect. 6.5.5) decreases by about 4.3 percent. Also, this perihelion increase causes a decrease of 0.1076 EOS, or about 3.2 km/s, to give $V_f \cong 71.4$ km/s at $t = t_f$.

The strong increase of the absolute value of the escape latitude compensates for the above loss of speed (small indeed), i.e. from about $-7.8°$ to $-33.4°$. This is a very good result because medium latitudes of the celestial sphere can be accessed

7.4 Three-Dimensional Motion Reversal

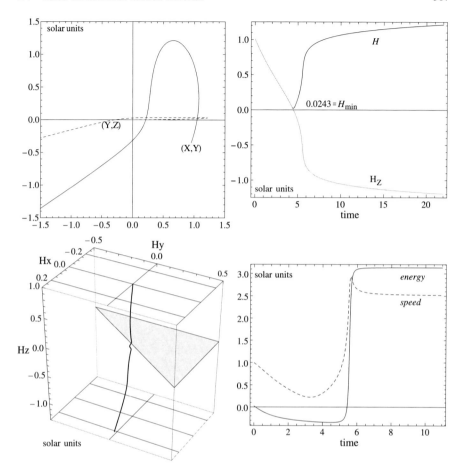

Fig. 7.10 Case 7.3. Sailcraft flight characterized by the three-**L** control with a **sinusoidal** transversal number in the trajectory arc-2, as in Table 7.6, where the amplitude of the second interval has been significantly increased. Solar units have been used throughout; *top left:* two projections of the 3D trajectory on the XY-plane and on the YZ-plane, respectively, of HIF; *top right*: behaviors of the invariant H and the third component of **H**; *bottom left*: path of the **H**-vector tip; *bottom right*: time profiles of energy and speed

by a small value of the normal number. By shifting more than 10 percent, with the other quantities fixed, the minimum in escape latitude can be $-37.7°$.

Finally, the differences $\mathbf{R}_f - \mathbf{R}_f^{(2D)}$ and $\mathbf{V}_f - \mathbf{V}_f^{(2D)}$ show that the current 3D flight is considerably different from the base profile. Nevertheless, the projection on the XY-plane looks like a 2D H-reversal trajectory, according to Theorem 7.2.

Now, what would happen if we delay the interval $[t_1, t_2]$ with respect to $t^{\circledast}$? Next case will analyze the ensuing effect on the trajectory.

end case 7.4 ∎

7 The Theory of Fast Solar Sailing

Table 7.7 Case 7.4. Main data of the 3D trajectory driven by the three-**L** control with a *sinusoidal* transversal number in the trajectory arc-2. In this case, the second interval has been moved *earlier* with respect to Case 7.3. Such data generate the profiles reported in Fig. 7.11. θ_f is the latitude of the terminal escape direction

$\mathbf{R}_0$	$(0.95, -0.35, 0)$ AU		
$\mathbf{V}_0$	$(2.15, 5.873, 0)/2\pi$ EOS		EOS $= $ (AU/year) $\cdot (2\pi)$
t_0 , t_1	0	$0.6617412 \cdot 2\pi$	time unit $=$ year$/2\pi$
$\mathbf{L}^{(1)}$	$(0.52, -0.28, 0.014)$		
t_1 , t_2	$0.6617412 \cdot 2\pi$	$0.7443296 \cdot 2\pi$	
$\mathbf{L}^{(2)}$	$(0.52, \mathcal{L}_t^{(2)}(t), \mathcal{L}_n^{(2)}(t))$		**Eqs.** (7.82) **and** (7.81)
t_2 , t_f	$0.7443296 \cdot 2\pi$	$7 \cdot 2\pi$	
$\mathbf{L}^{(3)}$	$(0.52, 0.28, 0.014)$		
$t^{\square}, \varphi(t^{\square})$	$0.7030354 \cdot 2\pi$	$175.558°$	
$\kappa(t^{\square}), \tau(t^{\square})$	1.21102	-2.06970	AU^{-1}
$t^{\square} - t^{\circledast}$		-0.0518917	$(-3.01659$ day$)$
$t^{\square} - t^r$		-0.08051825	$(-4.68074$ day$)$
R_p, t_p	0.215422 AU	5.55316	$(322.819$ day$)$
$\mathbf{R}_f$	$(-51.569, -57.317, -50.919)$ AU		
$\mathbf{V}_f$	$(-1.3403, -1.4768, -1.3312)$ EOS		
R_f, V_f	92.398 AU	2.3978 EOS	
θ_f	$-33.441°$		
$\mathbf{R}_f - \mathbf{R}_f^{(2D)}$	$(30.752, -7.3435, -50.919)$ AU		
$\mathbf{V}_f - \mathbf{V}_f^{(2D)}$	$(0.80334, -0.19142, -1.3312)$ EOS		

Case 7.5 $\boxed{t_1 < t^{\circledast} < t^{\square}}$ Now, let us move the interval $[t_1, t_2]$ of Case 7.3 *later* so that $t^{\circledast}$ falls in the open semi-interval $(t_1, t^{\square})$. We considered a shift equal to $-(t_2 - t_1)/10$, or about -3 days. All other inputs have been kept unchanged; thus, this case differs from the previous one by the opposite shift. Instants t_1 and t_2 are reported in Table 7.8. The top-left panel of Fig. 7.12 gives the first result; on the XY-plane, the oriented curvature changes sign twice. In addition, from the YZ projection, one sees that the sailcraft crosses the ecliptic plane twice. However, these features should *not* to be confused with the motion change, which takes place twice as well, but at two other times. Theorem 7.2 applies in the neighborhoods of these two points, which are marked by • in the XY-projection; the inset of the top-right panel of the figure shows clearly that the function $H_z(t)$ has two zeros separated by a temporal interval of about 8.75 days. The related instants have been denoted by t^r_{+-} and t^r_{-+} in Table 7.8, meaning that the motion changes from direct $(+)$ to retrograde $(-)$, or vice versa. In the inset, the vertical lines represent t_1, $t^{\circledast}$, $t^{\square}$, and t_2, in the order. The middle plot shows the time profiles of the $\mathbf{R}$ components and $H_z(t)$, with the

7.4 Three-Dimensional Motion Reversal

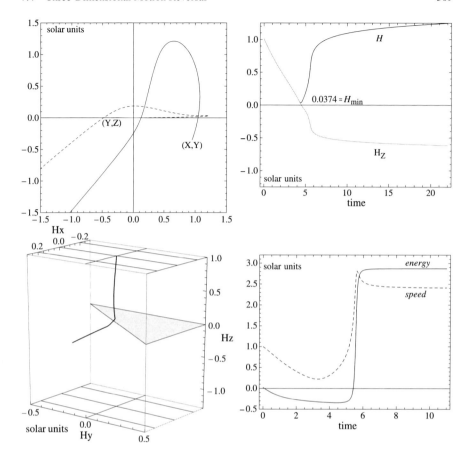

Fig. 7.11 Case 7.4. Sailcraft flight characterized by three-L control with a **sinusoidal** transversal number in the trajectory arc-2. In this case, the second interval has been moved *earlier* with respect to Case 7.3. *Top left:* two projections of the 3D trajectory on the XY-plane and on the YZ-plane, respectively, of HIF; *top right*: behaviors of the invariant H and the third component of **H**; *bottom left*: path of the **H**-vector tip; *bottom right*: time profiles of energy and speed

reversal points marked again. It is apparent that the ecliptic plane crossing times are later than the motion reversal ones. Between the two crossings, as reported in Table 7.8, the sailcraft achieves its minimum distance from the Sun: it is further reduced to $R_p \cong 0.1825$ AU, which entails an increase of 20 percent in the solar irradiance.

In the **H**-space, the two motion reversals result in a sort of semicircle about the origin, as plotted in the bottom-left panel of Fig. 7.12. After that, angular momentum evolves in the positive semispace. The bottom-right subplot shows that the pseudo-cruise speed has increased, due to the perihelion decrease, to about 16.486 AU/year, or 78.15 km/s. Notable is the fact that the escape direction turns to medium positive latitude, the first hitherto, namely with the *same* sign of $\mathcal{L}_n$. From the last two lines

310   7 The Theory of Fast Solar Sailing

Table 7.8 Case 7.5. Main data of the 3D sailcraft trajectory driven by the three-**L** control with a *sinusoidal* transversal number in the trajectory arc-2. In this case, the second interval has been moved *later* with respect to Case 7.3. Such data generate the profiles reported in Fig. 7.12 and discussed in the text. θ_f denotes the latitude of the terminal escape direction

$\mathbf{R}_0$	$(0.95, -0.35, 0)$ AU		
$\mathbf{V}_0$	$(2.15, 5.873, 0)/2\pi$ EOS		EOS $= $ (AU/year) $\cdot (2\pi)$
t_0 , t_1	0	$0.67825884 \cdot 2\pi$	time unit $=$ year$/2\pi$
$\mathbf{L}^{(1)}$	$(0.52, -0.28, 0.014)$		
t_1 , t_2	$0.67825884 \cdot 2\pi$	$0.76084724 \cdot 2\pi$	
$\mathbf{L}^{(2)}$	$(0.52, \mathcal{L}_t^{(2)}(t), \mathcal{L}_n^{(2)}(t))$		**Eqs.** (7.82) **and** (7.81)
t_2 , t_f	$0.76084724 \cdot 2\pi$	$7 \cdot 2\pi$	
$\mathbf{L}^{(3)}$	$(0.52, 0.28, 0.014)$		
$t^\square, \varphi(t^\square)$	$0.71955304 \cdot 2\pi$	$178.764°$	
$\kappa(t^\square), \tau(t^\square)$	1.094052	-38.32524	AU^{-1}
$t^\square - t^\circledast$		$+0.0518919$	(3.01660 day)
$t^\square - t^r_{+-}$		$+0.0506310$	(2.94330 day)
$t^\square - t^r_{-+}$		-0.0999081	(-5.80790 day)
R_p, t_p	0.182485 AU	5.54664	(322.44 day)
$\mathbf{R}_f$	$(6.40124, -92.6411, 39.9150)$ AU		
$\mathbf{V}_f$	$(0.175694, -2.40102, 1.04329)$ EOS		
R_f	101.077 AU		
V_f	2.6238 EOS	78.150 km/s	
θ_f	$23.26°$		
$\mathbf{R}_f - \mathbf{R}_f^{(2D)}$	$(88.723, -42.668, 39.9150)$ AU		
$\mathbf{V}_f - \mathbf{V}_f^{(2D)}$	$(2.31928, -1.11565, 1.04329)$ EOS		

of Table 7.8, one recognizes that this case is significantly different from the base profile.

end case 7.5 ∎

Case 7.6 $\boxed{t_2 < t^\circledast}$ This case consists of choosing the interval $[t_1, t_2]$ so that $t_2 < t^\circledast$; in the case in point $t_2 - t^\circledast = -(t_2 - t_1)/10$. The main results are reported in Table 7.9, while the evolution of trajectory-related quantities are plotted in Fig. 7.13. At first glance, the XY-projection in the top-left panel looks very similar to that one in the top-left plot of Fig. 7.12. However, there is no straight line (passing through the origin of the XY-plane) which is tangent to the 2D curve; in other words, one finds no velocity and position vectors parallel/antiparallel to one another, namely, there is no point where $\tilde{\mathbf{H}} = 0$. Thus, condition (b) of Theorem 7.2 is not satisfied,

7.4 Three-Dimensional Motion Reversal

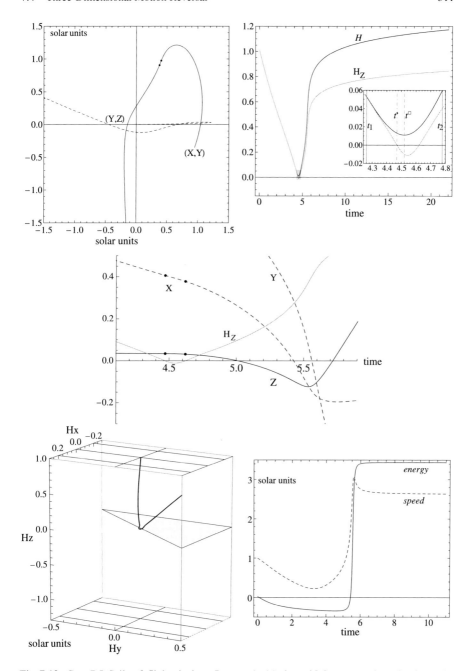

Fig. 7.12 Case 7.5. Sailcraft flight via three-**L** control with **sinusoidal** transversal number in arc-2. The second interval has been moved *later* with respect to Case 7.3. Solar units have been used; *top left*: trajectory projections on the XY-plane and YZ-plane, respectively; *top right*: behaviors of the invariant H and the Z-component of **H**; *center*: time profiles of the position components and H_z; *bottom left*: path of the **H**-vector tip; *bottom right*: time profiles of energy and speed

312   7 The Theory of Fast Solar Sailing

Table 7.9 Case 7.6. Main data of the 3D sailcraft trajectory driven by the three-**L** control with a *sinusoidal* transversal number in the trajectory arc-2. In this case, the second **L** control is applied wholly *before* the reversal time of the base profile. Such data generate the profiles in Fig. 7.13 discussed in the text. θ_f is the latitude of the terminal escape direction

$\mathbf{R}_0$	$(0.95, -0.35, 0)$ AU		
$\mathbf{V}_0$	$(2.15, 5.873, 0)/2\pi$ EOS		EOS $= (\mathrm{AU/year}) \cdot (2\pi)$
t_0, t_1	0	$0.6204469 \cdot 2\pi$	time unit $= \mathrm{year}/2\pi$
$\mathbf{L}^{(1)}$	$(0.52, -0.28, 0.014)$		
t_1, t_2	$0.6204469 \cdot 2\pi$	$0.7030353 \cdot 2\pi$	
$\mathbf{L}^{(2)}$	$(0.52, \mathcal{L}_\mathrm{t}^{(2)}(t), \mathcal{L}_\mathrm{n}^{(2)}(t))$		**Eqs. (7.82) and (7.81)**
t_2, t_f	$0.7030353 \cdot 2\pi$	$7 \cdot 2\pi$	
$\mathbf{L}^{(3)}$	$(0.52, 0.28, 0.014)$		
$t^\square, \varphi(t^\square)$	$0.6617411 \cdot 2\pi$	$166.440°$	
$\kappa(t^\square), \tau(t^\square)$	1.67240	8.26154	AU^{-1}
$t^\square - t^\circledast$		-0.3113509	(-18.09959 day)
$t^\square - t^r$		no reversal	
R_p, t_p	0.305295 AU	5.57263	(323.951 day)
$\mathbf{R}_f$	$(4.7791, -67.316, -35.33)$ AU		
$\mathbf{V}_f$	$(0.13832, -1.7381, -0.92802)$ EOS		
R_f, V_f	76.174 AU	1.9752 EOS	
θ_f	$-27.633°$		
$\mathbf{R}_f - \mathbf{R}_f^{(2D)}$	$(87.100, -17.343, -35.33)$ AU		
$\mathbf{V}_f - \mathbf{V}_f^{(2D)}$	$(2.2819, -0.45271, -0.92802)$ EOS		

and one expects that the 3D trajectory exhibits no motion reversal. This is confirmed by the top-right panel of Fig. 7.13, where both H and H_z are always positive. Note that the minimum of H is rather high compared to the previous cases; this is due to the choice of having the variable $\mathcal{L}_\mathrm{t}$ "too far" from $t^\circledast$. Thus, the perihelion is high. Table 7.9 confirms an increase up to $\gtrsim 0.3$ AU.

The bottom-left panel of the figure shows the full evolution of $\mathbf{H}$ that neither lays horizontally on nor crosses the XY-plane. Its final behavior shows clearly that the motion in the pseudo-cruise regime is always direct.

Finally, energy and speed are plotted in the bottom-right subplot of Fig. 7.13. This time, in the cruise regime, energy is *numerically* lower than speed, even though close to it. This happens when the pseudo-cruise speed (see Proposition 7.3) is equal to 2, or very close to 2, as (numerically) $\frac{1}{2} V_\mathrm{cruise}^2 = V_\mathrm{cruise}$. For faster sailing, cruise energy is always distinctly higher than cruise speed (numerically, in solar units). The escape latitude is negative, as in the first four cases, because there is one ecliptic plane crossing, i.e. an odd number of crossings.

end case 7.6 ∎

7.4 Three-Dimensional Motion Reversal

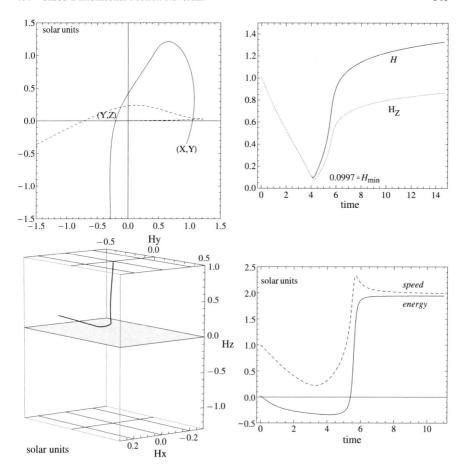

Fig. 7.13 Case 7.6. Sailcraft flight via the three-**L** control with a **sinusoidal** transversal number in the arc-2. In this case, the second **L** control is applied wholly *before* the reversal time of the base profile, as specified in Table 7.9. Solar units have been used throughout; *top left:* two projections of the 3D trajectory on the XY-plane and on the YZ-plane, respectively, of HIF; *top right:* behaviors of the invariant H and the third component of **H**; *bottom left:* path of the **H**-vector tip; *bottom right:* time profiles of energy and speed

Case 7.7 $\boxed{\mathcal{L}_n \text{ increased}}$ The above analyzed Case 7.5, i.e. $t_1 < t^{\circledast} < t^{\square}$, has been chosen as the case where to apply a notable increase to the normal number $\mathcal{L}_n^{(1)}$, i.e. from 0.014 to 0.13. From few numeric experiments, we saw that such value allows getting a high (negative) terminal latitude. The analysis of the progressive modifications, as $\mathcal{L}_n^{(1)}$ is changed by steps, is not reported here; a notable point is that, at $\mathcal{L}_n^{(1)} \geq 0.1105$, the number of reversal points is greater than 2.

The current case highlights three reversal points. We have added further subplots to Fig. 7.14. The reversal-motion markers are in the XY-projections of trajectory (top-left subplot) and hodograph (top-right subplot). They show the **H**-reversals tak-

314   7 The Theory of Fast Solar Sailing

Table 7.10 Case 7.7. Main data of the 3D sailcraft trajectory driven by the three-**L** control with a *sinusoidal* transversal number in the trajectory arc-2. In this case, the normal number has been increased from 0.014 to 0.13. Such data generate the profiles reported in Fig. 7.14 and discussed in the text. θ_f denotes the latitude of the terminal escape direction. There are three motion-reversal points and one XY-crossing point

$\mathbf{R}_0$	$(0.95, -0.35, 0)$ AU		
$\mathbf{V}_0$	$(2.15, 5.873, 0)/2\pi$ EOS		EOS $= $ (AU/year) $\cdot (2\pi)$
t_0, t_1	0	$0.67825884 \cdot 2\pi$	time unit $=$ year$/2\pi$
$\mathbf{L}^{(1)}$	$(0.52, -0.28, 0.13)$		
t_1, t_2	$0.67825884 \cdot 2\pi$	$0.76084724 \cdot 2\pi$	
$\mathbf{L}^{(2)}$	$(0.52, \mathcal{L}_t^{(2)}(t), \mathcal{L}_n^{(2)}(t))$		**Eqs. (7.82) and (7.81)**
t_2, t_f	$0.76084724 \cdot 2\pi$	$7 \cdot 2\pi$	
$\mathbf{L}^{(3)}$	$(0.52, 0.28, 0.13)$		
$t^\square, \varphi(t^\square)$	$0.71955304 \cdot 2\pi$	$178.764°$	
$\kappa(t^\square), \tau(t^\square)$	1.20457	-44.5786	AU^{-1}
$t^\square - t^\circledast$		$+0.0518919$	(3.01660 day)
$t^\square - t_1^r$		$+0.170977$	(9.939336 day)
$t^\square - t_2^r$		-0.0140751	(-0.8182206 day)
$t^\square - t_3^r$		-0.4283939	(-24.90358 day)
R_p, t_p	0.182485 AU	5.54665	(322.44 day)
$\mathbf{R}_f$	$(-29.6386, -48.6305, -83.5057)$ AU		
$\mathbf{V}_f$	$(-0.76854, -1.25169, -2.17413)$ EOS		
R_f	101.077 AU		
V_f	2.6238 EOS	78.150 km/s	
θ_f	$-55.706°$		
$\mathbf{R}_f - \mathbf{R}_f^{(2D)}$	$(52.683, 1.3426, -83.506)$ AU		
$\mathbf{V}_f - \mathbf{V}_f^{(2D)}$	$(1.3750, 0.033685, -2.1741)$ EOS		

ing place all before perihelion. In three disjoint neighborhoods, respectively, Theorem 7.2 applies as the two central subplots—on which also H_z is drawn—show clearly. The first two reversal times occur inside the second control interval; in particular $t^\circledast \in (t_1^r, t_2^r)$. The third one is earlier than perihelion by $\cong -34.715$ days. Even this 3D case confirms that every reversal event is located well in advance of perihelion, giving the sailcraft sufficient time to accelerate and to achieve the escape condition $E = 0$ before the perihelion as well.

The behavior of the **H**-vector tip is shown in the bottom-left panel of the figure. Starting from the top plane, this curve crosses the xy-plane three times, and then turn to the cruise regime.

With respect to Case 7.5, one will note that the pair (R_p, t_p) is unchanged. However, longitude and latitude of $\|\mathbf{R}_p\|$ are different; such values are important for

7.4 Three-Dimensional Motion Reversal

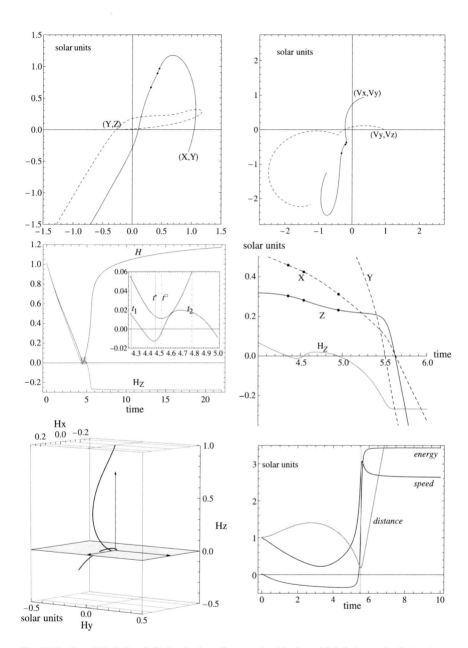

Fig. 7.14 Case 7.7. Sailcraft flight via three-**L** control with **sinusoidal** $\mathcal{L}_t$ in arc-2. $\mathcal{L}_n$ has been increased (see Table 7.10). Solar units have been used throughout; *top left*: trajectory projections on the XY and YZ-planes; *top right*: hodograph projections; *center left* behaviors of H and the third component of **H**; *center right*: time profiles of **R** components and H_z; *bottom left*: path of **H**-vector tip; *bottom right*: time profiles of distance, speed, and energy; the vertical line represents t_p

316   7 The Theory of Fast Solar Sailing

sailcraft-Earth communications. From the cruise regime viewpoint, both energy and speed manifest the same profiles.

What then is the main advantage in getting a higher normal number that, in turn, entails a lower sailcraft sail loading, namely, a more advanced design of sailcraft? The answer comes from the escape latitude value, which now amounts to $-55.7°$. This, for instance, means the possibility of exploring large deep-space zones along directions highly inclined with respect to the ecliptic plane, and without requiring planetary flybys.

end case 7.7 ■

7.4.4 About the Definition of 3D Fast Sailing

The definition of two-dimensional fast sailing (p. 277), the considerations in Sect. 7.4.1, Theorem 7.2 and Remark 7.4, and the numerical results in the previous section suggest us featuring the essential points of three-dimensional fast sailing mode in the following definition.

Definition 7.3 3D Fast Sailing (D3FS). Three-dimensional fast-sailing sailcraft motion is characterized by the following points:

1. the (heliocentric) trajectory is differentiable and regular everywhere (at least three times), and is neither *closed* nor *simply*-closed;
2. there exist pairs of direct-retrograde, or retrograde-direct, arcs separated by points where the third component H_z of the orbital angular momentum $\mathbf{H}$ vanishes. The related instants $t_j^r, t_j^r \in (t_0, t_f)$, $j = 1, \dots, n < \infty$ are the elements of a *non*-empty countable set;
3. in a sufficiently small neighborhood of each t_j^r, the function $H_z(t)$ is strictly monotonic;
4. the last retrograde or direct arc (if motion starts counterclockwise or clockwise, respectively) exhibits a strictly positive time derivative of the orbital energy;
5. this arc contains the $E = 0$ point and the perihelion, and the respective instants satisfy $t_{esc} < t_p$;
6. at the perihelion, speed is not stationary, but $\dot{V}_p > 0$;
7. sufficiently far from the perihelion, sailcraft eventually recedes from the Sun with quasi-constant velocity.

Let us comment on the above points in their essential aspects. There could be many families of $\mathbf{L}$ of a set compliant with Definition 7.3. The three-$\mathbf{L}$ employed above in several configurations is a working element, but not the only one. In this respect, one may make distinction between *fast* and *enhanced fast* SPS on a quantitative basis. The former family of trajectories provides cruise speeds $V_{\text{cruise}} \lesssim V_p$, whereas the latter family regards higher cruise speeds, namely, $V_{\text{cruise}} > V_p$. This will appear clearer in the next chapter.

7.5 General Comments 317

H_z replaces the role the H-function has in two dimensions from the viewpoint of motion reversal. Although the respective meanings are alike, at this point it should be clear that the trajectory profiles satisfying Definition 7.1 *not* always are XY-projections of 3D trajectories meeting Definition 7.3; such projections cannot in general be got by merely setting $\mathcal{L}_n(t) = 0$ in the $\mathbf{L}$ history generating a 3D fast profile. Nevertheless, all the above examples shows that H is always positive in trajectories beginning counterclockwise. This can be viewed as a state constraint, which should be satisfied by fast sailcraft.

Definition 7.3 does not exclude thrust maneuvers (see Sect. 5.5). In the next chapters, we will use such potential for increasing the cruise speed of an extra-solar mission and/or achieving several goals via one sailcraft launch.

7.5 General Comments

We deem that this chapter needs some final remarks due to its complexity. The rationale of this chapter is to prove that special sets of SPS trajectories exist. Sailcraft with sufficiently high lightness numbers could achieve terminal speeds considerably higher than the orbital speed of the departure planet, i.e. the Earth, at least in realistic scenarios of in-space propulsion regarding the first half of the 21st century. One can object that nuclear-electric propulsion, or even sophisticated multiple gravity-assist planetary maneuvers could achieve the same goal. NEP is indeed the other realistic candidate not only for missions to the outer solar system, but also for interstellar precursor missions. SPS is *not* in contrast to NEP. They could "co-exist" very well inasmuch as a key point to be grasped is that the current Physics does **not** allow us to devise a *unique* propulsion type practically efficacious for *any* space mission. In addition, particular near-term missions could be accomplished efficiently via gravity-assist strategies.

Plainly, such three different approaches have remarkable own characteristics. With regard to SPS, the objective to achieve high terminal speeds is not the only one. Fast-SPS will not require any planetary flyby or expendable mass (aside from very small mass quantities by some backup attitude control devices), but shall be required to perform any pre-perihelion path in a short time. Thus, the phrase "fast solar sailing" refers to *both* high final speed and short time for achieving a quasi-cruise regime. As a point of fact, if we achieved a high cruise speed after many and more years, then the high-terminal-speed advantage would be greatly degraded. Fast solar propulsion vehicles could be directed along high heliocentric latitudes for a wide variety of targets.

How should be light sailcraft controlled for getting the just mentioned features? Well, the sections of this chapter show that a high-lightness-number sailcraft has to lose most of its initial energy (that is high because the radial number is high) in order to be able to acquire an amount much higher than that lost. Such an acquisition is possible only flying by the Sun closely. There exist two large sets of trajectories: one entails at least one motion change, namely, from direct to retrograde motion, or vice versa; the other one is a full-direct (or full-retrograde) motion. From a thrust maneuver viewpoint, the former is the simplest one, for which the orbital angular momentum evolution is essential.

3D SPS has been "triggered" by numerical evidences from and issues in integrating the motion equations. However, historically, this was a sort of loop inasmuch as the conventional orbit frame of reference showed some limitations. Thus, the first basic thing has been to extend the definition of heliocentric orbital frame. Then, in order to keep the mathematics manageable and, at the same time, to deal with very simple attitude controls (another basic thing for a realistic fast mission), piecewise-constant lightness vectors have been first discussed deeply for two-dimensional motions. Subsequently, the three-dimensional motion has required piecewise-continuous controls. We have explored several cases where some of the key parameters have been chosen in order to discover/emphasize trajectory effects, a sort of "partial derivative" algorithm; this procedure has revealed many features "hidden" in the highly non-linear dynamics of a fast sailcraft.

Once the mission analyst is aware of such so complicated features, she/he can go forward to more complicated aspects of a realistic sailcraft flight. Out of such aspects, some new items will be the object of the remaining two chapters.

References

1. Borisenko, A. I., Tarapov, I. E. (1979), Vector and Tensor Analysis with Applications. New York: Dover.
2. Gray, A., Abbena, E., Salamon, S. (2006), Modern Differential Geometry of Curves and Surfaces with Mathematica^{TM} (3rd edn.). London: Chapman and Hall/CRC. ISBN 1584884487.
3. Ince, E. L. (1926), Ordinary Differential Equations. New York: Dover. ISBN 0-486-60349-0.
4. Kühnel, W. (2006), Student Mathematical Library: Vol. 16. Differential Geometry: Curves-Surfaces-Manifolds (2nd edn.). Providence: American Mathematical Society. ISBN 0-8218-3988-8.
5. McInnes, C. R. (2004), Solar Sailing: Technology, Dynamics and Mission Applications (2nd edn.). Berlin: Springer-Praxis, ISBN 3540210628, ISBN 978-3540210627.
6. Matloff, G. L., Vulpetti, G., Bangs, C., Haggerty, R. (2002), The Interstellar Probe (ISP): Pre-Perihelion Trajectories and Application of Holography. NASA/CR-2002-211730.
7. Matloff, G. L. (2005), Deep-Space Probes (2nd edn.). Chichester: Springer-Praxis, ISBN 3-540-24772-6.
8. Mewaldt, R. A., Kangas, J., Kerridge, S. J., Neugebauer, M. (1995), A small interstellar probe to the heliospheric boundary and interstellar space. Acta Astronautica, 35, supplement.
9. Vulpetti, G. (1992), Missions to the heliopause and beyond by staged propulsion spacecraft. In IAF Congress-43 in The 1st World Space Congress, Washington, DC, 28 August–5 September 1992, paper IAA-92-0240.
10. Vulpetti, G. (1996), 3D high-speed escape heliocentric trajectories by all-metallic-sail low-mass sailcraft. Acta Astronautica, 39, 161–170.
11. Vulpetti, G. (1996), The AURORA project: light design of a technology demonstration mission. In 1st IAA International Symposium on Realistic Near-Term Advanced Scientific Space Missions, Turin, Italy, 25–27 June 1996.
12. Vulpetti, G. (1997), Sailcraft at high speed by orbital angular momentum reversal. Acta Astronautica, 40(10), 733–758.

References

13. Vulpetti, G. (1998), A sailing mode in space: 3D fast trajectories by orbital angular-momentum reversal. In NASA-JPL Ninth Advanced Space Propulsion Research Workshop and Conference, Pasadena, 11–13 March 1998.
14. Vulpetti, G. (1999), General 3D H-reversal trajectories for high-speed sailcraft. Academy Transactions Note, AA, 44(1), 67–73.
15. Vulpetti, G. (2006), The sailcraft splitting concept. In Fourth IAA International Symposium on Realistic Near-Term Advanced Scientific Space Missions, Aosta, Italy, 4–6 July 2005, also published on Journal of the British Interplanetary Society, 59(2), 48–53, 2006.
16. Vulpetti, G., Johnson, L., Matloff, G. L. (2008), Solar Sails, a Novel Approach to Interplanetary Travel. New York: Springer/Copernicus Books/Praxis. ISBN 978-0-387-34404-1, doi:10.1007/978-0-387-68500-7.

Part IV
Advanced Aspects

Applying the theory developed in the previous chapters to potential scientific missions is the main purpose of this final part of this book. In Chap. 8, special lightness vector profiles, based on what learnt from the previous chapters, are studied in order to increase the cruise speed of the fast trajectories. This investigation has been embedded in the general problem of optimizing solar-photon sailcraft, in particular that one of getting a very high cruise speed via a wise employment of mass-to-area ratio particularly low, since potentially realizable via nanoscience/nanotechnology. In Chap. 9, overall quantitative effects from variable total solar irradiance and optical degradation are analyzed. Both topics are still in their infancy in the solar-sail literature.

Chapter 8
Approach to SPS Trajectory Optimization

Classical Tools for Studying Advanced Missions via Solar-Photon Sailing
Classical calculus of variations, Pontryagin principle, and NLP provide very good tools for optimizing SPS trajectories with respect to some index of performance. In addition, many robust procedures there exist since many years in the major numerical libraries. In this chapter, we start from the established Pontryagin principle, and arrive to Non-Linear Programming. This process will be driven by some key properties of the SPS dynamical system, *and* by the basic guideline of keeping the time history of the sail attitude control as less difficult as possible. As far as one may envisage nowadays, this is one of the key points for getting a *viable* SPS, especially regarding fast large-sail sailcraft.

With regard to symbology, we will denote vectors by boldfaced upright Greek letters (e.g. $\boldsymbol{\Phi}$, $\boldsymbol{\Lambda}$, etc.) as well as by bold Roman letters. Normal-type capital Roman and Greek letters can denote matrices as well. Partial derivatives of scalars or vectors with respect to a vector will be denoted, for example, by $\partial_\mathbf{X} f$ or $\partial_\mathbf{X} \mathbf{F}^\top$, respectively. We extend the zero-vector convention to matrices, i.e. $A = 0$ will mean that all entries vanish.

8.1 Current Optimization Problem: Variational Approach

Let us re-write Eqs. (7.15) as a set of $n = 6$ first-order ordinary differential equations

$$\dot{\mathbf{R}} = \mathbf{V} \tag{8.1a}$$

$$\dot{\mathbf{V}} = \frac{\mu_\odot}{R^2}\left[-(1 - \mathcal{L}_r)\bar{\mathbf{r}} + \mathcal{L}_t\bar{\mathbf{h}} \times \bar{\mathbf{r}} + \mathcal{L}_n\bar{\mathbf{h}}\right] = -\frac{\mu_\odot}{R^3}\mathbf{R} + \frac{\mu_\odot}{R^2}\,\Xi\mathbf{L} \tag{8.1b}$$

$$\Xi \equiv (\bar{\mathbf{r}} \quad \bar{\mathbf{h}} \times \bar{\mathbf{r}} \quad \bar{\mathbf{h}}) \tag{8.1c}$$

where matrix Ξ consists of three orthonormal *independent* vectors. The control $\mathbf{L}$ in the system (8.1a)–(8.1c) has constraints. Before setting them formally, it is advisable to make some additional considerations. In the previous chapters, we focused

G. Vulpetti, *Fast Solar Sailing*, Space Technology Library 30,
DOI 10.1007/978-94-007-4777-7_8, © Springer Science+Business Media Dordrecht 2013

323

324   8 Approach to SPS Trajectory Optimization

the attention on trajectories with $\mathcal{L} < 1$, where again $\mathcal{L}$ denotes the length of $\mathbf{L}$. However, $\mathbf{L}$-maneuvers after the perihelion are not forbidden, and in point of fact we will see that very fast trajectories can be obtained via a control satisfying the constraints

$$0 < \mathcal{L}(t) < 1, \quad t \in [t_0, t_m], \ t_m > t_p$$
$$1 \leqslant \mathcal{L}(t) \leqslant \mathcal{L}_{max}, \quad t > t_m \tag{8.2}$$

Here, t_p is again the perihelion time, and $\mathcal{L}_{max}$ is the highest design value of the lightness number for a given mission to be realized. $\mathcal{L}_{max} > 1$ stands for (very) advanced sailcraft.

Now, it is known that, in the optimization of many rocket-propulsion missions, the on-off control of the thrust magnitude is determined by a switching function, while the thrust direction is provided by the primer vector. Throttling is a process generally independent of thrust vectoring, and this reflects various techniques even though throttling a many-engine system can be used to get the desired thrust direction. From this viewpoint, SPS needs a further clarification. The value of the mission-dependent $\mathcal{L}_{max}$ could be required during the flight, but the direction of vector $\mathbf{L}$ is forced to become radial. Although this is not a completely strict statement, this feature can be easily checked via Eq. (6.136). (For instance, if one models the sailcraft dynamics by taking the aberration of light into account, the direction of maximum thrust is very close, but not equal, to the radial one.) However, conceptually the behavior of the maximum thrust does not change. Said differently, once the sail materials have been selected, the sailcraft sail loading could be calculated by $\mathcal{L}_{max}$; if Eq. (6.136) is employed in the sailcraft design, then $\mathcal{L}_{max}$ is just the value obtained by $\mathbf{n} = \mathbf{u} = (1\ 0\ 0)^{\top}$. $\mathcal{L}_{max}$ does not need to be achieved during the flight (e.g. the results from the optimization process may indicate that the sail direction is nowhere radial). We can summarize the constraints on $\mathbf{L}$ in the following form:

$$0 < \mathcal{L} = \sqrt{\mathcal{L}_r^2 + \mathcal{L}_t^2 + \mathcal{L}_n^2} \leqslant \mathcal{L}_{max}, \quad 0 < \mathcal{L}_r \leqslant \mathcal{L}_{max},$$
$$\mathcal{L}_t^2 + \mathcal{L}_n^2 = \left(\mathcal{L}^2 - \mathcal{L}_r^2\right)\tilde{u}(\mathcal{L} - \mathcal{L}_{max}), \quad \tilde{u}(x) = \begin{cases} 1, & x < 0 \\ 0, & x = 0 \end{cases} \tag{8.3}$$

Thus, the transversal components of the lightness vector at the highest design value of $\mathcal{L}$ are necessarily zero both, and take on small values (of any sign) in small intervals $[\mathcal{L}_{max} - \varepsilon, \mathcal{L}_{max}]$, where $0 < \varepsilon \ll \mathcal{L}_{max}$. We will see the impact of these constraints on the optimal problem below.

A set of optimization problems relevant to fast sailing can be expressed as a Bolza problem with the following objective function:

$$J = \eta(t_f, \mathbf{R}_f, \mathbf{V}_f) + \int_{t_0}^{t_f} w(t, \mathbf{R}, \mathbf{V}, \mathbf{L})\, dt \tag{8.4}$$
$$w(t, \mathbf{R}, \mathbf{V}, \mathbf{L}) \equiv w_0(t, \mathbf{R}, \mathbf{V}) + \mathbf{w}_1(t, \mathbf{R}, \mathbf{V}) \cdot \mathbf{L}(t)$$

8.1 Current Optimization Problem: Variational Approach

where, for the moment, we consider t_0 fixed, while t_f may or may not be specified. J has to be minimized. As particular cases, it includes flight-time, terminal speed, sail loading, and combinations of them. A quadratic or another non-linear index of performance for fast trajectory has not yet been considered hitherto. This is crucial choice because it changes the whole optimization problem, as it will be apparent below.

Note that the dynamical system is non-linear in the state variables, but linear in the control expressed by the lightness vector. The Hamiltonian of this system is defined as:

$$
\begin{aligned}
\mathbb{H} &= w + \boldsymbol{\Psi} \cdot \mathbf{V} + \boldsymbol{\Lambda} \cdot \left(-\frac{\mu_\odot}{R^3} \mathbf{R} + \frac{\mu_\odot}{R^2} \Xi \mathbf{L} \right) \\
&= w + \boldsymbol{\Psi}^\top \mathbf{V} + \boldsymbol{\Lambda}^\top \left(-\frac{\mu_\odot}{R^3} \mathbf{R} + \frac{\mu_\odot}{R^2} \Xi \mathbf{L} \right)
\end{aligned} \tag{8.5}
$$

In the second line, we employed the matrix symbology for all vector elements. Vector $\mathbf{C} \equiv (\boldsymbol{\Psi}, \boldsymbol{\Lambda})$ is the co-state or the adjoint of the sailcraft state $\mathbf{S} \equiv (\mathbf{R}, \mathbf{V})$. It is a time-varying Lagrange vector multiplier. The evolution of the adjoint quantities is given by

$$
\dot{\boldsymbol{\Psi}} = -\partial_{\mathbf{R}} \mathbb{H} = -\partial_{\mathbf{R}} \left(w - \frac{\mu_\odot}{R^3} \boldsymbol{\Lambda}^\top \mathbf{R} + \frac{\mu_\odot}{R^2} \boldsymbol{\Lambda}^\top \Xi \mathbf{L} \right) \tag{8.6a}
$$

$$
\dot{\boldsymbol{\Lambda}} = -\partial_{\mathbf{V}} \mathbb{H} = -\boldsymbol{\Psi} - \partial_{\mathbf{V}} \left(w + \frac{\mu_\odot}{R^2} \boldsymbol{\Lambda}^\top \Xi \mathbf{L} \right) \tag{8.6b}
$$

because the definition of Ξ includes $\mathbf{R}$ and $\mathbf{V}$ explicitly.

We now formally state the possible state constraints and the implied transversality conditions. Postponing the problem of optimizing the flight epoch later in this chapter, we assume the following boundary conditions:

$$
\mathbf{R}(t_0) = \mathbf{R}_0, \qquad \mathbf{V}(t_0) = \mathbf{V}_0 \tag{8.7a}
$$

$$
\boldsymbol{\beta}\left(t_f, \mathbf{R}(t_f), \mathbf{V}(t_f)\right) = 0 \tag{8.7b}
$$

in which $\boldsymbol{\beta}$ is an s-dimensional vector with $s \leqslant n = 6$. Thus, at the stopping time, the state is generally assigned through implicit-form constraints lower than n in number. The transversality conditions give the expression of the co-state at the terminal time:

$$
\mathbf{C}(t_f) = (\partial_{\mathbf{S}} \eta)_{t_f} + \left(\partial_{\mathbf{S}} \boldsymbol{\beta}^\top \mathbf{a}\right)_{t_f} \tag{8.8}
$$

of which only $n - s$ components are independent because $\mathbf{a}$ is an undetermined s-dimensional vector of constant Lagrange multipliers; $\mathbf{a}$ has to be found as a part of the solution.

The vector control may be discontinuous at some (corner) point(s); nevertheless, *in the current framework*, the Erdmann-Weierstrass conditions ensure that Hamiltonian and co-state are continuous also at corner points. Because we known that SPS does not allow any state discontinuity but mass (which has not yet been considered in the motion equations), we can assert that $\mathbf{S}$, $\mathbf{C}$, and $\mathbb{H}$ are continuous functions in $[t_0, t_f]$, the whole flight time.

However, we have to mention two circumstances of great general interest, for which the previous statement does not hold.

- *State Constraints.* An optimized trajectory may contain finite arcs belonging to some hyper-surface of the state space, the equation of which is pre-assigned. More generally, a subset of the state variables could be required to evolve only in certain regions of the n-dimensional space; such regions are described by the state-only inequality constraints $\phi_j(\mathbf{S}(t)) \geqslant 0$, $j = 1..r < n$, $t \in [t_e, t_x]$ (subscripts e and x stands for *entry* and *exit*, respectively). Plenty of literature (in the course of over half a century) has regarded state constraints; here, we like to signal only a few problems that such constraints induce. For the sake of simplicity, we will restrict to $r = 1$ because the generalization is formally straightforward. If the state $\mathbf{S}(t)$ is on a segment completely internal to the domain D_ϕ of $\phi(\mathbf{S}(t)) \geqslant 0$, then there is no problem in describing the spacecraft motion. The problem arises when the optimal state joins the boundary of D_ϕ at the entry time t_e, and has to stay on this boundary until time t_x, at which it exits. Of course, depending on the whole problem, it may possible that $t_e \equiv t_0$ and/or $t_x \equiv t_f$. There have been devised various methods for dealing with active state constraint(s). If one does not want to leave the variational formulation, then the key point is that time differentiating the equality $\phi(\mathbf{S}(t)) = 0$ eventually turns into an equation where the control vector appears explicitly (unless one has put some pathological constraint). For instance, if the constraint contains all components of $\mathbf{S}$, then the control could be made explicit in the first derivative:

$$\dot{\phi}(\mathbf{S}(t)) = \dot{\mathbf{S}}^\top \partial_{\mathbf{S}}\phi = \mathbf{f}^\top \partial_{\mathbf{S}}\phi = 0 \tag{8.9}$$

where the vector $\mathbf{f} = \mathbf{f}(\mathbf{S}, \mathbf{U}, t)$ denotes the general right-hand side of the motion equations. Time derivatives up to order $\bar{k}$ may be required for taking the control out; thus, $\bar{k} + 1$ relationships of the form $\phi^{(k)} = 0$ are built, the last of which can be viewed as the state/control condition to be satisfied for the state is on the boundary for from t_e to t_x. In general, the striking feature from state constraints is that the co-state $\mathbf{C}$ becomes discontinuous at the entry instant t_e, and additional multipliers have to be introduced; they are generally in different number for each constraint because the various orders $\bar{k}$ may be unequal.
- *State Discontinuity.* Perhaps, the most classical example in astrodynamics is the optimization of the ascent phase of a multi-stage rocket trajectory, where the launcher mass undergoes jumps at times given by a set of implicit relationships. In general, using the variational approach on the initially-established coordinates results in discontinuity of the adjoint state and of the Hamiltonian at the state jump instants. This increases the difficulty of getting a satisfactorily-converging solution of the full set of variables. Another way to face with the problem without rejecting the variational method consists of augmenting the dimensionality of the state and the co-state. Considering the optimization of an N_s-stage launcher

8.1 Current Optimization Problem: Variational Approach

carrying some payload from a ground site to a given orbit, the launcher mass m may be replaced by N_s variables, namely, the stage propellant masses. Such state variables, acting in separate burning phases, are continuous with their own initial values and time derivatives. Of course, there are N_s related adjoint variables. A very good example is given in [56] for the Ariane-5 launcher. A fast sailcraft may jettison a mass (which may be another spacecraft with no sail), and proceeds with a notable higher lightness number, whereupon is admissible some **L**-maneuver(s) resulting in higher speed.

We will include the above features in sailcraft trajectory optimization later, under a different optimization method. For now, we will go on discussing via Pontryagin's principle in order to highlight the major issues of the current problem.

As t_f is left free in general, this lack of information is balanced by the last (transversality) condition giving the optimal Hamiltonian at the terminal time:

$$\mathbb{H}_f^* \equiv \mathbb{H}(t_f, \mathbf{S}^*, \mathbf{C}^*, \mathbf{L}^*) = -(\partial_t \eta)_{t_f} - \left(\partial_t \boldsymbol{\beta}^\top \mathbf{a}\right)_{t_f} \tag{8.10}$$

where the right-hand side has to be calculated at $\mathbf{S}^*$. Thus, relationships (8.7a), (8.7b), (8.8)–(8.10) provide $2n + 1$ conditions for the current-problem equations.

In fast sailing, Eqs. (8.8)–(8.10) can be considerably simplified by the choice of the functions entering the definitions of η and w. However, the real difficulty one has to face with comes from the linearity of the hamiltonian and the structure of the orthogonal matrix Ξ.

The strong form of the minimum principle[1] states that

$$\mathbb{H}(t, \mathbf{S}^*, \mathbf{C}^*, \mathbf{L}^*) \leqslant \mathbb{H}(t, \mathbf{S}^*, \mathbf{C}^*, \mathbf{L}) \tag{8.11}$$

where ()* denotes optimality. Both the optimal control $\mathbf{L}^*$, if any, and the generic control $\mathbf{L}$ belong to the set of admissible controls. Inequality (8.11) is accompanied by the property that $\mathbf{S}^* \neq 0, \forall t \in [t_0, t_f]$.

After inserting the definition of $w(t)$, i.e. the second line in (8.4), into inequality (8.11) and re-arranging terms, the Hamiltonian is minimized if the function

$$(\partial_\mathbf{L} \mathbb{H})^\top \mathbf{L} = \left(\mathbf{w}_1^\top + \frac{\mu_\odot}{R^2} \boldsymbol{\Lambda}^\top \Xi \right) \mathbf{L} \equiv \mathbf{F}^\top \mathbf{L} \tag{8.12}$$

is minimized along the trajectory. When $\mathbf{w}_1$ is assigned, it is usually non-zero (but, at most, at a finite number of times) whereupon vector $\mathbf{F} \neq 0$ but at most in the instants of a set $\{t_j'\}$, $j = 1..n' < \infty$; then, in order to minimize the above dot-product, the control $\mathbf{L}$ has to be chosen antiparallel to vector $\mathbf{F}$ where $\|\mathbf{F}\| > 0$.

At this point, one might think to give $\|\mathbf{L}\|$ its maximum value $\mathcal{L}_{max}$ for obtaining the lowest value of the product at each point of the trajectory. However, in the

---

[1]Historically, the principle was formulated by L.S. Pontryagin by for maximization problems; subsequently, the "minimum version" arose straightforwardly because of the numerous practical minimization problems in the various areas of research and engineering.

case of SPS, the two "minimizing operations" are conflicting. As a point of fact, because of the constraints (8.3) and the related considerations, employing the maximum allowed value of $\mathcal{L}$ would preclude the flexibility of changing direction and so utilizing the non-conservative field via $\mathcal{L}_t \neq 0$ and $\mathcal{L}_n \neq 0$, which were shown to be so important in fast sailing (even though not yet optimized). Thus, the only information one may get is that

$$\mathbf{L}/\|\mathbf{L}\| = -\mathbf{F}/\|\mathbf{F}\|, \quad \|\mathbf{F}\| > 0 \tag{8.13}$$

but one does not know which $\mathcal{L}(t)$ renders the trajectory the best one with respect to the chosen objective function.

One may object that, once the sailcraft sail loading is dictated from other requirements (e.g. space-agency's budget and objectives, mission-implied technological developments, spinoffs, and so forth), the complete control can be carried out. Yes, but some indices of performance would be ruled out; for instance, maximization of the terminal speed for sail loading values higher than 2 g/m^2 would lose its meaning. Nevertheless, depending on the endpoints of $\mathbf{S}$, the radial mode may result to be the optimal one.

Pontryagin principle is not able to provide us with useful information about $\mathbf{L}$ if, for a finite interval of time, $\partial_\mathbf{L} \mathbb{H} = 0$. There may be values of the $\mathbf{L}$'s components inside their respective boundaries that could minimize the objective function. This would be the case for an optimal problem that admits singular arcs; this is not infrequent in process control and optimal designs. Here, the occurrence of singularity has to be ascribed to the Hamiltonian that is linear in the control.

In general, if the Hamiltonian is not linear in every component of the control, singularity could arise during a time interval $\mathcal{T}^{(s)} \subseteq [t_0, t_f]$ if the Hessian matrix $\partial_\mathbf{L}^2 \mathbb{H}$ were singular. Definitions of singular control are given—even differently—in many textbooks and papers; e.g. see [6, 13, 34, 46, 58, 66, 72, 76], and the many references inside.

As noted above, a further complication in SPS is due to the control vectorial. In order to understand and extract information about the control in singular problems, two additional (not equivalent) *necessary* conditions have been found: (1) the *generalized* Legendre-Clebsch condition, and (2) the Jacobson condition. The complexity of the minimization problem increases hugely; in particular, the first condition requires a sequence of time differentiations of $\partial_\mathbf{L} \mathbb{H}$ by employing the state and co-state equations (until the control re-appears explicitly), whereas the second condition entails a new $n \times n$ matrix differential equation. In the above-cited references, the interested reader can find plenty of work made by many investigators in the course of over fifty years, and the treatment of several important problems such as the singular problem order, the order(s) of singular subarc(s), the junction conditions, controls with non-singular and singular components, and others. It is beyond the scope of such book to apply such theoretical tools to fast sailing. At least to the author knowledge, no detailed analysis of *singular* SPS aspects has been done hitherto in literature. Therefore, this research area is quite open.

8.2 Current Optimization Problem: Non-Linear Programming

At this point, we have to do a number of observations we deem worthy of note for going on with a framework inside which an optimization of fast SPS trajectories makes sense from theoretical and practical viewpoints:

i. Several numerical methods for singular optimal problems have been proposed, investigated, and reviewed, e.g. (in chronological order) [6, 18–20, 25, 36, 44], and [15]. In particular, it is enlightening the concept on which the computational method in [25] and [6] relies. To the author knowledge, this concept is still valid even though it should not be considered as a sort of panacea for singular controls: it depends on the problems to be solved, of course. Because non-singular optimal problems, and related computer algorithms, are less difficult to deal with and implement than singular optimal algorithms, then it may be convenient to transform a singular problem into a non-singular problem sequence, the limit of which coincides with the original singular problem. Thus, the initial objective function J is modified as $\tilde{J} = J + \frac{1}{2}\varepsilon \int_{t_0}^{t_f} \mathbf{U}^\top \mathbf{U}\, dt$. In reality, the continuous parameter ε is realized through a discrete sequence of decreasing ε_k. When the parameter is given some very small value, the non-singular optimal problem solution is considered to be coincident with the searched singular optimal solution. *With regard to SPS, would-it be suitable to keep the advantage of the concept, i.e. the conversion into a non-singular problem, and avoid serious numerical convergence issues?*

ii. Suppose for a moment one has developed a full theoretical variational-based tool for SPS as formulated above, and also has implemented a code solving numerically the problem. Suppose now one likes to add planetary perturbations for more realistic flight design; in addition, it is probable that—in the course of years—actual sailcraft designs require more complicated state and/or state-control constraints, or various discontinuities in state, or something really more complicated such as the optical degradation. Though modern computer languages allow a powerful, flexible, and efficient architecture in coding, nevertheless most of the theoretical calculation has to be "updated" materially (albeit the theoretical framework is the same). *Would-it be suitable to have theoretical and computational tools less arduous to "render accordant" to the evolution of new mission concepts?*

iii. Finally, suppose one has achieved the best solution, really the optimal one, to a family of SPS trajectory problems. Query is the following: is the obtained attitude profile *viable*? One should not forget that controlling a very extended object in space is not an easy task. A difficult attitude control would mean additional mass and reliability lessening. *Would-it be suitable to get a generally suboptimal control, which is more realizable and less difficult to compute?*

8.2 Current Optimization Problem: Non-Linear Programming

In order to answer affirmatively to the three questions set above, we re-formulate the problem of optimization of SPS trajectories, especially with regard to fast sailing,

330   8 Approach to SPS Trajectory Optimization

in terms of Non-Linear Programming. We build the current optimization problem by steps marked by $\boxed{\sharp \twoheadrightarrow}$, where $\sharp = 1..7$. Whenever it is possible, we will use the same nomenclature(s) of Sect. 8.1 and of the previous chapters.

$\boxed{1 \twoheadrightarrow}$ The first step consists of transforming the current linear-in-control dynamical system into a system non-linear in control; this can be done via the vector equation of the type

$$\mathbf{L} = \mathbf{L}(\sigma, \mathbf{n}) \tag{8.14}$$

of which a meaningful example is given by Eq. (6.136); (vector $\mathbf{x}_s$ is a function of $\mathbf{n}$ only, because $\mathbf{u} = \mathbf{R}/R$ is a constant in EHOF). Question: since we are changing the formulation and the method of the optimization problem, why the need of converting it into a non-linear control system? The main reason is the "management" of the control constraints (8.3) in realistic way, namely, through an optical diffraction/absorption/re-radiation model, as we saw in detail in Chap. 6. On the other side, what we have learnt through the employment of the lightness vector will be taken into account explicitly in the computation of optimal profiles in the NLP sense; this is conceptually correct inasmuch as the acceleration equation is expressed via the transformation: from acceleration-in-EHOF to acceleration-in-HIF.

$\boxed{2 \twoheadrightarrow}$ The second step is to replace the Bolza functional (8.4) by an approximating objective function encompassing many problems of great interest in the SPS optimization. In the fast-sailing system of equations expressed in terms of $\mathbf{n}$ and σ, $\|\mathbf{n}\|$ is fixed, while an important aspect is to allocate a possible mass separation for enhancing the mission capabilities. One can handle several realistic criteria of the form

$$\mathfrak{J} = \mathfrak{J}\big(t_f, \mathbf{R}(t_f), \mathbf{V}(t_f), m(t_f)\big) \tag{8.15}$$

where the terminal time may be assigned or left free. Functions of this type cover various linear and non-linear criterion problems. One will note that the sailcraft state is now seven-dimensional; as a point of fact, not only NLP allows the mission analyst to include mass discontinuity in the problem with a difficulty considerably lower than in the variational approach, but also may easily include the option of a (rocket-based) sail attitude control subsystem that ejects mass.

$\boxed{3 \twoheadrightarrow}$ The third step regards the sailcraft motion equations. We should include realistic features modeled in the framework of ODEs. To such aim, we have to recall what discussed in Sect. 5.3, in particular Eqs. (5.16)–(5.17) on p. 138. Thus, we consider sailcraft trajectories for which sailcraft is sufficiently far from the planets for the motion is not planetocentric, but still sensitive to their gravitational perturbations to be non-negligible in high-precision flight computation. For the sake of clarity, we restore some symbols employed in Chap. 5. Taking the vector $(\mathbf{R}_{\odot\square}, \mathbf{V}_\square, m_\square)$ as the sailcraft's dynamical state perturbed by N_{ncb} planets ($\bullet$), the here-adopted equations of the motion in HIF are the following:

$$\dot{\mathbf{R}}_{\odot\square} = \mathbf{V}_\square \tag{8.16a}$$

8.2 Current Optimization Problem: Non-Linear Programming

$$\dot{\mathbf{V}}_\square = -\mu_\odot \frac{\mathbf{R}_{\odot\square}}{\|\mathbf{R}_{\odot\square}\|^3} + \frac{\mu_\odot}{\|\mathbf{R}_{\odot\square}\|^2} \Xi \mathbf{L}(\sigma, \mathbf{n}) + \sum_{\bullet=1}^{N_{ncb}} \left(\mathbf{K}_\square^{(\bullet)} - \mathbf{K}_\odot^{(\bullet)}\right) \tag{8.16b}$$

$$\dot{m}_\square = -\dot{m}_{(ac)}(t) - \Delta m\, \delta(t - t_{(sp)}) \tag{8.16c}$$

where again $\mathbf{R}_{ab} \equiv \mathbf{R}_b - \mathbf{R}_a$. Some remarks about Eqs. (8.16a)–(8.16c) are in order.

Remark 8.1 An optimal problem involving continuous state variables can be approximated by a discrete problem. The numerical integration of ODEs can be viewed as a sequence of logical stages, each of which processes the output of the previous stage, and applies the current control for computing the output at the subsequent time tag. (The treatment of the initial stage is plain.) Via the calculated full trajectory in the phase space, the final and the intermediate states are available for the objective function, while the terminal boundary conditions and the intermediate constraints (if any) can be updated according to the chosen optimization algorithm. Thus, though Eqs. (8.16a)–(8.16c) are not algebraic equations of evolution, NLP can be employed very well. This point will influence our choice of the specific NLP optimization method (Sect. 8.4).

Remark 8.2 Vector function $\mathbf{L}(\sigma, \mathbf{n})$ in Eq. (8.16b) means that the flight designer has chosen a model of the sunlight-sail interaction, which characterizes the role of the sail axis in EHOF and the sailcraft sail loading as the control quantities. Some of them may be piecewise-constant parameters, but some others may be piecewise-continuous functions, the parameters of which will undergo optimization. A meaningful example consists of a control sequence like the three-$\mathbf{L}$ smooth control analyzed in detail in Chap. 7. We will use a more sophisticated control too, on the wake of the properties precisely found in that chapter.

If we want to deal with a sailcraft perturbed by gravitational bodies, then the total system contains the acceleration equations of all involved bodies. Although the current motion model does not consider General Relativity, however we will use the set of JPL planetary ephemeris files, the main properties of which have been mentioned on p. 130. Let us remind the reader that time in Eqs. (8.16a)–(8.16c) is TDB, according to the definition of HIF made in Sect. 5.3 with regard to Eqs. (5.16).

Remark 8.3 Subscript (ac) in Eq. (8.16c) stands for some attitude control entailing mass ejection, but with a zero total momentum with respect to the sailcraft barycenter for not causing additional perturbation to the sailcraft motion. $\dot{m}_{(ac)}(t)$ is the mass flow rate, which may be assumed a non-zero constant in some small subinterval of the flight time, because the main attitude control system should be propellantless. However, either as backup or parallel subsystem in some crucial trajectory arc, the impact of (micro)rocket-based attitude control subsystems should be analyzed also as a possible trajectory perturbation. This rather long and specific task will not be pursued here. In the mission examples of the next sub-sections, the baseline is $\dot{m}_{(ac)} = 0$ throughout the flight.

The rightmost term in Eq. (8.16c) accounts for a possible splitting-up of the sailcraft. An example is a two-spacecraft sailcraft mission with very different payloads, e.g. one spacecraft (without sail) is released on an orbit partially internal to Mercury, while the other one (with the sail) becomes an extra-solar probe. Thus, with the above baseline, Δm is given by the mass of the first spacecraft. In general, separation time $t_{(sp)}$ is not known a-priori, but is one of the optimization outcomes.

$\boxed{4 \twoheadrightarrow}$ The fourth step is devoted to set up boundary conditions: equality and inequality constraints of the current NLP problem. In a more realistic trajectory computation, the sailcraft does not start from some arbitrary point of a solar system with no planets; instead, after a planetocentric flight carried out by either the launcher's upper stage or its sail (depending on σ), it achieves a point sufficiently far from the departure planet that the trajectory can be described "heliocentrically". In the family of fast trajectories, we consider the Earth-Moon as the departure system. Without computing the planetocentric phase explicitly, we will prefix the distance of the sailcraft from Earth-Moon barycenter (EMB), and leave its direction free. The sailcraft speed in the EMB-centered frame will depend on the launcher payload capability, while its direction may be fixed or left free. For example, this frame may have the axes parallel to those ones of HIF or may coincide with the heliocentric orbital frame of the EMB (EMB/OF). In addition, since EMB follows its continuously-evolving orbit about the Sun, trajectories towards a distant target can be particularly sensitive to the flight epoch, namely, the Julian Date of the "injection" of the sailcraft into the solar gravitational field. As a result, t_0 could be a parameter to be optimized. We can then write the initial conditions as follows:

$$\left\| \mathbf{R}_{\odot\square}(t_0) - \mathbf{R}_{\odot\circledcirc}(t_0) \right\| = r_{(outer)} \tag{8.17a}$$

$$0 \leqslant \left\| \mathbf{V}_{\square}(t_0) - \mathbf{V}_{\circledcirc}(t_0) \right\| \leqslant v_{(max)} \tag{8.17b}$$

$$m_{\square}(t_0) = m_0 \leftrightarrows \boxed{\text{S/C design}} \tag{8.17c}$$

where $\circledcirc$ denotes the EMB, and $r_{(outer)}$ is the value of the outer radius of the "sphere" of Earth's influence on the sailcraft corresponding to a small fraction (usually 0.01) of the two-body heliocentric acceleration; a value of 0.01776 AU will be employed accordingly to [62],[2] and numerical calculations of the restricted four-body problem. After injection, sailcraft is assumed to be driven mainly by solar gravity and radiation pressure, and perturbed by planets, including the Earth-Moon

---

[2]Qualitatively speaking, in the phase space, there is a transition region, or the weak stability boundaries (WSB), between gravitational capture and escape from the Moon. A region of chaotic behavior in the Earth-Moon system was originally studied in [3] and [4], and then in many published papers on celestial mechanics, e.g. [5]. In such studies, distances at which non-conventional orbit transfers may be realized are about 1.5 millions of kilometers. The nature of the WSB has been also the topic of university theses, e.g. in [12], and an extended study (where low-thrust propulsion is added) can be found in [71]. The value given here to $r_{(outer)}$, i.e. $\cong 2.7 \times 10^6$ km, stems also from the advantage that the sailcraft injection should be planned well farther than WSBs.

8.2 Current Optimization Problem: Non-Linear Programming

system as a whole. Constraint (8.17b) stems from the previous considerations plus the following: $v_{(max)}$ may be lower than what allowed by the launcher because, depending on t_0, a high hyperbolic excess may be incompatible with trajectories involving angular momentum reversal. In some cases, $v_{(max)} = 0$ is a good guess to optimize, or just a realistic choice.

In general, the injection-referred control parameters can be grouped as

$$\mathbb{U}_0 \equiv (t_0, \boldsymbol{\rho}_{\odot\square}, \boldsymbol{v}_\square)$$

$$\boldsymbol{\rho}_{\odot\square} = (\mathbf{R}_\square - \mathbf{R}_\odot)_{t_0}^{(\text{EMB/OF})}, \qquad \boldsymbol{v}_\square = (\mathbf{V}_\square - \mathbf{V}_\odot)_{t_0}^{(\text{EMB/OF})} \qquad (8.18)$$

Injection time, sailcraft position and velocity—resolved in EMB/OF at the injection time—give sailcraft trajectory optimization a great flexibility.

Note 8.1 Box in the line (8.17c) points out that sailcraft mass at epoch will be guessed through a mass breakdown model of sailcraft based on the design of the major systems; mass model and trajectory profiles "interact". The optimal trajectory computation is therefore an iterative process. During a real project, such "interaction" does not develop in a short time; instead, it is a long process where many requirements and constraints—relevant to the vehicle, its payload(s), and goal(s)—have to be taken into account.

With regard to the terminal constraints, in principle we have two intriguing deep-space mission kinds to be dealt with: (1) to flight close to massive object(s) of the Kuiper belt, e.g. the Haumean system [55], and (2) to point to (a set of) extra-solar targets. The success of the mission set of kind-1 presumably depends not only on accurate knowledge of the trans-neptunian object ephemerides, but also on the engineering of a hybrid propulsion like sailing+ion/propulsion.[3] Since we are dealing with fast trajectories with possible extension to farther targets, is the kind-2 that is more appropriate here; therefore,

$$R_{\odot\square}(t_f) = R_f, \qquad \psi(t_f) = \psi_f, \qquad \theta(t_f) = \theta_f \qquad (8.19)$$

where using heliocentric distance-longitude-latitude coordinates, or (R, ψ, θ) respectively, for position is more meaningful than referring to the Cartesian form.[4] We specified no terminal condition regarding velocity for two reasons. One is that the terminal speed enters the objective function in the current framework; the other one is that, in fast sailing, the last trajectory section is practically rectilinear, namely, the direction of the velocity practically coincides with the position direction, which already is specified in conditions (8.19).

---

[3]Though not applied to flyby or rendezvous of distant celestial bodies, the concept of such a hybrid system can be traced back to [75].

[4]In any case, internal computations in our trajectory computer code are accomplished via normalized Cartesian components in order to get the highest precision/accuracy after many tens of billions of floating-point operations.

334   8 Approach to SPS Trajectory Optimization

$\boxed{5 \rightarrow\!\!\!\rightarrow}$ In the fifth step, we set intermediate constraints; we begin with

$$R_p \leqslant \|\mathbf{R}_{\odot\square}(t_p)\| \leqslant R_p + \Delta R_p \tag{8.20a}$$

$$T_s(t) \leqslant T_s^{(max)} \tag{8.20b}$$

$$m_\square(t_{(sp)}^+) = m_\square(t_{(sp)}^-) - \Delta m, \quad t_{(sp)} > t_p \tag{8.20c}$$

$$0 < H(t) \ll H_0, \quad \forall t \in [t_1^*, t_2^*] \supseteq \mathcal{T}^{(H_z)} \tag{8.20d}$$

where $\mathcal{T}^{(H_z)}$ denotes the shortest time interval containing all instants at which $H_z = 0$. For instance, in Case 7.7 (p. 313), $\mathcal{T}^{(H_z)}$ is the interval from the first to the third zero of H_z in the inset of the center-left plot of Fig. 7.14. Inequalities (8.20a)–(8.20d) need some additional explanation.

- Inequality (8.20a) represents a corridor or lane outside of which either the sailcraft sail-system working is not guaranteed or the main objective of the mission is degraded. Below, we will see that the ratio $\Delta R_p / R_p$ may be somewhat low from trajectory viewpoint.
- Inequality (8.20b) appears obvious, and one may restrict its validity to a neighborhood of the perihelion. However, because a post-perihelion attitude maneuver cannot be excluded when the distance is not much different from R_p, the new incidence angle of sunlight may cause the sail temperature to increase beyond its upper bound. It is safer to extend the temperature constraint over the flight time, formally speaking, also because the maneuver time is not known a priori.
- Inequality (8.20c) refers to the mass separation, the instant of which is not known before the optimization begins; therefore either it should be optimized or it may be prefixed by inferring it from a set of optimizations without separation.
- Inequality (8.20d) is based on Proposition 7.4 on p. 299; dealing with sailcraft starting counterclockwise, this inequality (if set in the optimization problem) allows the mission analyst to employ the many properties found in Sect. 7.4.3 in order to guess the related trajectory. If not set, the code(s) may find different families of trajectories, depending on the chosen objective function.

Sailcraft sail loading has lower and upper bounds:

$$\sigma_{(min)} < \sigma \leqslant \sigma_{(max)} \tag{8.21}$$

From [43], we can set $\sigma_{(max)} = 2.1$ g/m^2. The **L** model employed in [43], though a flat-sail one, was more sophisticated than the current one.[5] For a sail with Aluminum as the reflective layer, the reported value of $\sigma_{(max)}$ could be considered that value above which no H-reversal trajectory can be obtained. With regard to $\sigma_{(min)}$, let us first model the whole sail system as follows:

$$m_S/A = \sigma_{BS} + (m_B + m_D)/A \tag{8.22}$$

---

[5]Finite-size Sun and limb-darkening in any sail orientation, sail's optical degradation, and Doppler-effect were considered in the lightness vector model, but there was no backscattering included.

8.2 Current Optimization Problem: Non-Linear Programming

where A is the actual sail area; m_S denotes the mass of the sail system, which we have split into three sub-systems: (1) the bare sail with mass $m_{BS} = A\sigma_{BS}$, (2) the booms or support sub-system having mass m_B, and (3) the deployment sub-system with mass m_D. The flight design feasibility study should issue which materials, and their thicknesses, should be adopted for the sail of the planned mission. Then, the calculation of the bare-sail loading σ_{BS} is straightforward; such value is practically independent of the rest of sailcraft. For various architectures, m_B scales closely to $\sqrt{A}$; to minimize the boom's mass per unit length (or the boom's specific mass) is a goal in designing a boom-based sail support sub-system [26]. This specific mass varies also with the chosen sail-supporting architecture. m_D may be assumed a pre-fixed fraction of m_B. These considerations induce us to take σ_{BS} as the smallest of the lower bounds of the σ. Such limit is achievable by no real sailcraft to design/optimize; therefore,

$$\sigma_{(min)} > \sigma_{BS} = \sum_{j=1}^{N_l} \rho_j \delta_j, \quad \delta_j > \delta_{j,(min)} \tag{8.23}$$

where ρ_j and δ_j denote the density and the thickness of the material forming the layer-j of the sail. $\delta_{j,(min)}$ is the minimum mean thickness for each material; it depends mainly on physical and construction reasons (see Sect. 6.3).

$\boxed{6 \rightarrow}$ Differently from the variational approach, for which a general switching function logically divides the whole trajectory in the phase space in sub-arcs where (generally distinct) controls are active, here the 3D trajectory is segmented in a number of arcs, which determines the overall number of degrees of freedom. The following control sequence is considered here:

Control Parameters:

1. The whole trajectory is divided in five thrusting arcs, and an additional terminal coasting arc if suitable for the objective function. The first three arcs, from the injection time to an instant shortly after the perihelion, look like the three-**L** profile detailed in Sect. 7.4. In particular, in order to keep as simple the implementation of the control as possible, one may begin with searching for optimal controls of the following form (if any):
   a. $\mathbf{n}^{(1)}$ constant in EHOF for $\Delta t^{(1)} \equiv t_1 - t_0$.
   b. $\Xi\mathbf{n}^{(2)}$ constant in HIF for $\Delta t^{(2)} \equiv t_2 - t_1$, containing the instant $t^{\circledast}$ of the base profile defined in Note 7.6 on p. 299.
   c. $\mathbf{n}^{(3)}$ constant in EHOF for $\Delta t^{(3)} \equiv t_3 - t_2$; perihelion time belongs to the interior of $\Delta t^{(3)}$.
2. The fourth arc and fifth arc are devoted to either mass release or **L**-maneuver, or both.
   i. $\mathbf{n}^{(4)}$ constant in EHOF for $\Delta t^{(4)} \equiv t_4 - t_3$.
   ii. $\mathbf{n}^{(5)}$ constant in EHOF for $\Delta t^{(5)} \equiv t_5 - t_4$, and $t_5 = t_f$ if there is no coasting arc.
3. A terminal coasting may be included as the arc No. 6, as mentioned. This may be appropriate if, for instance, the target distance is outside the heliosphere and the

336   8 Approach to SPS Trajectory Optimization

objective function does not change appreciably if the sail jettisoned at t_5. $\Delta t^{(j)}$, ($j = 1..5$) or $\Delta t^{(j)}$ ($j = 1..6$) (if coasting is requested) are to be optimized as well.

Each thrusting-arc time may or may not be restricted in an interval depending, for example, on the requirement of using the reversal mode instead of the full direct motion mode, or viceversa.

Remark 8.4 Because the thrust direction in EHOF has to be transformed via the EHOF-to-HIF rotation matrix Ξ, which varies with time, then $\mathbf{L}$ gives rise to a thrust that is piecewise-continuous vector function in HIF, in general. In order to avoid (strong) attitude re-orientation (but the initial one) during the pre-perihelion phase, continuity is required between thrusting arcs 1 and 2, and between arcs 2 and 3. Specifically, we set

$$(\Xi \mathbf{n})_{t_1} = \mathbf{n}^{(\mathrm{HIF})}(t_1)$$
$$\mathbf{n}^{(\mathrm{HIF})}(t_2) = (\Xi \mathbf{n})_{t_2}$$
(8.24)

where again $\mathbf{n}$ is resolved in EHOF, and $\mathbf{n}^{(\mathrm{HIF})}$ denotes the sail axis direction given directly in HIF. These equalities represent realistic situations to be added to the points on p. 335.

Because mass jettisoning and $\mathbf{L}$-maneuver could not be managed if both events were temporarily too close to each other, the fourth sailing arc should contain the jettisoning, whereas the $\mathbf{L}$-maneuver should be allocated in the fifth arc. The reverse sequence is not suitable, unless the target orbit of the released mass requires an adjustment of the main trajectory.

Depending on the chosen method for controlling $\mathbf{n}$, there may be bounds for the control angles. Avoiding large sunlight incidence angles (as emphasized in Chap. 6) could be another physics-driven need.

$\boxed{7 \twoheadrightarrow}$ We now are able to state the current NLP optimization problem formally.

Problem 8.1 Given the equations (8.16a)–(8.16c), describing the evolution (in the solar system) of the vector state $\mathbf{S} \equiv (\mathbf{R}_{\odot\square}, \mathbf{V}_{\square}, m_{\square})$ of a sailcraft driven from $\mathbf{S}_0$ to $\mathbf{S}_f$ (partially-fixed endpoints) by the control $\mathbf{U} \equiv \mathbb{U}_0 \cup \{\sigma, \Delta t^{(j)}, \mathbf{n}^{(j)}\}$—where $\mathbb{U}_0$ was defined in (8.18), and $j = 1..n_s$, $n_s = 5, 6$ according to the control sequence at step $\boxed{6 \twoheadrightarrow}$—find the control $\mathbf{U}^{(opt)}$ that minimizes the objective function (8.15), subject to state/control equality and inequality constraints as expressed by (8.17a)–(8.17c), (8.19), (8.20a)–(8.20d), (8.21), and (8.23).

8.3 The Photon-Acceleration Technological Equivalent

In this section, we will introduce two concepts useful for trajectory/sailcraft design: the thrust efficiency and the technological equivalent of SPS acceleration. The sec-

8.3 The Photon-Acceleration Technological Equivalent 337

ond one utilizes the first concept. The concept of SPS thrust efficiency may be even utilized for the analysis of the results in Sect. 8.4.

Suppose to have an ideal sailcraft with a *flat* and *perfectly*-reflecting sail at rest in a certain position $\mathbf{R}$. Then, with $\mathbf{n}$ parallel to $\mathbf{R}/R$, this sailcraft will have a lightness number $\mathcal{L}_{(ideal)} = \sigma_{(cr)}/\sigma$. Now, if another sailcraft at $\mathbf{R}$ exhibits an actual $\mathcal{L}$, then we define the following ratio

$$\eta_{(thrust)} = \frac{\mathcal{L}}{\mathcal{L}_{(ideal)}} = \mathcal{L}\frac{\sigma}{\sigma_{(cr)}} \tag{8.25}$$

as the thrust acceleration efficiency of the real sailcraft. Note that $\sigma = \sigma_{(cr)}$ does not mean that the lightness number is unity, simply because any *real* sail could in no case achieve the *ideal* performance in acceleration, even if flat. Thus, the critical (sailcraft sail) loading is the value of the loading for which a sailcraft exhibits thrust efficiency equal to lightness number. If, additionally, the sail were flat and perfectly reflecting, then one would get $\mathcal{L} = \eta_{(thrust)} = 1$. If $\sigma \neq \sigma_{(cr)}$, then one has $\eta_{(thrust)} \leqslant 1$ and $\mathcal{L} \neq \eta_{(thrust)}$.

Now, let us consider the reciprocal of the sailcraft sail loading, $A_\square \equiv A/m_\square$, which we name the sailcraft's specific area. Because A is already an effective value (i.e. the sail-system's part actually irradiated by sunlight), $A_\square$ has a role similar to the effective specific power w_0, defined in Problem 1.1 on p. 21, inasmuch as $\mathbf{L}$ is proportional to $A_\square$. One can also define the sail-system's specific area, say, $A_\diamond \equiv A/m_S$, where m_S has been modeled in Eq. (8.22). Using this relationship, the general one $m_\square = m_S + m_L$, where m_L is the gross payload of the sailcraft, and the just-defined specific areas, one can re-arrange terms to get

$$A_\square = A_\diamond \frac{A}{A + A_\diamond m_L} \quad \Leftrightarrow \quad A_\diamond = A_\square \frac{A}{A - A_\square m_L} \tag{8.26a}$$

$$A_\diamond = \frac{A}{m_{BS} + (1 + \eth)m_B\mathfrak{g}\sqrt{A}} \tag{8.26b}$$

$$\mathcal{L} = \hat{\sigma} A_\square \quad \Rightarrow \quad A_\square = \frac{\mathcal{L}}{\hat{\sigma}} \equiv \mathbb{G}\mathcal{L} \tag{8.26c}$$

$$A_\square = \frac{\mathcal{L}}{\sigma_{(cr)}\eta_{(thrust)}} \tag{8.26d}$$

The new symbols m_B, $\mathfrak{g}$, and $\eth$ denote: the boom's mass per unit length, a dimensionless factor depending on the sail architecture (e.g. $\mathfrak{g} = 2\sqrt{2}$ for a simple square sail), and the deployment-subsystem mass on the boom subsystem mass ratio, respectively. Quantity $\hat{\sigma}$ "synthesizes" the sunlight-sail interaction; it has the same units of the sailcraft sail loading. From an interpretation viewpoint, we may mean by it the "viability of converting" a given surface carrying the unit mass into (normalized) SPS acceleration. Conversely, if an SPS trajectory requires a certain lightness number, then $1/\hat{\sigma}$ may express the "difficulty in realizing" this SPS acceleration through a physical surface transporting the unit mass. $\mathbb{G} \equiv \hat{\sigma}^{-1}$, which depends on the sail materials as well, is here named the technological equivalent of photon-sail acceleration; $\mathbb{G} \cong (650.72/\eta_{(thrust)})$ m$^2/kg$, a value related to the Sun, of course.

Even though the direction of $\mathbf{L}$ is a function of time, in general, however the actual specific area should result in the same value: deviations may come from errors in determining actual lightness numbers through thrust measurements. Nevertheless, the design value shall correspond to the maximum value of the lightness number or $\mathcal{L}_{max}$ (see p. 324). In both the mentioned concepts, the role of the optical diffraction—which is the main component of the sail-light interaction—appears clearly through the flat-sail model of the lightness number; it is quite general though.

Example 8.1 In preliminary flight design, let us suppose that one likes a certain value $\mathcal{L}_{max}$ for designing a certain mission trajectory; the related sailcraft thrust efficiency amounts to

$$\eta_{min} \leqslant \eta_{(thrust)} < 1 \tag{8.27}$$

where the lowest allowed value η_{min} of thrust efficiency is estimated from the sail materials that one should employ. The range of the technological equivalent is determined by

$$650.72 < \mathbb{G} \leqslant 650.72/\eta_{min} \text{ m}^2/\text{kg} \tag{8.28}$$

Consequently, the specific area of the sailcraft (to be designed) has to satisfy the inequality:

$$650.72\mathcal{L}_{max} < A_\square \leqslant (650.72/\eta_{min})\mathcal{L}_{max} \text{ m}^2/\text{kg} \tag{8.29}$$

which tells the designers which system technologies should be employed for realizing the desired mission. Later in this chapter, we will see that an optimal specific area, or equivalently, sail loading is found for each considered mission.

8.4 Numerical Examples via Non-Linear Programming

We have to point out the method(s) utilized here for dealing with Problem 8.1. Here, the approach is to transform this constrained problem into a set of $N_u \geqslant 1$ unconstrained problems, the index of performance of which is expressed as the sum of the squares of functions. Such functions are the constraints suitably modified. This sum (which contains the original index of performance) has to be minimized. Thus, we are able to convert the original problem into a Non-Linear Least-Squares (NLLS) minimization, for which several reliable methods are not only known since decades, but also there are very-well implemented procedures in numerical libraries for computer. Why N_u is greater than 1 has been mentioned in Note 8.1 on p. 333. We have included the case $N_u = 1$ for dealing with feasibility analysis, where usually parametric runs are carried out.

8.4 Numerical Examples via Non-Linear Programming

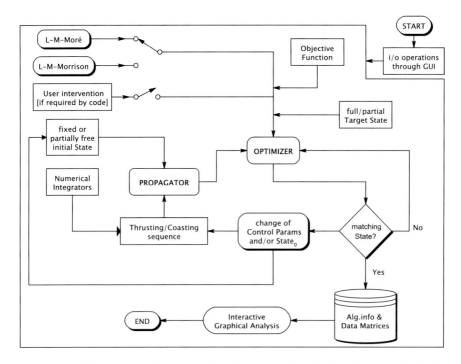

Fig. 8.1 Simplified scheme of the code adopted here for solving Problem 8.1 numerically. The propagator provides the force field, according to the user specifications, and integrates the ensuing motion equations driven by the control parameters selected according to the mission aims. The optimizer processes the state/control equality and inequality constraints in order to minimize the augmented objective function in the NLLS sense

Joining such considerations to those ones made in Remark 8.1 on p. 331 results in an algorithm for solving Problem 8.1, the flow diagram of which is shown in Fig. 8.1 in its basic blocks.

The flow is similar to many other schemes that are usually chosen for achieving convergence in an optimal control problem. Here, the main point consists of adopting (two different versions of) the Levenberg-Marquardt (LM) method for dealing with the current optimization problem (see below). The type of method known as the Levenberg-Marquardt, [33, 42], is an efficacious modification of Gauss-Newton method, and is detailed in modern literature, e.g. [2, 8, 30, 40, 82], and implemented as high-precision routines in widely-employed Fortran and C++ libraries, e.g. [59, 60, 74], and [24]; the compiled source codes of such routines are also embedded in many compilers for various platforms. Also, Maple[TM] and Mathematica[TM] have the LM algorithm included in their associated packages. LM method is also known as the Levenberg-Marquardt-Morrison (LMM) method, which remembers the contribution of the third author [52, 54]. Below, we will use the acronym LMM in order to distinguish this algorithm from that in [49–51].

Qualitatively, the core of the LM strategy for getting solutions to NLLS problems can be summarized as follows: if the current iterate is far from the local minimum, then the algorithm proceeds similarly to the *steepest descent* method; as the current iterate approaches the minimum, it behaves like the *Gauss-Newton* method. The update of the current search point is regulated by the Levenberg-Marquardt parameter, or the damping parameter. LM method is a sort of blend of the advantages of the steepest descent and Gauss-Newton methods, and it works very well in practice.

> In unconstrained minimization, there are two principal policies for trying to achieve global convergence upon iterations: (a) the line search methods (with or without derivatives), and (b) the *trust-region* methods.
>
> The line search without derivatives relies on some properties of the function to be minimized; as a point of fact, such function is not required to be differentiable. The golden-section method and Fibonacci method are classical examples of derivative-free line search.
>
> The derivative-based line search resorts to the concept of descent direction, i.e. a direction forming an obtuse angle with the local gradient of the function. The steepest-descent and Gauss-Newton methods are examples of multi-dimensional line search using derivatives.
>
> Differently, in the trust-region strategy, one builds a model approximating the function in a finite region containing the current iterate; this is just named the thrust region because one places confidence that this assumed model of the original function is well appropriate. Such a region is a function of the vector increment to be given to the current-iterate. To find the minimum of such function-model is the sub-problem to be solved for going to the next iterate. LM method resembles strongly to a trust-region strategy: many authors consider the LM method as a type of trust-region strategy.
>
> LM method is considered fast and *robust* in the sense of NLP.

For this chapter, we will use two different algorithms and implementations of type LM: (1) the LMM version, and (2) the Levenberg-Marquardt-Moré (LME) version [49–51]. In the double-precision routine DNOLF1 of LMM [21], only function values are used; Jacobian matrix is evaluated via the forward difference. The version by Moré focuses on the variable scaling already present in the original LMM algorithm. He modified the previous criterion into an adaptive scaling, and this resulted in a scale-invariant LM algorithm remarkably more robust than the original one. Again, partial derivatives are computed by the forward difference. This point is very important for our current minimization inasmuch as there are no analytical partial derivatives available for solving Problem 8.1.

Why using two methods for the optimization of sailcraft trajectories? This is a matter of many-year experience with the families of fast trajectories, which are rather sensitive to initial conditions, and contain arcs of notably different time scales of $\mathbf{L}$, especially when a low perihelion is required. In addition, the 2D and 3D

8.4 Numerical Examples via Non-Linear Programming

reversal-motion profiles are particularly sensitive to the control in the neighborhood of the reversal point(s). If either LMM or LME algorithm fails, nevertheless it provides precious information for the new guess and, often, for improving the assumed trajectory segmentation; for instance, if three arcs were assumed with specific attitude laws, the gained information could indicate that the optimal trajectory is made of four arcs. In addition, the solution of either may be used as the starting profile for the other method for getting a second optimal profile to be compared with the first one.

Before proceeding with the numerical results in the following subsections, we have to mention how we have manipulated the current equality and inequality constraints in order to build the vector function, say, F the *squared* norm of which shall be minimized.

We give the F's first component the objective function, i.e.

$$F_0 = \mathfrak{J}(t_f, \mathbf{R}(t_f), \mathbf{V}(t_f), m(t_f)) \tag{8.30}$$

Therefore, no modification is applied to the original expression (8.15).

Let us consider the generic scalar equality of the original problem, and define the simple function

$$\varepsilon(\zeta, \zeta_{(ref)}) = \zeta(\mathbf{X}, \mathbf{U}) - \zeta_{(ref)} \tag{8.31}$$

where ζ is the quantity of interest computed in the current iteration: the control $\mathbf{U}$ was defined in Problem 8.1, while $\mathbf{X}$ denotes the state calculated at the instant required in the original constraint. $\zeta_{(ref)}$ is the value prefixed for ζ. The k-th component of the function F is given the following function

$$F_k = \sqrt{a_k}\,\varepsilon(\zeta_k, \zeta_{(k,ref)}), \quad k = 1..N_{eq} \tag{8.32}$$

where a_k is a prefixed large positive number (the square root is for convenience, as we will see below).

Now, let us consider the generic scalar inequality of the original problem, and define a set of four functions $\iota_{(<,=)}, \iota_{(=,<)}, \iota_{(<,<)}$ and $\iota_{(=,=)}$ where the subscripts refer to the inclusion/exclusion of the bound(s); it is sufficient to show the first element

$$\iota_{(<,=)}(y, y_{min}, y_{max}) = \begin{cases} 0, & y_{min} < y \leqslant y_{max} \\ |y - y_{min} + o|, & y \leqslant y_{min} \\ |y - y_{max}|, & y > y_{max} \end{cases} \tag{8.33}$$

$$-\infty < y_{min} < y_{max} < \infty \tag{8.34}$$

where o is a small number, say, a factor ten of the 8-byte machine precision ($\cong 2.2 \times 10^{-16}$). Thus, the $(N_{eq} + j)$-th component of the function F can be given the following function:

$$F_{N_{eq}+j} = \sqrt{b_j}\,\iota_{(\square,\square)}(\xi_j(\mathbf{Y}, \mathbf{U}), \xi_{(j,min)}, \xi_{(j,max)}), \quad j = 1..N_{ineq} \tag{8.35}$$

where the two slots accept the inclusion/exclusion of the lower and/or upper bounds, and b_j is a prefixed large positive number (the square root is for convenience). $\mathbf{Y}$ is the state computed at the instant specified in the original constraint. $\xi_{(j,min)}$ and $\xi_{(j,max)}$ denote the assigned lower and upper bounds of ξ_j, which is the quantity of interest computed in the current iteration. For example, if the original inequality were $\xi_{(3,min)} < \xi_3 \leqslant \xi_{(3,max)}$, then $F_{N_{eq}+3} = \sqrt{b_3}\iota_{(<,=)}(\xi_3, \xi_{(3,min)}, \xi_{(3,max)})$, where ξ_3 is the value computed in the current iteration.

Thus, the augmented function is built; the NLLS unconstrained problem requires to minimize

$$\mathfrak{J} = F^{\top}F = F_0{}^2 + \sum_{k=1}^{N_{eq}} F_k{}^2 + \sum_{j=1}^{N_{ineq}} F_{N_{eq}+j}{}^2$$

$$= \mathfrak{J}^2 + \sum_{k=1}^{N_{eq}} a_k \left[\varepsilon(\zeta_k, \zeta_{(k,ref)})\right]^2 + \sum_{j=1}^{N_{ineq}} b_j \left[\iota_{(\square,\square)}(\xi_j(\mathbf{Y}, \mathbf{U}), \xi_{(j,min)}, \xi_{(j,max)})\right]^2$$

$$(8.36)$$

The definitions (8.31) and (8.33) allow treating equality and inequality constraints in $\mathfrak{J}$ in the same way. The squared components $a_k[.]^2 \equiv E_k$ and $b_j[.]^2 \equiv I_j$ act as severe penalties to the original objective function. Qualitatively, at the beginning of the search, $\mathfrak{J}$ is expected to be large because most of the constraints are violated to a great extent. During the search, if the quantities ζ and/or ξ deviate still more from their prefixed value or range, respectively, the penalties E_k and/or I_j increase further, thus forcing the algorithm to change the infeasible iterates drastically in order to decrease $\mathfrak{J}$. In general, a decrease of $\mathfrak{J}^2$ proceeds simultaneously with the (strong) decrease of E_k and I_j. When all E_k and I_j become of the same order of magnitude or less than $\mathfrak{J}^2$, then the constraints are almost satisfied and the converging value of original objective function gets meaningful.

In principle, in the limit at which every E_k and I_j vanish for a_k and b_k diverging, the obtained control, say, $\mathbf{U}^{(lim)}$ is a solution to Problem 8.1, namely, $\mathbf{U}^{(lim)} = \mathbf{U}^{(opt)}$. The reader is suggested consulting the Chap. 9 of [2] for the related lemmas and theorems. In practice, the tolerances on the equality constraints are determined by the values $\{a_k\}$, $k = 1..N_{eq}$, the choice of which also affects the algorithm speed towards convergence, and (sometimes) may prevent the convergence achievement.

8.4.1 Using Results from Scalar Scattering Theory

The optimal profiles, carried out in the sense of NLP in this and Sects. 8.4.2–8.4.3, are based on the SST through which the hemispherical optical quantities entering the lightness-vector equation (6.136) are computed, as anticipated in Remark 6.8 on p. 246.

8.4 Numerical Examples via Non-Linear Programming

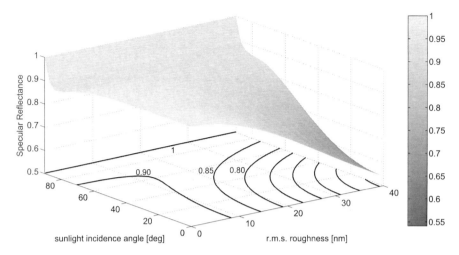

Fig. 8.2 Aluminium's *specular* reflectance, according to SST, as function of the light's incidence angle and the surface's root mean square roughness. Figure is a 3D re-representation of data coming from the algorithm in [77]. Courtesy of Elsevier

Figures 8.2, 8.3 and 8.4 are new plots of the Aluminum's results obtained in [77] by employing SST just with regard to the specular reflectance, the directional-hemispherical diffuse reflectance, and the total absorptance *vs.* the *rms* roughness and the incidence angle of (unpolarized) sunlight.

Figure 8.2 plots an extended set of the results obtained in [77]. The maximum value of the *rms* height is 40 nm, a value that should be considered as an upper bound for the Aluminum layer of high-quality sails. In any case, if roughness is too high, not only SST does not hold (and one has to resort to the diffraction theories in Sect. 6.5.4), but also diffuse reflection becomes dominant inducing a clear lessening of the thrust. The contours of the specular reflectance $\mathcal{R}_s$ are plotted and labeled on the plane $(\vartheta_\odot, \tilde{z})$. One should note that, as $\tilde{z}$ increases, the paraxial reflectance values become markedly lower than those ones at greater incidence angles. One can note the local "canyon" of $\mathcal{R}_s$ at large incidence angles (between 80°–90°). In a real mission analysis, sail's large-scale curvature and wrinkling (Sect. 6.5.7) should be analyzed carefully case by case, and compared with the results from either SST or VST.[6] In any case, mission analysts needs accurate measurements of reflectance made on *representative* samples of the materials of the sail to be flown. This is a key-point in sailcraft trajectory design.

Figure 8.3 shows the directional-hemispherical *diffuse* reflectance of Aluminum *vs.* $\vartheta_\odot$ and $\tilde{z}$. It is strictly increasing with $\tilde{z}$ for any fixed $\vartheta_\odot$, as the contour levels show clearly.

---

[6] Already in the Seventies, during their studies on the comet Halley rendezvous via solar sailing, JPL pointed out a thrust falling to zero at incidence over 60° for a (sufficiently) curved sail.

344   8 Approach to SPS Trajectory Optimization

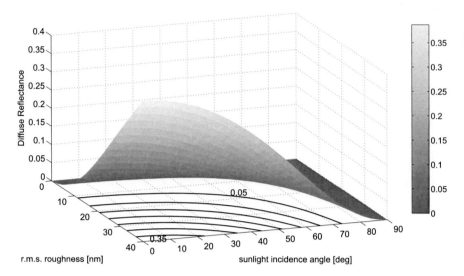

Fig. 8.3 Aluminium's directional-hemispherical *diffuse* reflectance, according to the scalar scattering theory, as function of the light's incidence angle and the surface's root mean square roughness. Figure is a 3D re-representation of data from the algorithm in from [77]. Courtesy of Elsevier

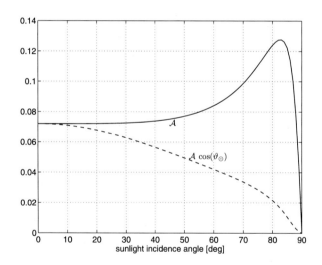

Fig. 8.4 Aluminum's total absorptance (*solid line*), according to SST, as function of the light's incidence angle. The absorptance times the cosine of the incidence angle is plotted as well (*dashed line*). No transparency has been assumed. Data comes from the algorithm in [77]. Courtesy of Elsevier

The absorptance $\mathcal{A}$ and the product $\mathcal{A}\cos\vartheta_\odot$ have been graphed in Fig. 8.4 as function of $\vartheta_\odot$; this product enters the temperature sail given by equation (6.99). No transparency has been assumed according to the original algorithm. The product

8.4 Numerical Examples via Non-Linear Programming

can be approximated very well by the following cubic polynomial in a useful range

$$\mathcal{A}\cos\vartheta_\odot = 0.0119258\vartheta_\odot^3 - 0.0401377\vartheta_\odot^2 + 0.0001553\vartheta_\odot + 0.0720440$$

$$0 \leqslant \vartheta_\odot \leqslant 70\pi/180 \tag{8.37}$$

where the incidence angle is expressed in radians. The maximum of the residuals is 3.1×10^{-4}, so the calculation of the sail temperature can be implemented very easily.

After the first applications of SST to aluminized-sail mission concepts ([77, 78], and [43]), other authors re-considered the problem of realistic sails (including a temperature limit), e.g. [14] and [47]. For their fundamental importance in Optics, applying VSTs to SPS should and could result in a meaningful evolution, as shown in Chap. 6, in general, and Sect. 6.5 with Sect. 6.7.2, in particular.

For the NASA study reported in [43], regarding the solar-sailing option of the Interstellar Probe concept, the author employed a computer code that knows the Aluminum's above-explained optical quantities as function of $\vartheta_\odot$ and $\tilde{z}$. This code underwent several extensions in the years, and has been further modified for this chapter, compliantly with the results of Sect. 6.7.2, in order to be applied to the flat-sail model given by Eq. (6.136). This new version is thus able to answer the following question in the framework of fast sailing:

Question 8.1 If the scattering of the solar light produces a total diffuse photon momentum with a component parallel to the sail plane, how much is the control, optimized in the sense stated above, affected?

8.4.2 Lightness Number Smaller than Unity

In this subsection, we will discuss an extra-solar mission example that satisfies Theorem 7.2, is compliant with the results plotted in Fig. 7.5, and satisfies Definition 7.3. Trajectory is optimized in the sense of Problem 8.1.

For the sake of explanation simplicity, conventionally we will often use phrases like "...the orbit of $\mathbf{W}$...", where $\mathbf{W}$ is any real three-dimensional vector, the time evolution of which is graphed in the space of all possible component values. Here, the tip of the vector describes an orbit as time goes by. Thus, in particular, the orbit of position is the usual trajectory, while the orbit of velocity coincides with hodograph. We will employ such convention for vectors other than $\mathbf{R}$ and $\mathbf{V}$. If $\mathbf{W}$ is constant in a certain time interval, then a dot represents such evolution.

The reasoning underlying the example of concept of mission driven by $\mathcal{L} < 1$ starts from analyzing a potential fast sailing with a sailcraft sail loading roughly in the middle of the range from the critical loading to the maximum practical value, 2.1 g/m^2, beyond which fast sailing is not allowed. The scientific goal is to send a sailcraft at 200 AU in the direction of the heliopause nose, namely, to longitude and latitude equal to 254.5°, and 7.5°, respectively, in a time significantly lower than the MHJT. Maximizing the terminal speed guarantees that any possible continuation of the mission (provided the sailcraft systems go on working) will be time-minimized. If one minimizes the time to 200 AU, then this value would be shorter than a few hours with respect to the max-V_f time, or less than 0.00002 in relative difference. Therefore, in the current context, minimizing time and maximizing terminal speed produce the same results.

Here the choice is to not jettison the sail, e.g. just after the Jupiter orbit, unless some very serious issue takes place; moreover, keeping the sail (in a simple orientation mode) might offer some additional opportunities of enhancing the scientific purposes. Nevertheless, very low temperatures may be a severe issue.

In this mission concept, we opt for utilizing conventional materials: a reflective thin layer of Aluminum deposited on a plastic support, the backside of which is covered by a thin layer of Chromium. The real challenge comes from the plastic support. First, its thickness should be one order of magnitude lower than that used in the IKAROS sail [27], and one fourth that of the NanoSail-D2 sail via the space-qualified CP1 [29]; such sails, though, are much, much smaller than fast sails. Second, the perihelion is limited by the polymer Glass Transition Temperature (GTT); for CP1, GTT is equal to 536 K, and is considerably lower than its decomposition temperature (over 800 K) [32]. In the temperature constraint, Inequality (8.20b), we set $T_s^{(max)} = 500$ K that, in turn, is much lower than the melting points of Aluminum and Chromium. This safety value affects the sailcraft sail loading. Let us see why. If $\mathcal{L}_{max}$ is sufficiently high, then in the deceleration/acceleration arc from t_0 to t_p, $\cos \vartheta_\odot$ is sufficiently less than one 1 for $T < T_s^{(max)}$ is satisfied, while $\mathbf{L}$ allows the sailcraft to flyby the Sun within the prefixed range of perihelion. On the other side, if $\mathcal{L}_{max}$ were smaller (i.e. technologically more favorable), $\cos \vartheta_\odot$ would be forced to be higher, namely, the sail temperature could violate the constraint. One may then raise the perihelion, but this would decrease the terminal speed. Thus, an optimal σ or, equivalently, $A_\square$—compliant with both the chosen sail materials and the required perihelion corridor—is expected.

The problem of sail temperature when sail draws close to the Sun has been addressed independently also in [31].

One may guess that very-large full-nanotech-enhanced polyimide sheets may be made for (also) achieving ultra-thin thickness (about 0.5 µm), higher GTT ($\gtrsim 600$ K), and high (> 0.6) hemispherical emittance so that one could avoid using Chromium or another emissive layer. Once aluminized, this lighter membrane would result in a two-layer sail utilizable in direct or reverse motion sailcraft flights with 0.15–0.20 AU perihelia, provided that environmental situations can still be met.

8.4 Numerical Examples via Non-Linear Programming

> Actually, we are on the way. As one can check from Ref. [32], currently produced by NeXolve Corporation (Huntsville, Alabama), there are two types of CP1: Clear LaRC^{TM} CP1, and *conductive* Black LaRC^{TM} CP1. The second one has been enhanced in its properties by using Carbon nanoparticles.

In Sect. 8.4.3, we will concern with an example of mission concept utilizing some promising results from Nanotechnology. Here, we will continue to analyze the current $\mathcal{L} < 1$ example of fast-trajectory concept in detail. Since the target distance (200 AU) should be well beyond that of the near heliopause, and along its nominal direction, we will refer to this mission concept also as the Mission to Near-Heliopause and Beyond (MNHB).

Note 8.2 Unless otherwise specified, most of data and figures will be expressed in solar units (S.U.), namely, $\mu_\odot = 1$, unit length $=$ AU, and unit time $=$ year$/2\pi$ or, equivalently, unit speed $=$ EOS. The value of the year, as coming from file DE405 (where 1 AU $= 1.49597870691 \times 10^{11}$ m), amounts approximately to 365.2569 days. However, we will often express values of speed in astronomical units per 365.25 days, namely, AU/year $= 4.740470$ km/s. The corresponding earth orbital speed, say, EOS$' = 1.00001889$EOS.

Table 8.1 reports sailcraft and sail data employed in this example of $\mathcal{L} < 1$ mission concept. As just mentioned, one can see that an issue (nowadays) to sail realization is the very low thickness (0.5 μm) of the substrate; supposing hopefully that this may be done, even this low value entails a mass of the polyimide layer (CP1, 115 kg), which is a high fraction (0.39) of the sailcraft injection mass. In order to get $\mathcal{L} > 0.7$, the spacecraft mass has been restricted to 80 kg. Because at least 1/4 of this mass should be reserved for science, one needs advanced systems/subsystems.

For these basic reasons, the international Aurora Collaboration (January 1994–December 2000) accomplished its research on sailcraft technology, trajectories and mission concepts by searching for innovative concepts about in-orbit removal of the plastic substrate [65]. Thus, the aim was to finally get a bilayer sail of 100-nm Aluminum and 20-nm Chromium, as a baseline. A sailcraft mission concept based on such all-metal sail was proposed for achieving the inner regions of the Solar Gravitational Lens (SGL) [78]. A historical overview on such collaboration and its main results can be found on pp. 156–157 of [79]. Extensive studies on SGL can be found in [37–39].

Table 8.2 reports the main figures regarding the optimal controls and the ensuing trajectory; the evolutions of state and control, and the other key variables, satisfying the specified equality and inequality constraints, are shown in Figs. 8.5–8.15.

Figure 8.5 shows the trajectory arc inside the box $[(-1.5 \ 1.5), (-1.5 \ 1.5), (-0.25 \ 0.5)]$ AU (black line), its projection on the XY-plane (medium-gray line), and the orbits of the inner planets (light-gray lines) in the same time interval. Gravitational perturbations due to these planets and Jupiter have been included in the sailcraft motion field. Slanted small segments denote time tags, t_0, t_1, t_2, t_3, and about

Table 8.1 MNHB concept: *sailcraft data* from trajectory optimization with $\mathcal{L} < 1$. Symbols defined in this chapter have been employed

Mass breakdown			
$m_\square$	297 kg	σ	1.8562 g/m^2
m_L	80 kg		
m_S	217 kg	σ_S	1.3562 g/m^2
m_{BS}	173 kg	σ_{BS}	1.081 g/m^2
$m_{BS}^{(substrate)}$	115 kg		
m_B	40 kg	$\mathfrak{m}_B$	35.4 g/m
m_D	4 kg		

Sail data			
$\sqrt{A}$	400.0 m	$\mathfrak{g}$	$2\sqrt{2}$
$A_\square$	538.720 m^2/kg	$A_\diamond$	737.327 m^2/kg
Al-CP1-Cr	95 nm		
	500 nm		
	15 nm		
Al-layer $\tilde{z}$	20 nm	Al-layer $\tilde{l}$	120 nm
$\mathcal{L}$	0.7182	$\eta_{(thrust)}$	0.8675
$\mathbb{G}$	750.1 m^2/kg		

30 days beyond. Together with the results graphed in other figures, Theorem 7.2 is satisfied for one point in the second time interval of control; the 3D motion reverses at t^r, as reported in Table 8.2. Subsequently, at $t^{c_1} = 361.15$ day, the sailcraft goes below the XY-plane, and re-emerges from negative latitude after about 37 days (at $t^{c_2} = 398.30$ day) for pointing towards the target direction. No close encounters with the inner planets take place.

Figure 8.6 shows the orbit of **L**. This curve starts from the small square; the variable part is almost entirely due to the second interval $t_2 - t_1$, where the transversal number strictly increases, and crosses the zero value, where also the normal number reaches a local maximum. Though slightly, the radial number first decreases, then increases. According to the theory developed in Sect. 7.4.3, the changes of $\mathcal{L}_t$ and $\mathcal{L}_n$ result in motion reversal under the conditions of Theorem 7.2. The isolated point, also reported in Table 8.2, on the right vertical plane represents the **L**-maneuver; this one, chiefly through the increase of the radial number, mitigates the gravity deceleration as the sailcraft moves away from the Sun.

The explicit time evolution of the components of **L** is shown in Fig. 8.7; the vertical segments represent the optimized time bounds of the control. The event $\mathcal{L}_t = 0$ takes place *before* the middle of the interval, or $t^\square < t^m$. After the post-perihelion **L**-maneuver, the effective solar gravity is further reduced, and the transversal number is still positive, i.e. the energy rate is positive. Note that the maneuver produces $\mathcal{L}_n < 0$, and therefore $(\mathbf{H} \times \dot{\mathbf{H}})_{\square, t_3}$ becomes radially inward, according to Eq. (7.19a). Time allocation of the maneuver has been optimized as well.

8.4 Numerical Examples via Non-Linear Programming

Table 8.2 MNHB concept: trajectory optimization with $\mathcal{L} < 1$. Optimal control parameters and variables have been reported for a mission to 200 AU along the direction of the closest heliopause. Trajectory data have been added, compliantly with symbols defined in this chapter. Subscripts $\odot$ and $\square$ in position/velocity symbols were dropped for simplicity. t^r, $t^\square$, and t^m denote the motion reversal time, the E-min and H-min time, and the middle time of the control interval from t_1 to t_2; t^{c1} and t^{c2} denote the XY-plane crossing times. The bottom lines give the additional transfer times to 400 AU, and to the minimum (theoretical) distance of SGL

5-Arc trajectory field Optimization index injection date t_0	Thrust + gravity $V(t_f), t_f$ free 2029.10.03 06:27:34	Sun	Inner planets	Jupiter
$\mathbf{R}(t_0)$	AU	0.9825873	0.1898193	-1.914095×10^{-5}
$\mathbf{V}(t_0)$	AU/year	-1.1851558	6.1665726	-4.018873×10^{-4}
$R(t_0), V(t_0)$	AU, EOS$'$	1.0007543	0.9994019	
$t_1 - t_0$	day	299.88793		
$t_2 - t_1$	day	25.80032		
$t_3 - t_2$	day	55.55077		
$t_4 - t_3$	day	365.00000		
$t_f - t_4$	year	11.5056456		
$t^r, t^\square, t^m$	day	306.01496	306.30165	312.78809
$\|\mathbf{H}\|^r, H^\square, H^m$	AU·EOS$'$	0.003363	0.003181	0.025487
t^{c1}, t^{c2}	day	361.14965	398.30472	
$\mathbf{R}_p$	AU	0.1796313	0.0282044	-0.0704428
R_p, t_p, V_p	AU, day, EOS$'$	0.19500	377.1992	2.67741
V_{cruise}	EOS$'$	2.5418		
$\mathbf{L}^{(1)}$	constant	0.534810	-0.248777	0.033144
$\mathbf{L}^{(2)}$	see Fig. 8.7			
$\mathbf{L}^{(3)}$	see Fig. 8.7			
$\mathbf{L}^{(4)}(t_3^+)$	impuls. man.	0.69983	0.15452	-0.053053
$\mathbf{L}^{(5)}(t_4)$		0.69938	0.15440	-0.053012
$\mathbf{L}^{(5)}(t_f)$		0.6994	0.15441	-0.053014
$R(t_f), \psi(t_f), \theta(t_f)$	AU, deg., deg.	200.00	254.50	7.50
$V(t_f)$	EOS$'$, AU/year	2.5409	15.965	
t_f	year	13.549		
$t^{(400\ \mathrm{AU})} - t_f$	year	12.53		
$t^{(550\ \mathrm{AU})} - t_f$	year	21.93		

The azimuth and elevation of $\mathbf{n}$ in EHOF, denoted here by α_n and δ_n, and the sunlight incidence angle $\vartheta_\odot$ are plotted in Fig. 8.8. Aside from the post-perihelion $\mathbf{L}$'s re-orientation maneuver, the incidence angle is practically constant. This means

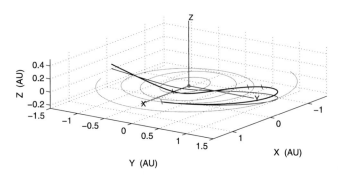

Fig. 8.5 3D optimal trajectory to the near heliopause via $\mathcal{L} < 1$. Only the trajectory arcs inside the Mars' orbit are shown (*black line*). The trajectory's final arc is practically rectilinear towards the heliopause nose. In addition to the orbits of the inner planets (*light-gray lines*), the trajectory projection on the HIF's XY-plane is shown (*medium-gray line*). It is apparent that motion reverses

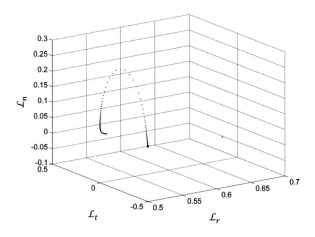

Fig. 8.6 Orbit of the optimal lightness vector for MNHB. The *small square* represents $\mathbf{L}(t_0)$. The variable $\mathbf{L}(t)$ from t_1 to t_3^- is arc-like, whereas the *isolated dot on the right* represents the practically constant lightness vector after the impulse-modeled **L**-maneuver

that the optimal control is realized in EHOF via **n** either constant or rotating about the x-axis (i.e. the $\mathbf{R}_{\odot\square}$ direction, or $\bar{\mathbf{r}}$). More precisely, in the variable case, we split vector $\mathbf{n}^\top = (n_x, 0, 0) + (0, n_y, n_z) = n_x \bar{\mathbf{r}} + \mathbf{n}_{yz}$, with constant $n_x = \cos\vartheta_\odot$, whereas $\mathbf{n}_{yz}$ has variable angular speed, say, ω_x about $\bar{\mathbf{r}}$. ω_x exhibits a local maximum at $t^\square$ with a value of $2.8°$/hour.

Figure 8.9 shows the fourth plot about the sailcraft control, i.e. the overall behavior of $\mathcal{L}$ and $\eta_{(thrust)}$ that satisfy equation (8.25). One can note the achievement of high efficiency after the post-perihelion **L**-maneuver, i.e. when $\mathcal{L}$ is large.

The evolution of the sail temperature T_s, together with R, is plotted in Fig. 8.10. It is apparent that the plastic substrate of the reflective layer undergoes a temperature up to 500 K around the perihelion, according to the constraint. On the other side,

8.4 Numerical Examples via Non-Linear Programming

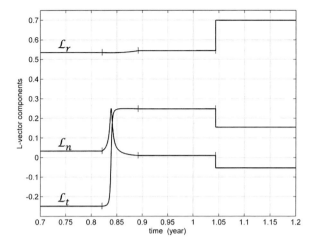

Fig. 8.7 Evolution of **L** for MNHB via $\mathcal{L} < 1$. The *vertical small segments* denote the optimized time bounds of the control. In the second interval, $\mathcal{L}_t$ crosses zero when $\mathcal{L}_n$ achieves its maximum, according to the theory. At the end of the third interval, a **L**-maneuver is applied, and optimized, for reducing the effect of the gravitational deceleration

Fig. 8.8 MNHB via $\mathcal{L} < 1$: evolution of the sail axis angles (α_n, δ_n) (i.e. azimuth and elevation in EHOF), and the sunlight incidence angle. The *inset* zooms in the second control interval

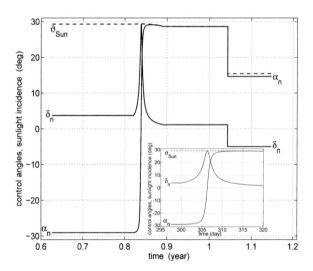

a sufficiently low temperature might cause severe issue(s) to the sail system as a whole. Although new polyimides such as CP1 and CP2 are operational at cryogenic temperatures [29], however more accurate investigations should be done because sail temperatures beyond the planetary range may be below 50 K. If hazards were predicted, then they could make up a valid reason for jettisoning a sail made of conventional materials.

Returning to the mission trajectory's dynamical quantities, the evolution of the invariant H and the Z-component of the orbital angular momentum are shown in

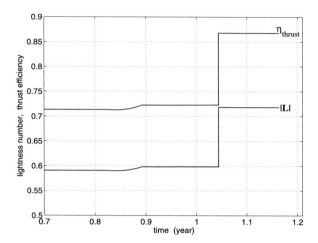

Fig. 8.9 MNHB via $\mathcal{L} < 1$: evolution of $\mathcal{L}$ and $\eta_{(thrust)}$. The two discontinuities correspond to the **L**-maneuver. Note that both quantities monotonically increase, but such that their ratio is constant according to Eq. (8.25)

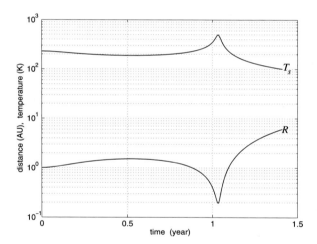

Fig. 8.10 MNHB via $\mathcal{L} < 1$: evolution of distance and sail temperature. Temperature stays in the 400–500 K interval for 13 days

Fig. 8.11. In the bottom zoomed subplot, the motion reversal and the minimum of H are recognized clearly. There is only one **H**-reversal point in the current trajectory profile.

The orbit of **H** is plotted in Fig. 8.12. Combining this one and Fig. 8.11, in particular, this mission trajectory features the peculiar property of high-$\mathcal{L}$ motion reversal in $\mathfrak{R}^3$, as expressed by Theorem 7.2.

The orbit of velocity, or the hodograph, in Fig. 8.13 looks like part of a treble clef, where the minimum distance from the origin does not correspond to the trajectory

8.4 Numerical Examples via Non-Linear Programming

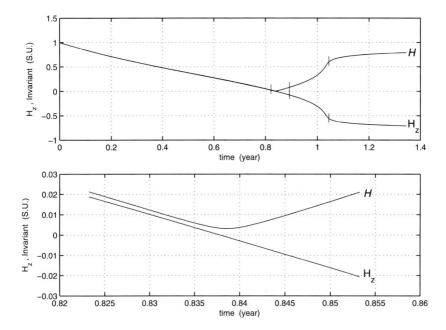

Fig. 8.11 MNHB via $\mathcal{L} < 1$: evolution of the invariant H and the third component of the angular momentum. The *vertical small segments* denote the optimized time bounds of the control intervals

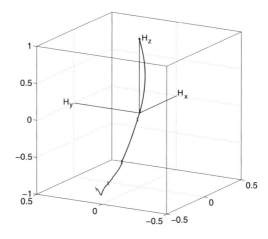

Fig. 8.12 MNHB via $\mathcal{L} < 1$: orbit of **H**. The *tilted small segments* denote the optimized time bounds of the control **L**. **H** crosses the XY-plane at $t^r \in [t_1, t_2]$, as reported in Table 8.2, but $\|\mathbf{L}\| \neq 0$

perihelion, but to V_{min} in expression (7.56). Similarly, the maximum distance is the value V_{max} in the same equation. A small (left) neighborhood of the terminal point represents the quasi-cruise phase of the flight. 3D hodographs like this one generalize the hodographs in Fig. 7.2. After projecting **V** on the Vx-Vy plane, here a point can be found such that $\tilde{\mathbf{V}} \times \dot{\tilde{\mathbf{V}}} = 0$, where $\tilde{(.)}$ mark velocity and acceleration on the projection plane. But we will not insist further on this aspect.

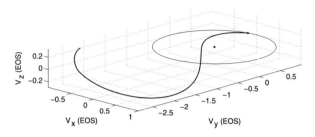

Fig. 8.13 Hodograph of the MNHB sailcraft. The *circle* represents the unit-speed hodograph, namely, it corresponds to the circular orbit of a particle with 1 EOS of speed. The *small square* denotes the flight start

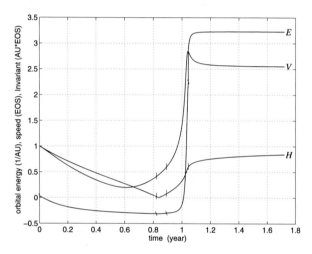

Fig. 8.14 Optimal sailcraft's E, H, and V exhibited by MNHB sailcraft via $\mathcal{L} < 1$. Vertical/slanting small segments represent the time bounds of the optimal trajectory arcs

Figure 8.14 shows the time histories of the sailcraft energy, speed, and the invariant-H again for the reader ease. The minima of E and H take place at the same time, according to Proposition 7.4 on p. 299. The insertion of the **L**-maneuver produces $V_\square(t_f) = 15.965$ AU/year. If one performs the optimization of MNHB under the same constraints and objectives, but without a post-perihelion maneuver, then the set of optimal parameters takes on appreciably different values because the maneuver makes the achievement of the target direction "easier" via three additional degrees of freedom. However, the optimal profile would entail $V_\square(t_f) = 15.194$ AU/year, or an increase of the flight time of 0.55 years, or also 1.11 years on arrival at the minimum theoretical distance of the SGL. Such differences appears sufficiently small to consider avoiding a trajectory profile with post-perihelion attitude re-orientation. From the injection time viewpoint, the no-reorientation trajectory profile regards a launch about a week later (on October 10, 2029).

8.4 Numerical Examples via Non-Linear Programming

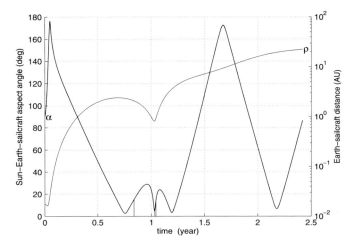

Fig. 8.15 Sun-Earth-sailcraft aspect angle, say, $\alpha_{\odot\square}^{\oplus}$ in MNHB. This is the angular distance between the Sun's barycenter and the sailcraft as it would be observed from the Earth barycenter. Also reported is $\rho_{\oplus\square}$, i.e. the Earth-sailcraft distance scaled on the right

Remark 8.5 Whatever $\mathcal{L} < 1$ profile is chosen for MNHB, with or without post-perihelion re-orientation maneuver, or full-direct motion, one of the key points is to design a sailcraft with an optimal sail loading for a given set of sail materials and architecture, as explained on p. 346, in order to also satisfy the strong constraints on perihelion and sail temperature.

The launch windows of the two profiles, as well as the opportunity of MNHB full-direct-flight (necessarily a few months in advance), are beyond the purposes of this chapter. However, we have to stress that more than one launch opportunity per year are admitted, and *every* year too. Why so a large advantage with respect to current rocket systems? Because fast sailing allows getting rid of any planetary flyby and, at the same time, achieving a high cruise speed appropriate for extra-solar robotic explorations.

Finally, in Fig. 8.15, we plotted two quantities denoted by $\alpha_{\odot\square}^{\oplus}$ and $\rho_{\oplus\square}$. The first is the angle between the Sun's barycenter and the sailcraft as it would be observed from the Earth barycenter; the second is the Earth-sailcraft distance. From left to right, the small vertical segments represent the values of $\alpha_{\odot\square}^{\oplus}$ at t^r, t_p, and t_3. Such values amount to: 15.018°, 5.652°, and 14.706°, respectively.[7] Note also that perihelion and maneuver events take place when the sailcraft is far from Earth by less than 1 AU. The triangular line on the right repeats almost periodically every 2.2 years during the flight. Such curves may be of interest in sailcraft-Earth communication.

---

[7]The mean angular radius of the Sun, as seen from 1 AU, is approximately 0.2665°.

356   8 Approach to SPS Trajectory Optimization

In the next section, the trajectory features and aspects so far discussed will help us to set up an example of fast sailing via a sailcraft with $\mathcal{L}_{max} > 2$. The gain of cruise speed will be appreciable.

8.4.3 Lightness Number Greater than Unity

The reasoning leading to the qualitative explanation of why actually there exists an optimal specific area or, equivalently, an optimal sail loading has been summarized on p. 346. The following question arises: how to proceed in order to get terminal speeds notably higher than those ones obtainable by $\mathcal{L}_{max} < 1$? It is plain that we should build sailcraft with $\mathcal{L}_{max} > 1$, in general, and hopefully not less than 2, in particular. In addition, it is essential to find sail-system and spacecraft materials which tolerate higher temperatures and solar-particle fluxes, so that (moderately) lower perihelia may become accessible, if requested. Sailcraft runs through the hazard distance range in a relatively short time; for instance, sailcraft trajectory described in Sect. 8.4.2 entails about 8.5 days for crossing the $R_\square$-range 0.250–0.195–0.250 AU. On the other side, if one wants to get a sufficiently higher lightness number, which should be effected after the perihelion (because the pre-perihelion arc is governed by $\mathcal{L} < 1$ in any case), we should not be restricted only to re-orienting the sail axis. There is a further advantage. If one accomplishes a re-orientation maneuver *after* detaching some mass, then the lessened sailcraft sail loading enables increasing the maneuver lightness number. The practical implementation of the two **L** maneuvers will require a certain time between them.

In Sect. 5.5, on p. 161, **L**-maneuver type (5) has been pointed out. Thus, mass separation is not only admitted, but can also cause a large variation of the lightness number, and without re-orienting the sail axis. Of course, this detachment could have a real meaning if the mass to be released were part of the overall sailcraft payload. In other words, at least two missions in one should be designed. Let us make these considerations quantitative.

The mission concept, which in general we will name the Extra-Solar Exploration Mission (ESEM), to be analyzed entails a sailcraft consisting of one sail system and two spacecraft with two different sets of objectives. In terms of mass, we can write simply

$$m_0 = m_S + m_L^{(1)} + m_L^{(0)} \equiv m^{(0)} \tag{8.38}$$

where $m^{(0)}$ denotes the mass of the full sailcraft at injection, $m_L^{(1)}$ is the extra-solar mission payload, and $m_L^{(0)}$ is the inner-solar-system (or Sun-bound) mission payload that includes its AOCS. Spacecraft-0 is that one to be separated from the extra-solar sailcraft of mass $m^{(1)} = m^{(0)} - m_L^{(0)}$. As a point of fact, once it is unfastened, spacecraft-0's orbital energy returns to the standard Keplerian expression because its $\mathcal{L}_r$ is zero. Depending on separation time, this energy could be negative.

8.4 Numerical Examples via Non-Linear Programming

Using the sail system mass model in Eqs. (8.22) and (8.26b), we write

$$m_S = A/A_\diamond = m_{BS} + (1+\eth)g\sqrt{A}m_B$$
$$m_L^{(1)} + m_S = m^{(0)} - m_L^{(0)} \tag{8.39}$$

As a result of the separation of spacecraft-0 from the mother sailcraft, lightness vector jumps from $\mathbf{L}^{(0)}$ to $\mathbf{L}^{(1)} \parallel \mathbf{L}^{(0)}$, but chiefly from $\mathcal{L}_{max}^{(0)}$ to a notably higher $\mathcal{L}_{max}^{(1)}$. In order to keep the formal problem mathematically simple and meaningful, let us suppose that the optical quantities of the sail do not change appreciably in the space environment; this means that the maximum value of the thrust efficiency is an invariant. Using the relationship (8.25), and renaming thrust efficiency by η for simplicity, we can write

$$\mathcal{L}_{max}^{(0)} \frac{\sigma^{(0)}}{\sigma_{(cr)}} = \eta_{max}^{(0)} = \eta_{max}^{(1)} = \mathcal{L}_{max}^{(1)} \frac{\sigma^{(1)}}{\sigma_{(cr)}} \tag{8.40}$$

Setting $1 < \zeta \equiv \mathcal{L}_{max}^{(1)}/\mathcal{L}_{max}^{(0)}$, and combining Eqs. (8.39)–(8.40) applied to a simple-square (i.e. not a square of squares) sail results in

$$m^{(0)} = m_L^{(0)} \zeta/(\zeta - 1) \tag{8.41a}$$
$$0 = \sigma_{BS} l_S^2 + (1+\eth)2\sqrt{2}m_B l_S + m_L^{(1)} - m_L^{(0)}/(\zeta - 1) \tag{8.41b}$$

Depending on the mission and the technology one could use, these non-linear equations may be viewed from various design viewpoints because of the sail architecture and the pre-perihelion trajectory arcs requiring $\mathcal{L}_{max}^{(0)} < 1$. The choice of sail architecture implies how the sail has to be kept deployed under the pressure of the solar radiation. As an example of preliminary boom design, utilizing the theory explained in Chap. 1 of [26], one may set up a model of the boom sub-system for two cases: (a) a truss loaded in compression, and (b) a truss loaded in bending. Each case produces one truss-architecture-dependent value of the boom's specific mass, which also depends on the required length of the booms. Among the results of the algorithm, the optimal value of the boom mass per unit length is of concern here. The related structure will be able to carry lateral and axial loads for keeping the sail as required by the mission. Real boom sub-systems are considerably more complicated, and are treated via advanced methods, e.g. the finite-element method and the control volume method. Anyway, a measured or calculated value of m_B has to satisfy the simple Eq. (8.41b) because is this one that contains the ζ-requirement. The overall process is iterative.

As pointed out previously, the pre-perihelion motion requires $\mathcal{L} < 1$; numerical experience shows that, in terms of sailcraft sail loading, the following inequalities can be set for a rapid pre-perihelion arc without inclining the sail too much (Chap. 6):

$$\sigma_{(cr)} < \sigma^{(0)} < \sigma_{max}^{(0)} = 1.9 \text{ g/m}^2 \tag{8.42}$$

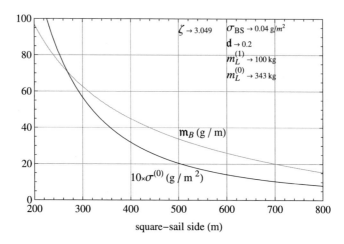

Fig. 8.16 Boom specific mass and pre-perihelion sailcraft sail loading vs sail side for the ESEM concept

Figure 8.16 shows m_B and $\sigma^{(0)}$ as function of the sail edge length for the set of parameters reported on the figure (note that $\sigma^{(0)}$ has been scaled up by a factor 10). It is apparent that, if one likes a high lightness number in the post-perihelion phase, much mass has to be separated; thus, $\zeta \sim m_L^{(0)}/m_L^{(1)}$. The value of the bare-sail loading is very low, below the best multi-layer sail consisting of metal and polyimide materials, but compliantly with the values of areal density obtained recently in laboratory for MWCNT sheet [87]. However, this is not enough; the specific mass of the sail-supporting booms has to be "agreeable" with the lightness of the sail material. The low mass of the main sail sub-systems guarantees that (1) the sailcraft gets a positive energy in the pre-perihelion arc, and (2) the lightness number after the Sun-bound spacecraft separation is sufficiently high to accelerate the sailcraft to a cruise speed notably higher than that achievable by $\mathcal{L}_{max} < 1$. The acceleration may be further increased by a conventional attitude maneuver after the mass release.

From Fig. 8.16, the strong constraint (8.42) may be satisfied in the sail side range approximately from 519 m to 576 m. Square sails of such sizes are difficult to control; however, they are sufficiently smaller than the well-known 800-m square sail[8] proposed in the 1970s by JPL for a (subsequently canceled) project of rendezvous with the Halley's comet in 1986. At that time, the motion-reversal sailing concept was not yet introduced; therefore, the JPL sailcraft—after a transfer from the Earth orbit down to a circular orbit of radius 0.3 AU—would had to rotate the orbit slowly for becoming retrograde. Subsequently, a motion-retrograde transfer would have delivered the sailcraft to rendezvous the comet.

In the wake of what said, the reader is invited to investigate the rendezvous of a sailcraft with the Halley's comet (1P/Halley) in (presumably) 2061 via the 3D motion reversal

---

[8]Such a figure has been often considered in the solar-sailing literature as a very meaningful comparison term, also considering the technology available in the Seventies.

8.4 Numerical Examples via Non-Linear Programming

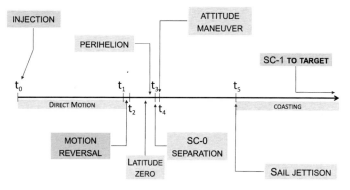

Fig. 8.17 Example of ESEM: sequence of basic events used in the actual computation of the optimal trajectory. Time intervals are approximatively to scale

mode. Projected orbital elements can be obtained via the JPL's Horizons Web Interface at http://ssd.jpl.nasa.gov. Nanotech materials and large space structures should be a standard at such future time. This study may be the topic of a Ph.D. thesis.

▶ The current ESEM concept relies on utilizing nanotechnology for strongly decreasing the mass of the otherwise heaviest component of a sailcraft: the sail system. The ratio between the sail system mass and the gross payload of $\mathcal{L}_{max} > 1$ sailcraft could then be lower than 1, a result unlikely achievable by conventional (though of high performance) materials. Per contra, the achievement of very large surfaces with high reflectance in the SPS electromagnetic regions and, possibly, almost transparent in the deep-space communication microwave region is yet far [83]. We are making the assumption that such unconventional materials may become available in the medium term. Thus, similarly to other areas of advanced space propulsion, the following questions can be reasonably set:

Question 8.2 Suppose that in the medium term Nanoscience/Nanotechnology advancements enable us to build large sailing structures; may then they be utilized for interstellar *precursor* missions?

Question 8.3 Conversely, might pure astrodynamical results such as the fast sailing theory contribute to stimulating nanotechnological research for ultra-light sail systems, and their likely spinoffs?

Figure 8.17 shows the sequence of the basic events in our example of ESEM concept. Their time allocations have been optimized again via the NLP-based code employed for the MNHB case. Note that, similarly to MNHB, both the **L**-maneuvers take place after the perihelion in order to maximize the terminal speed much after the sail jettisoning (occurring roughly half way between Saturn and Uranus orbits). Sailcraft data are reported in Table 8.3. One should note that the technological equivalent of the extra-solar sailcraft is lower than that of the MNHB sailcraft. This is due

Table 8.3 ESEM concept: *sailcraft data* from trajectory optimization with $\mathcal{L} > 1$. The bare-sail thickness has been expressed via the MWCNT sheet density, left indicated, because it depends on the densifying process; the resulting layer may be ~ 50 nm thick [87]. Note that the actual $\mathcal{L}^{(1)}/\mathcal{L}^{(0)} > \zeta$ because of the (rather) different $\vartheta_\odot$ in the pre-perihelion and post-perihelion arcs

Mass breakdown			
$m^{(0)}$	510.5 kg	$\sigma^{(0)}$	1.68482 g/m^2
$m^{(1)}$	167.5 kg	$\sigma^{(1)}$	0.552805 g/m^2
$m_L^{(0)}$	343 kg	$m_L^{(1)}$	100 kg
m_S	67.5 kg	σ_S	0.222772 g/m^2
m_{BS}	12.1 kg	σ_{BS}	0.04 g/m^2
m_B	46.2 kg	m_B	29.70 g/m
m_D	9.2 kg		

Sail data			
A	0.303 km^2	$\mathfrak{g}$	$2\sqrt{2}$
$A_\square^{(0)}$	593.5 m^2/kg	$A_\diamond$	4488.89 m^2/kg
$A_\square^{(1)}$	1808.9 m^2/kg		
MWCNT membranes	$\delta_{BS} = 0.04$ g/m^2/ρ_{BS}		
$\mathcal{R}_s$	0.8 constant		
$\mathcal{R}_d$	0.1 constant		
$\mathcal{A}$	0.1 constant		
sail $\tilde{z}$	20 nm	sail $\tilde{l}$	120 nm
$\mathcal{L}^{(0)}$	0.59409	$\eta^{(0)}$	0.65138
$\mathbb{G}^{(0)}$	998.99 m^2/kg		
$\mathcal{L}^{(1)}$	2.57110	$\eta^{(1)}$	0.92496
$\mathbb{G}^{(1)}$	703.51 m^2/kg		

to the current higher thrust efficiency in this ESEM concept. Before discussing the results in detail, a remark is in order.

Remark 8.6 In Table 8.3, the adopted values for specular reflectance, diffuse reflectance, and absorptance are supposed independent of the incidence angle of sunlight. Among all uncountable sets of admissible values for such quantities (for materials only envisaged today on so a large scale as solar sail), we chose rounded values close to the Aluminum's. In addition, in order to again apply the Rayleigh-Rice-based scattering detailed in Sect. 6.7, we have assumed:

i. the same surface parameters taken for the MNHB sail; this is possible because—as pointed out in Sect. 6.4.3—PSD is a property of the (sampled) surface, not of its materials.
ii. the following expression of the angle $\tilde{\beta}$ appearing in Eq. (6.136)

$$\tilde{\beta} = \pi/2 + 0.675\vartheta_\odot \tag{8.43}$$

8.4 Numerical Examples via Non-Linear Programming

which is the same function used for Aluminum on the basis of an estimated mean response from a high-index metal, according to the results in Sect. 6.5.3.1 and Sect. 6.7.2.

The above working points have been taken for making the current problem mathematically tractable and physically realistic.

Table 8.4 reports the values of the overall control quantities, the intermediate and target states, and additional trajectory-related information. Among the new results, we can note a number of features different from the corresponding Table 8.2 of the MNHB case. Some were expected, some others were not. First, the injection date is in advance of about 11 days, close to the autumnal equinox. Second, the reversal motion takes place about 10 hours before the minimum of the invariant H, but both are in the second half of the interval $[t_1, t_2]$.

Third, the perihelion distance is the same, but this time is above the ecliptic plane. ESEM's V_p is higher than both the perihelion and the terminal speeds of the MNHB flight. Since no **L**-maneuver has been accomplished hitherto, this value is ascribable only to the increased specific area. This seems in contrast to the optimization of MNHB; however, if the MNHB-sailcraft had a specific area of about 593.5 m^2/kg, this would entail a decrease of 27.4 kg of the sailcraft mass. This would be in contrast to the assumed sail made of Aluminum, polyimide, and Chromium: the ultra thin thickness of the substrate is already a challenge, as we pointed out there. Alternatively, if the MNHB-sail were unchanged, the sailcraft's payload would decrease by 27 on 80 kg, a too high fraction indeed.

Fourth, as indicated in the fourth block of lines, spacecraft-0 is left on an eccentric orbit with a perifocus of 0.19 AU and apofocus equal to 1.254 AU; in addition, orbital inclination is 140°, namely, the orbit is retrograde. The current ESEM case is not focused on a specific orbit of $m_L^{(0)}$; in a real case, the trajectory computation would have at least four additional constraints: $\mathbf{K}^{(0)}(t_3) = \mathbf{K}_3^{(0)}$, where $\mathbf{K}$ denotes the (vector of) Keplerian elements, and the separation time t_3 is left free. There would be yet many degrees of freedom for getting a very-high-speed extra-solar payload. This topic would deserve further investigation.

The other lines of Table 8.4 will be discussed as we proceed with commenting on the following figures.

Figure 8.18 shows the trajectory arc inside the box $[(-1.5\ 1.5), (-1.5\ 1.5), (-0.25\ 0.5)]$ AU, its projection on the XY-plane, and the orbits of the inner planets in the same time interval. Gravitational perturbations due to these planets and Jupiter have been included in the sailcraft motion field. Inclined small segments denote time tags, t_0, t_1, t_2, t_3, t_4, and about 20 days beyond. Together with the results graphed in other figures, Theorem 7.2 is satisfied for one point in the second time interval of control. The 3D motion reverses at t^r, as reported in Table 8.4. Subsequently, at $t^c \cong 297.39$ day, the sailcraft emerges from negative latitudes, and go towards its perihelion achieved 17 days later. No close encounters with the inner planets take place.

According to the theory, one local maximum, say, R^a of $R(t)$ takes place during the pre-perihelion phase; in this case, the point $R^a = 1.382$ AU is reached after

362   8 Approach to SPS Trajectory Optimization

Table 8.4 ESEM concept: trajectory optimization with $\mathcal{L} > 1$. Optimal control parameters and variables have been reported for a mission to 200 AU along the direction of the closest heliopause. Trajectory data have been added, compliantly with symbols defined in this chapter. Subscripts $\odot$ and $\square$ in position/velocity symbols were dropped for simplicity. The bottom lines give the additional transfer times to 400 AU, to the minimum distance of SGL (550 AU), to 750 AU, and to the target of the TAU mission devised by JPL in 1980s

6-Arc trajectory field Optimization index injection date t_0	Thrust + gravity $V(t_f)$, t_f free 2029.09.22 12:52:37	Sun	Inner planets	Jupiter
$\mathbf{R}(t_0)$	AU	1.0037679	0.0069857	-6.0405×10^{-6}
$\mathbf{V}(t_0)$	AU/year	-0.0354830	6.2602100	-4.1424×10^{-4}
$R(t_0)$, $V(t_0)$	AU, AU/year	1.0037922	6.2603106	
$t_1 - t_0, t_1$	day	244.971482	244.971482	
$t_2 - t_1, t_2$	day	14.366108	259.337590	
$t_3 - t_2, t_3$	day	56.490564	315.828154	
$t_4 - t_3, t_4$	day	4.999997	320.828151	
$t_5 - t_4, t_5$	day	182.132509	502.960660	
$t_f - t_5, t_f$	year, day	6.000993	2694.82335	
$t^r, t^\square, t^m$	day	252.886719	253.300906	252.154536
R^r, V^r	AU, AU/year	1.039149	2.95261	
t^a, R^a	day, AU	155.213555	1.381887	
$\|\mathbf{H}\|^r, H^\square, H^m$	AU·EOS$'$	0.0062197	0.005907	0.007985
t^c, R^c	day, AU	297.392668	0.49981	
$\mathbf{R}_p$	AU	0.1510906	-0.0110854	0.122755
R_p, t_p, V_p	AU, day, AU/year	0.19499	314.3498	17.3496
V_{cruise}	AU/year	30.738		
$\mathbf{L}^{(1)}$	constant	0.522173	-0.279950	-0.043651
$\mathbf{L}^{(2)}$	see Fig. 8.20			
$\mathbf{L}^{(3)}$	see Fig. 8.20			
$\mathbf{L}^{(4)}(t_3^+)$	$m_L^{(0)}$-separ.	1.58907	0.84638	-0.17201
$\mathbf{L}^{(5)}(t_4^+)$	re-orient. man.	2.57106	-0.014394	0.000313
(t_5^+)	sail jettis. (year)	1.377031		
SC-0 orbit	(a, e, i)	0.722138 AU	0.737277	140.1198°
	(Ω, ω, E)	70.6334°	65.7018°	13.2502°
$R(t_f), \psi(t_f), \theta(t_f)$	AU, deg., deg.	200.00	254.50	7.50
$V(t_f)$	EOS$'$, AU/year	4.89065	30.72886	
t_f	year	7.378024		
$t^{(400\ \text{AU})} - t_f$	year	6.51		
$t^{(550\ \text{AU})} - t_f$	year	11.39		
$t^{(750\ \text{AU})} - t_f$	year	17.9		
$t^{(1000\ \text{AU})} - t_f$	year	26.03		

8.4 Numerical Examples via Non-Linear Programming

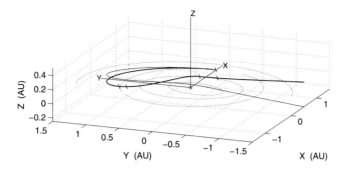

Fig. 8.18 3D optimal trajectory for the current concept of ESEM via $\mathcal{L} > 1$. Only the trajectory arcs inside the Mars' orbit are shown (*black line*). The trajectory's final arc is practically rectilinear towards the closest heliopause. In addition to the orbits of the inner planets (*light-gray lines*), the trajectory projection on the HIF's XY-plane is shown (*medium-gray line*)

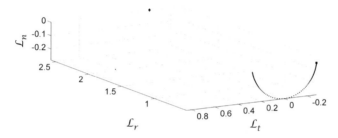

Fig. 8.19 Orbit of **L** inducing the ESEM trajectory of Fig. 8.18. The *two isolated dots* represent two **L**-maneuvers: the spacecraft-0 separation, and the sail re-orientation (*diamond dot*) described both in the text

approximately at $t^a = 155.21$ days after injection. More precise values are given in Table 8.4, where other distances at particularly meaningful times are reported as well.

Figure 8.19 shows the orbit of the lightness vector bringing about the ESEM trajectory graphed in Fig. 8.18, and summarized in Tables 8.3 and 8.4. The pre-perihelion arc profile is similar to that of Fig. 8.6. The square dot denotes the beginning of flight, the isolated dot represents the $m_L^{(0)}$ separation maneuver, whereas the diamond marker is the sail re-orientation maneuver. The quasi-circular control acts in the interval $[t_1, t_3]$ and is responsible for the motion reversal and the flyby of the Sun. Note that $\mathcal{L}_n$ evolves only in the negative subspace.

The component-wise evolution of the lightness vector can be analyzed from Fig. 8.20. Time tag t_3 (see Table 8.4) divides the plane $(t, \mathcal{L})$ in two regions: the left one where the sailcraft has $\mathcal{L} < 1$, and the right one where $\mathcal{L} > 1$ as the result of the spacecraft-0 separation, as also shown in Fig. 8.21 where the decimal logarithm of the sailcraft mass is plotted as well. The pre-separation behaviors of the **L** components are similar to those ones in Fig. 8.7 except that now $\mathcal{L}_n$ is always negative, and achieves the admissible local minimum at $t^\square$, according to the theory. At t_4, the

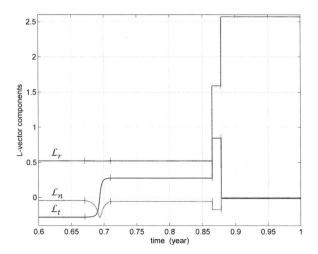

Fig. 8.20 Evolution of **L** in the interval 0.6–1 year for ESEM via $\mathcal{L} > 1$. The *vertical small segments* denote the optimized time bounds of the control. In the second interval, $\mathcal{L}_t$ crosses zero when $\mathcal{L}_n$ achieves a local minimum. The two jumps are due to spacecraft-0 separation and re-orientation attitude maneuver, in the order

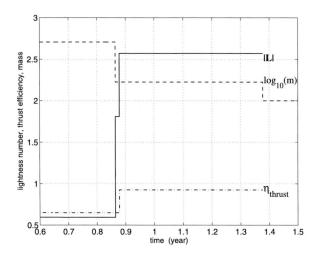

Fig. 8.21 ESEM via $\mathcal{L} > 1$: evolution of mass, lightness number, and thrust efficiency

sail re-orientation maneuver is performed. The two large increases of the lightness number are the direct consequence of having considered nanotechnological materials for the sail system. Like the MNHB, the re-orientation maneuver causing high $\mathcal{L}_r \lesssim \mathcal{L}$ produces a higher thrust efficiency, this time greater than 0.92, due to the large $\mathcal{L}$.

Figure 8.22 shows the evolution of the sail axis and the sunlight incidence angles; the temporal window zooms in the interval from t_1 to t_4, at which (we remind) sail

8.4 Numerical Examples via Non-Linear Programming

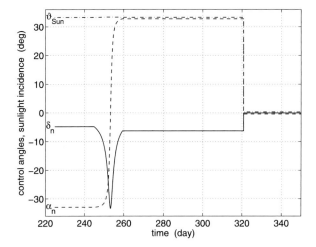

Fig. 8.22 ESEM via $\mathcal{L} > 1$: evolution of the sail axis angles (α_n, δ_n) (i.e. azimuth and elevation in EHOF), and the sunlight incidence angle

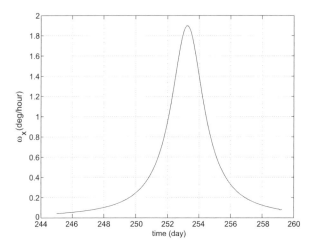

Fig. 8.23 ESEM via $\mathcal{L} > 1$: rotation of the sail axis **n** about $\bar{\mathbf{r}}$ with angular speed ω_x. See text for explanation

re-orientation is accomplished. From t_0 to t_4^-, incidence variations (less than $0.15°$) are hardly discernible in the plot; $\vartheta_\odot$ can be considered constant again. Repeating the reasoning made on p. 349, ω_x has been plotted in Fig. 8.23, and has a maximum at $t^\square$ of $1.9°$/hour.

Time evolutions of sailcraft mass, distance, and sail temperature are plotted in Fig. 8.24. Spacecraft-0 separation and sail jettisoning are clearly identified also in terms of sailcraft-Sun distance and sail temperature. As pointed out, the heaviest spacecraft is separated at t_3, or at 0.2039 AU, approximately 35.5 hours after peri-

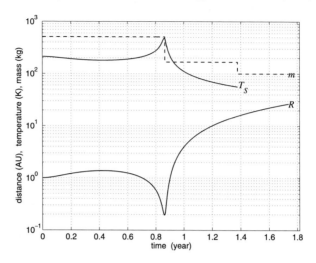

Fig. 8.24 ESEM via $\mathcal{L} > 1$: evolution of m, R, and T_s. Here, temperature curve ends when the sail is jettisoned, approximately at 15.5 AU, and after almost 503 days after sailcraft injection

helion. The optimized time t_5, reported in Table 8.4, of sail jettisoning is approximately 503 days (after sailcraft injection, or 188.6 days past the perihelion), when sailcraft is 15.54 AU from the Sun; this means a mean *radial* speed of approximately 29.7 AU/year.[9]

Note 8.3 The sail's current temperature profile should be well within the performances of Carbon-based materials. This would suggest employing a perihelion considerably lower than the current one. However, the impact on the other sailcraft systems and sub-systems may be prohibitive or require much additional mass for protection below 0.15 AU.

Note 8.4 The technique to be used for sail jettisoning is beyond the aims of this chapter. However, one should note that the sail continues to accelerate (very slightly) if left with the nominal t_5-orientation, while $m_L^{(1)}$, i.e. the spacecraft-1, decelerates (very slightly).

Figure 8.25 shows the evolutions of the invariant H and the component H_z; in particular, a neighborhood of the reversal point is zoomed in. The motion-reversal time precedes the time of the minimum of $H(t)$ (see Table 8.4). Like the example of MNHB, there is only one **H**-reversal point in the current trajectory profile of ESEM concept.

---

[9] Incidentally, a speed value of this order of magnitude might be appropriate for flyby flight of large Trans-Neptunian objects such as, for example, the Pluto-Charon or the Haumean system. However, too a high speed would reduce the close-encounter observation time.

8.4 Numerical Examples via Non-Linear Programming

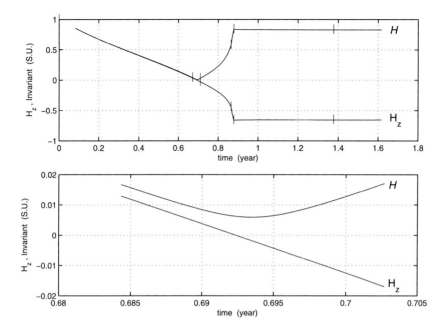

Fig. 8.25 ESEM via $\mathcal{L} > 1$: evolution of the invariant H and the third component of the angular momentum: *(top)* the *vertical small segments* denote the optimized time bounds of the control intervals; *(bottom)* zoom-in on both variables around the reversal point

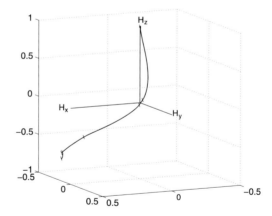

Fig. 8.26 ESEM via $\mathcal{L} > 1$: orbit of **H**. Units are AU times EOS. Time tags show clearly that one motion reversal takes place

Figure 8.26 displays the orbit of the orbital angular momentum. The small square denotes the start of flight, while time tags show clearly that one motion reversal takes place between t_1 and t_2, and with $\|\mathbf{H}\| \neq 0$ according to Theorem 7.2. The fourth and fifth tags seem to be in reverse order because of the sail jettisoning: only gravity acts on the interstellar payload for $t > t_5$.

Compared to Fig. 8.13, the ESEM sailcraft hodograph plotted in Fig. 8.27 exhibits additional interesting features that can be inferred by inspection using the

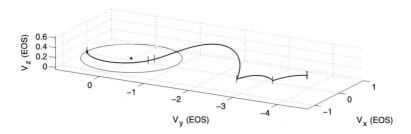

Fig. 8.27 ESEM via $\mathcal{L} > 1$: the sailcraft hodograph. The *small square* denotes the injection vector

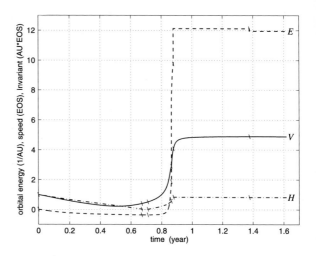

Fig. 8.28 ESEM via $\mathcal{L} > 1$: evolution of energy and speed. H-function history has been graphed again for a better visualization of the events

added time tags; again, the small black square refers to the injection (t_0). As usually, speed is the norm of (vector) velocity, i.e. $V = \|\mathbf{V}\|$. Some properties of the MNHB-related hodograph (Fig. 8.13) are shared with the current hodograph. The interval $[t_1, t_2]$ belongs to the trajectory acceleration arc, but V is still lower than 1 EOS. In this interval, lightness vector varies, and induces a subsequent quick increase of speed. Before $\dot{V}$ changes sign, spacecraft-0 is detached from the sailcraft at time t_3. This produces a further increase of the extra-solar vehicle speed. At t_4, the re-orientation causes an additional increase of speed; however, a part of this re-orientation is needed for steering the sailcraft towards the target direction. The last two time tags correspond to sail jettisoning and to the coasting end; they are practically coincident in the plot, revealing that the code-determined jettisoning distance brings about a negligible loss of speed during the coasting.

The remarkable output of this mission example is shown in Fig. 8.28, where the same units of Fig. 8.14 have been used for easier comparison. Again, the small segments stand for the time tags bounding the optimized trajectory arcs. The small decrease of energy at t_5^+ is due to sail jettisoning, namely, $\mathcal{L}_r(t) = 0, t > t_5$. V and H are continuous through, of course. At t_5, sailcraft distance is 15.541 AU, while its speed amounts to 30.805 AU/year. At 200 AU, the speed results in 30.729 AU/year,

8.5 Other Studies on 2D and 3D *H*-Reversal Motion

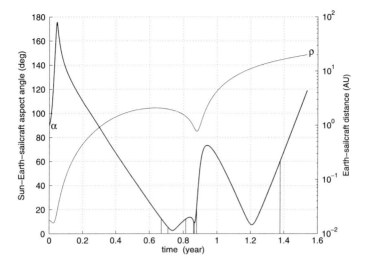

Fig. 8.29 Sun-Earth-sailcraft aspect angle, say, $\alpha_{\odot\square}^{\oplus}$ in the current example of ESEM. $\alpha_{\odot\square}^{\oplus}$ is the angular distance between the Sun's barycenter and the sailcraft as it would be observed from the Earth barycenter. Also reported is $\rho_{\oplus\square}$, i.e. the Earth-sailcraft distance scaled on the *right*. *Vertical segments* from the time axis to the aspect angle curve are at the time tags t_1, t_2, t^c, t_p, t_3, t_4, and t_5 in the order

which can be taken as constant for the prolonged mission time, as reported in the last lines of Table 8.4. In particular, the mass of this envisaged sailcraft, the sail/boom values, the payload mass, the propulsion time and cruise speed, and the times at various destinations can be compared with the corresponding numerical output of many other interstellar precursor mission and spacecraft concepts, issued in the past 30 years, by several individual authors or laboratory teams in USA, Europe and China.

Finally, as we did for the example of MNHB, we show the histories of the Sun-Earth-sailcraft aspect angle and the Earth-sailcraft distance in Fig. 8.29. Vertical segments from the time axis to the aspect angle curve are at the time tags t_1, t_2, t^c, t_p, t_3, t_4, and t_5, in the chronological order. The minimum aspect angle is 2.53°; thus, in all the mentioned operations (motion reversal, spacecraft separation, re-orientation maneuver, and sail jettisoning) sailcraft-Earth are in geometric visibility, namely, not hindered by the solar disk.

8.5 Other Studies on 2D and 3D *H*-Reversal Motion

We have been dealing with trajectories with two main arcs, one direct and one retrograde, the second one being unlimited. Thus, usually, the number of *H*-reversal points $\bar{n}$ is just one. We have understood in the previous chapters that, in the two dimensional space, this is a significant point in the definition and achievement of fast sailing, even though direct-direct motion solutions there exist and are of notable importance, as we discussed in Sect. 7.3. However, in Case 7.5 of Sect. 7.4.3, we have seen that 3D fast sailing trajectories could have $\bar{n} = 2$, though the related reversal times are close to one another.

One may have fast sailing with an odd $\bar{n} > 1$, e.g. Case 7.7 in Sect. 7.4.3, with both high cruise speed and high escape latitude.

When $\bar{n}$ is even, *other* very intriguing trajectories can arise. In particular, very recently, investigators in Italy and China have introduced new 2D and 3D **H**-reversal orbits of *non*-escaping heliocentric sailcraft, as we report below. They have employed the classical formalism [45] for force field and control, and Pontryagin's maximum principle in the related optimization problems. Here you are some basic items of such works. Due details can be found in the original papers.

In paper [73], the authors have employed a smart phase space for classifying all possible classes of two-dimensional trajectories of solar-photon sailcraft relatively to an ideal model of sail (i.e. flat and perfectly reflecting). They have divided the complete set of 2D trajectories into three subsets: (1) pseudo-hyperbolic trajectories, (2) angular momentum reversal trajectories, and (3) spiral trajectories. The employed phase space varies with the lightness number and the Sun angle.

In paper [48], cone angle is time varying, and the related program closes the trajectory. Orbits are *periodic*. The set of such orbits is referred to as H^2-reversal trajectory (H2RT) because each orbit consists of two arcs: one arc is in direct motion, and the other one is of reverse motion, but the number of motion-reversal is two. In each period, there is an attitude reorientation maneuver at the perihelion. The maximum lightness number in each element of the set is higher than $1/2$, compliantly with the results in Fig. 7.5. The H2RT set is found by minimizing the characteristic acceleration (which is proportional to $\mathcal{L}_{max}$ in a flight). The number of applications of such intriguing periodic orbits may be several in the future; one particularly interesting is the achievement of a near-helio-stationary condition.

In paper [84], the authors focused on three new potential applications of the reversal of the sailcraft's orbital angular momentum. These goals can be named as follows: (1) space observations via periodic double-motion-reversal 2D (2D-DMR) and 3D orbits, (2) heliocentric transfer between a counterclockwise orbit and a clockwise, even non-coplanar, orbit, and (3) collision with potential Earth-threatening small (counterclockwise) celestial bodies because the sailcraft-asteroid relative speed would be considerably high. Each set of trajectories could find significant applications in the future.

The author of this book notifies the reader that H2RT and 2D-DMR represent the same set of orbits, obtained by the respective research teams independently, and in the same period of time. This is a special historical situation regarding advanced SPS.

In papers [85, 86], the authors employ a combination of sailcraft speed at the sail jettisoning time and flight time as the index of performance (or objective function), and use an ideal sail. Pontryagin's maximum principle is utilized for getting the optimal control history of the sail orientation. The final distance is specified (but not the direction), and the (maximum) lightness number is prefixed (not optimized)

8.6 Objections and Concluding Remarks   371

because there is no sailcraft mass breakdown model nor constraints related to the sailcraft's real systems. Apart from the combined objective function, the overall approach is similar to that adopted in [64], where only time minimization is considered. The conclusions in these papers are similar, namely, the *unconstrained* direct solar flyby exhibits less flight-time than the totally **L**-*constant* motion-reversal solar flyby. Though that is plain, however the (itemized) answer to such conclusion will be given in next section, where major considerations about motion-reversal trajectories are emphasized. Meanwhile, we have to point out a meaningful result from papers [85, 86]: the fully **L**-*constant* motion-reversal trajectory is a local optimal solution in the framework of the Pontryagin principle, enforcing the so important aspect of attitude control simplicity.

8.6 Objections and Concluding Remarks

This section is devoted to comments on the set of trajectories explained in this book, and also is the continuation of Remark 7.3.

The components of the lightness vector represent an efficacious, simple and natural control space, and we have been employing it. The fundamental property is that sailcraft trajectory dynamics can be based entirely on the $\mathcal{L}_r$-positive semispace *without* requiring a model of the TSI-sail interaction because only the vector acceleration in the sailcraft frame intervenes explicitly. If one uses the sail axis (control) angles, then a model of this interaction has to be specified. Of course, if the model is not sufficiently accurate, many of the calculated properties may result unrealistic.

It is the *actualization* of mission trajectories that requires an explicit model of the TSI-sail interaction, (i.e. the optical diffraction, chiefly, as we know from Chap. 6), and the implementation of the technological tools necessary for a real flight. These two aspects—trajectory dynamics and interaction modeling—are not restricted to SPS, of course; in particular, one should note similarity of the SPS motion equations, as described via the lightness vector formalism, with the equations of Gauss in Celestial Mechanics. However, in SPS, they take on a special meaning because— differently from rocket propulsion—no expenditure of vehicular mass is involved, and thrust may be made comparable to or higher than the local solar gravitational acceleration for all the time the sailcraft stays in the solar system or until sail jettisoning is commanded.

From a pure trajectory optimization viewpoint, one may wonder whether the dynamical output (maximum terminal speed, minimum time to target, some combination of both, and so forth) may be further improved by time-varying the lightness vector *also* in the deceleration trajectory arc. Even though an over-simplified model of sail (i.e. an inclined perfectly-reflecting sail) has been often employed in literature, nevertheless a more realistic optimization of sailcraft trajectory could lead to a rather different value of the objective function, in general. For simplicity of explanation, apart from the perturbations to the general lightness vector, this type of control may be identified as the deceleration-arc **L**-variable control mode; in other

372 8 Approach to SPS Trajectory Optimization

words, the *whole* control history would be free to be optimized. Of course, this consideration is completely general. In particular, in literature, it has been applied to the direct-motion flight, particular cases of which have been dealt with in Sect. 7.3. As a result, one could be induced to conclude that there would be no need to resort to the angular-momentum reversal mode for getting fast sailing. However, there are some basic objections to this "conclusion". Let us discuss them point by point below. They may be viewed also as a critical fast-SPS-related summary, which is appropriate at this point of the book.

(i) Direct-motion and motion-reversal modes for fast sailing are not conflicting with one another. They share the common strategy of loosing energy initially, and of subsequently gaining much more energy than what would be obtainable via the plain control (p. 269). Without such control policy a close solar flyby would not be possible; it stemmed from the studies on the **L**-constant motion-reversal in the 1992–1997 time frame.

(ii) The motion reversal trajectory properties are due to the lightness vector components that satisfy certain general relationships (carried out in Chap. 7), which in turn give them their physical meaning. The motion reversal via either piecewise-constant **L**-profile or constant-variable-constant **L**-profile has the great advantage to reduce the sail control problem considerably, a big problem especially for large sails. An unconstrained optimization could result in a fine mathematical exercise if other key systems of the vehicle were considered as ideal systems.

(iii) If one tries to optimize a fast SPS trajectory *globally* in the sense of the optimal control theory, one should not forget that the maximum principle is not able to provide solutions to singular control problems, as we discussed in Sect. 8.1. For the reasons detailed on p. 329, NLP for the *constrained* optimization of both MNHB and ESEM concepts has been employed in this book for carrying out many quantitative results.

(iv) The optimization of fast sailcraft trajectories might risk to become a sort of misconception. In the past decades, most of the literature on SPS contained papers where the sail is quite ideal, namely, no optical diffraction, no absorption, no perturbing accelerations apart from the gravitational ones (not always taken into account). As a consequence, any claimed global optimal results should be re-calculated because the control profiles and their properties could change appreciably; to what extent this takes place may result (much) more complicated to analyze than that one might expect. The mission concepts discussed in this chapter have been focused first on the physics of photon sails (which is mainly governed by diffraction of light), and then constrained optimizations have been carried out.

(v) The existence of 3D motion-reversal *increase* the number of ESEM opportunities a year, every year, in general. In particular, multiple reversals in the same trajectory (Chap. 7) should be investigated more deeply in order to find aspects for exotic applications in Space. New investigations could extend the results obtained in Sect. 7.4.

(vi) Anyway, if future developments of the sail attitude control methods (and their implementations) were to result in wider performance, a full time-varying (realistic) control of the deceleration phase is expected to *enlarge* further the potentialities of both the motion-reversal and direct-motion trajectory families and, therefore, of the whole set of fast-sailing trajectories.

(vii) Fast SPS, and the motion-reversal concept in particular, relies on a solid mathematical base with three main theorems and many propositions. The theory lends itself to extensions in a natural way by, for instance, applying basic theorems of Group Theory to the sailcraft motion equations. Although such an advanced topic is outside the purposes of this book, however the following point might stimulate investigators to further fruitful results:

> Despite its formal simplicity, Eqs. (7.26) or (7.27) are not solvable in closed form for $\mathcal{L}_t \neq 0$. Very shortly, via symmetry determination, one can see that equations (7.26) entail the Abel differential equation of the *first* kind with rational coefficients; to it, at the time of this writing, the exact general solution is not known.

In the course of many years, this fact has affected the author's research for getting the properties of fast solar sailing, as we have seen in the previous chapters. Thus, this should not be viewed as a big drawback inasmuch as most of the properties found and explained in this book have been obtained by looking over SPS trajectory calculation from different viewpoints.

References

1. Abbott, A. S., McIntyre, J. E. (1968), Dynamic Programming: Vol. X. Guidance, Flight Mechanics and Trajectory Optimization. NASA CR-1009.
2. Bazaraa, M. S., Sherali, H. D., Shetty, C. M. (2006), Nonlinear Programming, Theory and Algorithms (3rd edn.). New York: Wiley. ISBN-10: 0-471-48600-0, ISBN-13: 978-0-471-48600-8.
3. Belbruno, E. A. (1987), Lunar capture orbits, a method for constructing Earth–Moon trajectories and the lunar GAS mission. In Proceedings of AIAA/DGLR/JSASS Inter. Propl. Conf. AIAA paper No. 87-1054.
4. Belbruno, E. A., Miller, J. (1993), Sun-perturbed Earth-to-Moon transfers with ballistic capture. Journal of Guidance, Control, and Dynamics, 16, 770–775.
5. Belbruno, E. A., Gidea, M., Topputto, F. (2010), Weak stability boundary and invariant manifolds. SIAM Journal on Applied Dynamical Systems, 9, 1061–1089.
6. Bell, D. J., Jacobson, D. H. (1975), Singular Optimal Control Problems. London: Academic Press. ISBN 0-12-085060-5.
7. Bertsekas, D. P. (1996), Constrained Optimization and Lagrange Multiplier Methods. New York: Athena Scientific. ISBN 1-886529-04-3.
8. Bertsekas, D. P. (2004), Nonlinear Programming (2nd edn.). New York: Athena Scientific. ISBN 1-886529-00-0.
9. Bertsekas, D. P. (2007), Dynamic Programming and Optimal Control. New York: Athena Scientific. ISBN 1-886529-08-6 (Vols. I and II, 3rd edn., Two-Volume Set).
10. Bertsekas, D. P. (2011), Approximate Dynamic Programming, update June 2011. http://web.mit.edu/dimitrib/www/dpchapter.pdf.
11. Bryson, H. (1969), Applied Optimal Control: Optimization, Estimation, and Control. Waltham: Blaisdell.

12. Ceccaroni, M. (2008), The weak stability boundary. Mathematics Thesis, Faculty of Mathematical, Physical and Natural Sciences, Roma-3, Rome, Italy.
13. Chachuat, B. (2009), Optimal Control, Lectures 25-27: Maximum Principles. McMaster University, Hamilton, ON, Canada.
14. Dachwald, B. (2004), Solar sail performance requirements for missions to the outer solar system and beyond. In 55th IAC 2004, Vancouver, Canada. Paper IAC-04-S.P.11.
15. Diehl, M., Glineur, F., Jarlebring, E., Michiels, W. (Eds.) (2010), Recent Advances in Optimization and Its Applications in Engineering. Berlin: Springer, ISBN 978-3-642-12597-3, e-ISBN 978-3-642-12598-0.
16. Denn, M. M. (1969), Optimization by Variational Methods. New York: McGraw-Hill. Library of Congress catalog card number 73-75167 16395.
17. Dreyfus, S. E. (1967), Dynamic Programming and the Calculus of Variations. New York: Academic Press.
18. Edge, E. R., Powers, W. F. (1976), Shuttle ascent trajectory optimization with function space quasi-Newton techniques. AIAA Journal, 14, 1369–1379.
19. Edge, E. R., Powers, W. F. (1976), Function-space quasi-Newton algorithms for optimal control problems with bounded controls and singular arcs. Journal of Optimization Theory and Applications, 20, 455–479.
20. Flaherty, J. E., O'Malley, R. E. Jr. (1977), On the computation of singular controls. IEEE Transactions on Automatic Control, AC-22, 640–648.
21. Fujitsu Scientific Subroutine Library II (2001), Minimization of the Sum of Squares of Functions. Fujitsu Ltd.
22. Grive, I., Nash, S. G., Sofer, A. (2009), Linear and Nonlinear Optimization (2nd edn.). Philadelphia: SIAM. ISBN 978-0-898716-61-0.
23. Hull, D. G. (2011), Optimal Control Theory for Applications. Berlin: Springer. ISBN 978-1-4419-2299-1.
24. IMSL Fortran Numerical Library (2010), User's Guide Math Library, version 7.0. Visual Numerics, Inc.
25. Jacobson, D. H., Gershwin, S. B., Lele, M. M. (1970), Computation of optimal singular controls. IEEE Transactions on Automatic Control, AC-15, 67–73.
26. Jenkins, C. H. M. (2006), AIAA Progress in Astronautics and Aeronautics: Vol. 212. Recent Advances in Gossamer Spacecraft. Berlin: Springer. ISBN 1-56347-777-7.
27. JAXA (2010), IKAROS Project. http://www.jspec.jaxa.jp/e/activity/ikaros.html.
28. Johnson, L., Young, R. M., Montgomery, E. E. IV (2007), Status of solar sail propulsion: moving toward an interstellar probe. AIP Conference Proceedings, 886, 207–214.
29. Johnson, L., Whorton, M., Heaton, A., Pinson, R., Laue, G., Adams, C. (2011), NanoSail-D: a solar sail demonstration mission. Acta Astronautica, 68(5–6), 571–575.
30. Kelley, C. T. (1999), Iterative Methods for Optimization. Philadelphia: SIAM (electronic version).
31. Koblik, V. V., Polyakova, E. N., Sokolov, L. L., Shmirov, A. S. (1996), Controlled solar sailing transfer flights into near-sun orbits under restriction on sail temperature. Cosmic Research, 34, 572–578.
32. Laue, G. (2012), Data-sheets on polyimides. ManTech International Corporation. http://www.mantechmaterials.com.
33. Levenberg, K. (1944), A method for the solution of certain nonlinear problems in least squares. Quarterly of Applied Mathematics, 2, 164–168.
34. Locatelli, A. (2001), Optimal Control: An Introduction. Basel: Birkhäuser. ISBN 3-7643-6408-4.
35. Luenberger, D. G., Ye, Y. (2010), Linear and Nonlinear Programming (3rd edn.). Berlin: Springer. ISBN 978-1-4419-4504-4, e-ISBN 978-0-387-74503-9.
36. Luus, R. (1992), On the application of iterative dynamic programming to singular optimal control problems. IEEE Transactions on Automatic Control, 37, 1802–1806.
37. Maccone, C. (2009), Deep Space Flight and Communications—Exploiting the Sun as a Gravitational Lens. Berlin: Praxis-Springer. ISBN 978-3-540-72942-6.

References

38. Maccone, C. (2010), Realistic targets at 1000 AU for interstellar precursor missions. Acta Astronautica, 67, 526–538.
39. Maccone, C. (2011), Interstellar radio links enhanced by exploiting the Sun as a gravitational lens. Acta Astronautica, 68, 76–84.
40. Madsen, K., Nielsen, H. B., Tingleff, O. (2004), Methods for Non-Linear Least Squares Problems. Informatics and Technical Modeling (2nd edn.). Technical University of Denmark.
41. Mangad, M., Schwartz, M. D. (1968), The Calculus of Variations and Modern Applications: Vol. IV. Guidance, Flight Mechanics and Trajectory Optimization. NASA CR-1003.
42. Marquardt, D. W. (1963), An algorithm for least squares estimation of nonlinear parameters. SIAM Journal on Applied Mathematics, 11, 431–441.
43. Matloff, G. L., Vulpetti, G., Bangs, C., Haggerty, R. (2002), The Interstellar Probe (ISP): Pre-Perihelion Trajectories and Application of Holography. NASA/CR-2002-211730.
44. Maurer, H. (1976), Numerical solution of singular control problems using multiple shooting techniques. Journal of Optimization Theory and Applications, 18, 235–257.
45. McInnes, C. R. (2004), Solar Sailing: Technology, Dynamics and Mission Applications (2nd edn.). Berlin: Springer-Praxis. ISBN 3540210628, ISBN 978-3540210627.
46. McIntyre, J. E. (1968), The Pontryagin Maximum Principle: Vol. VII. Guidance, Flight Mechanics and Trajectory Optimization. NASA CR-1006.
47. Mengali, G., Quarta, A., Dachwald, B. (2007), Refined solar sail force model with mission application. Journal of Guidance, Control, and Dynamics, 30(2), 512–520.
48. Mengali, G., Quarta, A. A., Romagnoli, D., Circi, C. (2010), $\mathbf{H}^2$-Reversal trajectory: a new mission application for high-performance solar sails. Advances in Space Research doi:10.1016/j.asr.2010.11.037.
49. Moré, J. J. (1978), The Levenberg-Marquardt algorithm: implementation and theory. In G. A. Watson (Ed.), Lecture Notes in Mathematics: Vol. 630. Numerical Analysis (pp. 105–116). Berlin: Springer.
50. Moré, J. J., Garbow, B. S., Hillstrom, K. E. (1980), User guide for MINPACK-1. Argonne National Laboratory Report ANL-80-74.
51. Moré, J. J., Sorensen, D. C. (1983), Computing a trust-region step. SIAM Journal on Scientific and Statistical Computing, 4, 553–572.
52. Morrison, D. D. (1960), Methods for nonlinear least squares problems and convergence proofs. In JPL Seminar Proceedings.
53. NASA HelioWeb, Heliocentric Trajectories for Selected Spacecraft, Planets, and Comets. http://cohoweb.gsfc.nasa.gov/helios/.
54. Osborne, M. R. (1976), Nonlinear least squares—the Levenberg algorithm revisited. Journal of the Australian Mathematical Society, 19(Series B), 343–357.
55. Poncy, J., Baig, J. F., Feresin, F., Martinot, V. (2011), A preliminary assessment of an orbiter in the Haumean system: how quickly can a planetary orbiter reach such a distant target? Acta Astronautica, 68(5–6), 622–628.
56. Ponssard, C., Graichen, K., Petit, N., Laurent-Varin, J. (2009), Ascent optimization for a heavy space launcher. In Proc. of the European Control Conference, Budapest, 23–26 August 2009. ISBN 978-963-311-369-1.
57. Pontryagin, L. S., et al. (1961), The Mathematical Theory of Optimal Processes (Vol. 4), translation of a Russian book, reprinted by Gordon and Breach Science (1986). ISBN 2-88124-077-1.
58. Powers, W. F. (1980), On the order of singular control problems. Journal of Optimization Theory and Applications, 32(4).
59. Press, W. H., Teukolsky, S. A., Vetterling, W. T., Flannery, B. P. (1992), Numerical Recipes in Fortran (2nd edn.). Cambridge: Cambridge University Press. ISBN 0-521-43064-X.
60. Press, W. H., Teukolsky, S. A., Vetterling, W. T., Flannery, B. P. (2007), Numerical Recipes: The Art of Scientific Computing (3rd edn.). Cambridge: Cambridge University Press. ISBN-10: 0521880688.
61. Ross, I. M. (2009), A Primer on Pontryagin's Principle in Optimal Control. Collegiate Publishers. ISBN 978-0-9843571-0-9.

376   8 Approach to SPS Trajectory Optimization

62. Roy, A. E. (2005), Orbital Motion (4th edn.). Bristol: Institute of Physics Publishing. ISBN 0-7503-10154.
63. Ruszczyński, A. P. (2006), Nonlinear Optimization. Princeton: Princeton University Press. ISBN 978-0-691-11915-1, ISBN 0-691-11915-5.
64. Sauer, C. G. Jr. (1999), Solar Sail Trajectories for Solar-Polar and Interstellar Probe Missions. AAS 99-336, pp. 1–16.
65. Scagione, S., Vulpetti, G. (1999), The aurora project: removal of plastic substrate to obtain an all-metal solar sail. Acta Astronautica, 44(2–4), 147–150.
66. Shakhverdyan, A. S., Shakhverdyan, S. V. (2004), Theory of singular optimal controls with applications to space flight mechanics. Cosmic Research, 42(3), 289–299. Translated from Kosmicheskie Issledovaniya, 42(3), 302–312.
67. Sewell, G. (2005), The Numerical Solution of Ordinary and Partial Differential Equations (2nd edn., pp. 345). New York: Whiley InterScience. ISBN 978-0-471-73580-9.
68. Stengel, R. F. (1994), Optimal Control and Estimation. New York: Dover. ISBN 0-486-68200-5.
69. Tanaka, K., Soma, E., Yokota, R., Shimazaki, K., Tsuda, Y., Kawaguchi, J., IKAROS demonstration team (2010), Develpment of thin film solar array for small solar power demonstrator IKAROS. In 61st International Astronautical Congress. IAC-10.C3.4.3.
70. Tsuda, Y., Mori, O., Sawada, H., Yamamoto, T., Saiki, T., Endo, T., Kawaguchi, J. (2011). Flight status of IKAROS deep space solar sail demonstrator. Acta Astronautica, 69(9–10), 833–840.
71. Topputto, F. (2007), Low-thrust non-Keplerian orbits: analysis, design, and control. Ph.D. Thesis, Polytechnic of Milan, Dept. of Aerospace Engineering.
72. Vapnyarskii, I. B. (2001), Optimal singular regime. SpringerLink http://eom.springer.de/o/o068480.htm.
73. Wokes, S., Palmer, P., Roberts, M. (2008), Classification of two dimensional fixed sun angle solar sail trajectories. Journal of Guidance, Control, and Dynamics, 31, 1249 (2008).
74. Vetterling, W. T., Teukolsky, S. A., Press, W. H., Flannery, B. P. (1994), Numerical Recipes Example Book (Fortran) (2nd edn.). Cambridge: Cambridge University Press. ISBN 0-521-43721-0.
75. Vulpetti, G. (1992), Missions to the heliopause and beyond by staged propulsion spacecraft, In The 1st World Space Congress, Washington, DC, 28 August–5 September 1992. Paper IAA-92-0240 at IAF Congress-43.
76. Volker, M. (1996), Singular optimal control—the state of the art. University of Kaiserslautern, Kaiserslautern, Germany. ftp://www.mathematik.uni-kl.de/pub/Math/Papers/AGTM-reports/report_169.ps.gz.
77. Vulpetti, G., Scaglione, S. (1999), The Aurora project: estimation of the optical scail parameters. Acta Astronautica, 44(2–4), 123–132.
78. Vulpetti, G. (2000), Sailcraft-based mission to the solar gravitational lens. In STAIF-2000, Albuquerque, NM, 30 January–3 February 2000.
79. Vulpetti, G., Johnson, L., Matloff, G. L. (2008), Solar Sails, A Novel Approach to Interplanetary Travel. New York: Springer/Copernicus Books/Praxis. ISBN 978-0-387-34404-1, doi:10.1007/978-0-387-68500-7.
80. Vulpetti, G. (2011), Reaching extra-solar-system targets via large post-perihelion lightness-jumping sailcraft. Acta Astronautica, 68(5–6). doi:10.1016/j.actaastro.2010.02.025.
81. Yagasaki, K. (2004), Sun-perturbed earth-to-moon transfers with low energy and moderate flight time. Celestial Mechanics & Dynamical Astronomy, 90(3–4), 197–212. doi:10.1007/s10569-004-0406-8.
82. Yuan, Y. X., Yuan, Y. Y. X., Yuan, Y. X. (1998), Advances in Nonlinear Programming. Dordrecht: Kluwer Academic, ISBN 0792350537, ISBN 9780792350538.
83. Vulpetti, G., Santoli, S., Mocci, G. (2008), Preliminary investigation on carbon nanotube membranes for photon solar sails. Journal of the British Interplanetary Society, 61(8), 2849.
84. Zeng, X., Baoyin, H., Li, J., Gong, S. (2011), New applications of the H-reversal trajectory using solar sails. In Cornell University Library. arXiv:1103.1470v1 [astro-ph.SR].

85. Zeng, X., Li, J., Baoyin, H., Gong, S. (2011), Trajectory optimization and applications using high performance solar sails. Theoretical and Applied Mechanics, 1, 033001. doi:10.1063/2.1103301.
86. Zeng, X., Alfriend, K. T., Li, J. (2012), Optimal solar sail trajectory analysis for interstellar missions. AAS 12-234.
87. Zhang, M., Fang, S., Zakhidov, A. A., Lee, S. B., Aliev, A. E., Williams, C. D., Atkinson, K. R., Baughman, R. H. (2005), Strong, transparent, multifunctional, carbon nanotube sheets. Science, 309, 1215–1219.

Chapter 9
Advanced Features in Solar-Photon Sailing

Should Space Environment Be Taken into Account in Trajectory Calculation?
What solar UV-light and solar wind may induce in a metallic sail has been described in previous chapters. The variability of TSI/SSI has been summarized in Chap. 2. Solar-wind's features have been emphasized in Chap. 3, including its turbulent character. Some effects of UV-light and solar-wind ions directly on a sail have been considered quantitatively in Chap. 4. Thrust acceleration models may be extended by taking the fluctuating interplanetary environment into account through different ways.

TSI variations change the level of thrust directly, while ultraviolet SSI may alter the optical properties of the sail's reflective material. Such modifications may be modeled as a function of the UV-energy absorbed by unit area of this layer as sailcraft moves. One says that UV-light causes an optical degradation.

Solar wind is a very small perturbation to sailcraft motion from the dynamical pressure viewpoint. However, as we saw in Sect. 4.3, a fraction of the solar-wind ions can be stopped throughout the reflective layer, and may bring about some degradation if the absorbed dose is high during the flight.

By degradation, we mean that the sail's absorptance increases, while the reflectance decreases with respect to the undamaged-material's values; here, degradation is meant as a *permanent* modification. This has the twofold effect of changing thrust components and increasing sail temperature, which is an issue in general. Degradation may be induced by UV-light *and* ion bombardment at the same time.

Such topics are only hinted here. Further studies—devoted to a detailed analysis of the quantitative impact of UV-light and solar-wind particles on sailcraft trajectories/orbits—are necessary also in view of new materials for sail making.

9.1 TSI-Variable Fast Trajectories

To the author knowledge, at the time of this writing only three papers [21, 23, 24] in the specialized literature, and a summary sheet [22], have been devoted to the complicated problem of taking the variability of TSI into account in SPS sailcraft trajectory computations and, thus, in mission analysis. Even the cited references should be regarded as preliminary items. We like to signal some key aspects, which should be enlarged in the future. The open issues, mainly related to the prediction of TSI for astrodynamical purposes (i.e. an application of solar physics), should result in advanced investigation for more realistic sailcraft flight design.

We will follow a *macroscopic* quantitative approach to such problem, which is oriented to sail temperature and sailcraft thrust perturbation directly. A *microscopic* quantitative approach, where many particle phenomena are detailed, can be found in [9, 10], and [11].

At least hitherto, there are three main approaches to deal with the problem of analyzing the effects on heliocentric sailcraft trajectories from variable TSI:

(a) Paper [24] regards the trajectory analysis of IKAROS by including the estimation of TSI force; the multi-plane solar sail model has been employed. This model of sail geometry takes into account sail deformation and non-uniform optical properties. Short period orbit determination is discussed; it is found that estimating specular reflectance is more suitable than estimating sail area. The authors point out that this one may not be the general case, and additional studies are required.

(b) Paper [21] shows that a nominal heliocentric circular orbit of radius 0.659 AU, for high-performance warning mission via SPS, is perturbed by the TSI fluctuations much more than by the gravitational perturbations due to the inner planets and Jupiter. Also, a sailing transfer orbit between Earth and Mars is highly perturbed by TSI fluctuations. These orbits have been envisaged in the 2003–2004 time-frame (where a good rendezvous opportunity took place). Why using examples in the past time? Because we have *realizations* of the fluctuating irradiance process observed via satellites, namely, the time-series of TSI. In other words, in this preliminary work, no TSI *prediction* model (unreliable still today) has been employed.

(c) Paper [23] is a continuation of [21], and goes into details of the Mars-sailcraft rendezvous. It turns out that irradiance-fluctuation perturbations are large, namely, if the TSI-constant optimal control sequence is not re-optimized under the TSI-variable environment, then rendezvous fails.

We now apply what summarized at points (b–c) to the optimization case detailed in Sect. 8.4.2. The guiding principle of the current analysis is to set the trajectory case in some (obviously past) period where reliable TSI measurements are available. It is expected that any trajectory change depends on the particular time series occurred in the chosen period. However, our goal here is to solely derive preliminary quantitative results by employing actual TSI data.

We have moved the MNHB trajectory back to October 3, 2009, i.e. exactly 20 years in advance; we have TSI daily means from the PMOD composite (Chap. 2) for two years after injection when the Sun-sailcraft distance is more than 16 AU. A sail jettisoning has been introduced because the final thrusting (which was included in the original mission concept) is uninfluential in the fluctuations context.

9.1 TSI-Variable Fast Trajectories

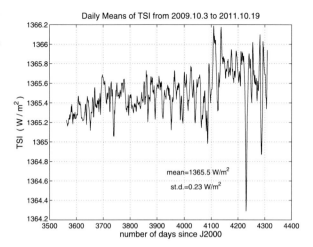

Fig. 9.1 History of TSI daily means between October 2009 and October 2011. Data come from the PMOD composite (Chap. 2). Courtesy of C. Fröhlich

Thus, we took the optimal control found for the first four trajectory arcs of the MNHB flight (Table 8.2), and wrote an input file to the code by changing the injection date, namely, replacing 2029.10.03 06:27:34, i.e. $\Delta JD = JD_{inj} - J2000 = 10867.76914$ TT, with 2009.10.03 06:00:00, or $\Delta JD = 3562.75$ TT. Coasting time was set equal to the interval $t_f - t_4 \cong 11.51$ year as in Table 8.2. For simplicity, we denote such trajectory profile by $P^{(ref)}$. We then integrated $P^{(ref)}$, and obtained the following main differences with respect to the original MNHB profile, say, $P^{(o)}$:

$$\Delta R_p = -0.0001 \text{ AU}$$
$$\Delta R(t_4) = 0.0055 \text{ AU}, \quad \Delta \psi(t_4) = 0.101°, \quad \Delta \theta(t_4) = -0.002°$$
$$\Delta R(t_f) = -0.86 \text{ AU}, \quad \Delta \psi(t_f) = 0.16°, \quad \Delta \theta(t_f) = -0.04° \quad (9.1)$$
$$\Delta V(t_f) = -0.09 \text{ AU/year}$$

where the employed the same nomenclature of Table 8.2. In the framework of a fast flight to the near heliopause and interstellar medium, these values (due to the different planetary positions during the two time frames) can be considered null differences. Therefore, $P^{(ref)}$ can be taken as the trajectory profile to be used in comparing TSI-constant and TSI-variable trajectory computations. We remind the reader that $P^{(ref)}$ has been computed with a constant TSI equal to 1366 W/m^2.

Figure 9.1 shows the daily means of TSI, taken from the PMOD composite in Fig. 2.4, from 2009.10.3 to 2011.10.19, which is the period chosen for the numerical analysis. As reported, the averaged value of total irradiance is equal to 1365.5 W/m^2. This is a low-mean and low-fluctuation period of the recent solar activity, in the early rising part of cycle-24. The first natural question is:

Question 9.1 Is a small irradiance decrease—such as 0.5 W/m^2 < 0.037 percent of the mean—appreciable in computing a TSI-constant fast-sailing trajectory?

382   9 Advanced Features in Solar-Photon Sailing

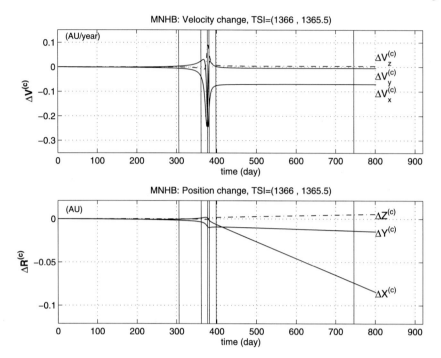

Fig. 9.2 TSI-constant MNHB profiles comparison: histories of the change in position and velocity of the 1365.5-constant profile with respect to the 1366-constant profile. *Vertical lines* denote the time tags t^r, t^{c_1}, t_p, t^{c_2}, t_3, and t_4, in the order, according the nomenclature of Table 8.2

The sailcraft motion equations have been integrated by using the $P^{(ref)}$ input file with the new value of TSI, i.e. 1365.5 W/m². The new sailcraft state (and related quantities) were tagged at the same intervals of the $P^{(ref)}$ values, namely, close to 12 hours. (Such value has been chosen for making meaningful the subsequent comparison with the variable case.) Figure 9.2 summarizes the answer to Question 9.1 by showing the components of the following vector differences:

$$\Delta \mathbf{R}^{(c)}(t) \equiv \mathbf{R}^{(1365.5)}(t) - \mathbf{R}^{(1366)}(t) \quad \text{(AU)}$$
$$\Delta \mathbf{V}^{(c)}(t) \equiv \mathbf{V}^{(1365.5)}(t) - \mathbf{V}^{(1366)}(t) \quad \text{(AU/year)} \quad (9.2)$$

where the superscript (c) denotes that TSI has been kept constant at the indicated values in W/m². We denote the state $(\mathbf{R}^{(1365.5)}(t), \mathbf{V}^{(1365.5)}(t))$ by $P^{(c)}$.

Already during the deceleration arc, small velocity differences between the two profiles appear, and they amplify in the interval from the first ecliptic crossing time t^{c_1} to the perihelion time t_p. $\Delta \mathbf{V}^{(c)}(t)$ is practically restored at t_3 in the y/z-components, but only partially in the x-component; the sail attitude maneuver at t_3 does not attenuate the divergence (since it is not inserted for this purpose). That causes a detour with respect to the nominal direction of the near heliopause as expressed by the progressive increase of $\Delta \mathbf{R}^{(c)}(t)$ in the bottom part of Fig. 9.2.

9.1 TSI-Variable Fast Trajectories 383

At t_f, the difference in direction amounts to $0.24°$, whereas $\Delta V(t_f) = +0.025$ AU/year, and $\Delta R(t_f) = +0.31$ AU (because $\Delta R_p = -0.0006$ AU). Within the framework of this MNHB flight concept, or alike, such differences are negligible. Not the same conclusion may be inferred if, for instance, one of the mission aims were to closely flyby an object of the Kuiper belt.

Question 9.2 Is the standard deviation of TSI daily means in Fig. 9.1 so low that one might deem useless the analysis of the full variable case?

For answering this question, let us describe how we calculated the following state difference histories:

$$\Delta \mathbf{R}^{(v)}(t) \equiv \mathbf{R}^{(v)}(t) - \mathbf{R}^{(1366)}(t) \quad \text{(AU)}$$
$$\Delta \mathbf{V}^{(v)}(t) \equiv \mathbf{V}^{(v)}(t) - \mathbf{V}^{(1366)}(t) \quad \text{(AU/year)}$$

$$(9.3)$$

where the superscript (v) denotes that the TSI time series of Fig. 9.1 has been used in integrating the motion equations. The state $(\mathbf{R}^{(v)}(t), \mathbf{V}^{(v)}(t))$ is indicated by $P^{(v)}$. Tags are every twelve hours again, but during the numerical integration we employed the linear interpolation between two consecutive daily means of TSI in the process of advancing the solution by one step. In practice, all frequencies higher than ~ 50 μHz in Fig. 2.14 have not been included in the numerical integration, compliantly with the fact of using TSI values averaged over one day in this analysis. In employing such data set, the implicit assumption has been made that irradiance variations sufficiently more rapid than 1 day (e.g. 2–3 hours or less) do not affect appreciably the motion of this sailcraft, for which $A_\square \cong 539$ m²/kg (see Table 8.1). Here, "appreciably" signifies that if the motion equations are integrated (under such short-time fluctuations) even for a long time, then the resulting final state is very close to the nominal one. This assumption may be verified a-posteriori by TSI *hourly* means compatible with the utilized daily means.

There is another hypothesis behind the current analysis. We applied TSI observations, which are adjusted at 1 AU, to the sailcraft at any distance from the Sun without time delay/advance, namely, using the $1/R^2$ scaling law solely. More precisely, the following relationship has been employed

$$I^{\text{HIF}}(t) = I^{\text{HIF}}(\text{TDB}) = I^{\text{HIF}}(T_{eph}) \cong \text{TSI(TT)}/R^2 \qquad (9.4)$$

where R is expressed in AU; we have applied what discussed in Sect. 5.1 about the time scales.

Figure 9.3 uses the same scales in Fig. 9.2. This gives an immediate overall comparison between the two cases: qualitatively, the two sets of results are quite similar; however, the TSI-fluctuating case shows greater dispersions. Quantitatively, we report the following terminal divergences at the same t_f value: the direction difference amounts to $0.32°$, whereas $\Delta V(t_f) = +0.033$ AU/year, and $\Delta R(t_f) = +0.42$ AU (because $\Delta R_p = -0.0008$ AU). Thus, the considered actual realization of TSI would not have required a re-calculation of the mission control sequence. In contrast

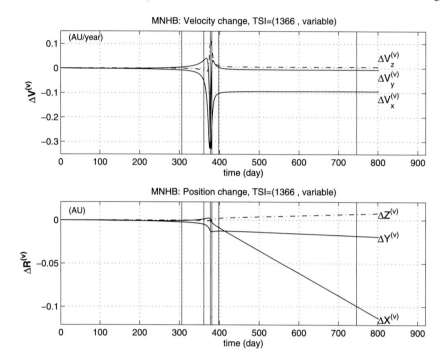

Fig. 9.3 TSI-variable MNHB profiles comparison: histories of the change in position and velocity of the TSI-variable profile with respect to the 1366-constant profile. *Vertical lines* denote the time tags t^r, t^{c_1}, t_p, t^{c_2}, t_3, and t_4, in the order, according the nomenclature of Table 8.2

to the Mars-sailcraft rendezvous detailed in [23], where non-considering fluctuating TSI in trajectory calculation could cause missing the rendezvous, an extra-solar mission concept towards a "broad target" should not require particular "worry" from the viewpoint of the target missing problem. Instead, as the nominal perihelion may decrease, a care should be needed if the mission entails a particularly low perihelion.

At this point, a question arises naturally: how to evaluate and/or compute realistic SPS trajectories by including TSI variations explicitly, and so evaluating how much a mission flight may be affected by them? The problem appears much complicated from both the solar-physics and astrodynamical viewpoints. Differently from reconstructing TSI, predicting TSI (and SSI)—remembering what has been summarized in Chap. 2—is still unreliable if we mean to predict, for instance, time series similar to what we have employed here, which are the outcome of a sophisticated processing of many-satellite measurements. Let us very shortly discuss and/or re-emphasize some items stemming from Chap. 6 and [21, 24], and [23].

1. An accurate model of thrust is a fundamental ingredient for the SPS realization in general, and fast-sailing in particular. After the feasibility mission analysis, new measurements of the sail's optical quantities should be carried out.
2. One may think of utilizing an onboard radiometer subsystem. This might help the process of trajectory-oriented determination of short TSI series, namely, this

subsystem may be part of the Navigation, Guidance, and Control (NGC) system. On the other hand, this entails additional mass: one needs a larger sail for keeping the (design's maximum value of the) lightness number unchanged. This would not be the only complication: data from the sailcraft radiometer subsystem (even if two radiometers are activated for simultaneous measurements) may be insufficient for providing the NGC system with reliable TSI values. However, measuring TSI at various heliographic longitudes and latitudes might be part of the solar-physics package.

3. TSI appears not only to have slow cyclic and secular behaviors, but also rapid fluctuations with different spectral laws. Sailcraft acceleration is linear in the lightness vector, which in principle acts for all the required time, and depends on TSI. A classical view regards sail axis angles, ballast mass displacements, mass to be separated, RCD parameters (depending on the sailcraft design) as deterministic inputs (controls) to a non-linear system (i.e. the sailcraft motion equations) perturbed by environmental disturbances due to irradiance fluctuations and particle bombardment (which could modify the sail's optical quantities). The ensuing (sailcraft) state enters the sensor subsystem that produces (noised) measurements; part of these ones are a feedback to the controller. For this complicated dynamical system, Dynamic Programming is appropriate as optimization method, e.g. [2, 14, 16], and [17]. The various aspects of this vast area of future SPS investigations may be topics for Ph.D. dissertations.

4. Sail is a large TSI sensor. Its thrust acceleration, as non-gravitational, could be sensed by onboard accelerometers, especially if the lightness number is close or higher than 1. However, as done in the IKAROS mission (for which the lightness number is < 0.001), radiation-pressure acceleration may be calculated mainly via Orbit Determination Process (ODP) [18]. Historically, the "solar radiation constant" was one of the solve-for parameters in the widely-used *batch* ODP named the Goddard Trajectory Determination System (GTDS), e.g. [4]. Instead, because TSI is a time-varying quantity, another view might be to consider its modeled evolution as a part of augmented motion equations.

The **L**-fluctuating scenario is not the only one potentially capable to modify the "tranquillity" of the deterministic description of SPS. Next section is devoted to a short explanation of that.

9.2 Modeling Modifications of Optical Parameters

We have left another complicated problem for the last; even this one is strongly dependent on the mission flight and the payload objectives. Again, in this chapter, the following considerations should be regarded as quantitative hints for subsequent more fruitful investigations.

In the chapter opening, we defined the here-used concept of degradation for solar-sail reflective layers. In what follows, we will employ the concept of energy fluence,

386   9 Advanced Features in Solar-Photon Sailing

closely related to that of absorbed dose, which is defined in Radiology and radio-logical protection [6]. Here, we are interested in the UV and solar-wind ion energy absorbed by the reflective layer of a solar sail. We call the following time-dependent quantity

$$F^{(UV)}(t) = \int_{t_0}^{t} dt' \int_{\lambda_{min}^{(UV)}}^{\lambda_{max}^{(UV)}} \mathcal{A}_\lambda I_\lambda(R(t'), t') \, d\lambda \tag{9.5a}$$

$$I_\lambda(R(t), t) = \frac{SSI(t)}{R^2(t)} \cos \vartheta_\odot(t) \tag{9.5b}$$

the *absorbed fluence* of UV radiation. It is the energy per sail's unit area absorbed from the mission epoch t_0 to the time t. Again, $I_\lambda(R, t)$ denotes the spectral irradiance, $\mathcal{A}_\lambda$ is the spectral absorptance, and $\vartheta_\odot$ is the sunlight incidence angle. The integral is calculated along the sailcraft path $R(t)$. Note that we have written the spectral solar irradiance as function of time, as it is. $F^{(UV)}(t)$ has units J/m^2, and its value at t depends on the past trajectory history.

Inserting Eq. (9.5b) into (9.5a), and using the definitions (6.44), one can write

$$\begin{aligned} F^{(UV)}(t) &= \int_{t_0}^{t} \frac{\cos \vartheta_\odot(t')}{R^2(t')} \, dt' \mathcal{A}^{(UV)} \int_{\lambda_{min}^{(UV)}}^{\lambda_{max}^{(UV)}} SSI(t') \, d\lambda \\ &= \int_{t_0}^{t} \mathcal{A}^{(UV)} \mathbb{I}(t') \frac{\cos \vartheta_\odot(t')}{R^2(t')} \, dt' \equiv \int_{t_0}^{t} \dot{F}^{(UV)}(t') \, dt' \end{aligned} \tag{9.6}$$

where $\mathcal{A}^{(UV)}$ denotes the total absorptance in the UV region, and $\mathbb{I}(t)$ is the SSI integrated over the same interval of ultraviolet wavelengths. The integral (9.6) is not simple because the ultraviolet absorptance is a function of time *through* the total fluence accumulated at t. However, this may be not restricted to the ultraviolet band.

It is important to realize that massive-particle and photon bombardment may affect the sail's *whole* response in absorptance, in general, because they may modify the surface of the reflective layer and the substrate as well, as the deposited energy increases more and more. Here, the absorptance regarding the TSI region causing thrust is supposed dependent on the linear combination of fluence contributions

$$\mathcal{A} = \mathcal{A}\big(f_{(UV)} F^{(UV)}(t) + f_{(SW)} F^{(SW)}(t) \big) \equiv \mathcal{A}(F(t)) \tag{9.7}$$

where $F^{(SW)}$ is the solar-wind fluence, i.e. the energy deposited by the wind ions in the sail materials. $f_{(\blacklozenge)}$, where $\blacklozenge \in \{UV, SW\}$, denotes the radiation effect scale. In the current framework, we will consider $f_{(\blacklozenge)}$ prefixed numbers. We will refer to $F(t)$ as the weighted fluence on the sail; it is a dimensionless quantity.

9.2 Modeling Modifications of Optical Parameters

Let us then define the SW fluence. The energy flux, or fluence rate, of the wind impinging on the sail may be expressed here as

$$\dot{F}_{in}^{(SW)} = \sum_{i=1}^{N_{ions}} \frac{n_p(1)}{R^2} \mathfrak{I}^{(i)} \frac{1}{2} m^{(i)} \left\| \boldsymbol{v}^{(bulk)} - \mathbf{V} \right\|^2 \left(\boldsymbol{v}^{(bulk)} - \mathbf{V} \right) \cdot \mathbf{n}$$

$$\equiv W_{in} \sum_{i=1}^{N_{ions}} \mathfrak{I}^{(i)} m^{(i)} \tag{9.8}$$

where $m^{(i)}$ is the mass of the ion ith-species with $m^{(1)} = m_p$, $n_p(1)$ is the proton number density at 1 AU (Chap. 4), $\boldsymbol{v}^{(bulk)}$ denotes the bulk velocity, and $\mathfrak{I}^{(i)}$ is the ion abundance relative to the proton one, set equal to 1. N_{ions} is the number of ion species in the model. Again, R is the Sun-sailcraft distance in AU, $\mathbf{V}$ is the sailcraft velocity, and $\mathbf{n}$ denotes the sail axis orientation. All ions are endowed with the same $\boldsymbol{v}^{(bulk)}$ (p. 67). (The fact that the r.h.s. of Eq. (9.8) is an invariant may be employed for calculating it easier.)

In Chap. 4, we saw that solar-wind protons are almost all stopped in the Aluminum layer, whereas α-particles penetrate the substrate; all the more, the solar-wind heavier ions do that. Each species undergoes a certain backscattering in interacting with the reflective layer. Let us denote the fraction of ion energy deposited in this layer by $f^{(i)}$, and the backscattering yield by $b^{(i)}$; therefore, the following effective abundance (in the current scope) may be defined for each ion species:

$$\mathfrak{I}_{eff}^{(i)} = \left(1 - b^{(i)} \right) f^{(i)} \mathfrak{I}^{(i)} \tag{9.9}$$

Thus, the flux of the solar-wind energy absorbed by the reflecting layer can be expressed compactly as

$$\dot{F}^{(SW)} = W_{in} \sum_{i=1}^{N_{ions}} \mathfrak{I}_{eff}^{(i)} m^{(i)} \tag{9.10}$$

and, consequently, the absorbed fluence related to the kinetic energy of SW-ions is defined by

$$F^{(SW)}(t) = \int_{t_0}^{t} \dot{F}^{(SW)} \, dt' \tag{9.11}$$

Each of above-defined absorbed fluences can be related to the energy absorbed by unit mass, or dose, of the material of concern (the reflective material, here) during the time interval from t_0 to t:

$$D_{abs}^{(\blacklozenge)} \equiv \Delta E_{abs}^{(\blacklozenge)} / \Delta m^{(rl)} = F^{(\blacklozenge)} / \rho_s \delta_s, \quad \blacklozenge = \text{UV}, \text{SW} \tag{9.12}$$

where ρ_S and δ_S denote the density and the thickness of the reflective layer (rl). The SI Unit for dose is J/kg, or gray (Gy). Apart from the plain difference of targets and aims, this definition of dose (appropriate in the discussion of the current model of degradation) differs from that employed in Biology and Radiology [6] in some important physical items:

a. Here, fluence regards the amount of energy that is actually absorbed by the specimen, not its mean value; therefore, it encompasses any time-dependent contributions, including the stochastic ones.
b. Dose and fluence depend not only on the kind of radiation and material, but also are tied to the movement of the material (i.e. the sail) in the radiation environment, and its relative orientation.
c. The sail material is assumed to be unable to spontaneously eliminate (like many biological processes do) the external substance(s) absorbed, i.e. the solar-wind ions. Thus, as the operational life of a sail increases, changes in absorptance accumulate.

Once the two contributions to fluence have been clarified, one may evaluate the degradation as the difference

$$\delta \mathcal{A}^{(\mathrm{UV})} = \delta \mathcal{A} = \mathcal{A}_{(a)} - \mathcal{A}_{(u)} \equiv \mathfrak{A}(\mathrm{F}(t))$$
$$0 \leqslant \mathfrak{A}(\mathrm{F}(t)) \leqslant 1 - \mathcal{A}_{(u)} \tag{9.13}$$

where $\mathrm{F}(t)$ was defined in (9.7); we will discuss about the function $\mathfrak{A}(\mathrm{F})$ below. The subscripts (u) and (a) stand for the undamaged-material and the actual values, respectively. The decrease of the reflectance of the layer under photon and massive-particle bombardment may be formally defined by means of a reflection degradation coefficient $f^{(re)}$

$$\mathcal{R}_{(a)} = \left(1 - f^{(re)}\right)\mathcal{R}_{(u)} \tag{9.14a}$$
$$\mathcal{R}_{(a)} + \mathcal{A}_{(a)} = \mathcal{R}_{(u)} + \mathcal{A}_{(u)} = 1 \tag{9.14b}$$

Constraints (9.14b) allows relating the reflection degradation coefficient to the function $\mathfrak{A}(\mathrm{F})$:

$$f^{(re)} = (\mathcal{A}_{(a)} - \mathcal{A}_{(u)})/\mathcal{R}_{(u)} = \mathfrak{A}(\mathrm{F})/\mathcal{R}_{(u)} \tag{9.15}$$

With regard to the hemispherical emittance, we suppose that there is no change in the function $\mathcal{E}^{(f)}(T)$; however, recalling Eq. (6.99), an absorptance increase may cause a non-negligible increase of the sail temperature (i.e. such that to violate the maximum-temperature constraint). The new radiative equilibrium temperature depends on how much the emittance varies with temperature; if T_s violates the constraint, then the sail attitude control has to be modified so to increase $\vartheta_\odot$. This is object of trajectory optimization.

9.2 Modeling Modifications of Optical Parameters 389

Question 9.3 How is the real implication of Eqs. (9.6)–(9.11)?

Question 9.4 How much does degradation affect a fast trajectory numerically?

Even though the answer to the first question contribute to the results one aims at achieving in practice, however one should have some numerical indications on how things might behave by using some further simplifications. For this reason, we will try to answer Question 9.3 at the end of this section. Meanwhile, in order to arrange a numerical method for a preliminary reply to Question 9.4, we assume for the moment that the integrals defining the two energy fluences be usual Riemann integrals by inputting mean values of the fluctuating quantities. Essentially, we integrate the motion equation and compute the sailcraft state and integrals (numerically) at the current time $t > t_0$. Through the function $\mathfrak{A}(F)$ (that we will choose below), the optical changes of the reflective-layer material (Aluminum, here) are then calculated. These new values are inserted in the lightness vector equation and in the radiative sail equilibrium equation. Then, the sailcraft state can be computed at $t + \Delta t$; and so on.

As in the previous examples of fast trajectories in Chap. 8, we employed some high-precision integration methods, i.e. those ones by Adams-Bashforth-Moulton (variable stepsize, variable order), modified Runge-Kutta-Shank (variable stepsize), and Bulirsch-Stoer (variable stepsize). Each integrator has been implemented as a Fortran-95 module, the module subprogram part of which is arranged into three nested levels (as usually): the *driver* routine, the *stepper* routine, and the *algorithm* routine. In order to deal with the current model of optical sail degradation, the driver routine has been modified.

In the current model, things are to be traced back to the function $\mathfrak{A}(F)$, which is a matter of experiments. In these ones, usually ion and UV-photon beams bombard given specimens for sufficiently long periods; absorptance is periodically measured as function of either the absorbed dose or of the particle fluence. References [1, 3], and [5] are some examples of experimental investigation of the space environmental effects on candidate solar sail materials, while Ref. [7] is an example of recent investigation of space environment effects on polymers. One can find extensive research in these topics which are of utmost importance for a long working of spacecraft.

Nevertheless, here we will follow a different approach, essentially numerical and conservative, essentially because what is particularly important in fast-trajectory design is to know whether a considerable change, if any, in the optical quantities of the sail may or may not jeopardize the mission (or, in this case, the mission concept). First, we choose the following simple function for expressing the absorptance change:

$$\mathfrak{A}(F) = (1 - A_{(u)})\big[1 - \exp(-F(t))\big] \qquad (9.16)$$

where $F(t)$ was defined in (9.7). The adopted function satisfies the constraints in (9.13), and is similar to the profile of the Al-Kapton-related experimental data, e.g. some data from [3] and polymer-related data in [7]. In the current framework, we have to determine the radiation effect scales $f_{(UV)}$ and $f_{(SW)}$ that define $F(t)$. We adopt the following criterion for ultraviolet energy: if the sail material is exposed to one prefixed fluence amount, e.g. $F_1^{(UV)} = 1$ GJ/m^2, of ultraviolet-region photons, what is the scale that produces a change of absorptance equal to the unperturbed absorptance value? We will use a similar criterion for solar-wind ions with $F_1^{(SW)} = 0.4$ MJ/m^2; this because the UV energy nominally absorbed by Aluminum is at least a factor $\sim 2,500$ higher than solar-wind ion energy (per unit time) at the same distance from the Sun. Therefore, one has

$$(1 - \mathcal{A}_{(u)})\left[1 - \exp\left(-f_{(UV)} F_1^{(UV)}\right)\right] = \mathcal{A}_{(u)} \qquad (9.17a)$$

$$(1 - \mathcal{A}_{(u)})\left[1 - \exp\left(-f_{(SW)} F_1^{(SW)}\right)\right] = \mathcal{A}_{(u)} \qquad (9.17b)$$

Searching for real solutions of the equation $(1 - \mathcal{A}_{(u)})(1 - \exp(-f F)) = \mathcal{A}_{(u)}$ results in

$$f = F^{-1} \ln \frac{1 - \mathcal{A}_{(u)}}{1 - 2\mathcal{A}_{(u)}} \qquad (9.18)$$

Inserting the above values of $F_1^{(UV)}$ and $F_1^{(SW)}$ for F, and $\mathcal{A}_{(u)} = 0.072$ for Aluminum,[1] one gets the following approximations

$$f_{(UV)} = 0.08 \text{ m}^2/\text{GJ}, \qquad f_{(SW)} = 200 \text{ m}^2/\text{GJ} \qquad (9.19)$$

Therefore, here we have utilized the following functions for modeling the absorptance and reflectance changes in the optimization code:

$$\mathfrak{A}(F) = 0.928\left[1 - \exp\left(-0.08 F^{(UV)} - 200 F^{(SW)}\right)\right] \qquad (9.20a)$$

$$f^{(re)} = \mathfrak{A}(F)/(\mathcal{R}_{s,(u)} + \mathcal{R}_{d,(u)}) \qquad (9.20b)$$

$$\mathcal{R}_{s,(a)} = (1 - f^{(re)})\mathcal{R}_{s,(u)} \qquad (9.20c)$$

$$\mathcal{R}_{d,(a)} = (1 - f^{(re)})\mathcal{R}_{d,(u)} \qquad (9.20d)$$

where both fluences are expressed in GJ/m^2; specular reflectance and diffuse hemispherical reflectance have been assumed to change by the same $f^{(re)}$.

The case for MNHB has been re-optimized with TSI kept constant at 1366 W/m^2; this has been done in order to highlight the effects, if any, from the sail's optical degradation without additional perturbations to the nominal profile computed in Chap. 8. We will report only the main differences via one table and six figures, which are discussed below.

Table 9.1 is the corresponding one to Table 8.2 on p. 349. It contains the new controls that cause the trajectory shown in Fig. 9.4. Plot ranges and 3D-viewpoint are the same of Fig. 8.5: even on a first inspection, trajectory appears to be modified

---

[1] In the current simplified framework, the solar irradiance has been assumed constant on average in the ultraviolet band (Fig. 2.9); then, combining data from Fig. 8.4 and Fig. 1 of [15], we took the value 0.072 for the undamaged-material's ultraviolet-band absorptance in a broad range of incidence angles.

9.2 Modeling Modifications of Optical Parameters

Table 9.1 MNHB concept: trajectory re-optimized under optical sail degradation. Optimal control parameters and variables have been reported for a mission to 200 AU along the direction of the closest heliopause. Subscripts $\odot$ and $\square$ in position/velocity symbols were dropped for simplicity. $t^r, t^\square$, and t^m denote the motion reversal time, the E-min and H-min time, and the middle time of the control interval from t_1 to t_2; t^{c1} and t^{c2} denote the XY-plane crossing times. The bottom lines give the additional transfer times to 400 AU, and to the minimum (theoretical) distance of SGL. To be compared with Table 8.2 (p. 349)

5-Arc trajectory field Optimization index injection date t_0	Thrust + gravity $V(t_f), t_f$ free 2029.10.07 12:53:46	Sun	Inner planets	Jupiter
$\mathbf{R}(t_0)$	AU	0.9675824	0.2478517	1.713321×10^{-2}
$\mathbf{V}(t_0)$	AU/year	-1.6362687	6.0704743	-3.931775×10^{-4}
$R(t_0), V(t_0)$	AU, EOS$'$	0.9989693	1.000628	
$t_1 - t_0$	day	315.29556		
$t_2 - t_1$	day	34.58021		
$t_3 - t_2$	day	66.395		
$t_4 - t_3$	day	365.00000		
$t_f - t_4$	year	13.35041		
$t^r, t^\square, t^m$	day	320.46971	320.56662	332.58567
$\|\mathbf{H}\|^r, H^\square, H^m$	AU·EOS$'$	0.0015307	0.0015062	0.0421923
t^{c1}, t^{c2}	day	385.3523	430.1883	
$\mathbf{R}_p$	AU	0.2187208	0.0505459	-0.0521966
R_p, t_p, V_p	AU, day, EOS$'$	0.23047	407.59346	2.39552
V_{cruise}	EOS$'$	2.21303		
$\mathbf{L}^{(1)}(t_0)$	opt. degrad.	0.551469	-0.245765	0.020677
$\mathbf{L}^{(2)}$	see Fig. 9.6			
$\mathbf{L}^{(3)}$	see Fig. 9.6			
$\mathbf{L}^{(4)}(t_3^+)$	impuls. man.	0.61758	0.16946	-0.08641
$\mathbf{L}^{(5)}(t_4)$		0.61338	0.16724	-0.08528
$\mathbf{L}^{(5)}(t_f)$		0.61332	0.1672	-0.08526
$R(t_f), \psi(t_f), \theta(t_f)$	AU, deg., deg.	200.00	254.5	7.50
$V(t_f)$	EOS$'$, AU/year	2.2112	13.893	
t_f	year	15.489		
$t^{(400\ \text{AU})} - t_f$	year	14.4		
$t^{(550\ \text{AU})} - t_f$	year	25.19		

in its starting point (the injection date is different), in elongation beyond the Mars orbit, and in perihelion. In particular, at the injection, the azimuth (97.5°) and elevation (76.3°) of the sailcraft in the EMB/OF results to be notably different from those

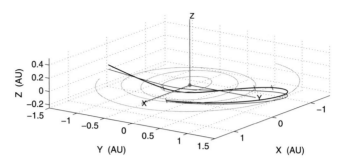

Fig. 9.4 MNHB optimal trajectory (*black line*) re-optimized under optical sail degradation. The trajectory's final arc is practically rectilinear towards the heliopause nose. In addition to the orbits of the inner planets (*light-gray lines*), the trajectory projection on the HIF's XY-plane is shown (*medium-gray line*). It is apparent that motion reverses. To be compared with Fig. 8.5 on p. 350

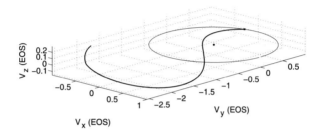

Fig. 9.5 Hodograph of the MNHB with sail under optical degradation, to be compared with Fig. 8.13 on p. 354. Again, the *circle* represents the hodograph of a gravity-only body circling the Sun at 1 AU. The sailcraft curve shape is kept, but its terminal point is closer to the origin

ones of the non-degradation case, i.e. 90° and 0.004°, respectively. In addition, the injection date is forward-shifted by about 4.5 days.

The re-optimized hodograph is shown in Fig. 9.5. It is very similar to that of Fig. 8.13; however, the local maximum and the terminal point are lower, namely, sailcraft has slowed down.

Time evolutions of the **L** components are shown in Fig. 9.6. Plot window is the same as in Fig. 8.7 on p. 351, so the comparison between the two figures is straightforward. It is quite apparent that the optimal control under material's progressive modification is moved forward in time essentially because sailcraft moves more distant from the Sun for keeping the temperature rise (due to the absorptance increase) sufficiently low with respect to GTT. Another item is the higher variability of the radial and transversal lightness numbers (compared to the non-degradation case) before the **L**-maneuver. This one takes place about 8.7 days after the perihelion, and is lower in magnitude; as a point of fact, past the perihelion, the reflectance has already decreased because fluence is high. Equivalently, though the design value of $\mathcal{L}_{max}$ is approximately 0.72, the actual highest value in the flight is $\mathcal{L} \lesssim 0.65$.

9.2 Modeling Modifications of Optical Parameters

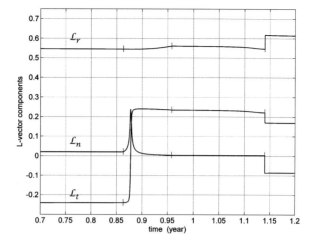

Fig. 9.6 Evolution of **L** for MNHB with optical degradation. The *vertical small segments* denote the re-optimized time bounds of the control. To be compared with Fig. 8.7 on p. 351

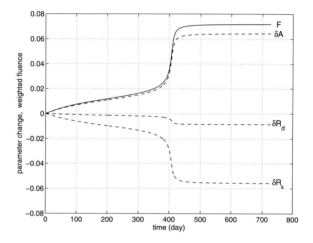

Fig. 9.7 MNHB: variations of the sail's optical quantities as the (weighted) fluence increases along the trajectory of Fig. 9.4

How the sail's optical quantities vary is shown in Fig. 9.7, where the weighted fluence F(t) is also plotted. The high accumulation takes place approximately from 350 to 450 days since injection, namely, in the interval $t_p \pm 50$ days. Subsequently, fluence saturates. Correspondingly, so behave the variations of the optical quantities. These changes decrease the lightness vector components and sail temperature appreciably such that sailcraft is forced to increase perihelion state: $\delta R_p = 0.0355$ AU, $\delta t_p = 30.4$ day, and $\delta V_p = -1.77$ AU/year.

Total fluence and sail temperature are plotted in Fig. 9.8. Maximum of temperature is 516 K, namely, 20 K less than the CP1's GTT. Temperatures over 500 K have to be withstood by the sail for 4.72 days. The constraint of 20 K below the CP1 glass transition temperature has been chosen arbitrarily by the author, who does not know what really happens chemically to a very large membrane of aluminized CP1 at so high temperatures for some days. Nevertheless, the values carried out in this analysis indicate quantitatively problems, if any, to be dealt with.

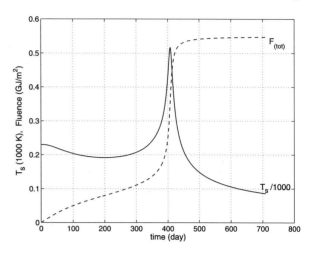

Fig. 9.8 MNHB: sail temperature profile resulting from the change in the optical quantities. With respect to Fig. 8.10 on p. 352, the peak of temperature is closer to, but not exceeding, GTT

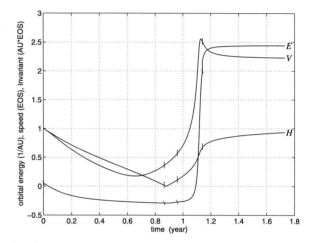

Fig. 9.9 Dynamical output of the current MNHB concept including optical degradation. These plots should be compared with Fig. 8.14 (p. 354)

The dynamical output of this mission concept has been summarized in Fig. 9.9, which is similar to Fig. 8.14 on p. 354. However, terminal speed (as also written in Table 9.1) comes down to 2.21 EOS, or 13.89 AU/year. Flight time to 200 AU is approximately 15.49 years. This is still a fine fast mission.

The previous results depend on the chosen functions given by Eqs. (9.20a)–(9.20d). Aside from a refinement of the method, an area of investigation would consist of processing experimental data from UV, electron and ion bombardment experiments—which are essential to the whole evaluation problem of space environment impact onto a sail—under international standards. Even for such problem, collecting, compare, and processing data from *many* laboratories is an arduous task. Sometimes, some information conflicts with some other one; in other cases, a laboratory has not approved documents on such experimental data for public release; also, the experimental setup and specimen material(s) may be not sufficiently detailed in papers. After a reliable procedure of experimental data processing, flight

9.2 Modeling Modifications of Optical Parameters

designs like MNHB might result to be too conservative (and this would be really a happy news). Anyway, the previous algorithm lends itself to be adjusted for a better numerical simulation of the interplanetary environmental effects on a fast sailcraft.

The very fast ESEM trajectory under the described space environment was not analyzed. One may expect that Carbon Nanotube (CNT)-based materials for fast sailing (at least) should not exhibit issues regarding high temperature and/or optical/chemical degradation(s). For the moment, this is an author's conjecture (and hope).

Finally, what remains to do is to answer Question 9.3. As mentioned above, we have assumed that the integrals at the base of the current model of sail's optical degradation are deterministic Riemann integrals.

Suppose that—through a deterministic analysis similar to or more sophisticated than that employed above—a certain set of alike SPS trajectories (open or closed) is not sufficiently sensitive to sail's optical degradation; in other words, either controlling the transfer flight to target or keeping the operational orbit can be accomplished almost nominally, i.e. as if either the sail's optical quantities were environment-unchangeable or degradation acted as small perturbations. In this second case, one could apply perturbation theory to the sailcraft motion equations, and there would be no additional complication(s).

Instead, suppose that the mentioned preliminary analysis shows that SPS trajectories are appreciably affected by optical degradation. What does *appreciably* means? Qualitatively, this may be expressed as follows: given a set of mission objectives, one computes the optimal trajectory profile without any intrinsic optical variation of the sail materials; let us call its control the *baseline* profile. Then, by employing a deterministic model of optical degradation with data coming from experiments, the trajectory is re-optimized; let us name this control the *unbiased* profile (because the above deterministic effect has been included). Finally, let us use the baseline control to propagate the sailcraft state via the motion equations with optical degradation; let us call it the *biased* profile. Well, if (1) constraints are strongly violated in biased profile, and (2) the unbiased profile confirms the achievement of the mission goals, but with acceptable decrease of the objective function, then a better model of sailcraft motion should probably be considered in order to compute a realistic profile. What would such improved/refined motion equations come from?

Given for granted that any optical-quantities change may be ascribed to the absorbed environmental fluence, let us recall that both SSI, wind ion density, and bulk speed fluctuate. If we think to conceptually model any long-term and/or cyclic feature(s) in such quantities, what remains is their stochastic behavior. As a straightforward result, the integrals in (9.6) and (9.11) are to be meant in the mean square (ms) sense for every $t \in [t_0, t_f]$ (e.g. [8, 13]). More exactly, one of the theorems of the mean square calculus (Theorem 3.9 on p. 68 of [8]) allows us to state that, if the above quantities—formally denoted by $\dot{F}^{(UV)}(t)$ and $\dot{F}^{(SW)}(t)$—are ms Riemann integrable for every $t \in [t_0, t_f]$, then $F^{(UV)}(t)$ and $F^{(SW)}(t)$—given by Eqs. (9.6) and (9.11)—are random variables of t, and continuous in ms on $[t_0, t_f]$. If, in addition, these integrands are ms continuous, then $F^{(UV)}(t)$ and $F^{(SW)}(t)$ are ms differentiable with $dF^{(UV)}/dt = \dot{F}^{(UV)}$ and $dF^{(SW)}/dt = \dot{F}^{(SW)}$.

Remark 9.1 The problem of the variability and partial stochastic character of TSI will require wide investigations in order to establish whether the most general SPS motion is describable via (though complicated) ordinary differential equations or an augmented system of differential equations for which the stochastic aspect is no longer negligible.

Note 9.1 The author does not have data from the IKAROS trajectory regarding the topics of this chapter. Presumably, some related issues/items might be not-detectable, even via ODP, simply because the lightness number of IKAROS is too low. However, future papers by JAXA might deal with these aspects.

9.3 Conclusion

In the course of many decades, the theory of motion of sail-based space vehicles has been augmenting in complexity as conceptual as practical. As for any space system, also solar-photon sailcraft needs physical principles describable by mathematical tools in order to be understood, and then engineered. However, often, before reaching a full maturity in technology level, one has to conceive of appropriate trajectories through which advanced payload aims can be set and achieved. This is an iterative process that, in the case for solar-photon sailing, can stimulate plenty of scientific and/or utilitarian objectives for spaceflight. Why is this possible? Because Astrodynamics of solar sailing, like Celestial Mechanics, is a scientific discipline in itself. Understanding it deeply will help the design of missions more and more advanced for pushing space exploration considerably forward. Both scientific exploration and pacific utilization of Space are essential for a civilization who aims at expanding beyond its birth planet.

The author likes to finish this book by shortly featuring solar-photon sailing from the viewpoint of basic physical aspects:

– Solar-Photon Sailing is a continuous-thrusting mode for space travel based on the Solar-Radiation Pressure *and* Diffraction of Light. Sailing relying only on light's absorption and/or diffuse transmission is conceptually admissible, but restricted in control and/or low in acceleration. One may substitute the adjective 'stellar' for 'solar' with a wider meaning.
– The center of mass of a general sail undergoes three fields: two are conservative, and one is non-conservative. When sailcraft's specific area is sufficiently high, thrust acceleration components become of the same order of magnitude of the gravity acceleration, or more. They are able to bring about completely new trajectory and hodograph families with no expenditure of mass.
– In general, sailcraft's orbital energy and orbital angular momentum per unit mass are not motion constants or piecewise constants. Instead, there exists an invariant the evolution of which can be controlled and is fundamental to the formation of exotic trajectories for scientific research and utilitarian purposes in Space.

References

1. Albarado, T. L., Hollerman, W. A., Edwards, D., Hubbs, W., Semmel, C. (2005), electron exposure measurements of candidate solar sail materials. Journal of Solar Energy Engineering, 127(1). http://dx.doi.org/10.1115/1.1823495.
2. Bellman, R. E., Dreyfus, S. E. (1962), Applied Dynamic Programming. Princeton: Princeton University Press.
3. Edwards, D. L., Hubbs, W., Stanaland, T., Hollerman, A., Altstatt, R. L. (2002), Characterization of space environmental effects on candidate solar sail material. Proceedings of SPIE, 4823, 67. http://dx.doi.org/10.1117/12.451455.
4. NASA GSFC (1989), Goddard Trajectory Determination System (GTDS), Mathematical Theory, Rev. 1, FDD/552-89/00l, CSC/TR-89/S00l.
5. Hollerman, W. A., Bergeron, N. P., Moore, R. J. (2005), Proton survivability measurements for candidate solar sail materials. In Nuclear Science Symposium Conference. New York: IEEE. doi:10.1109/NSSMIC.2005.1596590.
6. International Commission on Radiological Protection (2008), ICRP Publication 103: Recommendations of the ICRP. Amsterdam: Elsevier. ISBN 0-7020-3048-1, 978-0-7020-3048-2.
7. Ishizawa, J., Mori, K. (2009), Space environment effects on cross-linked ETFE polymer, session 1: radiation and charging effects. In ISMSE 2009, 14–18 September 2009.
8. Jazwinski, A. H. (1970), Mathematics in Science and Engineering: Vol. 64. Stochastic Processes and Filtering Theory. New York: Academic Press.
9. Kezerashvili, R. Ya., Matloff, G. L. (2007), Solar radiation and the beryllium hollow-body sail, 1: the ionization and disintegration effects. Journal of the British Interplanetary Society, 60, 169–179.
10. Kezerashvili, R. Ya., Matloff, G. L. (2008), Solar radiation and the beryllium hollow-body sail, 2: diffusion, recombination and erosion processes. Journal of the British Interplanetary Society, 61, 47–57.
11. Kezerashvili, R. Ya., Matloff, G. L. (2009), Microscopic approach to analyze solar-sail space-environment effects. Advances in Space Research, 44(7), 859–869.
12. Liberzon, D. (2012), Calculus of Variations and Optimal Control Theory. Princeton: Princeton University Press. ISBN 978-0-691-15187-8.
13. Maybeck, P. S. (1982), Stochastic Models, Estimation and Control, Vol. 2. New York: Academic Press. ISBN 0-12-480702-X.
14. Maybeck, P. S. (1982), Stochastic Models, Estimation and Control, Vol. 3. New York: Academic Press. ISBN 0-12-480703-8.
15. Rakić, A. D. (1995), Algorithm for the determination of intrinsic optical constants of metal films: application to aluminum. Applied Optics, 34, 22.
16. Siouris, G. M. (1996), An Engineering Approach to Optimal Control and Estimation Theory. New York: Wiley. ISBN 0-471-12126-6.
17. Spall, J. C. (2003), Introduction to Stochastic Search and Optimization: Estimation, Simulation, and Control. New York: Wiley-Interscience. ISBN 0-471-33052-3.
18. Tsuda, Y., Mori, O., Sawada, H., Yamamoto, T., Saiki, T., Endo, T., Kawaguchi, J. (2011), Flight status of IKAROS deep space solar sail demonstrator. Acta Astronautica, 69(9–10), 833–840.
19. van Kampen, N. G. (2007), Stochastic Processes in Physics and Chemistry (3rd edn.). North-Holland Personal Library. Amsterdam: Elsevier.
20. Vulpetti, G. (2002), Sailcraft trajectory options for the interstellar probe: mathematical theory and numerical results. In The Interstellar Probe (ISP): Pre-Perihelion Trajectories and Application of Holography. NASA/CR-2002-211730.
21. Vulpetti, G. (2010), Effect of the total solar irradiance variations on solar-sail low eccentricity orbits. Acta Astronautica, 67(1–2). doi:10.1016/j.actaastro.2010.02.004.
22. Vulpetti, G. (2010), Impact of total solar irradiance fluctuations on solar-sail mission design. SciTopics, November 5.

23. Vulpetti, G. (2011), Total solar irradiance fluctuation effects on sailcraft-Mars rendezvous. Acta Astronautica, 68(5–6). doi:10.1016/j.actaastro.2010.01.010.
24. Yamaguchi, T., Mimasu, Y., Tsuda, Y., Funase, R., Sawada, H., Mori, O., Morimoto, M. Y., Takeuchi, H., Yoshikawa, M. (2010), Trajectory analysis of small solar sail demonstration spacecraft IKAROS considering the uncertainty of solar radiation pressure. Transactions of the Japan Society for Aeronautical and Space Sciences, Aerospace Technology Japan, 8, 37–43.

Index

A

Absorbed fluence
  ms differentiable, 395
  dose, 387
  ion abundance, 387
  ion backscattering, 387
  ion energy deposited, 387
  radiation effect scales, 386, 390
  UV radiation, 386
  weighted fluence, 386
Acceleration ratio, 15
Active mass, 4
Adatoms, 178
Alfvén waves, 69
Aluminum's
  hemispherical diffuse reflectance (SST),
    344
  specular reflectance (SST), 343
  total absorptance (SST), 344
Aluminum's dielectric function, 210
Annihilation, 13
Antimatter, 13
Architecture
  deployed structural, 83
  deployment, 83
Astronautics
  Ackeret, J., 10
  Bussard, R.W., 10
  Cleaver, A.V., 10
  Esnault-Pelterie, R., 10
  Goddard, R.H., 10
  Koelle, H.H., 10
  Oberth, H., 10
  Ruppe, H.O., 10
  Sanger, E., 10
  Shepherd, L.R., 10
  solar sailing

  Garwin, 59
  Tsiolkovsky, 59
  Zander, 59
  Stuhlinger, E., 10
  Tsiolkovsky, K.E., 10
  Zander, F.A., 10
Atomic clock, 126
Aurora Collaboration, 347

B

BE, 36
Beam divergence, 24
Beam spread, 25
Bidirectional reflectance, 187
BIPM, 126
BIPS, 133
Blackbody, 55, 57
Boltzmann constant, 56
Boost, 4, 7
Bow-shock, 70

C

Charging time, 107
Chromosphere, 50
CIE, 35
CIR, 72, 73
  pulse duration, 72
Classical and relativistic concepts, a remark,
    168
Coefficient of diffuse momentum, 42
Coefficient of emissive momentum, 223
Comparisons
  orthonormal frames
    Frenet-Serret *vs* HOF, 153
Comparisons:1-burn rocket + sail, 286, 287
Comparisons:2-burn rocket, 285, 287
Coordinate time, 4, 129, 168

G. Vulpetti, *Fast Solar Sailing*, Space Technology Library 30,
DOI 10.1007/978-94-007-4777-7, © Springer Science+Business Media Dordrecht 2013

400 Index

Corona, 50
Coronal holes, 66
  polar, 67
CP1, 85, 346
CP2, 85
Criterion for neglecting thrust perturbation(s), 227
Cyclotron, gyro radius, 96

D
DE405, 132, 133
  basic features, 130
DE405-SOFA drift, 132
DE405/LE405, 129
Deployed structural, 83
Deployment, 83
Deposition process
  CVD, 179
  destination, 177
  film deposition rate, 178
  PVD, 178, 179
    evaporation technique, 178
    reactive sputtering, 179
    sputtering technique, 178
  source, 177
  TSD, 179
Differential directed surface, 37
Diffraction of light, 166
  BRDF, 183, 186
    main phenomena, 185
  BRDF split, 213
  BSDF, 183
  BSSRDF, 185
  BTDF, 183, 188
  CCBSDF, 183, 187
  defect scattering, 197
  diffracted radiance, 203
  diffuse reflectance, 43, 213
    B-theory, 214
    BK-theory, 215
    gHS-theory, 218
    mBK-theory, 217
    RR-theory, 215
  directional-hemispherical reflectance, 187
  directional-hemispherical transmittance, 188
  Fraunhofer diffraction, 203
  Fresnel diffraction, 203
  inverse scattering problem, 216
  material scattering, 197
  scalar scattering theory (SST), 204
  spectral bidirectional reflectance, 187
  specular lobe, 212
  specular reflectance, 213

TIS, 194
topographic scattering, 197
vector scattering theory (VST), 204
Vinci, Grimaldi, 203
wave phase difference, 202
Diffraction theories
  comments (see comparisons), 206
  comparisons
    Beckmann-Kirchhoff theory, 217
    Born approximation, 214
    generalized Harvey-Shack theory, 218
    modified Beckmann-Kirchhoff theory, 218
    Rayleigh-Rice vector theory, 215
    scalar Kirchhoff, 204
    summary of features, 206
    vector Kirchhoff, 204
Disturbing/perturbing function, 138
Doppler effect, 170
DSFG, 27
Dynamic Programming, 385

E
Earth
  lightness vector *wrt*, 154, 156
  radiance, 156
  surface temperature, 156
Ecliptic obliquity at J2000, 137
Effective jet speed, 10
Ejection beam, 5
Energetics, 16
Energy conservation, 7
Energy-momentum, 7
ERA, 126
Escape trajectories, 20
ESEM concept
  L evolution, 364
  boom specific mass, 357, 358
  evolution of E and V, 368
  evolution of H and H_z, 367
  hodograph, 368
  lightness number evolution, 364
  mass discontinuities, 364
  optimal control table, 362
  orbit of **H**, 367
  orbit of **L**, 363
  questions, 359
  sail axis and incidence angles, 365
  sail axis rotation, 365
  sail loading inequalities, 357
  sail temperature and distance, 366
  sail-system mass model, 357
  sailcraft data table, 360
  sailcraft mass model, 356

ESEM concept (*cont.*)
  sequence of basic events, 359
  Sun-Earth-S/C aspect angle, 369
  thrust efficiency evolution, 364
  trajectory, 363
Exhaust, 5
Exhaust speed, 9
  spread, 24
External momentum flux, 65
Extreme ultraviolet imaging telescope, 66

F
Facula, 51
Field-free space, 7, 9
Filling factors, 54
Finite-burn losses, 16
Forecasting TSI, 54
Four-vector, 7
  four-momentum, 7
Frenet-Serret equations, 149

G
Generalized Lorentz factor, 169
Geocenter, 127
Geocentric Celestial Reference System
  (GCRS), 128
Geometric-optics approximation, 168
GeoSail, 87
Golden rule, 215
GR, 127
Gravity, 7, 14, 16
  acceleration, 14
  losses, 16
Gray-atmosphere, 167
GTT, 346, 392

H
HCS, 71
Heliographic reference frame, 96
Heliopause, 70
Heliosheath, 70
Heliosphere, 70
  bow shock, 70
  heliopause, 70
  termination shock, 70
Heliospheric fluid, 168
Heliospheric magnetic field, 66
HMF, 69, 96
  open lines, 71
  spiral angle, 96
HPS, 71

I
IAU, 127, 131

ICRS, 128, 135
Ideal diffuser, 41
IERS, 126
IF, 4, 8
IKAROS, ix, 62, 84, 86, 162, 181, 346, 380,
  396
  lightness number, 385
  low-gain antennas, 86
  multi-plane sail model, 380
  Reflectance Control Device, 157
Inert mass, 5
Infinitesimal source, 41
Infinitesimal spectral irradiance, 170
Infinitesimal surface, 41
Input momentum, 174
International Celestial Reference Frame
  (ICRF), 128
Interstellar wind, 70
Ion propulsion, 23
Irradiance variations, 383
ISO, 35
Isotropic source of light, 40
IUPAP, 35

J
Jerk, 160
Jet, 5
Jet speed, 6

K
Kapton, 85

L
Lambertian surface, 40, 43
Larmor frequency, 96
Leap second, 127
Light, 37
  statistics, 37
    Bose-Einstein statistics, 37
    poissonian, 37
    sub-poissonian, 37
    super-poissonian, 37
Limb brightening, 167
Limb darkening, 166

M
Masking/shadowing, 190
Mass breakdown, 11, 18
  e, 18
  m, 11
Maxwellian distribution, 102
Mean human job time, 21
Mean solar radiance, 167
Measured spectral irradiance, 57

Membrane
backside BTDF, 226
crease, 82, 180
radiance from backside, 225
seam, 180
taut region, 83
tear, 82
true, 180
wrinkle, 82, 180
material wrinkles, 180
structural wrinkles, 180
wrinkled region, 83
model examples, 232
radial, 231
uniform, 231
Metric, 7, 172
MHJT, 346
Micro-sailcraft, 88
Minimum of thrust, 11
Minkowski spacetime, 136
MNHB, 380
nominal profiles, 347, 350–355
tables, 348, 349
TSI-constant, 382
TSI-variable, 384
MNHB re-optimized, 390
energy, speed, 394
fluence, temperature, 394
lightness vector, 393
optical changes, 393
trajectory, 392
weighted fluence, 393
MNHBre-optimized
hodograph, 392
Models, 52
Momentum flux source, 61
MWCNT monolayer membranes, 86
Mylar, 85

N
Nano-sailcraft, 88
Nanoprobes, 88
NanoSail-D2, ix, 86, 346
Near heliopause, 21, 27, 29
direction, 30
NEP Rocket, 23
Network, 50
Neutrinos, 12
NIP, 23
Non-gravitational acceleration, 139
Non-Linear Programming
LME, 340
LMM, 339, 340
NLLS, 342

Number spectral radiance, 169
Numerical integrators, 389

O
Observational frequency, 170
ODP, 385
Onboard, 384
Optical degradation, 379
adopted functions, 390
definition, 379
hemispherical emittance, 388
how much in fast sailing?, 389
modeled absorptance change, 389
question on the meaning, 389
reflection degradation coefficient, 388
Optics
evanescent waves, 217
Fresnel polarization reflectances, 211
polarization of light, 207
high refraction index, 209
Note, 208
polarization factors, 208
two options, 209

P
Perveance, 27
utilization, 27
Photoemission, 109
Photon momentum on sail
incidence approximation, 175
input momentum approximation, 176, 234
irradiance approximation, 175
output momentum, 234
reflected 2nd-order momentum, 212
reflected 4th-order momentum, 211
total 1st-order reaction, 236
Photon pressure, 174
Photon rocket, 6, 11, 13, 19
Photon statistics, 36
sub Poissonian, 36
super Poissonian, 36
Photon-induced space sailing, 36
Photosphere, 48, 50
Physics of damage, 118
displacement energy, 118
recoil cascade, 118
sail
Aluminum sputtering, 120
atom sputtering, 119
ion backscattering, 119
proton backscattering yield, 120
sail damage, 119
sputtering yield, 119
surface binding energy, 118

Index

403

Physics of damage (*cont.*)
 vacancy/interstitial-atom pair, 118
 vacancy/lattice-site binding energy, 118
Pizzo's model, 68
Plages, 51
Planck constant, 56
Plasma parameter, 95
PLEPH, 131
Poisson equation, 105
Power/propulsion system
 efficiencies, 26
PPU, 25
Proper time, 4, 126
PTE, 225

Q
Quiet Sun, 50

R
Radiation pressure, 81
Radiometric quantities, 35, 37
 definitions of, 37
  irradiance, 39
  radiance, 39
  radiant exitance, 38
  radiant intensity, 39
  radiant power, 38
  source of light, 37
  spectral irradiance, 39
  spectral radiance, 39
  spectral radiant exitance, 38
  spectral radiant intensity, 38
  spectral radiant power, 37
Rayleigh criterion, 194
Reference frames, 126
RLL coordinates, 142
Rocket equation, 9, 12
Rocket-vehicle equation
 differential form, 15
 integral form, 16
Rocketship, 4, 7
Roughness, 189
RTE, 224

S
Sail
 advanced reflective layer
  bias, 158
  control, 158
 bias-surface acceleration, 158
 control-surface acceleration, 159
Sail charging
 Al-layer potentials, 116
 arcing problem, 115

balance equation
 four-current model, 114
 three-current model, 113
bidirectional photoelectron yield, 110
electron shielding factor, 113
equilibrium, floating potential, 114
no particle propagation, 115
photoemission
 core electrons contribution, 111
 differential cross section, 110
 potential of ionization, 112
 total cross section, 110
thick-sheath approximation, 106
thin-sheath approximation, 106
total photoelectron flux, 110, 112
Sail support subsystem
 boom materials, 116
 longerons, diagonals, battens, 117
Sail wake, 107
Sailcraft, 82
 2D control regions
  fast motion, 272
  fourth quadrant, 273
  plunging region, 272
  slow motion, 272
 2D deceleration, 148
 H-function, 258
 acceleration's technological equivalent, 338
  example, 338
 Earth-spacecraft communication, 86
 elemental jerk, 159
 Extended Heliocentric Orbital Frame, 259
 H-reversal point
  differential properties, 276
 in the solar wind, 100
 invariant H, 261
 ion bombardment on sail, 117
 L-maneuver, 157, 159
 L-maneuver modes
  mass separation, 161
  optical-parameters variability, 161
  switching optical values, 161
  varying sail orientation, 160
  varying sunlight incidence, 160
 lightness normal number, 140
 lightness number, 140
 lightness radial number, 140
 lightness transversal number, 140
 lightness vector, 139, 143
 Modified Heliocentric Orbital Frame, 257
 motion equations, 135
  2D Cartesian explicit, 264
  2D compact, 264
  explicit properties, 261

404   Index

Sailcraft (*cont.*)
  motion reversal
    2D E and H examples, 267
    2D hodograph examples, 266
    2D trajectory examples, 264
    non-regular points, 269
    remark, 260
  normalized jerk, 160
  onboard accelerometers, 385
  orbital angular momentum, 256
    time derivative, 147
  orbital energy, 145
    time derivative, 145
  payload, 87
  planetocentric motion, 153
  questions about angular momentum, 147
  sail as TSI sensor, 385
  sail loading, 337
  sail specific area, 337
  sail system, spacecraft, 82
  sail-attached frame, 157
  thermal control, 85
  three-**L** control
    3D case-1, 300
    3D case-2, 302
    3D case-3, 305
    3D case-4, 305
    3D case-5, 308
    3D case-6, 310
    3D case-7, 313
    definition, 297
    properties of E and H, 299
  thrust, 236
  thrust acceleration, 139
  total acceleration, 267
  trajectory
    angle φ, 258
    angle ϑ, 262
    continuous planar, 257
    curvature, 149, 150
    energy minimum, 275
    jerk, 151
    non-regular reversal problem, 270
    oriented curvature, 149
    plain *vs* reversal mode, 269
    prefixed-perihelion problem, 271
    prefixed-time to perihelion, 272
    rectilinear, 257
    regular direct-retrograde arcs, 264
    torsion, 149–151
    zero along-track accel, 275
  vector work equation, 141
Sailcraft sail loading, 82, 140

Scattered exitance, 43
Scattering amplitude, 205
Scattering and diffraction, Note, 203
Scattering modes, 197
Scattering theories, 206
  Beckmann-Kirchhoff, 206
  Born approximation, 206
  Generalized Harvey-Shack, 206
  Modified Beckmann-Kirchhoff, 206
  Rayleigh-Rice, 206
Schwarzschild
  line element, 168
  radius of the Sun, 168
Self-collision times, 98
SGL, 347
Ship Frame, 4
SI second, 126
Simultaneous scalings, 132
SOF, 166, 170
SOFA, 127
SOFA-2006, 130
SOHO, 67
Solar dynamo, 53
  Babcock-Leighton, 53, 54
Solar images, 67, 68
Solar irradiance, 44
  A. Secchi, 44
  ACRIMSAT, 45
  C. G. Abbot, 44
  ENVISAT, 46
  measurements, 44
  NIMBUS-7, 44, 45
  PICARD, 47
  S. P. Langley, 44
  SDO, 46, 66
  SOHO, 45
  SOHO/VIRGO, 45
  SORCE, 46
Solar magnetic regions, 52
Solar sail
  aerodynamic center, 235
  all metal, 86
  architecture, 82, 230
    deployment, 83
    striped, 226
    striped, quadrant shape, 229
    structural, 83
    suspension, 226
    suspension, quadrant shape, 230
  assumptions, 197
  attitude control
    overview of types, 84
  bilayer sail, 85
  center of pressure, 84, 235

Index

Solar sail *(cont.)*
curved-sail sunlight incidence, 233
deployment, 84
examples of wrinkles, 181
flat sail
  D example-1, 243
  D example-2, 244
  D example-3, 244
  D example-4, 245
  diffuse-reflection momentum, 238
  Gaussian and isotropic PSD, 241
  input momentum, 237
  L explicit, 239
  re-radiation momentum, 238
  thrust per unit area, 238
  total diffuse momentum, 240
  vector **D**, 242
  vector **G**, 241
large-scale curvature, 228
materials, 85
monolayer sail, 85
sail frame transformations, 232, 235
sail total height deviation, 233
segmentation, 180
sheet (see membrane), 180
Sun-sail scene, 182
surface, 176
three-layer sail, 85
topology levels, 182
Solar shell model, 49
Solar spectral radiance, 170
Solar wind, 67, 99
as perturbation, 379
averages over cycle-23, 94
collision mean free path, 99
comparison of characteristic lengths, 100
concise features, 68
Debye length, 95
dynamical pressure histogram, 89
electron, proton collision frequency, 98
electron collision frequency, 99
electron plasma frequency, 97
fast streams, 67
heavy ions, 67
kinetic approach, 69
magnetohydrodynamic approach, 69
mean, mode, st. dev., 76
mean inter-particle distance, 94
object inside, 93
Parker's theory, 96, 97
proton gyro radius, 98
proton plasma frequency, 97
proton's inertial length, 97

skewness, kurtosis, 76
slow wind, 67
speed, dynamical pressure, 75–77
turbulence, 73, 74
  fully developed state, 74
  inertial range, 74
  intermittency, 74
  Reynolds number, 74
Ulysses, 68
Source frequency, 170
Source of light, 37
Spacecraft, 4, 84, 94, 95, 101
ACE, 72
differential charging, 101
SOHO, 66, 72
STEREO, 72
Ulysses, 96
Spacecraft charging, 103
absolute charging, 103
differential charging, 103
internal charging, 103
net-current equation, 103
sign convention, 105
Spaceship, 4
Specific area, 337
Specific mass, 26
Specific power, 11, 17, 22
Spread factor, 13
SPS, 396
1st basic theorem, 144
2D Fast Sailing
  cruising regime, 285
  definition, 277
  direct motion, 288
  direct motion (plots), 290
  direct motion (table), 291
  speed amplification arc, 283
  sufficient conditions, 282
  two launch opportunities, 292
2nd basic theorem, 262
3D Fast Sailing
  definition, 316
3D Motion Reversal
  L-constant inability, 294
  definition, 293
  different-point reversal, 297
  numerical issues, 295
3rd basic theorem, 295
L general constraints, 324
acceleration-limited, 152, 256
basic lemma, 260
flight optimization
  Bolza-type objective function, 324
  boundary conditions, 325

SPS (*cont.*)
 co-state evolution, 325
 Erdmann-Weierstrass conditions, 325
 generalized Legendre-Clebsch, 328
 Hamiltonian, 325
 Jacobson condition, 328
 Lagrange multipliers, 325
 linear-control Hamiltonian, 328
 NLP control parameters, 335
 NLP step-1, 330
 NLP step-2, 330
 NLP step-3, 330
 NLP step-4, 332
 NLP step-5, 333
 NLP step-6, 335
 NLP step-7, 336
 Non-Linear Programming, 329
 non-singular optimal **L**, 328
 Pontryagin's Principle, 327
 Pontryagin's strong form, 327
 singular arcs, 328
 singular Hessian matrix, 328
 state constraints, 326
 state discontinuity, 326
 three questions, 329
 transversality conditions, 325, 327
general motion reversal
 H2RT, 370
 other studies, 369
 some results, 370
Kuiper belt object flyby, 383
lightness number > 1, 356
motion equations
 mass change, 330
 NLP remarks, 331
 no mass rate, 323
planetocentric motion, 155
propositions
 along-track acceleration, 153
 blackbody sail, 201
 zero-normal-number trajectory, 152
 question on TSI-constant, 381
 question on TSI-variable, 383
 sailing excess, 286
 solar units, 347
SPS preliminary items, 61
Sputtering, 178
SR-related aberration, 172
SRIM, 117, 120
SSI, 57, 59, 171, 379, 384, 386, 395
SSI variations, 58
Stefan–Boltzmann constant, 56
Sunlight

space utilization, 60
Sun's barycenter orbit, 133, 134
Sunspot, 51
Supergranulation, 50
Surface, 190–194
 atoms per unit area, 177
 auto-correlation function, 192
 auto-correlation length, 192
 diffuse reflectance, 199
 diffuse transmittance, 199
 Gaussian probability density, 191
 images by AFM, 191
 isotropic, 196
 joint probability density, 193
 masking/shadowing, 190
 microfacet model, 188, 189
  the two basic conditions, 189
 models, 190
  deterministic, 190
  predictable stochastic, 190
  regular stochastic, 190
 PSD, 195, 197
  area PSD, 195
  profile PSD, 195
  spatial frequencies, 195, 202
 random height, 191
 random roughness, 191
 rms roughness or height, 191
 rms slope, 193
 rough surface, 194
 roughness, 184
 smooth, 194
 smooth and clean, 194
 smoothness criterion, 194
 specular reflectance, 199
 specular transmittance, 199
 sub-surface scattering, 189
 surface-to-bulk atoms ratio, 177
Swarm of tiny spacecraft, nanoprobes, 88

T
TCG, 129
TDB, 136
Technological equivalent of photon-sail
  acceleration, 338
Termination shock, 70
 Voyager-1, Voyager-2, 70
Thermal light, 36
Thermionic emission, 104
Thrust, 7, 11
 blackbody surface, 201
 dominant-transmission surface, 200
 Lambertian surface, 201
 no-transmission surface, 201

Index

Thrust (*cont.*)
  simplified momentum balance, 199
Thrust acceleration, 14, 15, 18, 20, 22–24, 337
  efficiency, 337
Thrust change (see sailcraft **L**-maneuver), 161
Thrust model
  bottom-up approach, 182
  induced thermal radiation, 221
  re-radiation momentum, 222
  sail temperature equation, 221
  total re-radiation momentum, 222
Thrusting arc
  long, 21
  terminal distance, 22
  terminal speed, 21
Time and reference frames, 125
  BCRS, 128
  BCRS, GCRS, 129
  day, 126
  ET, 130
  GOF, 154
  HOF, 139
  HOF time evolution, 150
  ICRF, 128
  ICRF1, 128
  ICRF2, 128
  ICRS, 128
  ICRS/Sun, 137
  J2000, 127
  JD, 126
  LB (def. constant), 129
  LG (def. constant), 127
  LVLHF, 155
  MHOF, 257
  SOF, 139
  TAI, 126, 130
  TCB, 129, 130, 166
  TCG, 127, 130
  TDB, 130
  Teph, 129
  TT, 126, 130
  UT1, 126
  UTC, 127
Time transformations

$T_{eph} - TT$, 131
  TDB-TCB, 131
  TT-TCG, 131
Total-energy distribution, 102
TSI, 44, 47, 48, 62, 171, 379, 380, 384, 385, 396
  fluctuations, 62
  frequency domain, 62
  histogram, 89
  PMOD composites
    window, 381
  prediction issue, 384
  TSI-variable trajectory, 380
TSI composites, 47
  ACRIM, 47
  extended PMOD-composite, 49
  IRMB, 47
  PMOD, 47, 48
Tsiolkovsky's equation, 10
TT-TDB, 132

U
UV light, 379

V
Vector spaces, 135
Vlasov equation, 105

W
Wien's displacement law, 52
Wind-based sail concepts
  common problems, 79
  e-sail, 78, 81
  magsail, 77, 80
  plasma sail, 78, 80
  two-sail spacecraft, 77, 80
Work function, 104, 109, 110

X
XUV/EUV relative changes, 57

Z
ZTM, 5, 8